Combinatorial games are finite, two-person, full-information games such as chess, checkers, go, domineering, dots-and-boxes, nim, and many others. This volume, arising from a workshop held at MSRI in July 1994, represents a significant addition to the literature of combinatorial games.

It includes expository articles by some of the masters in the field; studies of the classical games of chess and go from the point of view of combinatorial game theory; reports on computer advances such as the solution of nine-men morris and pentominoes; and new theoretical approaches, including extensions of the traditional framework to games with many players, or lacking perfect information, or involving loops. The book closes with an updated and commented list of unsolved problems by R. K. Guy and a comprehensive bibliography by A. Fraenkel.

Mathematical Sciences Research Institute
Publications

29

GAMES OF NO CHANCE

Mathematical Sciences Research Institute
Publications

Volumes 1 through 27 are available from Springer-Verlag

Games of No Chance

Combinatorial Games at MSRI, 1994

Edited by

Richard J. Nowakowski

Dalhousie University

CAMBRIDGE
UNIVERSITY PRESS

Richard J. Nowakowski
Department of Mathematics
Dalhousie University
Halifax, NS
Canada B3H 3J5
rjn@cs.dal.ca

Mathematical Sciences Research
Institute
1000 Centennial Drive
Berkeley, CA 94720

The Mathematical Sciences Research Institute wishes to acknowledge
support by the National Science Foundation.

PUBLISHED BY THE PRESS SYNDICATE OF THE UNIVERSITY OF CAMBRIDGE
The Pitt Building, Trumpington Street, Cambridge, United Kingdom

CAMBRIDGE UNIVERSITY PRESS
The Edinburgh Building, Cambridge CB2 2RU, UK www.cup.cam.ac.uk
40 West 20th Street, New York, NY 10011-4211, USA www.cup.org
10 Stamford Road, Oakleigh, Melbourne 3166, Australia
Ruiz de Alarćon 13, 28014 Madrid, Spain

First published 1996
First paperback edition 1998
Reprinted 1999

Library of Congress Cataloging-in-Publication Data is available.

A catalog record for this book is available from the British Library.

ISBN 0 521 57411 0 hardback
ISBN 0 521 64652 9 paperback

Transferred to Digital Printing 2004

Games of No Chance
MSRI Publications
Volume 29, 1996

Contents

Coda

Games of No Chance
MSRI Publications
Volume **29**, 1996

Preface

Combinatorial Game Theory, as an academic discipline, is still in its infancy. Many analyses of individual games have appeared in print, starting in 1902 with C. L. Bouton's analysis of the game of Nim. (For exact references to the works mentioned here, please see A. Fraenkel's bibliography on pages 493–537 of this volume.) It is was not until the 1930's that a consistent theory for impartial games was developed, independently, by R. Sprague and P. M. Grundy, later to be expanded and expounded upon by R. K. Guy and C. A. B. Smith. (Guy is still going strong, as evidenced by his energy at this Workshop.) J. H. Conway then developed the theory of partizan games, which represented a major advance. Based on this theory, D. Knuth wrote his modern moral tale, *Surreal Numbers*. The collaboration of E. R. Berlekamp, J. H. Conway and R. K. Guy gave us *Winning Ways*, which "set to music", or at least to "rhyme", the theory so far. In the process, many more games were discovered and analyzed: but more were discovered than solved!

This Workshop, held from 11 to 21 July 1994, gave evidence of the growing interest in combinatorial games on the part of experts from many fields: mathematicians, computer scientists, researchers in artificial intelligence, economists, and other social scientists. Players, some of whom make their living from games, also attended. Visitors such as D. Knuth and H. Wilf dropped in for a few hours or days and gave impromptu lectures. There was much cross-fertilization of ideas, as could be expected from a meeting of people from such varied backgrounds. One major paper by A. Fraenkel (pages 417–431) was conceived and essentially written during the Workshop, being inspired by V. Pless's talk.

But the Workshop was not all seminars. There were books, games and puzzles on display. Two official tournaments, Dots-and-Boxes and Domineering, attracted a lot of participants, and carried $500 first prizes (funded by E. R. B.) The final matches were shown over closed-circuit TV, so that the spectators could have a running commentary! (See pages 79–89.) Neither game is completely solved: Dots-and-Boxes is played by many school children, yet still holds mysteries for adults.

The articles in this volume are divided into four groups. Part I is introductory. Part II contains papers on some of the "classical" games, such as chess and

Go. Part III studies many other games, of greatly varying degrees of difficulty. Part IV contains articles that push the traditional theory in new directions: for example, by considering games more general than the strict definition of combinatorial games allows (see pages 1 and 363). The book closes with a list of unsolved problems by R. K. Guy, and a Master Bibliography by A. Fraenkel. The increasing role of computers can be witnessed throughout, in areas ranging from the solution of particular games to the use of the computer in teaching humans.

Many thanks must go the staff of MSRI, who helped make the Workshop a success. The facilities were wonderful. Thanks are due also to the Workshop chairs, E. R. Berlekamp and R. K. Guy. Together with the rest of the organizing committee (J. H. Conway, N. D. Elkies, A. S. Fraenkel, J. G. Propp, K. Thompson, and myself), they put together a wonderful and rich program.

In the preparation of this book, we are especially grateful to Silvio Levy, who essentially rewrote two of the articles, edited all the others, found good placements for the more than 200 figures, redrew some of them, and arranged the typesetting.

Richard J. Nowakowski

All Games Bright and Beautiful

What is a combinatorial game? The usual definition is a game in which

(i) there are two players moving alternately;
(ii) there are no chance devices and both players have perfect information;
(iii) the rules are such that the game must eventually end; and
(iv) there are no draws, and the winner is determined by who moves last.

In this section, two master expositors lead us through many examples of such games, and introduce the theory that has been developed to deal with them (pages 13–78). As an appetizer, you may prefer to read first J. H. Conway's charming study (pages 3–12) of a game that does not satisfy this definition, because it may well be endless. Or, if you already know the basic theory and are dying for action, turn to the reports of the Workshop tournament finals, on pages 79–89; or download the Gamesman's Toolkit, described on pages 93–98. Have fun!

Games of No Chance
MSRI Publications
Volume 29, 1996

The Angel Problem

JOHN H. CONWAY

ABSTRACT. Can the Devil, who removes one square per move from an infinite chessboard, strand the Angel, who can jump up to 1000 squares per move? It seems unlikely, but the answer is unknown. Andreas Blass and I have proved that the Devil *can* strand an Angel who's handicapped in one of several ways. I end with a challenge for the solution the general problem.

1. Introduction

The Angel and the Devil play their game on an infinite chessboard, with one square for each ordered pair of integers (x, y). On his turn, the Devil may eat any square of the board whatsoever; this square is then no longer available to the Angel. The Angel is a "chess piece" that can move to any uneaten square (X, Y) that is at most 1000 king's moves away from its present position (x, y)—in other words, for which $|X - x|$ and $|Y - y|$ are at most 1000. Angels have wings, so that it does not matter if any intervening squares have already been eaten.

The Devil wins if he can strand the Angel, that is, surround him by a moat of eaten squares of width at least 1000. The Angel wins just if he can continue to move forever.

What we have described is more precisely called an *Angel of power* 1000. The *Angel Problem* is this:

Determine whether an Angel of some power can defeat the Devil.

Berlekamp showed that the Devil can beat an Angel of power one (a chess King) on any board of size at least 32×33. However, it seems that it is impossible for the Devil to beat a Knight, and that would imply that an Angel of power two (which is considerably stronger than a Knight) will win. But can you prove it?

Well, nobody's managed to prove it yet, even when we make it much easier by making the Angel have power 1000 or some larger number. The main difficulty seems to be that the Devil cannot ever make a mistake: once he's made some moves, no matter how foolish, he is in a strictly better position than he was at the start of the game, since the squares he's eaten can only hinder the Angel.

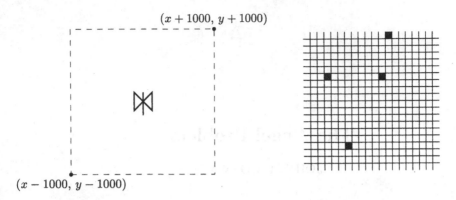

Figure 1. Left: An Angel about to fly; he can fly to any point in the square. Right: A Devil eating a square; three have already been eaten.

2. What's Wrong with Some Potential Strategies

Most people who think about this problem produce trial strategies for the Angel of the following general type. They advise the Angel to compute a certain "potential function" that depends on the eaten squares, and then to make a move that minimizes (or maximizes) this function.

The Devil usually finds it easy to beat such strategies, by finding some feature to which this function is too sensitive. Suppose, for instance, that the function is very sensitive to eaten squares close to the Angel. Then the Devil should build a horseshoe-shaped trap a light-year across, far to the north of the starting position. This will, of course, cause the Angel to fly southwards as fast as possible, but no matter. When the trap is set up, the Devil just goads the Angel into it by repeatedly eating the square just to the south of him. The Angel feels the pain of these "jabs at his behind" so strongly that he blindly stumbles right into the trap!

If the Angel switches to a new function that places great emphasis on eaten squares that are very far away, the Devil might build a mile-wide horseshoe and then scare the Angel into it by eating just a single square a megaparsec to the south of him.

The exact details of the Devil's strategy will of course depend on exact structure of the Angel's potential function. It seems to be impossible to get the function just right, and if it isn't just right, the Devil builds a trap that the potential function is insensitive to, and then steers the Angel into it using something to which the function is over-sensitive! A friend of mine, after hearing my talk of horseshoe-shaped traps, dreamed up a wonderful potential function that cleverly searched for horseshoe-shaped traps of all sizes and avoided them. I defeated it trivially by having the Devil use a modest little horseshoe that easily frightened the Angel into a trap of another shape!

3. Fools Rush On Where Angels Fear to Tread

About twenty years ago in Angell Hall, Ann Arbor, Andreas Blass and I spent some time with the Angel. It is reasonable to think that the Angel should be able to win by moving northward as fast as he can, combined with occasional zigzags to the east or west to avoid any obvious traps. So Blass and I made the following definition:

DEFINITION. A *Fool* is an Angel who is required always strictly to increase his y coordinate. So a Fool can make precisely those Angel moves from (x, y) to (X, Y) for which $Y > y$.

THEOREM 3.1. *The Devil can catch a Fool.*

PROOF. If the Fool is ever at some point P, he will be at all subsequent times in the "upward cone" from P, whose boundary is defined by the two upward rays of slope $\pm\frac{1}{1000}$ through P. Then we counsel the Devil to act as follows (Figure 2): he should truncate this cone by a horizontal line AB at a very large height H above the Fool's starting position, and use his first few moves to eat one out of every M squares along AB, where M is chosen so that this task will be comfortably finished when the Angel reaches a point Q on the halfway line that's distant $\frac{1}{2}H$ below AB (we'll arrange H to be exactly divisible by a large power of two).

At subsequent times, the Devil knows that the Fool will be safely ensconced in the smaller cone QCD, where CD is a subinterval of AB of exactly half its length, and for the next few of his moves, he should eat the second one of every M squares along the segment CD. He will have finished this by the time the Fool reaches a point R on the horizontal line $\frac{1}{4}H$ below AB. At later times, the

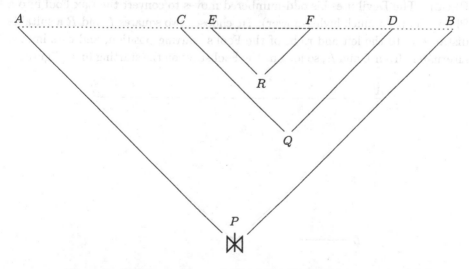

Figure 2. A Fool travelling north with the Devil eating along AB.

Fool will be trapped inside a still smaller cone REF, with $EF = \frac{1}{2}CD = \frac{1}{2}AB$, and the Devil should proceed to eat the third one of every M squares along the segment EF of AB.

If he proceeds in this way, then by the time the Fool reaches the horizontal line at distance $H' = 2^{-M}H$ below AB, the Devil will have eaten every square of the subsegment of AB that might still be reached by the Fool. The Devil should then continue, by eating the first out of every M squares on the segment $A'B'$ just below this one, a task which will be finished before the Fool reaches the horizontal line distant $\frac{1}{2}H'$ below $A'B'$, when he should start eating the second of every M squares on the portion $C'D'$ of $A'B'$ that is still accessible, and so on. We see that if we take H of the form 1000×2^N, where $N > 1000M$, then before the Fool crosses the horizontal line that is 1000 units below AB, the Devil will have eaten all squares between this line and AB that the Fool might reach, and so the Fool will be unable to move.   □

4. Lax Fools and Relaxed Fools

The kind of Fool we've been considering up to now is more precisely called a Plain Fool, since he may make only those Angel moves for which Y is strictly greater than y. Blass and I also showed how the Devil can beat some not-quite-so foolish Fools.

DEFINITION. A *Lax Fool* is a piece that can make precisely those Angel moves for which $Y \geq y$.

THEOREM 4.1. *The Devil can catch a Lax Fool.*

PROOF. The Devil uses his odd-numbered moves to convert the Lax Fool into a Plain Fool (of a much higher power). He chooses two squares L and R a suitable distance D to the left and right of the Fool's starting position, and eats inward alternately from L and R, so long as the Fool stays on the starting line (Figure 3).

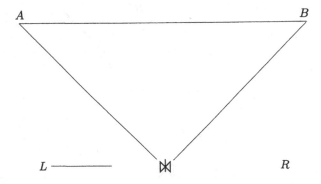

Figure 3. A Lax Fool travelling north.

Whenever the Fool makes an upwards move, the Devil takes the horizontal line through his new position as a new starting line.

Suppose that the Fool stays on the starting line for quite some time. Then he can use the four moves he has between two consecutive bites of the Devil at the left end of this line to move at most 4000 places. So, if we take $D = 4,000,000$, the Devil will have eaten 1000 consecutive squares at the left end of this line before the Fool can reach them. We now see that the Fool can stay on the starting line for at most 8,000,000 moves, since in 8,000,001 moves the Devil can eat every square between L and R.

If the Devil adopts this strategy, then the Fool must strictly increase his y coordinate at least once in every 8,000,000 moves. If we look at him only at these instants, we see him behaving like a Plain Fool of power 8,000,000,000, since each of his moves can have carried him at most a 1000 places. So if the Devil uses the strategy that catches an 8,000,000,000-power Plain Fool on his even-numbered moves, he'll safely trap the 1000-power Lax Fool.   □

Suppose now that the Angel does occasionally allow himself to decrease his y-coordinate, but promises never to make any sequence of moves that decreases it by more than a million.

DEFINITION. A *Relaxed Fool* (of laxity 1,000,000) is an Angel who promises that if he's ever at a position (X, Y), he will never be later at a position (x, y) with $y > Y - 1,000,000$.

THEOREM 4.2. *The Devil can catch a Relaxed Fool (of any laxity).*

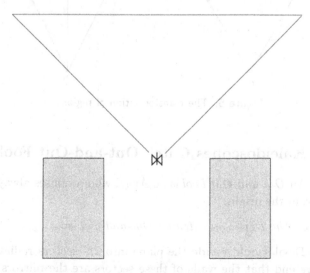

Figure 4. Funnelling the Relaxed Fool.

PROOF. This time, the Devil eats away sunken caissons of depth 1,000,000 at a suitable distance to the Left and Right of the Fool (see Figure 4). □

5. Variations on the Problem

Because our problem was inspired by Berlekamp's proof that the Devil can catch a chess King on all large enough boards, we've been using the chess King metric, in which the unit balls are squares aligned with the axes. However, the arguments that one meets early on in topology courses show that (provided one is allowed to change the number 1000) this really doesn't matter. It makes no difference to our problem if we suppose that the Angel is positioned at any time on an arbitrary point of the Euclidean plane, and can fly in one move to any other point at most 1000 units away, while the Devil may eat out all points of an arbitrary closed unit disc in any one of his moves.

In a similar way, we can prove that our problem is invariant under certain distortions. This happens for instance when the distorted image of every ball of radius R contains an undistorted ball of radius $\frac{1}{2}R$, and is contained in one of radius $2R$.

Suppose, for example, that the Angel promises always to stay inside the circular sector of Figure 5, and also that he will always increase his r-coordinate (his distance from the vertex O of this sector). Then the Devil can catch him, by distorting this sector into the triangle of the same figure, in such a way that the r-coordinate turns into the y-coordinate. But the transformation turns the Angel into a Plain Fool.

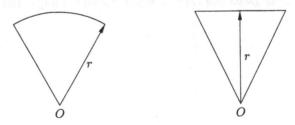

Figure 5. The transformation of regions.

6. Kaleidoscopes Catch Out-and-Out Fools.

DEFINITION. An *Out-and-Out Fool* is an Angel who promises always to increase his distance from the origin.

THEOREM 6.1. *The Devil can catch an Out-and-Out Fool.*

PROOF. The Devil should divide the plane into $2N$ sectors radiating from the origin, and pretend that the walls of these sectors are the mirrors of a kaleidoscope, as in Figure 6. There will then usually be $2N$ images of the Fool, one

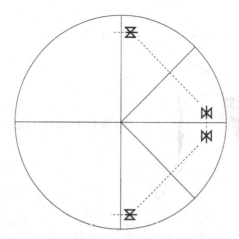

Figure 6. The kaleidoscope strategy.

in each sector. The Devil correspondingly distributes his moves between $2N$ demons, bidding each one of these to take care of the Fool's image in his own sector, which behaves like a Plain Fool of power $2000N$ confined as in Figure 5, and so can be caught by a slightly distorted Plain Fool strategy. □

Of course, the Devil can also catch the Relaxed Out-and Out Fool who merely promises never to make any sequence of moves that reduces his distance from the origin by more than 1,000,000 units.

7. An Extremely Diverting Strategy

THEOREM 7.1. *There is a strategy for the Devil that, when used, has the following property. For each point P of the plane and each distance D, no matter how the Angel moves there will be two times at the latter of which the Angel will be at least D units nearer to P than he was at the former.*

PROOF. It suffices to restrict P to the integral points of the plane, and the distances D to the integers. So we need only deal with countably many pairs (P, D), say (P_0, D_0), (P_1, D_1), (P_2, D_2), ... The Devil uses his odd-numbered moves to defeat an Angel that behaves like a Relaxed Out-and-Out Fool who has promised never to move more than D_0 units closer to P_0, his moves at twice odd numbers to catch one who promises never to move more than D_1 units closer to P_1, his moves at four times odd numbers to catch one who promises never to move more than D_2 units closer to P_2, and so on. Any sequence of moves by the real Angel that does not satisfy the conclusion of the theorem will lead to him being caught. □

It can be shown that, if the Devil employs our diverting strategy, there will be some very wiggly points on the Angel's trajectory, in the sense that he must cross any line through such a point at least 1000 times—no matter how he plays.

Angel's footprints

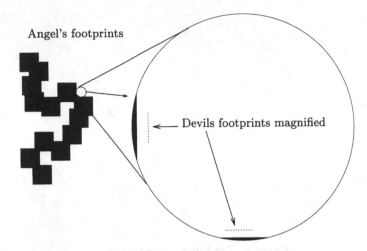

← — Devils footprints magnified

Figure 7. The Angel's large and the Devil's small footprints.

8. The Angel Is His Own Worst Enemy

The Devil only eats one square for each move of the Angel: we shall show that there is a sense in which the Angel himself eats millions!

THEOREM 8.1. *It does no damage to the Angel's prospects if we require him after each move to eat away every square that he could have landed on at that move, but didn't.*

PROOF. Consider the 2001-by-2001 array of squares that is accessible to the Angel from his starting position. We first show that the Devil, without harming his own prospects, can ensure that the Angel only returns to this array a bounded number of times.

If not, there will be indefinitely many times at which the Angel lands on some square of this array that he has already visited previously. At each such time, the Devil essentially has a free move, and he can use these moves to eat away a moat of width 1000 all around the array.

Now suppose that the Angel has any winning strategy whatsoever, say W. Then we shall convert W into another strategy that does not require him to land on any square on which he could have landed at a previous move. When he is at any point P, the Angel considers all the circumstances under which his winning strategy W causes him to repeatedly reenter the 2001-by-2001 array of squares accessible from P. He imagines himself in one of those circumstances that makes him re-enter this array the maximal number of times, and then "short-circuits" W by making right now the moves he would make on and after the last of these times. (He is in a better position now than he would be then, because the Devil has eaten fewer squares.)   □

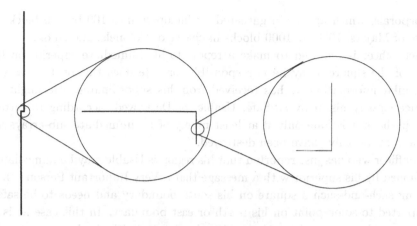

Figure 8. The straight and narrow path.

So the Angel is "really" burning out the path of all points within 1000 of each of his landing points as he moves. The Devil's consumption seems meager by comparison (see Figure 7).

9. Can the Angel Really Win?

If you've got this far, you've probably begun to wonder whether it's really quite safe to trust your fortune to the Angel. But please consider very carefully before selling your soul to the Devil! The Blass–Conway "Diverting Strategy" does indeed force the Angel to travel arbitrarily large distances in directions he might not have wanted to. But it only does so at unpredictable times in the middle of much larger journeys. So for example, at some stage in a megaparsec's journey north we can divert the Angel (say to the east) for a light-year or so; in the course of this smaller journey we might make him move south for an astronomical unit, with a mile's westward wandering in that, inside which we might divert him a foot or so to the north, and so on. If he moves like this, then (see Figure 8) we have indeed produced a point about which his path spirals, but the spiral is a very loose one indeed, and an observer at a suitably large scale would only see a very slight wobble on an almost entirely straight Northward path.

So, although we occasionally meet some unbelievers, and even after producing our Diverting Strategy, Blass and I remain firmly on the side of the Angel.

10. Kornering the Devil?

So does Tom Korner, who thinks the Devil might be defeated by the following kind of "military strategy". The Angel seconds from their places in Heaven a large number of military officers. He then puts a Private in charge of each individual square, and groups these by hundreds into 10-by-10 blocks in charge of Corporals, which are again gathered by hundreds into 100-by-100 blocks in charge of Majors, 1000-by-1000 blocks in charge of Colonels, and so on.

Each officer is required to make a report to his immediate superior on the status of the square array he is responsible for. He does this on the basis of the similar information he has received from his subordinates. He might for instance classify his array as Safe, Usable, or Destroyed, according to certain rules; perhaps it is Safe only if at least ninety of its immediate sub-arrays are safe, and at most five have been destroyed.

An officer who has just reported that his region is Usable may be immediately telephoned by his superior with a message that a Very Important Person is now entering such-and-such a square on his south boundary and needs to be safely transported to some point on his north or east boundary. In this case he is to plot a suitable route and telephone his subordinates with similar requests. An officer of the n-th rank whose region is Safe may face a stronger demand: perhaps he is only told that the VIP might arrive at any time in the next 10^n time units.

Korner thinks that perhaps the rules can be so designed that these orders can always be safely executed, and that at any time all officers of sufficiently high rank will make Safe reports, so that on a large enough scale the Angel can move completely freely. So do I, but neither of us has been able to turn these ideas into an explicit strategy for the Angel. All we can say is that a "military strategy" of this type could allow for Blass–Conway diversionary tactics, and is not so easy to hoodwink as the potential-function strategies, and so there might well be one that solves our problem.

Join the Fun!

This problem has been alive too long! So I offer $100 for a proof that a sufficiently high-powered Angel can win, and $1000 for a proof that the Devil can trap an Angel of any finite power.

JOHN H. CONWAY
DEPARTMENT OF MATHEMATICS
PRINCETON UNIVERSITY
FINE HALL
WASHINGTON ROAD
PRINCETON, NJ 08544

Games of No Chance
MSRI Publications
Volume **29**, 1996

Scenic Trails Ascending from Sea-Level Nim to Alpine Chess

AVIEZRI S. FRAENKEL

ABSTRACT. *Aim*: To present a systematic development of part of the theory of combinatorial games from the ground up. *Approach*: Computational complexity. Combinatorial games are completely determined; the questions of interest are efficiencies of strategies. *Methodology*: Divide and conquer. Ascend from Nim to chess in small strides at a gradient that's not too steep. *Presentation*: Informal; examples of games sampled from various strategic viewing points along scenic mountain trails illustrate the theory.

1. Introduction

All our games are two-player perfect-information games (no hidden information) without chance moves (no dice). Outcome is (lose, win) or (draw, draw) for the two players, who play alternately. We assume throughout *normal* play, i.e., the player making the last move wins and his opponent loses, unless *misère* play is specified, where the outcome is reversed. A draw is a dynamic tie, that is, a position from which neither player can force a win, but each has a nonlosing next move.

As we progress from the easy games to the more complex ones, we will develop some understanding of the *poset* of tractabilities and efficiencies of game strategies: whereas, in the realm of existential questions, tractabilities and efficiencies are by and large linearly ordered, from polynomial to exponential, for problems with an unbounded number of alternating quantifiers, such as games, the notion of a "tractable" or "efficient" computation is much more complex. (Which is more tractable: a game that ends after four moves, but it's undecidable who

---

Invited one-hour talk at MSRI Workshop on Combinatorial Games, July, 1994. Part of this paper was prepared while visiting Curtin University, Perth, WA. Thanks to my host Jamie Simpson for everything, and to Brian White and Renae Batina for their expert help in preparing the overhead pictures of the mountain scenes displayed during the oral presentation of this paper.

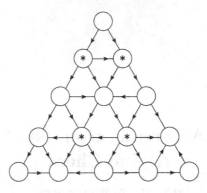

Figure 1. Beat Doug on this DAG (directed acyclic graph).

wins [Rabin 1957], or a game that takes an Ackermann number of moves to finish but the winner can play randomly, having to pay attention only near the end
[Fraenkel, Loebl and Nešetřil 1988]?)

When we say that a computation or strategy is polynomial or exponential,
we mean that the time it takes to evaluate the strategy is a polynomial or
exponential function in a most succinct form of the input size.

In Section 2 we review the classical theory (impartial games without draws,
no interaction between tokens). On our controlled ascent to chess we introduce
in Section 3 draws, on top of which we then add interaction between tokens in
Section 4. In Section 5 we review briefly partizan games. In Section 6 we show
how the approach and methodology outlined in the abstract can help to understand some of the many difficulties remaining in the classical theory (numerous
rocks are strewn also along other parts of the trails ascending towards chess).
This then leads naturally to the notion of tractable and efficient games, taken
up in Section 7, together with some more ways in which a game can become
intractable or inefficient.

This paper is largely expository, yet it contains material, mainly in parts of
Sections 6 and 7, not published before, to the best of our knowledge. The present
review is less formal than [Fraenkel 1991]: the emphasis here is on examples that
illustrate part of the theory. A fuller and more rigorous treatment is to appear
in [Fraenkel ≥ 1997].

2. The Classical Theory

Here we will learn to play games such as "Beat Doug" (Figure 1).

Place one token on each of the four starred vertices. A move consists of
selecting a token and moving it, along a directed edge, to a neighboring vertex
on this acyclic digraph. Tokens can coexist peacefully on the same vertex. For
the given position, what's the minimum time to:

(a) compute who can win;

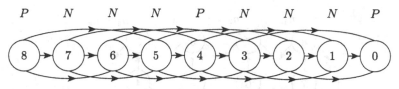

Figure 2. The game graph for Scoring, with initial score $n = 8$ and maximal step $t = 3$. Positions marked N are wins and P are losses for the player about to move.

(b) compute an optimal next move;
(c) consummate the win?

Following our divide and conquer methodology, let's begin with a very easy example, before solving Beat Doug. Given $n \in \mathbb{Z}^+$ (the initial score) and $t \in \mathbb{Z}^+$ (the maximal step size), a move in the game *Scoring* consists of selecting $i \in \{1, \ldots, t\}$ and subtracting i from the current score, initially n, to generate the new score. Play ends when the score 0 is reached.

The *game graph* $G = (V, E)$ for Scoring is shown in Figure 2, for $n = 8$ and $t = 3$. A position (vertex) $u \in V$ is labeled N if the player about to move from it can win; otherwise it's a P-position. Denoting by $\mathcal{P}$ the set of all P-positions, by $\mathcal{N}$ the set of all N-positions, and by $F(u)$ the set of all (direct) *followers* or *options* of any vertex u, we have, for any acyclic game,

$$u \in \mathcal{P} \quad \text{if and only if} \quad F(u) \subseteq \mathcal{N}, \tag{2.1}$$

$$u \in \mathcal{N} \quad \text{if and only if} \quad F(u) \cap \mathcal{P} \neq \varnothing. \tag{2.2}$$

As suggested by Figure 2, we have $\mathcal{P} = \{k(t+1) : k \in \mathbb{Z}^0\}$ and $\mathcal{N} = \{\{0, \ldots, n\} - \mathcal{P}\}$. The winning strategy consists of dividing n by $t + 1$. Then $n \in \mathcal{P}$ if and only if the remainder r is zero. If $r > 0$, the unique winning move is from n to $n - r$. Is this a "good" strategy?

INPUT SIZE: $\Theta(\log n)$ (*succinct* input).
STRATEGY COMPUTATION: $O(\log n)$ (linear scan of the $\lceil \log n \rceil$ digits of n).
LENGTH OF PLAY: $\lceil n/(t + 1) \rceil$.

Thus the computation time is linear in the input size, but the length of play is exponential. This does not prevent the strategy from being good: whereas we dislike computing in more than polynomial time, the human race relishes to see some of its members being tortured for an exponential length of time, from before the era of the Spanish matadors, through soccer and tennis, to chess and Go! But there are other requirements for making a strategy tractable, so at present let's say that the strategy is *reasonable*.

Now take k scores $n_1, \ldots, n_k \in \mathbb{Z}^+$ and $t \in \mathbb{Z}^+$, where $n_j \leq n$ for each j. A move consists of selecting one of the current scores and subtracting from it some $i \in \{1, \ldots, t\}$. Play ends when all the scores are zero. Figure 3 shows an example. This is a *sum* of Scoring games, itself also a Scoring game. It's easy to

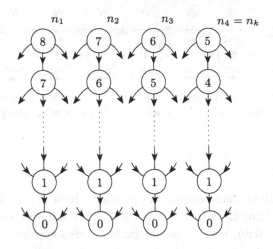

Figure 3. A Scoring game consisting of a sum of four Scoring games. Here $k = 4$, $n_1 = 8$, $n_2 = 7$, $n_3 = 6$, $n_4 = 5$, and $t = 3$.

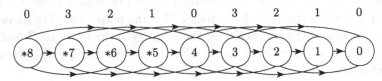

Figure 4. A game on a graph, but not a game graph.

see that the game of Figure 3 is equivalent to the game played on the digraph of Figure 4, with tokens on vertices 5, 6, 7 and 8. A move consists of selecting a token and moving it right by not more than $t = 3$ places. Tokens can coexist on the same vertex. Play ends when all tokens reside on 0. What's a winning strategy?

We hit two snags when trying to answer this question:

(i) The sum of N-positions is in $\mathcal{P} \cup \mathcal{N}$. Thus a token on each of 5 and 7 is seen to be an N-position (the move $7 \to 5$ clearly results in a P-position), whereas a token on each of 3 and on 7 is a P-position. So the simple P-, N-strategy breaks down for sums, which arise frequently in combinatorial game theory.

(ii) The game graph has exponential size in the input size $\Omega(k + \log n)$ of the "regular" digraph $G = (V, E)$ (with $|V| = n + 1$) on which the game is played with k tokens. However, this is not the game graph of the game: each tuple of k tokens on G corresponds to a single vertex of the game graph, whose vertex-set thus has size $\binom{k+n}{n}$—the number of k-combinations of $n+1$ distinct objects with at least k repetitions. For $k = n$ this gives $\binom{2n}{n} = \Theta(4^n / \sqrt{n})$.

The main contribution of the classical theory is to provide a polynomial strategy despite the exponential size of the game graph. On G, label each vertex u with the least nonnegative integer not among the labels of the followers of u (see top of Figure 4). These labels are called the *Sprague–Grundy* function of G, or the g-function for short [Sprague 1935–36; Grundy 1939]. It exists uniquely on every finite acyclic digraph. Then for $u = (u_1, \ldots, u_k)$, a vertex of the game graph (whose very construction entails exponential effort), we have

$$g(u) = g(u_1) \oplus \cdots \oplus g(u_k), \qquad \mathcal{P} = \{u : g(u) = 0\}, \qquad \mathcal{N} = \{u : g(u) > 0\},$$

where $\oplus$ denotes *Nim-sum* (summation over GF(2), also known as exclusive or). To compute a winning move from an N-position, note that there is some i for which $g(u_i)$ has a 1-bit at the binary position where $g(u)$ has its leftmost 1-bit. Reducing $g(u_i)$ appropriately makes the Nim-sum 0, and there's a corresponding move with the i-th token. For the example of Figure 4 we have

$$g(5) \oplus g(6) \oplus g(7) \oplus g(8) = 1 \oplus 2 \oplus 3 \oplus 0 = 0,$$

a P-position, so every move is losing.

Is the strategy polynomial? For Scoring, the remainders $r_1, \ldots, r_k$ of dividing $n_1, \ldots, n_k$ by $t+1$ are the g-values, as suggested by Figure 4. The computation of each r_j has size $O(\log n)$. Since $k \log n < (k + \log n)^2$, the strategy computation (items (a) and (b) at the beginning of this section) is polynomial in the input size. The length of play remains exponential.

Now consider a general nonsuccinct digraph $G = (V, E)$, by which we mean that the input size is not logarithmic. If the graph has $|V| = n$ vertices and $|E| = m$ edges, the input size is $\Theta((m + n) \log n)$ (each vertex is represented by its index of size $\log n$, and each edge by a pair of indices), and g can be computed in $O((m + n) \log n)$ steps (by a "depth-first" search; each g-value is at most n, of size at most $\log n$). For a sum of k tokens on the input digraph, the input size is $\Theta((k + m + n) \log n)$, and the strategy computation for the sum can be carried out in $O((k + m + n) \log n)$ steps (Nim-summing k summands of g-values). Note also that for a general digraph the length of play is only linear rather than exponential, as on a succinct (logarithmic input size) digraph.

Since the strategy for Scoring is polynomial for a single game as well as for a sum, we may say, informally, that it's a *tractable* strategy. (We'll see in Section 7 that there are further requirements for a strategy to be truly "efficient".)

Our original Beat Doug problem is now also solved with a tractable strategy. Figure 5 depicts the original digraph of Figure 1 with the g-values added in. Since $2 \oplus 3 \oplus 3 \oplus 4 = 6$, the given position is in $\mathcal{N}$. Moving $4 \to 2$ is a unique winning move. The winner can consummate his win in polynomial time.

Unfortunately, however, the strategy of classical games is not very robust: slight perturbations in various directions can make the analysis considerably more difficult. We'll return to this subject in Sections 6 and 7.

We point out that there is an important difference between the strategies of Beat Doug and Scoring. In both, the g-function plays a key role. But for the latter, some further property is needed to yield a strategy that's polynomial, since the input graph is (logarithmically) succinct. In this case the extra ingredient is the periodicity modulo $(t + 1)$ of g.

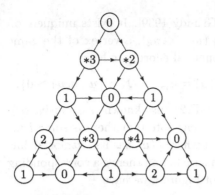

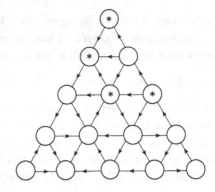

Figure 5. The beaten Doug.   **Figure 6.** Beat Craig in this cyclic digraph.

3. Introducing Draws

In this section we learn how to beat Craig efficiently. The four starred vertices in Figure 6 contain one token each. The moves are identical to those of Beat Doug, and tokens can coexist peacefully on any vertex. The only difference is that now the digraph $G = (V, E)$ may have cycles, and also loops (which correspond to passing a turn). In addition to the P- and N-positions, which satisfy (2.1) and (2.2), we now may have also Draw-positions D, which satisfy

$$u \in \mathcal{D} \quad \text{if and only if} \quad F(u) \cap \mathcal{P} = \varnothing \text{ and } F(u) \cap \mathcal{D} \neq \varnothing,$$

where $\mathcal{D}$ is the set of all D-positions.

Introducing cycles causes several problems:

- Moving a token from an N-position such as vertex 4 in Figure 7 to a P-position such as vertex 5 is a nonlosing move, but doesn't necessarily lead to a win. A win is achieved only if the token is moved to the leaf 3. The digraph might be embedded inside a large digraph, and it may not be clear to which P-follower to move in order to realize a win.

- The partition of V into $\mathcal{P}, \mathcal{N}$ and $\mathcal{D}$ is not unique, as it is for $\mathcal{P}$ and $\mathcal{N}$ in the classical case. For example, vertices 1 and 2 in Figure 7, if labeled P and N, would still satisfy (2.1) and (2.2), and likewise for vertices 8 and 9 (either can be labeled P and the other N).

Both of these shortcomings can be remedied by introducing a proper counter function attached to all the P-positions [Fraenkel $\geq$ 1977].

For handling sums, we would like to use the g-function, but there are two problems:

- The question of the existence of g on a digraph G with cycles or loops is NP-complete, even if G is planar and its degrees are ≤ 3, with each indegree ≤ 2 and each outdegree ≤ 2 [Fraenkel 1981] (see also [Chvátal 1973; van Leeuwen 1976; Fraenkel and Yesha 1979]).

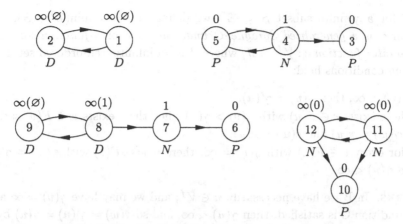

Figure 7. P-, N-, D- and γ-values for simple digraphs.

- The strategy of a cyclic game isn't always determined by the g-function, even if it exists.

This is one of those rare cases where two failures are better than one! The second failure opens up the possibility that perhaps there's another tool that always works, and if we are optimistic, we might even hope that it is also polynomial. There is indeed such a generalized g-function γ. It was introduced in [Smith 1966], and rediscovered in [Fraenkel and Perl 1975]; see also [Conway 1976, Ch. 11]. Of the possible definitions of γ, we give a simple one below; see [Fraenkel and Yesha 1986] for two other definitions and a proof of the equivalence of all three.

The γ-function is defined the same way as the g-function, except that it can assume not only values in $\mathbb{Z}^0 = \{n \in \mathbb{Z} : n \geq 0\}$, but also *infinite* values of the form $\infty(K)$, as we now explain. We have $\gamma(u) = \infty(K)$ if K is the set of finite γ-values of followers of u, and some follower v of u has the following properties: $\gamma(v)$ is also infinite, and v has no follower w with $\gamma(w)$ equal to the least nonnegative integer not in K. Figure 7 shows γ-values for some simple digraphs. Every finite digraph with n vertices and m edges has a unique γ-function that can be computed in $O(mn \log n)$ steps. This is a polynomial-time computation, though bigger than the g-values computation.

We also associate a (nonunique) counter function to every vertex with a finite γ-value, for the reasons explained above. Here is a more formal definition of γ. Given a digraph $G = (V, E)$ and a function $\gamma : V \to \mathbb{Z}^0 \cup \{\infty\}$, set

$$V^f = \{u \in V : \gamma(u) < \infty\}$$

and

$$\gamma'(u) = \operatorname{mex} \gamma\big(F(u)\big),$$

where, for any finite subset $S \subseteq \mathbb{Z}^0$, we define $\operatorname{mex} S = \min(\mathbb{Z}^0 - S)$. The function γ is a *generalized Sprague–Grundy function*, or *γ-function* for short, with *counter function* $c : V^f \to J$, where J is an infinite well-ordered set, if the following conditions hold:

A. If $\gamma(u) < \infty$, then $\gamma(u) = \gamma'(u)$.
B. If there exists $v \in F(u)$ with $\gamma(v) > \gamma(u)$, then there exists $w \in F(v)$ satisfying $\gamma(w) = \gamma(u)$ and $c(w) < c(u)$.
C. If, for every $v \in F(u)$ with $\gamma(v) = \infty$, there is $w \in F(v)$ with $\gamma(w) = \gamma'(u)$, then $\gamma(u) < \infty$.

REMARKS. In B we have necessarily $u \in V^f$; and we may have $\gamma(v) = \infty$ as in C. If condition C is satisfied, then $\gamma(u) < \infty$, and so $\gamma(w) = \gamma'(u) = \gamma(u)$ by A. Condition C is equivalent to the following statement:

C′. If $\gamma(u) = \infty$, then there is $v \in F(u)$ with $\gamma(v) = \infty(K)$ such that $\gamma'(u) \notin K$.

To keep the notation simple, we write throughout $\infty(0)$ for $\infty(\{0\})$, $\infty(0,1)$ for $\infty(\{0,1\})$, and so on.

To get a strategy for sums, define the *generalized Nim-sum* as the ordinary Nim-sum augmented by:

$$a \oplus \infty(L) = \infty(L) \oplus a = \infty(L \oplus a), \qquad \infty(K) \oplus \infty(L) = \infty(\varnothing),$$

where $a \in \mathbb{Z}^0$ and $L \oplus a = \{l \oplus a : l \in L\}$. For a sum of k tokens on a digraph $G = (V, E)$, let $\boldsymbol{u} = (u_1, \ldots, u_k)$. We then have $\gamma(\boldsymbol{u}) = \gamma(u_1) \oplus \cdots \oplus \gamma(u_k)$, and

$$\left.\begin{aligned}
\mathcal{P} &= \big\{\boldsymbol{u} : \gamma(\boldsymbol{u}) = 0\big\}, \\
\mathcal{N} &= \big\{\boldsymbol{u} : 0 < \gamma(\boldsymbol{u}) < \infty\big\} \cup \big\{\boldsymbol{u} : \gamma(\boldsymbol{u}) = \infty(K) \text{ and } 0 \in K\big\}, \\
\mathcal{D} &= \big\{\boldsymbol{u} : \gamma(\boldsymbol{u}) = \infty(K) \text{ and } 0 \notin K\big\}.
\end{aligned}\right\} \qquad (3.1)$$

Thus a sum consisting of a token on vertex 4 and one on 8 in Figure 7 has γ-value $1 \oplus \infty(1) = \infty(1 \oplus 1) = \infty(0)$, which is an N-position (the move $8 \to 7$ results in a P-position). Also one token on 11 or 12 is an N-position. But a token on both 11 and 12 or on 8 and 12 is a D-position of their sum, with γ-value $\infty(\varnothing)$. A token on 4 and 7 is a P-position of the sum.

With k tokens on a digraph, the strategy for the sum can be computed in $O((k + mn) \log n)$ steps. It is polynomial in the input size $\Theta((k + m + n) \log n)$, since $k + mn < (k + m + n)^2$. Also, for certain succinct "linear" graphs, γ provides a polynomial strategy. See [Fraenkel and Tassa 1975].

Beat Craig is now also solved with a tractable strategy. From the γ-values of Figure 8 we see that the position given in Figure 6 has γ-value $0 \oplus 1 \oplus 2 \oplus \infty(2,3) = 3 \oplus \infty(2,3) = \infty(1,0)$, so by (3.1) it's an N-position, and the unique winning move is $\infty(2,3) \to 3$. Again the winner can force a win in polynomial time, and can also delay it arbitrarily long, but this latter fact is less interesting.

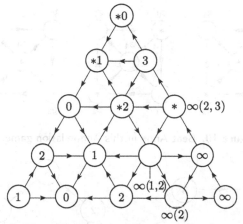

Figure 8. Craig has also been beaten.

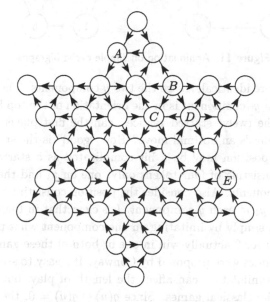

Figure 9. Beat an even bigger Craig.

As a homework problem, beat an even bigger Craig: compute the labels P, N, D for the digraph of Figure 9 with tokens placed as shown, or for various other initial token placements.

4. Adding Interactions between Tokens

Here we learn how to beat Anne. On the five-component digraph depicted in Figure 10, place tokens at arbitrary locations, but at most one token per vertex. A move is defined as in the previous games, but if a token is moved onto an occupied vertex, both tokens are annihilated (removed). The digraph has cycles, and could also have loops (passing positions). Note that the three components

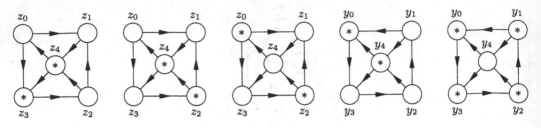

Figure 10. Beat Anne in this Ann-ihilation game.

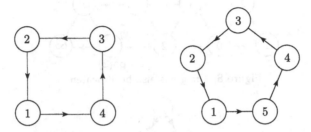

Figure 11. Annihilation on simple cyclic digraphs.

having z-vertices are identical, as are the two y-components. The only difference between a z- and a y- component is in the orientation of the top horizontal edge. With tokens on the twelve starred vertices, can the first player win or at least draw, and if so, what's an optimal move? How "good" is the strategy?

The indicated position may be a bit complicated as a starter. So consider first a position consisting of four tokens only: one on z_3 and the other on z_4 in two of the z-components. Also consider the position consisting of a single token on each of y_3 and y_4 in each y-component. It's clear that in these cases player 2 can at least draw, simply by imitating on one component what player 1 does on the other. Can player 2 actually win in one or both of these games?

Annihilation games were proposed by Conway. It's easy to see that on a finite acyclic digraph, annihilation can affect the length of play, but the strategy is the same as for the classical games. Since $g(u) \oplus g(u) = 0$, the winner doesn't need to use annihilation, and the loser cannot be helped by it. But the situation is quite different in the presence of cycles. In Figure 11 (left), a token on each of vertices 1 and 3 is clearly a D-position for the nonannihilation case, but it's a P-position when played with annihilation (the second move is a winning annihilation move). In Figure 11 (right), with annihilation, a token on each of 1 and 2 is an N-position, whereas a token on each of 1 and 3 is a D-position. The theory of annihilation games is discussed in depth in [Fraenkel and Yesha 1982]; see also [Fraenkel 1974; Fraenkel and Yesha 1976; 1979; Fraenkel, Tassa and Yesha 1978]. Ferguson [1984] considered misère annihilation play.

The *annihilation graph* is a certain game graph of an annihilation game. The annihilation graph of the annihilation game played on the digraph of Figure 11 (left) consists of two components. One is depicted in Figure 12, namely, the com-

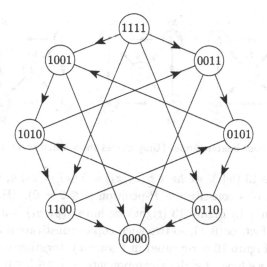

Figure 12. The "even" component G^0 of the annihilation graph G of the digraph of Figure 11 (left).

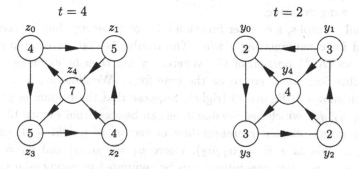

Figure 13. The γ-function.

ponent $G^0 = (V^0, E^0)$ with an even number of tokens. The "odd" component G^1 also has 8 vertices. In general, a digraph $G = (V, E)$ with $|V| = n$ vertices has an annihilation graph $\boldsymbol{G} = (\boldsymbol{V}, \boldsymbol{E})$ with $|\boldsymbol{V}| = 2^n$ vertices, namely all n-dimensional binary vectors. The γ-function on $\boldsymbol{G}$ determines whether any given position is in $\mathcal{P}$, $\mathcal{N}$ or $\mathcal{D}$, according to (3.1); and γ, together with its associated counter function, determines an optimal next move from an N- or D-position.

The only problem is the exponential size of $\boldsymbol{G}$. We can recover an $O(n^6)$ strategy by computing an *extended γ-function* $\underline{\gamma}$ on an induced subgraph of $\boldsymbol{G}$ of size $O(n^4)$, namely, on all vectors of weight ≤ 4 (at most four 1-bits). In Figure 13, the numbers inside the vertices are the $\underline{\gamma}$-values, computed by Gaussian elimination over GF(2) of an $n \times O(n^4)$ matrix. This computation also yields the values $t = 4$ for Figure 13 (left) and $t = 2$ for Figure 13 (right): If $\underline{\gamma}(\boldsymbol{u}) \geq t$, then $\gamma(\boldsymbol{u}) = \infty$, whereas $\underline{\gamma}(\boldsymbol{u}) < t$ implies $\gamma(\boldsymbol{u}) = \underline{\gamma}(\boldsymbol{u})$.

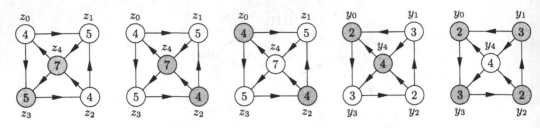

Figure 14. Poor beaten Anne. (Gray circles show initial token positions.)

Thus for Figure 13 (left), we have $\underline{\gamma}(z_3, z_4) = 5 \oplus 7 = 2 < 4$, so $\gamma(z_3, z_4) = 2$. Hence two such copies constitute a P-position ($2 \oplus 2 = 0$). (How can player 2 consummate a win?) In Figure 13 (right) we have $\underline{\gamma}(y_3, y_4) = 3 \oplus 4 = 7 > 2$, so $\gamma(y_3, y_4) = \infty$, in fact, $\infty(0, 1)$, so two such copies constitute a D-position. The position given in Figure 10 is repeated in Figure 14, together with the γ-values. From left to right we have: for the z-components, $\gamma = 2 \oplus 3 \oplus 0 = 1$; and for the y-components, $\infty(0, 1) + 0 = \infty(0, 1)$, so the γ-value is $\infty(0, 1) \oplus 1 = \infty(0, 1)$. Hence the position is an N-position by (3.1). There is, in fact, a unique winning move, namely $y_4 \to y_3$ in the second component from the right. Any other move leads to drawing or losing.

For small digraphs, a counter function c is not necessary, but for larger ones it's needed for consummating a win. The trouble is that we computed γ and c only for an $O(n^4)$ portion of $\mathbf{G}$. Whereas γ can then be extended easily to all of $\mathbf{G}$, this does not seem to be the case for c. We illustrate a way out on the digraph shown in Figure 13 (right). Suppose that the beginning position is $\mathbf{u} = (y_0, y_1, y_2, y_3)$, which has γ-value 0, as can be seen from Figure 13.

With $\mathbf{u}$ we associate a *representation* of vectors of weight ≤ 4, each with γ-value 0, in this case, $\tilde{\mathbf{u}} = (\mathbf{u}_1, \mathbf{u}_2)$, where $\mathbf{u}_1 = (y_0, y_2)$ and $\mathbf{u}_2 = (y_1, y_3)$, with $\mathbf{u} = \mathbf{u}_1 \oplus \mathbf{u}_2$. Representations can be computed in polynomial time, and a counter function c can be defined on them. Suppose player 1 moves from $\mathbf{u}$ to $\mathbf{v} = (y_1, y_2)$. The representation $\tilde{\mathbf{v}}$ is obtained by carrying out the move on the representation $\tilde{\mathbf{u}}$: $F_{0,3}(\mathbf{u}_1)$ (i.e., moving from y_0 to y_3) gives $\mathbf{u}_3 = (y_2, y_3)$, so $\tilde{\mathbf{v}} = (\mathbf{u}_3, \mathbf{u}_2)$, with $\mathbf{u}_3 \oplus \mathbf{u}_2 = \mathbf{v}$. Now player 2 would like to move to some $\mathbf{w}$ with $\gamma(\mathbf{w}) = 0$ and $c(\tilde{\mathbf{w}}) < c(\tilde{\mathbf{u}})$, namely, $\tilde{\mathbf{w}} = (\mathbf{u}_2)$. However, we see that $\mathbf{u}_2$ is a predecessor (immediate ancestor) of $\mathbf{v}$, rather than a follower of $\mathbf{v}$. Player 2 now pretends that player 1 began play from $\mathbf{u}_2$ rather than from $\mathbf{u}$, so arrived at $\mathbf{v}$ with representation $F_{3,2}(\mathbf{u}_2) = \tilde{\mathbf{v}}$. This has the empty representation as follower, so player 2 makes the final annihilation move. Followers of representations can always be chosen with c-value smaller than the c-value of their grandfather, and they always correspond to either a follower or predecessor of a position. Since the initial counter value has value $O(n^5)$, player 2 can win in that many moves, using an $O(n^6)$ computation.

This method can easily be extended to handle sums. Thus, according to our definition, we have a tractable strategy for annihilation games. Yet clearly it would be nice to improve on the $O(n^6)$ and to simplify the construction of the counter function.

Is there a narrow winning strategy that's polynomial? A strategy is *narrow* if it uses only the present position u for deciding whether u is a P-, N-, or D-position, and for computing a next optimal move. It is *broad* [Fraenkel 1991] if the computation involves any of the possible predecessors of u, whether actually encountered or not. It is *wide* if it uses any ancestor that was actually encountered in the play of the game. Kalmár [1928] and Smith [1966] defined wide strategies, but then both immediately reverted back to narrow strategies, since both authors remarked that the former do not seem to have any advantage over the latter. Yet for annihilation games we were able to exhibit only a broad strategy that is polynomial. Is this the alpine wind that's blowing?

Incidentally, for certain (Chinese) variations of Go, for chess and some other games, there are rules that forbid certain repetitions of positions, or modify the outcome in the presence of such repetitions. Now if all the history is included in the definition of a move, then every strategy is narrow. But the way Kalmár and Smith defined a move—much the same as the intuitive meaning—there is a difference between a narrow and wide strategy for these games. We also mention here the notion of "positional strategy" [Ehrenfeucht and Mycielski 1979; Zwick and Paterson ≥1996]. See also [Beck 1981; 1982; 1985; Chvátal and Erdős 1978].

As a homework problem, compute the label $\in \{P, N, D\}$ of the stellar configuration marked by letters in "Simulation of the SL comet fragments' encounter with Jupiter" (Figure 15), where J is Jupiter, the other letters are various fragments of the comet, and all the vertices are "space-stations". A move consists of selecting Jupiter or a fragment, and moving it to a neighboring space-station along a directed trajectory. Any two bodies colliding on a space-station explode and vanish in a cloud of interstellar dust. Note the six space-stations without exit, where a body becomes a "falling star". Is the given position a win for the (vicious) player 1, who aims at making the last move in the destruction of this subsystem of the solar system, or for the (equally vicious) player 2? Or is it a draw, so that a part of this subsystem will exist forever? And if so, can it be arranged for Jupiter to survive as well? (The encounter of the Shoemaker–Levy comet with Jupiter took place during the MSRI workshop.)

Various impartial and partizan variations of annihilation games were shown to be NP-hard, Pspace-complete or Exptime-complete [Fraenkel and Yesha 1979; Fraenkel and Goldschmidt 1987; Goldstein and Reingold 1995]. We mention here only briefly an interaction related to annihilation. Electrons and positrons are positioned on vertices of the game Matter and Antimatter (Figure 16). A move consists of moving a particle along a directed trajectory to an adjacent station—if not occupied by a particle of the same kind, since two electrons (and two positrons) repel each other. If there is a resident particle, and the incoming particle is of opposite type, they annihilate each other, and both disappear from the play. It is not very hard to determine the label of any position on the

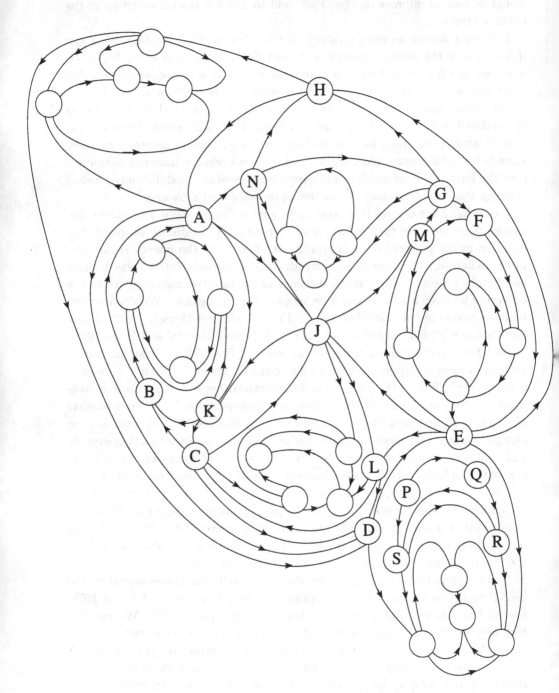

Figure 15. Interstellar encounter with Jupiter.

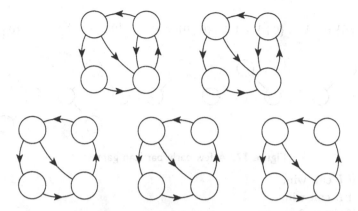

Figure 16. Matter and Antimatter.

given digraph. But what can be said about a general digraph? About succinct digraphs? Note that the special case where all the particles are of the same type, is the generalization of Welter played on the given digraph. Welter [Conway 1976, Ch. 13] is Nim with the restriction that no two piles have the same size. It has a polynomial strategy, but its validity proof is rather intricate. In Nim we are given finitely many piles. A move consists of selecting a pile and removing any positive number of tokens from it. The classical theory (Section 2) shows that the P-positions for Nim are simply those pile collections whose sizes Nim-add to zero.

5. Partizan Games

A game is *impartial* if both players have the same set of moves for all game positions. Otherwise the game is *partizan*. Nim-like games are impartial. Chess-like games are partizan, because if Gill plays the black pieces, Jean will not let him touch the white pieces.

In this section we shall refer to partizan games simply as games. The following two inductive definitions are due to Conway [1976; 1977; 1978a].

(i) If M^L and M^R are any two sets of games, there exists a game $M = \{M^L \mid M^R\}$. All games are constructed in this way.

(ii) If LE and RI are any two sets of numbers and no member of LE is $\geq$ any member of RI, then there exists a number $\{LE \mid RI\}$. All numbers are constructed in this way.

Thus the numbers constitute a subclass of the class of games.

The first games and numbers are created by putting $M^L = M^R = LE = RI = \varnothing$. Some samples are given in Figure 17, where L (Left) plays to the south-west and R (Right) to the south-east. If, as usual, the player first unable to move is the loser and his opponent the winner, then the examples suggest the following statements:

$$\{\mid\} = 0 \qquad \{\mid 0\} = -1 \qquad \{0 \mid \} = 1 \qquad \{0 \mid 0\} = * \qquad \{0 \mid 1\} = \tfrac{1}{2} \qquad \{0 \mid *\} = \uparrow$$

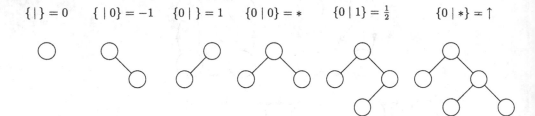

Figure 17. A few early partizan games.

$M > 0$ if L can win,

$M < 0$ if R can win,

$M = 0$ if player 2 can win,

$M \parallel 0$ if player 1 can win (for example in $*$).

We shall in fact define $>$, $<$, $=$, $\parallel$ by these conditions. The relations can be combined as follows:

$M \geq 0$ if L can win as player 2,

$M \leq 0$ if R can win as player 2,

$M \rhd 0$ if L can win as player 1,

$M \lhd 0$ if R can win as player 1.

Alternative inductive definitions of $\leq$ and $\parallel$ can be given. Let $x = \{x^L \mid x^R\}$. Then

$x \leq y$ if and only if $x^L \lhd y$ for all x^L and $x \lhd y^R$ for all y^R,

$x = y$ if and only if $x \leq y$ and $y \leq x$,

$x \parallel y$ (x is *fuzzy* with y) if and only if $x \not\leq y$ and $y \not\leq x$.

One can also provide a consistency proof of both definitions. This enables one to prove many properties of games in a simple manner. For example, define $-M = \{-M^R \mid -M^L\}$. Then $G - G = 0$ (player 2 can win in $G - G$ by imitating the moves of player 1 in the other game). Also:

$$x \leq y \lhd z \qquad \text{or} \qquad x \lhd y \leq z \Rightarrow x \lhd z$$

(consider the game $z - x = (z - y) + (y - x)$, in which L can win as first player), and

$$x^L \lhd x \lhd x^R$$

(clearly L can win as player 1 in $x - x^L$ and in $x^R - x$).

If x is a number, then $x^L < x < x^R$.

Most important is that a sum of games simply becomes a *sum*, defined by:

$$M + H = \{M^L + H, M + H^L \mid M^R + H, M + H^R\}.$$

Consider for example the game of Domineering [Conway 1976, Ch. 10; Berlekamp, Conway and Guy 1982, Ch. 5], played with dominoes each of which covers precisely two squares of an $n \times n$ chessboard. The player L tiles vertically, R

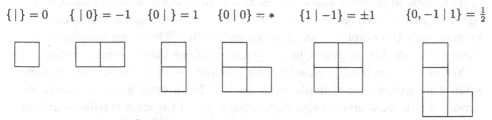

$\{\,|\,\} = 0$ $\{\,|\,0\} = -1$ $\{0\,|\,\} = 1$ $\{0\,|\,0\} = *$ $\{1\,|-1\} = \pm 1$ $\{0, -1\,|\,1\} = \tfrac{1}{2}$

Figure 18. A few values of a Domineering game.

horizontally. After Domineering is played for a while, the board may break up into several parts, whence the game becomes a sum of these parts. The values of several small configurations are given in Figure 18.

Since the relation between the value of a game M and 0 determines the strategy for playing it, computing the value of M is fundamental to the theory. In this direction the *Simplicity Theorem* is helpful: Let $x = \{x^L \mid x^R\}$ be a game. If there exists a *number* z such that

$$x^L \lhd z \rhd x^R$$

and no option of z satisfies this relation, then $x = z$. Thus $\{-1 \mid 99\} = 0$. There also exist algorithms for computing values associated with games that are not numbers. Partizan games with possible draws are discussed in [Li 1976; Conway 1978b; Shaki 1979; Fraenkel and Tassa 1982; Berlekamp, Conway and Guy 1982, Ch. 11].

Sums of partizan games are Pspace-complete [Morris 1981], even if the component games have the form $\{a \,\|\, \{b\,|\,c\}\}$ with $a, b, c \in \mathbb{Z}$ [Yedwab 1985; Moews 1994].

6. Sticky Classical Games

As we mentioned in the penultimate paragraph of Section 2, the classical theory is nonrobust (the same holds, even more so, for nonclassical theories). Let's examine a few of the things that can go wrong.

First consider misère play. The *pruned game graph* G, that is, the game graph from which all leaves have been pruned, gives the same full information for misère play as the game graph gives for normal play: its P-, N-positions determine who can win. But in the nontrivial cases, G is exponentially large, so of little help. The game graph G is neither the sum of its components nor of its pruned components, which might explain why nobody seems to have a general theory for misère play, though important advances have been made for many special subclasses and particular games. See, for example, [Berlekamp, Conway and Guy 1982, Ch. 13; Sibert and Conway 1992; Plambeck 1992; Banerji and Dunning 1992].

The classical theory is also very sensitive to interaction between tokens, which again expresses itself as the inability to decompose the game into sums. We mention only two examples: annihilation and Welter. There's an involved special theory for each of these games, but no general unifying theory seems to be known.

Yet another problem is posed by the succinctness of the input size of many interesting games—for example, *octal games*. In an octal game, n tokens are arranged as a linear array (input size: $\Theta(\log n)$), and the array is reduced and/or subdivided according to rules encoded in octal [Guy and Smith 1956; Berlekamp, Conway and Guy 1982, Ch. 4; Conway 1976, Ch. 11]. It's not known whether there's an octal game with finitely many nonzero octal digits that doesn't have a polynomial strategy, though for many such octal games no polynomial strategy has (yet?) been found. Some of them have such a huge period or preperiod that their polynomial nature asserts itself only for impractically high values of n. See [Gangolli and Plambeck 1989].

Of course all these and many more problems bestow upon the class of combinatorial games its distinctive, interesting and challenging flavor.

Let us now examine in some more detail the difficulties presented by a particular classical game, played with just two piles: Wythoff's game [Wythoff 1907; Yaglom and Yaglom 1967]. In this game there's a choice between two types of moves: you can either take any positive number of tokens from a single pile, just as in Nim, or you can remove the same (positive) number of tokens from both piles.

The first few g-values are depicted in Table 1. The structure of the 0's is well-understood, and can be computed in polynomial time [Fraenkel 1982; Fraenkel and Borosh 1973]. But the nonzero values appear at positions that are not yet well-understood. Some structure of the 1-values is portrayed in [Blass and Fraenkel 1990]; see [Pink 1991] for further developments. But no polynomial construction of all the nonzero g-values seems to be known—not even whether such a construction is likely to exist or not.

Why is this the case? The experts say that it's due to the nondisjunctive move of taking from both piles. To test this opinion, let's consider a game, to be called $(k, k+1)$-Nimdi (for reasons to become clear later); see Table 2 for the first few g-values. In this game a move consists of either taking any positive

0	1	2	3	4	5	6	7	8	9	10	11
1	2	0	4	5	3	7	8	6	10	11	9
2	0	1	5	3	4	8	6	7	11	9	10
3	4	5	6	2	0	1	9	10	12	8	7
4	5	3	2	7	6	9	0	1	8	13	14
5	3	4	0	6	8	10	1	2	7	12	15
6	7	8	1	9	10	3	4	5	13	0	2

Table 1. (k, k)-Wythoff.

number from a single pile, or else k from one and $k + 1$ from the other, for an arbitrary $k \in \mathbb{Z}^+$.

It won't take long for the reader to see that these values are exactly the same as those for *Nim*. The same holds if "for an arbitrary k" is replaced by "for a fixed k", say $k = 2$. So Table 2 also gives the g-values for $(2,3)$-Nimdi. Let's now consider the same game, but with $(2,3)$ replaced by $(1,3)$. In other words, a move consists of taking any positive number of tokens from a single pile, or else 1 token from one pile and 3 from the other. The first few g-values are listed in Table 3.

The next empty entry, for $(2,3)$, should be $2 \oplus 3 = 1$, according to the Nim-sum rule. However, the true value is 4. The reason is that $(2,3)$ has a follower $(2,3) - (1,3) = (1,0)$, which already has value 1. In other words, the g-value 1 of Nim has been "short-circuited" in $(1,3)$-Nimhoff! Note that in Wythoff, taking (k,k) short-circuits 0-values, but in $(k, k+1)$-Nimdi, no g-values have been short-circuited.

More generally, given piles of sizes $(a_1, \ldots, a_n)$ and a move set of the form $S = (b_1, \ldots, b_n)$, where $a_i \in \mathbb{Z}^+$ and $b_i \in \mathbb{Z}^0$ for $i \in \{1, \ldots, n\}$ such that $b_1 + \ldots + b_n > 0$, the moves of the game *Take* are of two types:

(i) Taking any positive number m from a single pile, so

$$(a_1, \ldots, a_n) \rightarrow (a_1, \ldots, a_{i-1}, a_i - m, a_{i+1}, \ldots, a_n).$$

(ii) Taking $b_1, \ldots, b_n$ from all the piles, so

$$(a_1, \ldots, a_n) \rightarrow (a_1 - b_1, \ldots, a_n - b_n).$$

Under what conditions is Take a Nimdi (Nim-in-disguise) game, i.e., a game whose g-values are exactly those of Nim? Blass and Fraenkel (to appear) have proved that Take is a Nimdi-game if and only if S is an odd set in the following sense: let m be the nonnegative integer such that $a_i 2^{-m}$ is an integer for all i but $a_j 2^{-m-1}$ is not an integer for some j. Then S is *odd* if $\sum_{k=1}^{n} a_k 2^{-m}$ is an odd integer. The divide and conquer methodology now suggests that we approach Wythoff gradually, by short-cicuiting only retricted sets of g-values of Nim, thus

0	1	2	3	4	5	6	7	8	9	10	11
1	0	3	2	5	4	7	6	9	8	11	10
2	3	0	1	6	7	4	5	10	11	8	9
3	2	1	0	7	6	5	4	11	10	9	8

Table 2. The first few g-values of $(k, k+1)$-Nimdi.

0	1	2	3	4	5	6	7	8	9	10	11
1	0	3	2	5	4	7	6	9	8	11	10
2	3	0									

Table 3. The first few g-values of $(1,3)$-Nimhoff.

generating a family of non-Nimdi games. Several such case studies of Nimhoff games are included in [Fraenkel and Lorberbom 1991].

For example, consider *Cyclic Nimhoff*, so named because of the cyclic structure of the g-values, where the restriction is $0 < \sum_{i=1}^{n} b_i < h$, where $h \in \mathbb{Z}^+$ is a fixed parameter. Thus $h = 1$ or $h = 2$ is Nim; $n = 2$, $h = 3$ is the fairy chess king-rook game; and $n = 2$, $h = 4$ is the fairy chess game king-rook-knight game. For a general cyclic Nimhoff game with pile sizes $(a_1, \ldots, a_n)$, we have

$$g(a_1, \ldots, a_n) = h(\lfloor a_1/h \rfloor \oplus \cdots \oplus \lfloor a_n/h \rfloor) + (a_1 + \cdots + a_n) \bmod h.$$

This formula implies that cyclic Nimhoff has a polynomial strategy for every fixed h. Note the combination of Nim-sum and ordinary sum, somewhat reminiscent of the strategy of Welter.

As another example, consider 2^k-*Nimhoff*, where k is any fixed positive integer. In this game we can remove 2^k tokens from two distinct piles, or remove a positive number of tokens from any single pile. Define the k-*Nim-sum* by $a \circledk b = a \oplus b \oplus a^k b^k$. In other words, the k-Nim-sum of a and b is $a \oplus b$, unless the k-th bits of a and b are both 1, in which case the least significant bit of $a \oplus b$ is complemented. The k-Nim-sum is not a generalization of Nim-sum, but it is associative. For 2^k-Nimhoff with n piles we then have

$$g(a_1, \ldots, a_n) = a_1 \circledk \cdots \circledk a_n.$$

Now that we have gained some understanding of the true nature of Wythoff's game, we can exploit it in at least two ways:

1. Interesting games seem to be obtained when we adjoin to a game its P-positions as moves! For example, consider W^2, which is Nim with the adjunction of (k, k) and Wythoff's P-positions as moves. We leave it as an exercise to compute the P-positions of W^2, and to iterate other games in the indicated manner.

2. A generalization of Wythoff to more than two piles has long been sought. It's now clear what has to be done: for three-pile Wythoff, the moves are to either take any positive number of tokens from a single pile, or take from all three, say k, l, m (with $k + l + m > 0$), such that $k \oplus l \oplus m = 0$. This is clearly a generalization of the usual two-pile Wythoff's game. Initial values of the P-positions [Chaba and Fraenkel] are listed in Table 4, namely the cases $j = 0$ (one of the three piles is empty—the usual Wythoff game) and $j = 1$ (one of the three piles has size 1). Recall that for two-pile Wythoff the golden section plays an important role. The same holds for three-pile Wythoff, except that there are many "initial disturbances", as in so many other impartial games.

A rather curious variation of the classical theory is *epidemiography*, motivated by the study of long games, especially the Hercules–Hydra game; see the survey article [Nešetřil and Thomas 1987]. Several perverse and maniacal forms of the malady were examined in [Fraenkel and Nešetřil 1985; Fraenkel, Loebl and Nešetřil 1988; Fraenkel and Lorberbom 1989]. The simplest variation is a mild case of *Dancing Mania*, called *Nimania*, sometimes observed in post-*pneumonia* patients.

In Nimania we are given a positive integer n. Player 1 begins by subtracting 1 from n. If $n = 1$, the result is the empty set, and the game ends with player 1 winning. If $n > 1$, one additional copy of the resulting number $n - 1$ is adjoined, so at the end of the first move there are two (indistinguishable) copies of $n - 1$ (denoted $(n - 1)^2$). At the k-th stage, where $k \geq 1$, a move consists of selecting a copy of a positive integer m of the present position, and subtracting 1 from it. If $m = 1$, the copy is deleted. If $m > 1$, then k copies of $m - 1$ are adjoined to the resulting $m - 1$ copies.

D^0	j	B^0	C^0	D^0	j	B^0	C^0	D^1	j	B^1	C^1	D^1	j	B^1	C^1
0	0	0	0	31	0	50	81	0	1	1	1	34	1	50	84
1	0	1	2	32	0	51	83	1	1	3	4	35	1	53	88
2	0	3	5	33	0	53	86	4	1	5	9	36	1	54	90
3	0	4	7	34	0	55	89	6	1	6	12	37	1	55	92
4	0	6	10	35	0	56	91	7	1	7	14	38	1	57	95
5	0	8	13	36	0	58	94	3	1	8	11	39	1	59	98
6	0	9	15	37	0	59	96	8	1	10	18	40	1	61	101
7	0	11	18	38	0	61	99	9	1	13	22	41	1	62	103
8	0	12	20	39	0	63	102	5	1	15	20	42	1	63	105
9	0	14	23	40	0	64	104	12	1	16	28	43	1	65	108
10	0	16	26	41	0	66	107	10	1	17	27	44	1	66	110
11	0	17	28	42	0	67	109	11	1	19	30	45	1	67	112
12	0	19	31	43	0	69	112	15	1	21	36	46	1	69	115
13	0	21	34	44	0	71	115	16	1	23	39	47	1	71	118
14	0	22	36	45	0	72	117	13	1	24	37	48	1	72	120
15	0	24	39	46	0	74	120	18	1	25	43	49	1	74	123
16	0	25	41	47	0	76	123	14	1	26	40	50	1	76	126
17	0	27	44	48	0	77	125	20	1	29	49	51	1	78	129
18	0	29	47	49	0	79	128	21	1	31	52	52	1	80	132
19	0	30	49	50	0	80	130	19	1	32	51	53	1	82	135
20	0	32	52	51	0	82	133	23	1	33	56	54	1	83	137
21	0	33	54	52	0	84	136	24	1	34	58	55	1	85	140
22	0	35	57	53	0	85	138	25	1	35	60	56	1	86	142
23	0	37	60	54	0	87	141	26	1	38	64	57	1	87	144
24	0	38	62	55	0	88	143	27	1	41	68	58	1	89	147
25	0	40	65	56	0	90	146	28	1	42	70	59	1	91	150
26	0	42	68	57	0	92	149	29	1	44	73	60	1	93	153
27	0	43	70	58	0	93	151	30	1	45	75	61	1	94	155
28	0	45	73	59	0	95	154	31	1	46	77	62	1	96	158
29	0	46	75	60	0	97	157	32	1	47	79	63	1	97	160
30	0	48	78	61	0	98	159	33	1	48	81	64	1	99	163

Table 4. Initial P-positions in three-pile Wythoff.

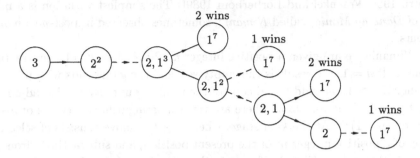

Figure 19. Player 1 can win Nimania for $n = 3$ in 13 moves. Solid arrows indicate Player 1's moves, and dashed arrows those of Player 2.

Since the numbers in successive positions decrease, the game terminates. Who wins? For $n = 1$ we saw above that player 1 wins. For $n = 2$, player 1 moves to 1^2, player 2 to 1, hence player 1 again wins. For $n = 3$, Figure 19 shows that by following the lower path, player 1 can again win. Unlike the cases $n = 1$ and $n = 2$, however, not all moves of player 1 for $n = 3$ are winning.

An attempt to resolve the case $n = 4$ by constructing a diagram similar to Figure 19 is rather frustrating. It turns out that for $n = 4$ the loser can delay the winner so that play lasts over 2^{44} moves! We have proved, however, the following surprising facts:

(i) Player 1 can win for every $n \geq 1$.

(ii) For $n \geq 4$, player 1 cannot hope to see his win being consummated in any reasonable amount of time: the smallest number of moves is $\geq 2^{2^{n-2}}$, and the largest is an Ackermann function.

(iii) For $n \geq 4$, player 1 has a *robust* winning strategy: most of the time player 1 can make random moves; only near the end of play does player 1 have to pay attention (as we saw for the case $n = 3$).

In view of (ii), where we saw that the length of play is at least doubly exponential, it seems reasonable to say that Nimania is not tractable, though the winning strategy is robust.

7. What's a Tractable or Efficient Strategy?

We are not aware that these questions have been addressed before in the literature. Since "Nim-type" games are considered to have good strategies, we now abstract some of their properties in an attempt to define the notions of a "tractable" or "efficient" game, in a slightly more formal way than the way we have used them above.

The subset T of combinatorial games with a *tractable* strategy has the following properties. For normal play of every $G \in T$, and every position u of G:

(a) The P-, N- or D-label of u can be computed in polynomial time.

(b) The next optimal move (from an N- to a P-position; from D- to a D-position) can be computed in polynomial time.

(c) The winner can consummate the win in at most an exponential number of moves.

(d) The set T is closed under summation, i.e., $G_1, G_2 \in T$ implies $G_1 + G_2 \in T$ (so (a), (b), (c) hold for $G_1 + G_2$).

The subset $T' \subseteq T$ for which (a)–(d) hold also for misère play is the subset of *efficient* games.

REMARKS. (i) Instead of "polynomial time" in (a) and (b) we could have specified some low polynomial bound, so that some games complete in P (see, for example, [Adachi, Iwata and Kasai 1984]), and possibly annihilation games, would be excluded. But the decision about how low that polynomial should be would be largely arbitrary, and we would lose the closure under composition of polynomials. Hence we preferred not to do this.

(ii) In (b) we could have included also a P-position, i.e., the requirement that the loser can compute in polynomial time a next move that makes play last as long as possible. In a way, this is included in (c). A more explicit enunciation on the speed of losing doesn't seem to be part of the requirements for a tractable strategy.

(iii) In Section 2 we saw that, for Scoring, play lasts for an exponential number of moves. In general, for succinct games, the loser can delay the win for an exponential number of moves. Is there a "more natural" succinct game for which the loser cannot force an exponential delay? There are some succinct games for which the loser cannot force an exponential delay, such as Kotzig's Nim [Berlekamp, Conway and Guy 1982, Ch. 15] of length $4n$ and move set $M = \{n, 2n\}$. This example is somewhat contrived, in that M is not fixed, and the game is not primitive in the sense of [Fraenkel, Jaffray, Kotzig and Sabidussi 1995, Section 3]. Is there a "natural" nonsuccinct game for which the loser can force precisely an exponential delay? Perhaps an epidemiography game with a sufficiently slowly growing function f (where at move k we adjoin $f(k)$ new copies; see [Fraenkel, Loebl and Nešetřil 1988; Fraenkel and Lorberbom 1989]), played on a general digraph, can provide an example.

(iv) There are several ways of compounding a given finite set of games—moving rules and ending rules. See, for example, [Conway 1976, Ch. 14]. Since the sum of games is the most natural, fundamental and important among the various compounds, we only required in (d) closure under game sums.

(v) One might consider a game efficient only if both its succinct and nonsuccinct versions fulfil conditions (a)–(d). But given a succinct game, there are often many different ways of defining a nonsuccinct variation; and given a nonsuccinct game, it is often not so clear what its succinct version is, if any. Hence this requirement was not included in the definition.

A panorama of the poset of strategy efficiencies can be viewed by letting Murphy's law loose on the tractability and efficiency definitions. Just about any perverse game behavior one may think of can be realized by some perturbation of (a)–(d). We have already met misère play, interaction between tokens and succinctness. These tend to affect (d) adversely. Yet we are not aware that misère play has been proven to be NP-hard, nor do we know of any succinct game that has been proven NP-hard; the complexity of so many succinct games is still open! But there are many interesting games involving interaction between tokens that have been proven Pspace-complete, Exptime-complete, or even Expspace-complete.

We have also seen that epidemiography games violate (c), and that Whythoff's game is not known to satisfy (d). The same holds for Moore's Nim [Moore 1909–10; Berlekamp, Conway and Guy 1982, Ch. 15], so for both Wythoff's game and Moore's Nim we only have, at present, a reasonable strategy in the sense of Section 2. For partizan games (d) is violated conditionally, in the sense that sums are Pspace-complete. (It's not more than a curiousity that *certain* sums of impartial Pspace-hard games are polynomial. Thus geography is Pspace-complete [Schaefer 1978]. Yet given two identical geography games, player 2 can win easily by imitating on one copy what the opponent does on the other.)

Incidentally, geography games [Fraenkel and Simonson 1993; Fraenkel, Scheinerman and Ullman 1993; Bodlaender 1993; Fraenkel, Jaffray, Kotzig and Sabidussi 1995] point to another weakness of the classical theory. The input digraph or graph is not succinct, the game does not appear to decompose into a sum, and the game graph is very large. This accounts for the completeness results of many geography games.

While we're at it, we point out another property of geography games. Their *initial position*, as well as an "arbitrary" mid-position, are Pspace-complete, since one is as complex as the other: the initial position has the same form as any mid-position. In fact, this is the case for their prototype, quantified Boolean satisfiability [Stockmeyer and Meyer 1973]. On the other hand, for board games such as checkers, chess, Go, Hex, Shogi or Othello, the completeness result holds for an *arbitrary position*, i.e., a "midgame" or "endgame" position, carefully concocted in the reduction proof. See, for example, [Fraenkel, Garey, Johnson, Schaefer and Yesha 1978; Lichtenstein and Sipser 1978; Fraenkel and Lichtenstein 1981; Robson 1984a; Robson 1984b; Adachi, Kamekawa and Iwata 1987; Iwata and Kasai 1994]. But for the initial position, which is rather simple or symmetric, the decision question who can win might be easier to solve.

For poset games with a largest element and for Hex it's in fact easy to see that player 1 can win, but the proof of this fact is nonconstructive. Yet one of these games, von Neumann's Hackendot, has been given an interesting polynomial strategy by Úlehla [1980]. See also Bushenhack in [Berlekamp, Conway and Guy 1982, Ch. 17]. Still unresolved poset games include chomp [Gale 1974] and a power-set game [Gale and Neyman 1982]; see also [Fraenkel and Scheinerman 1991].

The outcome of certain games can be made to depend on open problems in mathematics. See, for example, [Jones 1982; Jones and Fraenkel ≥ 1996]. Lastly, but far from exhaustively, we mention a game in which "the loser wins". Let $Q = x_1 - x_2x_4 - x_2 - x_4 - 1$. Two players alternately assign values to x_1, x_2, x_3, x_4 in this order. Player 1 has to select x_1 as a composite integer > 1 and x_3 as any positive integer, and player 2 selects any positive integers. Player 2 wins if $Q = 0$; otherwise player 1 wins. It should be clear that player 2 can win, since $Q = x_1 - (x_2 + 1)(x_4 + 1)$ and x_1 is composite. But if player 1 picks x_1 as the product of 2 large primes of about the same size, then player 2 can realize his win in practice only if he can crack the RSA public-key cryptosystem [Rivest, Shamir and Adleman 1978]. Thus, in practice, the loser wins! Also note that player 2 has an efficient probabilistic method, such as those of [Solovay and Strassen 1977; Rabin 1976; 1980], to determine with arbitrarily small error that player 1 did not cheat, that is, that he indeed selected a *composite* integer.

References

1. [Adachi, Iwata and Kasai 1984]   A. Adachi, S. Iwata and T. Kasai, "Some combinatorial game problems require $\Omega(n^k)$ time", *J. Assoc. Comput. Mach.* **31** (1984), 361–376.

2. [Adachi, Kamekawa and Iwata 1987] H. Adachi, H. Kamekawa and S. Iwata, "Shogi on $n \times n$ board is complete in exponential time" (in Japanese), *Trans. IEICE* **J70-D** (1987), 1843–1852.

3. [Banerji and Dunning 1992] R. B. Banerji and C. A. Dunning, "On misere games", *Cybernetics and Systems* **23** (1992), 221–228.

4. [Beck 1981] J. Beck, "On positional games", *J. Combin. Theory* A**30** (1981), 117–133.

5. [Beck 1982]   J. Beck, "Remarks on positional games, I", *Acta Math. Acad. Sci. Hungar.* **40**(1–2) (1982), 65–71.

6. [Beck 1985] J. Beck, "Random graphs and positional games on the complete graph", *Ann. Discrete Math.* **28** (1985), 7–13.

7. [Berlekamp, Conway and Guy 1982] E. R. Berlekamp, J. H. Conway and R. K. Guy, *Winning Ways for Your Mathematical Plays*, Academic Press, London, 1982.

8. [Berlekamp and Wolfe 1994] E. Berlekamp and D. Wolfe, *Mathematical Go: Chilling Gets the Last Point*, A. K. Peters, Wellesley, MA, 1994.

9. [Blass and Fraenkel 1990]   U. Blass and A. S. Fraenkel, "The Sprague–Grundy function of Wythoff's game", *Theor. Comput. Sci.* (*Math Games*) **75** (1990), 311–333.

10. [Bodlaender 1993] H. L. Bodlaender, "Complexity of path forming games", *Theor. Comput. Sci.* (*Math Games*) **110** (1993), 215–245.

11. [Chvátal 1973]   V. Chvátal, "On the computational complexity of finding a kernel", Report No. CRM-300, Centre de Recherches Mathématiques, Université de Montréal (1973).

12. [Chvátal and Erdős 1978]   V. Chvátal and P. Erdős, "Biased positional games", pp. 221–229 in *Algorithmic Aspects of Combinatorics* (edited by B. Alspach et al.), Ann. Discrete Math. **2**, North-Holland, Amsterdam, 1978.

13. [Conway 1976]   J. H. Conway, *On Numbers and Games*, Academic Press, London, 1976.

14. [Conway 1977]   J. H. Conway, "All games bright and beautiful", *Amer. Math. Monthly* **84** (1977), 417–434.

15. [Conway 1978a]   J. H. Conway, "A gamut of game theories", *Math. Mag.* **51** (1978), 5–12.

16. [Conway 1978b]   J. H. Conway, "Loopy games", pp. 55–74 in *Proc. Symp. Advances in Graph Theory* (edited by B. Bollobás), Ann. Discrete Math. **3**, North-Holland, Amsterdam, 1978.

17. [Ehrenfeucht and Mycielski 1979]   A. Ehrenfeucht and J. Mycielski, "Positional strategies for mean payoff games", *Internat. J. Game Theory* **8** (1979), 109–113.

18. [Ferguson 1984]   T. S. Ferguson, "Misère annihilation games", *J. Combin. Theory* **A37** (1984), 205–230.

19. [Fraenkel 1974]   A. S. Fraenkel, "Combinatorial games with an annihiliation rule", pp. 87–91 in *The Influence of Computing on Mathematical Research and Education* (edited by J. P. LaSalle), Proc. Symp. Appl. Math. **20**, Amer. Math. Soc., Providence, RI, 1974.

20. [Fraenkel 1981]   A. S. Fraenkel, "Planar kernel and Grundy with $d \leq 3$, $d_{out} \leq 2$, $d_{in} \leq 2$ are NP-complete", *Discrete Appl. Math.* **3** (1981), 257–262.

21. [Fraenkel 1982]   A. S. Fraenkel, "How to beat your Wythoff games' opponents on three fronts", *Amer. Math. Monthly* **89** (1982), 353–361.

22. [Fraenkel 1991]   A. S. Fraenkel, "Complexity of games", pp. 111–153 in *Combinatorial Games* (edited by R. K. Guy), Proc. Symp. Appl. Math. **43**, Amer. Math. Soc., Providence, RI, 1991.

23. [Fraenkel $\geq$ 1997]   A. S. Fraenkel, *Adventures in Games and Computational Complexity*, to appear in Graduate Studies in Mathematics, Amer. Math. Soc., Providence, RI.

24. [Fraenkel and Borosh 1973]   A. S. Fraenkel and I. Borosh, "A generalization of Wythoff's game", *J. Combin. Theory* **A15** (1973), 175–191.

25. [Fraenkel, Garey, Johnson, Schaefer and Yesha 1978]   A. S. Fraenkel, M. R. Garey, D. S. Johnson, T. Schaefer and Y. Yesha, "The complexity of checkers on an $n \times n$ board – preliminary report", pp. 55–64 in *Proc. 19th Annual Symp. Foundations of Computer Science*, IEEE Computer Soc., Long Beach, CA, 1978.

26. [Fraenkel and Goldschmidt 1987]   A. S. Fraenkel and E. Goldschmidt, "Pspace-hardness of some combinatorial games", *J. Combin. Theory* **A46** (1987), 21–38.

27. [Fraenkel, Jaffray, Kotzig and Sabidussi 1995]   A. S. Fraenkel, A. Jaffray, A. Kotzig and G. Sabidussi, "Modular Nim", *Theor. Comput. Sci. (Math Games)* **143** (1995), 319–333.

28. [Fraenkel and Lichtenstein 1981] A. S. Fraenkel and D. Lichtenstein, "Computing a perfect strategy for $n \times n$ chess requires time exponential in n", *J. Combin. Theory* A31 (1981), 199–214. Preliminary version appeared as pp. 278–293 in *Automata, Languages and Programming*, Acre, Israel, 1981 (edited by S. Even and O. Kariv), Lecture Notes Comp. Sci. 115, Springer, Berlin, 1981.

29. [Fraenkel, Loebl and Nešetřil 1988] A. S. Fraenkel, M. Loebl and J. Nešetřil, "Epidemiography, II. Games with a dozing yet winning player", *J. Combin. Theory* A49 (1988), 129–144.

30. [Fraenkel and Lorberbom 1989] A. S. Fraenkel and M. Lorberbom, "Epidemiography with various growth functions", *Discrete Appl. Math.* 25 (1989), 53–71.

31. [Fraenkel and Lorberbom 1991] A. S. Fraenkel and M. Lorberbom, "Nimhoff games", *J. Combin. Theory* A58 (1991), 1–25.

32. [Fraenkel and Nešetřil 1985] A. S. Fraenkel and J. Nešetřil, "Epidemiography", *Pacific J. Math.* 118 (1985), 369–381.

33. [Fraenkel and Perl 1975] A. S. Fraenkel and Y. Perl (1975), "Constructions in combinatorial games with cycles", pp. 667–699 in *Proc. Intern. Colloq. on Infinite and Finite Sets* (edited by A. Hajnal et al.), vol. 2, Colloq. Math. Soc. János Bolyai 10 North-Holland, Amsterdam, 1975.

34. [Fraenkel and Scheinerman 1991] A. S. Fraenkel and E. R. Scheinerman, "A deletion game on hypergraphs", *Discrete Appl. Math.* 30 (1991), 155–162.

35. [Fraenkel, Scheinerman and Ullman 1993] A. S. Fraenkel, E. R. Scheinerman and D. Ullman, "Undirected edge geography", *Theor. Comput. Sci. (Math Games)* 112 (1993), 371–381.

36. [Fraenkel and Simonson 1993] A. S. Fraenkel and S. Simonson, "Geography", *Theor. Comput. Sci. (Math Games)* 110 (1993), 197–214.

37. [Fraenkel and Tassa 1975] A. S. Fraenkel and U. Tassa, "Strategy for a class of games with dynamic ties", *Comput. Math. Appl.* 1 (1975), 237–254.

38. [Fraenkel and Tassa 1982] A. S. Fraenkel and U. Tassa, "Strategies for compounds of partizan games", *Math. Proc. Camb. Phil. Soc.* 92 (1982), 193–204.

39. [Fraenkel, Tassa and Yesha 1978] A. S. Fraenkel, U. Tassa and Y. Yesha, "Three annihilation games", *Math. Mag.* 51 (1978), 13–17.

40. [Fraenkel and Yesha 1976] A. S. Fraenkel and Y. Yesha, "Theory of annihilation games", *Bull. Amer. Math. Soc.* 82 (1976), 775–777.

41. [Fraenkel and Yesha 1979] A. S. Fraenkel and Y. Yesha, "Complexity of problems in games, graphs and algebraic equations", *Discrete Appl. Math.* 1 (1979), 15–30.

42. [Fraenkel and Yesha 1982] A. S. Fraenkel and Y. Yesha, "Theory of annihilation games – I", *J. Combin. Theory* B33 (1982), 60–86.

43. [Fraenkel and Yesha 1986] A. S. Fraenkel and Y. Yesha, "The generalized Sprague–Grundy function and its invariance under certain mappings", *J. Combin. Theory* A43 (1986), 165–177.

44. [Gale 1974] D. Gale, "A curious Nim-type game", *Amer. Math. Monthly* 81 (1974), 876–879.

45. [Gale and Neyman 1982] D. Gale and A. Neyman, "Nim-type games", *Internat. J. Game Theory* 11 (1982), 17–20.

46. [Gangolli and Plambeck 1989] A. Gangolli and T. Plambeck, "A note on periodicity in some octal games", *Internat. J. Game Theory* **18** (1989), 311–320.

47. [Goldstein and Reingold 1995] A. S. Goldstein and E. M. Reingold, "The complexity of pursuit on a graph", *Theor. Comput. Sci. (Math Games)* **143** (1995), 93–112.

48. [Grundy 1939] P. M. Grundy, "Mathematics and Games", *Eureka* **2** (1939), 6–8. Reprinted in *Eureka* **27** (1964), 9–11.

49. [Guy and Smith 1956] R. K. Guy and C. A. B. Smith, "The G-values of various games", *Proc. Camb. Phil. Soc.* **52** (1956), 514–526.

50. [Iwata and Kasai 1994] S. Iwata and T. Kasai, "The Othello game on an $n \times n$ board is PSPACE-complete", *Theor. Comput. Sci. (Math Games)* **123** (1994), 329–340.

51. [Jones 1982] J. P. Jones, "Some undecidable determined games", *Internat. J. Game Theory* **11** (1982), 63–70.

52. [Jones and Fraenkel $\geq$ 1996] J. P. Jones and A. S. Fraenkel, "Complexities of winning strategies in polynomial games", *J. Complexity*, in press.

53. [Kalmár 1928] L. Kalmár, "Zur Theorie der abstrakten Spiele", *Acta Sci. Math. Univ. Szeged* **4** (1928), 65–85.

54. [van Leeuwen 1976] J. van Leeuwen, "Having a Grundy-numbering is NP-complete", Report No. 207, Computer Science Dept., Pennsylvania State Univ., 1976.

55. [Li 1976] S.-Y. R. Li, "Sums of Zuchswang games", *J. Combin. Theory* A**21** (1976), 52–67.

56. [Lichtenstein and Sipser 1978] D. Lichtenstein and M. Sipser, "Go is Polynomial-Space hard", *J. Assoc. Comput. Mach.* **27** (1980), 393–401. Earlier draft appeared as pp. 48–54 in 19*th Annual Symposium on Foundations of Computer Science*, Ann Arbor, MI, 1978, IEEE Computer Soc., Long Beach, CA, 1978.

57. [Moews 1994] Chapter 5 in [Berlekamp and Wolfe 1994].

58. [Moore 1909–10] E. H. Moore, "A generalization of the game called nim", *Ann. of Math.* **11**, ser. 2 (1909–10), 93–94.

59. [Morris 1981] F. L. Morris, "Playing disjunctive sums is polynomial space complete", *Internat. J. Game Theory* **10** (1981), 195–205.

60. [Nešetřil and Thomas 1987] J. Nešetřil and R. Thomas (1987), "Well quasi ordering, long games and combinatorial study of undecidability", pp. 281–293 in *Logic and Combinatorics* (edited by S. G. Simpson), Contemp. Math. **65**, Amer. Math. Soc., Providence, RI, 1987.

61. [Pink 1991] N. Pink, "Wythoff's game and some properties of its Grundy scheme", preprint, 1991.

62. [Plambeck 1992] T. E. Plambeck, "Daisies, Kayles, and the Sibert–Conway decomposition in misère octal games", *Theor. Comput. Sci. (Math Games)* **96** (1992), 361–388.

63. [Rabin 1957] M. O. Rabin, "Effective computability of winning strategies", pp. 147–157 in *Contributions to the Theory of Games* (edited by H. W. Kuhn and A. W. Tucker), Ann. of Math. Studies **39**, Princeton Univ. Press, Princeton, 1957.

64. [Rabin 1976] M. O. Rabin, "Probabilistic algorithms", pp. 21–39 in *Algorithms and Complexity: New Directions and Recent Results* (edited by J. F. Traub), Academic Press, New York, 1976.

65. [Rabin 1980] M. O. Rabin, "Probabilistic algorithm for testing primality", *J. Number Theory* **12** (1980), 128–138.

66. [Rivest, Shamir and Adleman 1978] R. L. Rivest, A. Shamir and L. Adleman, "A method for obtaining digital signatures and public-key cryptosystems", *Comm. ACM* **21** (1978), 120–126.

67. [Robson 1984a] J. M. Robson, "Combinatorial games with exponential space complete decision problems", pp. 498–506 in *Math. Foundations of Computer Science* 1984 (edited by M. P. Chytil and V. Koubek), Lecture Notes in Computer Science **176**, Springer, Berlin, 1984.

68. [Robson 1984b] J. M. Robson, "*N* by *N* checkers is Exptime complete", *SIAM J. Comput.* **13** (1984), 252–267.

69. [Schaefer 1978] T. J. Schaefer, "On the complexity of some two-person perfect-information games", *J. Comput. System Sci.* **16** (1978), 185–225.

70. [Shaki 1979] A. S. Shaki, "Algebraic solutions of partizan games with cycles", *Math. Proc. Camb. Phil. Soc.* **85** (1979), 227–246.

71. [Sibert and Conway 1992] W. L. Sibert and J. H. Conway, "Mathematical Kayles", *Internat. J. Game Theory* **20** (1992), 237–246.

72. [Smith 1966] C. A. B. Smith, "Graphs and composite games", *J. Combin. Theory* **1** (1966), 51-81. Reprinted in slightly modified form in *A Seminar on Graph Theory* (edited by F. Harary), Holt, Rinehart and Winston, New York, 1967.

73. [Solovay and Strassen 1977] R. Solovay and V. Strassen, "A fast Monte-Carlo test for primality", *SIAM J. Comput.* **6** (1977), 84–85. Erratum, ibid. **7** (1978), 118.

74. [Sprague 1935–36] R. Sprague, "Über mathematische Kampfspiele", *Tôhoku Math. J.* **41** (1935–36), 438–444.

75. [Stockmeyer and Meyer 1973] L. J. Stockmeyer and A. R. Meyer (1973), "Word problems requiring exponential time", pp. 1-9 in *Fifth Annual ACM Symp. on Theory of Computing*, Assoc. Comput. Mach., New York, 1973.

76. [Úlehla 1980] J. Úlehla, "A complete analysis of von Neumann's Hackendot", *Internat. J. Game Theory* **9** (1980), 107–113.

77. [Wythoff 1907] W. A. Wythoff, "A modification of the game of Nim", *Nieuw Arch. Wisk.* **7** (1907), 199–202.

78. [Yaglom and Yaglom 1967] A. M. Yaglom and I. M. Yaglom, *Challenging Mathematical Problems with Elementary Solutions* (translated by J. McCawley, Jr.; revised and edited by B. Gordon), vol. 2, Holden-Day, San Francisco, 1967.

79. [Yedwab 1985] L. J. Yedwab, "On playing well in a sum of games", M.Sc. Thesis, MIT, 1985. Issued as MIT/LCS/TR-348, MIT Laboratory for Computer Science, Cambridge, MA.

80. [Zwick and Paterson ≥ 1996] U. Zwick and M. S. Paterson, "The complexity of mean payoff games on graphs". Preprint at http://www.math.tau.ac.il/~zwick. Preliminary version to appear in *Proc. COCOON* 1995, Xi'an, China. Final version submitted to *Theor. Comput. Sci.* (*Math Games*).

AVIEZRI S. FRAENKEL
DEPARTMENT OF APPLIED MATHEMATICS AND COMPUTER SCIENCE
WEIZMANN INSTITUTE OF SCIENCE
REHOVOT 76100, ISRAEL
fraenkel@wisdom.weizmann.ac.il
http://www.wisdom.weizmann.ac.il/~fraenkel/fraenkel.html

Games of No Chance
MSRI Publications
Volume **29**, 1996

What Is a Game?

RICHARD K. GUY

ABSTRACT. This introduction to the theory of combinatorial games relies
on many examples to illustrate basic definitions and results.

This article skims the surface of the vast subject of combinatorial games.
It often makes reference to the foundational books *Winning Ways for Your
Mathematical Plays*, abbreviated WW [Berlekamp et al. 1982], and *On Numbers
and Games*, abbreviated ONAG [Conway 1976]. Other references that should
be consulted are [Fraenkel 1980; Guy 1983; Guy 1991], and the other articles in
this volume. See also Fraenkel's master bibliography on pages 493–537.

1. What We Mean by a Combinatorial Game

Our games are unlike those of "classical" game theory, that find application
in economics, management, and military strategy. Our games usually, though
perhaps not quite always, satisfy the following conditions:

1. There are just two players, often called Left and Right. There can be no
   question of *coalitions*.
2. There are several, usually finitely many, *positions*, and often a particular
   starting position.
3. There are clearly defined *rules* that specify the two sets of *moves* that Left
   and Right can make from a given position to its *options*.
4. Left and Right move alternately, in the game as a whole.
5. In the *normal play* convention a player unable to move *loses*.
6. The rules are such that play will always come to an end because some player
   will be unable to move. This is called the *ending condition*. There are no
   games that are drawn by repetition of moves.

This is a slightly revised reprint of the article of the same name that appeared in *Combinatorial Games*, Proceedings of Symposia in Applied Mathematics, Vol. 43, 1991. Permission
for use courtesy of the American Mathematical Society.

7. Both players know what is going on; there is *complete information*. There is no occasion for *bluffing*.

8. There are no *chance moves*: no dealing of cards; no rolling of dice.

Think about games that you know. How far do they satisfy these eight conditions?

Ludo, Snakes-and-Ladders and *Backgammon* all have complete information, but contain chance moves, since they all use dice.

Battleships, Kriegspiel, Three-Finger Morra and *Scissors-Paper-Stone* have no chance moves but the players do not have complete information about the disposition of their opponents' pieces or fingers. Moreover, in the finger games, the players move simultaneously rather than alternately.

Tic-Tac-Toe (*Noughts-and-Crosses*) fails 5 because a player unable to move is not necessarily the loser, since ties are possible. *Chess* also fails 5, and contains positions that are *tied* by stalemate (in which the last player does *not* win) and positions that are *drawn* by infinite play (of which perpetual check is a special case). The words "tied" and "drawn" are often used interchangeably, though with slight transatlantic differences, for games that are neither won nor lost. We suggest that *drawn* be used for cases when this happens because play is drawn out indefinitely and *tied* for cases when play definitely ends but the rules do not award a win to either player.

Monopoly fails on several counts. As in Ludo, there are chance moves and there may be more than two players. Players don't have complete information about the arrangement of the cards and the game could, theoretically, go on for ever.

In *Poker* much of the interest derives from the incompleteness of the information, the chance moves and the possibility of coalitions.

Bridge is a two-person game, each "person" being a team of two, but the players do not even have complete information about their own cards.

Nim is played with heaps of beans. When it is your turn to move, choose a heap and remove as many beans from it as you wish; perhaps the whole heap, but at least one bean. *Grundy's Game* is also played with heaps of beans. A move now is to split a heap into two heaps of *unequal* size, so that heaps of one or two cannot be split. *Wythoff's Game* is played with just two heaps. A move is to remove any number of beans from one heap, or *equal* numbers of beans from both heaps. This last option is an example of a *nondisjunctive* move: it doesn't satisfy the condition for the *sum* of two games, which we will define later. Nim, Grundy's Game, and Wythoff's Game each satisfy all of our eight conditions, together with the additional one that the options from any given position are the same for each player, regardless of whose turn it is to move. Such games are called *impartial*; those games in which the options for the two players are not all alike are called *partizan*.

Dots-and-Boxes is won by the player scoring the larger number of boxes, so that it doesn't satisfy the normal play convention. However, it can almost always be treated as an impartial game, satisfying the normal play convention. Part of its theory uses that of *Kayles* and of *Dawson's Kayles*. These two games are played with rows of skittles: in Kayles a move removes just one skittle, or two adjacent ones, so that the game repeatedly splits into a sum of smaller games. Similarly in Dawson's Kayles, in which the move is to remove any skittle, provided that its immediate neighbors, if any, are also removed.

Sylver Coinage is an impartial game that uses the *misère play convention* that the last person to play *loses*. Misère games are usually very difficult to analyze, though a recent breakthrough was made by William Sibert and John Conway [1992], who have found the complete analysis of *Misère Kayles*.

Go is a good example of a "hot" partizan game, i.e., one in which, in the great majority of positions, each player is eager to make the next move. Good progress is being made with analyzing the concluding stages of the game, using Elwyn's generalization of the idea of "overheating" [Berlekamp and Wolfe 1994; Berlekamp et al. 1982, 170–174].

We will often refer to a move as being "good" if it wins, and "bad" if it won't. In theory it usually suffices to find any good move, or to show than no good move exists. But in real life games there are many other criteria for choosing between your various options. If you're *losing*, then all your options are bad in the above sense, but in practice they're not all equal, and you might prefer one that makes the situation too complicated for your opponent to analyze (the *Enough Rope Principle*).

It is hard to draw the line between mathematics and psychology. There are even cases where you should prefer a bad move to a good one! Your opponent might be learning how to play a game with which you're already familiar. In this case you'll probably be able to win a few times despite the bad moves you deliberately make so as not to give away your strategy. Or one move, theoretically the best, might gain you only a dollar, while another, which loses a dollar, might win you a hundred if your opponent fails to find the subtle winning reply. Or you may be a baby-sitter, whose job is much more peaceful if your opponent wins. Or a card-sharp who's losing while the stakes are low, in anticipation of winning later when the stakes are higher.

2. Game Graphs and Trees

A game may be visualized as a digraph: the nodes are the positions and the arcs are the options. The arcs may be thought of as colored, say

bLue,   Red, or   grEen

according as the option is available

to Left only,   to Right only, or   to Either player.

Alternatively, we may distinguish between different plays of the game, that is, different dipaths in the digraph, by duplicating the nodes as necessary and

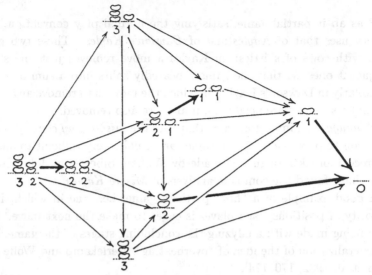

Figure 1. The game graph for the Nim position $\{3, 2\}$. From the leftmost position the next player can win by adopting the strategy indicated by the heavy arrows.

representing the game by a rooted tree. The root is the starting position and the arcs are directed away from the root.

Figures 1 and 2 show the game graph and the game tree for the position $\{3,2\}$ in a game of Nim: two heaps, one with three beans, the other with two. Nim is an example of an impartial game, in every position of which the same set of options is available to either player: think of the arcs in Figures 1 and 2 as being colored green. Nim is played with a number of heaps of beans. The typical option, for either player, is to choose a heap and remove from it as many beans as you wish: the whole heap maybe, but at least one bean.

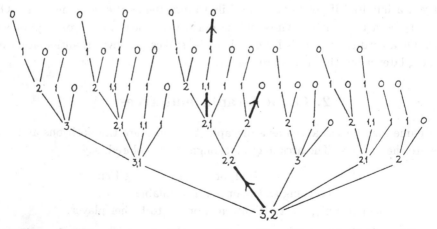

Figure 2. The game tree for the same Nim position. The root is $\{3, 2\}$, and the arcs are directed upwards.

Notice the difference between the *complete analysis* of a game, and a *winning strategy*. Figure 2 is a complete analysis: for a winning strategy it suffices to describe the four black arrows.

You may mentally identify three seemingly different aspects of the same idea.

- A *game*, i.e., the whole digraph or tree representing the game. For example, *the* Game of Chess, as opposed to *a* (particular) game of chess. We refer to the latter as a *play* of the game: compare *le jeu d'échecs, une partie d'échecs*.
- A *position* in a game; a particular node of the digraph, perhaps the root of the tree. For example, the standard opening position in Chess, ready for a play of the game.
- The *ordered pair* of sets of options available to the two players from a given position. For example, from the opening position for Chess, these are the options for the two players:

$$\{\text{White options} = \{\text{Pa3, Pa4}, \dots, \text{Ph3, Ph4, Na3, Nc3, Nf3, Nh3}\},$$
$$\text{Black options} = \{\text{Pa6, Pa5}, \dots, \text{Ph6, Ph5, Na6, Nc6, Nf6, Nh6}\}\}.$$

A position from which neither player has any option—for example, the rightmost in Figure 1, or any zero in Figure 2—is a *terminal position*, at which the game ends. The *outcome* is then specified by the rules. It may be a win for Left, or a win for Right, possibly accompanied by some score or payoff. The rules may not specify a winner, so that the game may end in a *tie*. For present purposes we will adopt the *normal play convention* that the winner is the player who has just made the last move: equivalently, *last player winning*; if you can't move, you lose. We won't say very much about the *misère play convention*, which accords the win to a player unable to move: *last player losing*. Analysis is far more difficult in this case.

To ensure that we *have* a last player, our games must end. We assume that they satisfy the *ending condition*: that there is no infinite sequence of options. Notice that this condition prohibits *all* infinite sequences, not merely those in which Left and Right make alternate moves. In order to give *values* to our games, we need to consider the possibility of several consecutive moves by the same player. This can occur in the play of the *sum* of two or more games, as we shall see.

A game that does not satisfy the ending condition is called a *loopy game*. Its digraph will contain a directed circuit or an infinite directed path. The outcome may be a *draw*: note that we distinguish between a *tied* game and one *drawn out* by infinite play. Chess exhibits both kinds of outcome: stalemate is a tie, but perpetual check, repetition of moves, or insufficient mating material are equivalent to draws.

3. The Formal Definition of a Game

This is deceptively simple: each game is an ordered pair of sets of games:

$$G = \{\{G^{L_1}, G^{L_2}, \ldots\} \mid \{G^{R_1}, G^{R_2}, \ldots\}\}.$$

To avoid proliferation of braces, we write this more compactly as $G = \{G^L \mid G^R\}$, where we must remember that G^L and G^R are *sets* of Left and Right options, which may, for example, be infinite, or empty. Indeed the definition is inductive, and the empty set is the basis for the induction, which starts with the *Endgame* $\{\varnothing \mid \varnothing\} = \{ \mid \}$, in which neither player has an option, and which we will denote by 0 (*zero*).

Here, and from now on, we use several symbols familiar in elementary arithmetic, with the strong implication that we can manipulate games in the same way that we manipulate numbers in ordinary arithmetic. Some games behave like numbers and we call them numbers, but to justify the manipulations takes more space than we have here, so turn to pages 71–96 of ONAG if you would like more detail and further examples.

It's helpful to attach ordinal numbers, or *birthdays*, to games, and to introduce the idea of *simplicity* [WW, 23–27]. When a move is made in a game, it becomes *simpler* in the sense that we arrive at a position with an earlier birthday. All definitions and proofs are inductive in that they are assumed to have been made for all simpler games. The basis, as we have already stated, is the simplest game of all, the Endgame, born on day zero.

On day one we have two sets, the empty set and the set $\{0\}$ consisting of the Endgame; so that we can visualize 2^2 games. Their game trees (in which Left's moves slope up to the left and Right's moves slope up to the right), together with their names, are shown in Figure 3.

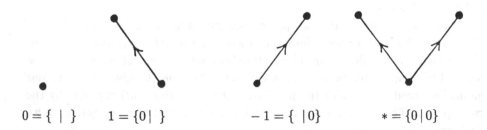

$$0 = \{ \mid \} \qquad 1 = \{0 \mid \} \qquad -1 = \{ \mid 0\} \qquad * = \{0 \mid 0\}$$

Figure 3. The four simplest games, born on days zero and one.

We quote from p. 72 of ONAG:

> The simplest game of all is the *Endgame*, 0. I curteously offer you the first move in this game, and call upon you to make it. You lose, of course, because 0 is defined as the game in which it is never legal to make a move.
>
> In the game $1 = \{0 \mid \}$, there is a legal move for Left, which ends the game, but at no time is there any legal move for Right. If I play Left, and you Right,

and you have the first move again (only fair, since you lost the previous game) you will lose again, being unable to move even from the initial position. To demonstrate my skill, I shall now start from the same position, make my legal move to 0, and call upon you to make yours.

Of course you are now beginning to suspect that Left always winds, so for our next game, -1, you may play as Left and I as Right! For the last of our examples, the new game $* = \{0\,|\,0\}$, you may play whichever role you wish, provided that for this privilege you allow me to play first.

In summary:

The Endgame is the prototype of games in which the next player loses, since no option is available: a *second-player win*.

The game 1 is a *Left win*, no matter who starts: if Louise starts, she goes to $\{\ |\ \} = 0$ and Richard has no option and loses; if Richard starts, he has no option and loses even more quickly.

The game -1 is a *Right win*, no matter who starts.

The game $\{0\,|\,0\} = *$ ("Star") is the simplest game that is not a number [WW, 40]. It is a *first-player win*.

4. The Four Outcome Classes

If we adopt the normal play convention, every game belongs to just one of four outcome classes [ONAG, Theorem 50] that are exemplified by the four games we've just seen. The terminology and notation are displayed in Figure 4.

If, in a game G		Right starts	
		and L has a winning strategy	and R has a winning strategy
Left starts	and R has a winning strategy	ZERO $G = 0$ 2nd wins	NEGATIVE $G < 0$ R wins
	and L has a winning strategy	POSITIVE $G > 0$ L wins	FUZZY $G \,\|\, 0$ 1st wins

Figure 4. The four outcome classes.

It is convenient to combine these outcome classes and symbols in pairs, in correspondence with the two rows and two columns of Figure 4:

If L has a winning strategy provided R starts, we write $G \geq 0$ (column 1).

If R has a winning strategy provided L starts, we write $G \leq 0$ (row 1).

If L has a winning strategy provided L starts, we write $G \vartriangleright 0$ (row 2).

If R has a winning strategy provided R starts, we write $G \vartriangleleft 0$ (column 2).

5. The Negative of a Game

A device to breathe new life into an otherwise one-sided contest is to allow a novice opponent, when he feels he is losing, to turn the board around, to reverse the roles of the two players, to handicap his more skilled adversary, by asking her to defend what appears to him to be an inferior position. This replaces the game by its negative. Formally, the *negative* of G,

$$-G = \{ -G^R \mid -G^L \}$$

is defined inductively [WW, 35]. Remember that $-G^R$, for example, is short for the set $\{-G^{R_1}, -G^{R_2}, \ldots\}$, whose members are simpler games than $-G$, and hence have been defined earlier. As with all our definitions, the empty set forms the basis for the induction.

6. Sums of Games

There are many ways of playing two or more games simultaneously, but often the most natural is what we call the *sum*, or *disjunctive compound* [ONAG, 75; WW, 33]. Nim, for example, is the sum of a number of games of one-heap Nim. In the sum of two or more component games, the player whose turn it is to move selects one component and makes a legal move in it:

$$G + H = \{ G^L + H, G + H^L \mid G^R + H, G + H^R \}.$$

Once again this is an inductive definition: $G^R + H$, for example, represents the set of options $\{G^{R_1} + H, G^{R_2} + H, \ldots\}$ each of which is a simpler game than $G + H$, so that addition there is already defined.

It's not hard to see that sums are commutative and associative, that $G+0 = G$, and [ONAG, Theorem 51] that $G + (-G) = 0$. In that last sentence we've used zero in two quite different senses. In $G + 0 = G$ we intended 0 to mean the Endgame, $\{ \mid \}$. In $G + (-G) = 0$ we intended "$= 0$" to mean "is a zero game", that is "belongs to the (very large!) equivalence class of games for which the second player has a winning strategy". Check that $1 + (-1) = 0$ and that $* + * = 0$, so that we can speak of the games $1 + (-1)$ and $* + *$ as having the same *value*, 0, as the Endgame, even though their *forms* are different.

More generally, we will say that two games are *equivalent*, and have the same *value*, and write $G = H$, if the game $G + (-H)$ is a second-player win. With the above definitions of sum, negative and zero, the set of all games forms a commutative group. Moreover, games form a partially ordered set, and we write $G > H$ just if $G - H > 0$, that is, if Left can win the sum $G + (-H)$, no matter who starts. On the other hand, if $G - H \parallel 0$, we say that G and H are incomparable and write $G \parallel H$. Our notation is justified by theorems such as the following, proved in [ONAG, 76]. If $G \geq 0$ and $H \geq 0$, then $G + H \geq 0$. If H is a zero game (that is, a win for the second player), then $G + H$ has the same outcome as G. If $H - K$ is a zero game, then $G + H$ and $G + K$ have the same outcome.

7. The Games Born on Day Two

As day two dawns we have four games to play with, and so $2^4 = 16$ sets of games. There are 16 choices for Left's options and 16 for Right's, giving a potential of 256 games on day two. However, things are not *quite* that complicated, in that for each player, some options are clearly preferable to others. The four games born on day one can be arranged in the lattice (in the poset sense, rather than the geometrical sense) of Figure 5, in which Left's preferences are placed higher, and Right's are lower.

Figure 5. The lattice of games born on day one.

The only set of options for which there is any doubt in either player's mind about the best move, is the incomparable pair $\{0, *\}$. So, for a player's options we need consider only six possibilities: the empty set, the four singletons, and this incomparable pair. Among the resulting 6^2 possibilities for games born on day two, just 22 are inequivalent and 18 are new. They are shown in Figure 6, the four quarters of which should be compared with those of Figure 4. These contain the zero game; six negative games; six positive; and nine fuzzy ones. The six sets of Right options run from left to right in increasing order of desirability from Right's point of view; Left's run correspondingly downwards.

Figure 7 shows the game trees for these twenty-two day-two games: the lowest node in each case is the root, arcs sloping up to the left are blue, those sloping up to the right are red. The trees for Star and sixteen of the day-two games have been condensed into digraphs in Figure 8; arcs labelled E are green and represent options available to both Left and Right.

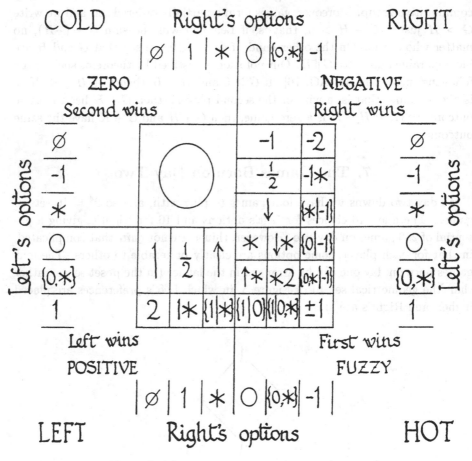

Figure 6. The twenty-two games born on day two.

The twenty-two games are exhibited as a poset in Figure 9. If two games are connected by an arc, or, transitively by a path of arcs, then the higher game has a greater value than the lower, as in Figure 5.

Examples and exercises. $\qquad$ $1+1 = 2,$ $\qquad$ $\frac{1}{2}+\frac{1}{2} = 1,$ $\qquad$ $*+* = 0,$
$\uparrow = \{0|*\} = \{0,*|*\} > 0$ ("Up is positive"), $\quad$ $\uparrow* = \{0,*|0\} = \uparrow+*$ ("Upstar"),
$0\|*2 = \{0,*|0,*\}$ ("0 is incomparable with Star-two"), $\qquad$ $1* = \{1|1\} = 1+*,$
$\{1|0\} > *,$ $\qquad$ $\downarrow* = \{0|0,*\}\|0$ ("Downstar is incomparable with zero"),
$\{1|0\} > \uparrow*,$ $\qquad$ $\{1|0\} > *2,$ $\qquad$ $\{1|0\} > \downarrow = \{*|0\},$ $\qquad$ $\{1|0\} > \downarrow*,$
$\{1|0\} > \pm1 = \{1|-1\},$ $\qquad$ $\{1|0\}\|0,$ $\qquad$ $\{1|0\}\|\uparrow,$ $\qquad$ $\{1|*\} > 0,$
$\{1|*\} > \uparrow,$ $\qquad$ $\{1|*\}\|\uparrow*,$ $\qquad$ $\{1|0,*\} > \downarrow,$ $\qquad$ $\{1|0,*\}\|\uparrow.$

Two important ways of classifying games are as *hot* or *cold* and as *partizan* or *impartial*. We will shortly make a brief attempt to distinguish hot from cold. Impartial games are those in which the two sets of Left and Right options are

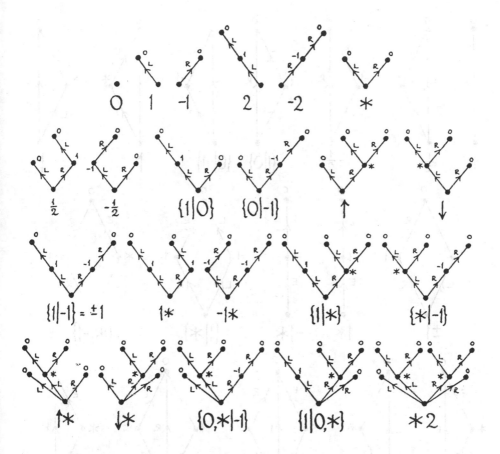

Figure 7. The game trees for the twenty-two games born on day two.

the same; in partizan games the two sets are different, in general. As illustrative examples, we describe two partizan games: Domineering is a hot game; Blue-Red Hackenbush is a cold one.

8. Domineering

The game of *Domineering*, also called *Crosscram*, is played as follows (see also [ONAG 74–75, 120–121; WW 117–120, 137–140], and this volume's articles starting on pages 85 and 311). Left and Right alternately place dominoes so that they exactly cover two squares of a checkerboard. Left orients her dominoes North-South and Right puts his East-West. Dominoes mustn't overlap each other or the edge of the board. A player loses who can find no appropriately oriented space for a domino. After a while, the available space may separate into disconnected regions, and the game becomes the sum of smaller games. Many

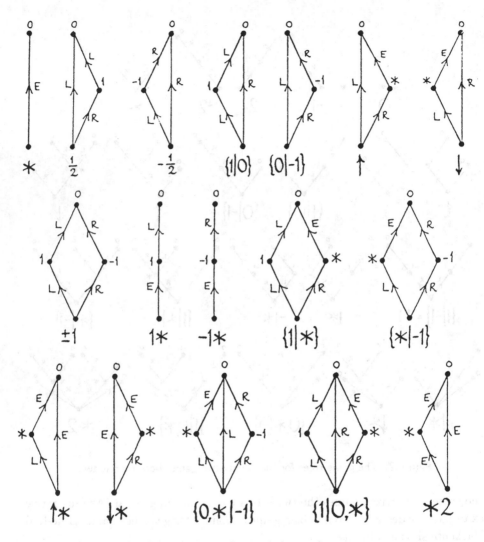

Figure 8. Digraphs for Star and sixteen day-two games.

of the games we have already seen are realized by small Domineering "boards". Check the values in Figure 10.

9. Hot and Cold Games

Domineering is an example of a *hot game*. These are the interesting games in which there's an advantage in having the move: the first player wins. If $G = \{G^L | G^R\}$, then the various differences $G^L - G$ and $G - G^R$ are the (Left and Right) *incentives* of G. These are always negative if G is a number. Numbers

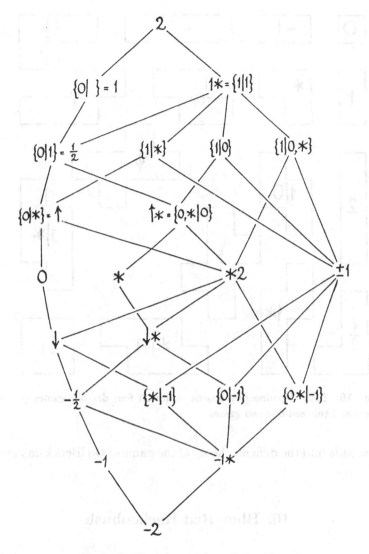

Figure 9. The poset of day-two games.

are *cold games* and the Number Avoidance Theorem tells you that you should

> Never move in a
> Number, unless there's
> Nothing else to do.

An earnest of the theory of hot games can be found in the work of Milnor [1953] and Hanner [1959]. For recent developments, see [WW, 141–182] and Berlekamp [1988], who is currently generalizing the theory of "overheating" and

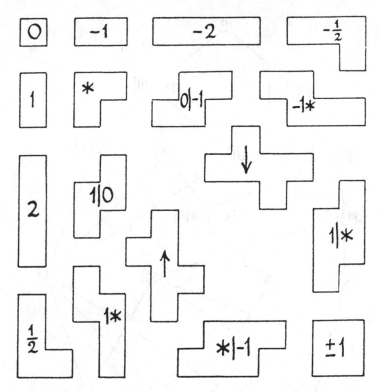

Figure 10. Small Domineering boards realize all four day-one games (top left corner) and thirteen day-two games.

making inroads into the difficult theory of the game of Go [Berlekamp and Wolfe 1994].

10. Blue–Red Hackenbush

This good example of a cold game is perhaps best played on a blackboard, using an eraser. Start with a *picture*, for example Figure 11, which is a graph, some of whose nodes are on the ground (the dotted line), and whose edges are either blue or red [ONAG, 86–91; WW, 3–8]. A Left (resp. Right) move is to delete a blue (resp. red) edge, together with any edges of either color that are no longer connected to the ground.

If Right deletes the dog's neck, for example, the head disappears as well. If Left deletes the body, no other edges disappear, but the picture breaks into the sum of two separate pictures. The aim, as usual, is to be the last player, the person whose move leaves no edges of the opponent's color.

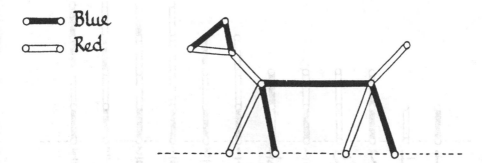

Figure 11. A Blue–Red Hackenbush picture.

11. When is a Game a Number?

Although the values of Blue–Red Hackenbush pictures may be hard to calculate (in the technical sense; WW, 210–212), they are all numbers. A game is a *number* [ONAG, 81] exactly if all its options are numbers and no Left option is greater than or equal to any Right option. The game $\pm 1 = \{1 \mid -1\}$ is *not* a number, since $1 \geq -1$; Star is not a number, because $0 \geq 0$; and $\uparrow = \{0 \mid *\}$ is not a number, because $*$ is not. Examples of numbers are

0	born on day 0
1 and -1	born on day 1
$\frac{1}{2}, -\frac{1}{2}, 2$ and -2	born on day 2
$\frac{1}{4}, -\frac{1}{4}, \frac{3}{4}, -\frac{3}{4}, 1\frac{1}{2}, -1\frac{1}{2}, 3$ and -3	born on day 3

On day ω *all* the remaining real numbers are born (as Dedekind sections of the binary rationals), as well as the first infinite ordinals, $\omega = \{0, 1, 2, \ldots \mid \}$ and $-\omega = \{ \mid 0, -1, -2, \ldots\}$ and infinitesimals such as $1/\omega = \{0 \mid 1, \frac{1}{2}, \frac{1}{4}, \ldots\}$.

The value of a game may be thought of as "the number of moves advantage to Left". For example, -2 is two moves advantage to Right. The first four Blue–Red Hackenbush values in Figure 12 are clear. Deletion of the blue edge in the fifth reduces the picture to 0, while deletion of the red edge leaves 1, so the fifth value is $\{0 \mid 1\} = \frac{1}{2}$.

Check that if you play a game comprising two copies of this and a single separate red edge, Figure 13, then the second player wins (although if Left starts, Right *can* make a bad reply). Check the remaining values in Figure 12.

Not only is the value of every Blue–Red Hackenbush picture a number, but every number can be represented by a Hackenbush string! To see this, work backwards from *Berlekamp's Rule* for evaluating Hackenbush strings [Berlekamp 1974]:

The *sign* is determined by the color of the edge touching the ground (+ for blue, − for red). Move up the string until there is a change in edge color. This first *pair* of differently colored consecutive edges represents the

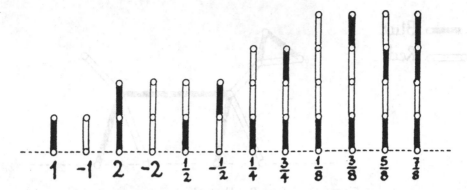

Figure 12. Values of some Blue–Red Hackenbush strings.

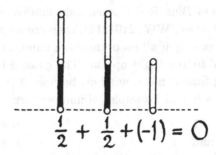

Figure 13. Two halves make a whole.

binary point. The number of edges *below* this pair gives the *integer part* of the number. Above the pair, label each edge with a binary digit, 1 or 0, according as its color agrees or disagrees with the color of the edge touching the ground, and adjoin an extra digit 1 at the end if the string is finite.

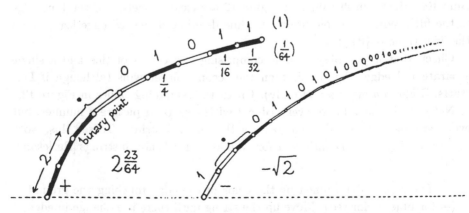

Figure 14. Berlekamp's Rule for Hackenbush strings.

The rule is illustrated in Figure 14; also use it to check the values in Figure 12. Infinitesimal numbers and infinite ordinals can also be represented by (infinite) Hackenbush strings [WW, 309–313].

12. Simplifying Games

Games can be simplified in two ways [ONAG, 109–112; WW, 62–64]: by *deleting dominated options*, or by *replacing reversible options*.

We used the first of these methods implicitly when we made our catalog of day-two games. If, in a game $G = \{A, B, C, \ldots \mid Z, Y, X, \ldots\}$, we have $B \geq A$, we say that B *dominates* A. Then A may be deleted, provided that B is retained. Similarly, if $Y \leq Z$, then Y dominates Z, and Z can be deleted if Y is retained.

Replacing reversible options is more subtle. A right option X is *reversible* if Left has a reply X^L that is at least as good for her as the original game, that is, if $X^L \geq G$. Then X may be replaced by the list of *all* the Right options, $X^{LR} = \{X^{LR_1}, X^{LR_2}, \ldots\}$, of X^L. Similarly, the Left option C is reversible if Right has a reply C^R at least as good for him as G is, that is if $C^R \leq G$, and then C may be replaced by the list of all the Left options C^{RL} of C^R.

For example, in the game

$$G = \{0, * \mid *\},$$

which is a more precise description of the game labelled "↑" in Figure 10, neither Left option dominates the other, since $* \parallel 0$, but the Left option $*$ is reversible, because $*^R = 0 \leq G$. (To see that $G \geq 0$, note that if Right starts, his only option is $* = \{0 \mid 0\}$ and Left plays to 0 and wins.) So the Left option $*$ may be replaced by all the Left options of $*^R = 0 = \{ \ \mid \ \}$; that is, it may be replaced by all the members of the empty set, that is, it may be deleted, and G simplifies to $\{0 \mid *\} = \uparrow$.

For another example, consider the game $H = \{\uparrow \mid \uparrow\}$ and examine each option for reversibility (Figure 15).

Check that H satisfies the *upstart equality* [ONAG, 77; WW, 73],

$$\{0 \mid \uparrow\} = \uparrow + \uparrow + * = \Uparrow * \quad (\text{``double up star''}).$$

References

[Berlekamp 1988] E. R. Berlekamp, "Blockbusting and domineering", *J. Combin. Theory* A**49** (1988), 67–116.

[Berlekamp 1974] E. R. Berlekamp, "The Hackenbush number system for compresssion of numerical data", *Inform. and Control* **26** (1974), 134–140.

[Berlekamp and Wolfe 1994] E. Berlekamp and D. Wolfe, *Mathematical Go: Chilling Gets the Last Point*, A. K. Peters, Wellesley, MA, 1994.

[Berlekamp et al. 1982] E. R. Berlekamp, J. H. Conway, and R. K. Guy, *Winning Ways For Your Mathematical Plays*, Academic Press, London, 1982.

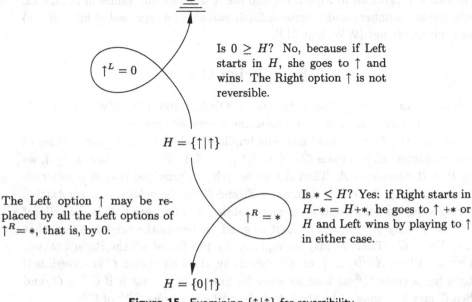

Figure 15. Examining $\{\uparrow|\uparrow\}$ for reversibility.

Is $0 \geq H$? No, because if Left starts in H, she goes to $\uparrow$ and wins. The Right option $\uparrow$ is not reversible.

$\uparrow^L = 0$

$H = \{\uparrow|\uparrow\}$

The Left option $\uparrow$ may be replaced by all the Left options of $\uparrow^R = *$, that is, by 0.

$\uparrow^R = *$

Is $* \leq H$? Yes: if Right starts in $H-* = H+*$, he goes to $\uparrow +*$ or H and Left wins by playing to $\uparrow$ in either case.

$H = \{0|\uparrow\}$

[Conway 1976] J. H. Conway, *On Numbers And Games*, Academic Press, London, 1976.

[Fraenkel 1980] A. S. Fraenkel, "From Nim to Go", pp. 137–156 in *Combinatorial Mathematics, Optimal Designs, and Their Applications* (edited by J. Srivastava), *Ann. Discrete Math.* **6**, North-Holland, Amsterdam, 1980.

[Guy 1983] R. K. Guy, "Graphs and games", pp. 269–295 in *Selected Topics in Graph Theory*, vol. 2 (edited by L. W. Beineke and R. J. Wilson), Academic Press, London, 1983.

[Guy 1991] R. K. Guy (editor), *Combinatorial Games*, Proc. Symp. Appl. Math. **43**, Amer. Math. Soc., Providence, RI, 1991.

[Hanner 1959] O. Hanner, "Mean play of sums of positional games", *Pacific J. Math.* **9** (1959), 81–99.

[Milnor 1953] J. Milnor, "Sums of positional games", pp. 291–301 in *Contributions to the Theory of Games*, vol. 2 (edited by H. W. Kuhn and A. W. Tucker), Ann. of Math. Stud. **28**, Princeton University Press, Princeton, 1953.

[Sibert and Conway 1992] W. L. Sibert and J. H. Conway, "Mathematical Kayles", *Internat. J. Game Theory* **20** (1992), 237–246.

RICHARD K. GUY
MATHEMATICS AND STATISTICS DEPARTMENT
UNIVERSITY OF CALGARY
2500 UNIVERSITY AVENUE
ALBERTA, T2N 1N4
CANADA

Games of No Chance
MSRI Publications
Volume **29**, 1996

Impartial Games

RICHARD K. GUY

In memory of Jack Kenyon, 1935-08-26 to 1994-09-19

ABSTRACT. We give examples and some general results about impartial
games, those in which both players are allowed the same moves at any
given time.

1. Introduction

We continue our introduction to combinatorial games with a survey of impartial games. Most of this material can also be found in WW [Berlekamp et al. 1982], particularly pp. 81–116, and in ONAG [Conway 1976], particularly pp. 112–130. An elementary introduction is given in [Guy 1989]; see also [Fraenkel 1996a], pp. 13–42 in this volume.

An *impartial game* is one in which the set of Left options is the same as the set of Right options. We've noticed in the preceding article the impartial games

$$\{ \ | \ \} = *0 = 0, \qquad \{0\,|\,0\} = *1 = * \quad \text{and} \quad \{0, *\,|\,0, *\} = *2.$$

that were born on days 0, 1, and 2, respectively, so it should come as no surprise that on day n the game

$$*n = \{*0, *1, *2, \ldots, *(n{-}1) \mid *0, *1, *2, \ldots, *(n{-}1)\}$$

is born. In fact any game of the type

$$\{*a, *b, *c, \ldots \mid *a, *b, *c, \ldots\}$$

has value $*m$, where $m = \operatorname{mex}\{a, b, c, \ldots\}$, the least nonnegative integer *not* in the set $\{a, b, c, \ldots\}$. To see this, notice that any option, $*a$ say, for which $a > m$,

---

This is a slightly revised reprint of the article of the same name that appeared in *Combinatorial Games*, Proceedings of Symposia in Applied Mathematics, Vol. 43, 1991. Permission for use courtesy of the American Mathematical Society.

is reversible, both as a Left option and as a Right option, because $*m$ is an option of $*a$,

$$*m \text{ is both } \geq \text{ and } \leq \{*a, *b, *c, \ldots \mid *a, *b, *c, \ldots\}$$

so that $*a$ may be replaced by the options of $*m$, namely $0, *, *2, \ldots, *(m-1)$.

This is the inductive step that proves the Sprague–Grundy theorem [Sprague 1935–36; Grundy 1939], which states that every position in an impartial game (or, which is the same, every impartial game) is equivalent to a *nim-heap* (see page 20 ff. in Fraenkel's article in this volume).

Since the Left and Right options are the same, $*n$ is its own negative, $*n + *n = 0$. Also, we need only write one set of options, and may define the *nimber*

$$*n = \{0, *, *2, \ldots, *(n-1)\}.$$

This exactly parallels John von Neumann's definition of ordinal numbers.

2. Examples of Impartial Games

We all know that the game of *Nim* is played with several heaps of beans. A move is to select a heap, and to remove any positive number of beans from it, possibly the whole heap. Any position in Nim is therefore the sum of several one-heap Nim games. The value of a single heap of n beans is $*n$.

It's easy to see how to win a game of Nim if there's only one (nonempty) heap: take the whole heap! But it's worth pausing for a moment to note exactly what your options are. They are to move to *any* smaller sized heap: they correspond exactly to the options in the definition of $*n$. It's also fairly easy to play well at two-heap Nim: if the heaps are unequal in size, remove enough beans from the larger to make the heaps equal. If the two heaps are already equal, then hope that it is your opponent's turn to move. From then on, use the *Tweedledum and Tweedledee Principle*: whatever your opponent does to one heap, you copy in the other. For more than two heaps, the theory is more tricky. It was discovered by Bouton [1902]. Imagine each heap to be partitioned into distinct powers of two. For example, Figure 1 shows four heaps of 27, 23, 22 and 15 beans partitioned in this way.

We can pair off and then ignore heaps of equal size, so concentrate on the columns with an *odd* number of parts, the ones, fours and sixteens. A good move would be to take $16 + 4 + 1 = 21$ from the 23 heap. If you leave a position for your opponent in which each power of two occurs *evenly* often, he will have to change the parity in at least one column, and then you will be able to restore it. Notice that finding a good move does *not* depend on there being appropriate powers of two all in the same row (heap). You could also take $16 - 4 + 1 = 13$ from the 27 heap, or $16 + 4 - 1 = 19$ from the 22 heap. Find three good moves from the Nim position $\{23, 19, 13, 12, 11\}$.

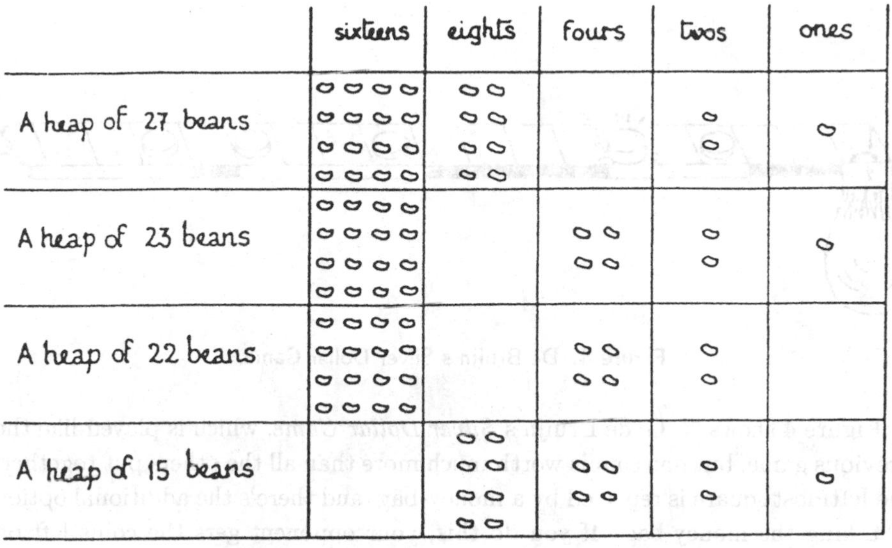

	sixteens	eights	fours	twos	ones
A heap of 27 beans	◯ ◯ ◯ ◯ ◯ ◯ ◯ ◯ ◯ ◯ ◯ ◯ ◯ ◯ ◯ ◯	◯ ◯ ◯ ◯ ◯ ◯ ◯ ◯		◯ ◯	◯
A heap of 23 beans	◯ ◯ ◯ ◯ ◯ ◯ ◯ ◯ ◯ ◯ ◯ ◯ ◯ ◯ ◯ ◯		◯ ◯ ◯ ◯	◯ ◯	◯
A heap of 22 beans	◯ ◯ ◯ ◯ ◯ ◯ ◯ ◯ ◯ ◯ ◯ ◯ ◯ ◯ ◯ ◯		◯ ◯ ◯ ◯	◯ ◯	
A heap of 15 beans		◯ ◯ ◯ ◯ ◯ ◯ ◯ ◯	◯ ◯ ◯ ◯	◯ ◯	◯

Figure 1. How to look at a Nim position.

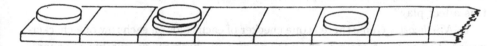

Figure 2. A game of Nimble.

Nimble is played with coins on a strip of squares (Figure 2). Take turns, moving just one coin to the left. You can jump onto or over other coins, even clear off the strip. You can have any number of coins on a square. The last player wins. Can you analyze this game? Suppose that you are not allowed to jump off the strip, so that the game ends when all coins are stacked on the left hand square. Can your analysis be modified to cope with this variant?

In the game shown in Figure 3, you're allowed at most one coin on a square, and you're not allowed to jump over other coins. A move is to *slide* a coin leftwards as far as you like, but not onto or over the next coin, and not off the end of the strip. The analysis is now more cunning: the black marks on the side of the strip may give you a hint.

Figure 3. A coin-sliding game.

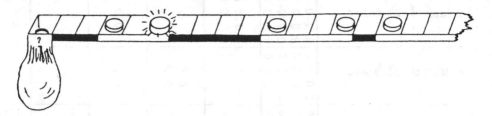

Figure 4. De Bruijn's Silver Dollar Game.

Figure 4 shows N. G. de Bruijn's *Silver Dollar Game*, which is played like the previous game, but one coin is worth much more than all the others put together, the leftmost square is replaced by a money-bag, and there's the additional option of taking the money-bag. If you do this, your opponent gets the coins left on the strip. When you've solved that game, consider the variant in which the additional option is to slide a coin *and* take the money-bag, all in one move.

Poker Nim is played like Nim, but with poker chips in place of beans; as well as removing chips from a heap, you may instead *add* chips to a heap. How does this affect play?

In *Northcott's Game* there is one checker of each color on each row of a checker-board (Figure 5). A move is to slide one of your checkers any number of squares in its own row, without jumping over your opponent's checker and without going off the board. This *looks* like a *partizan* game, and many people can't see any

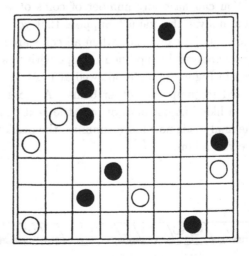

Figure 5. Northcott's Game.

point to it, and slide the checkers aimlessly. Indeed, the game doesn't appear to satisfy the *ending condition*, but there is a winner, and she can force the game to end. Remember that the aim is to be the last player to move.

Lasker's Nim is played like ordinary Nim, but with the additional option that you are allowed to split a heap into two smaller, nonempty heaps.

3. Nim-Addition

(See ONAG, pp. 50–51; WW, pp. 60–61.) We know that $*n + 0 = *n$ and that $*n + *n = 0$, and it's not hard to see that addition of impartial games, indeed of any of our games, is commutative and associative. Let's calculate

$$*2 + * = \{0, *\} + \{0\} = \{0+*, *+*, *2+0\} = \{*, 0, *2\} = *3.$$

Add $*$, or $*2$, to each side and obtain $*2 = *3 + *$ and $* = *3 + *2$. In general,

$$*a + *b = \{0, *, *2, \ldots, *(a-1)\} + \{0, *, *2, \ldots, *(b-1)\}$$
$$= \{0+*b, *+*b, \ldots, *(a-1)+*b, *a+0, *a+*, \ldots, *a+*(b-1)\},$$

and we can build a nim-addition table (Table 1) by noting that the options of an entry are just the earlier entries in the same row and the earlier entries in the same column. Each entry in Table 1 is the least nonnegative integer not appearing as an earlier entry in the same row or column. For instance, $*5 + *6 = *3$, because 3 is the first number not in the set $\{5, 4, 7, 6, 1, 0, 6, 7, 4, 5, 2\}$, i.e., the first six entries in row 5 and the first five entries in column 6. In the usual language, 3 is the *nim-sum* of 5 and 6, which is sometimes written $5 \overset{*}{+} 6 = 3$.

0	1	2	3	4	5	6	7	8	9	10	11	12	13	14	15
1	0	3	2	5	4	7	6	9	8	11	10	13	12	15	14
2	3	0	1	6	7	4	5	10	11	8	9	14	15	12	13
3	2	1	0	7	6	5	4	11	10	9	8	15	14	13	12
4	5	6	7	0	1	2	3	12	13	14	15	8	9	10	11
5	4	7	6	1	0	3	2	13	12	15	14	9	8	11	10
6	7	4	5	2	3	0	1	14	15	12	13	10	11	8	9
7	6	5	4	3	2	1	0	15	14	13	12	11	10	9	8
8	9	10	11	12	13	14	15	0	1	2	3	4	5	6	7
9	8	11	10	13	12	15	14	1	0	3	2	5	4	7	6
10	11	8	9	14	15	12	13	2	3	0	1	6	7	4	5
11	10	9	8	15	14	13	12	3	2	1	0	7	6	5	4
12	13	14	15	8	9	10	11	4	5	6	7	0	1	2	3
13	12	15	14	9	8	11	10	5	4	7	6	1	0	3	2
14	15	12	13	10	11	8	9	6	7	4	5	2	3	0	1
15	14	13	12	11	10	9	8	7	6	5	4	3	2	1	0

Table 1. Nim-addition table. The stars have been omitted; i.e., the entries are nim-values instead of nimbers.

Contrast the two equations $*5 + *6 = *3$ and $5 \overset{*}{+} 6 = 3$. In the first the summands are nimbers, i.e., values of impartial games, and the addition is a game sum. In the second the summands are *nim-values* and the addition is nim-addition.

Nim-addition is perhaps better known as addition without carry in base 2, or vector or coordinatewise addition over GF(2), or XOR (exclusive or): it is reassuring that it also follows from the more general idea of game sum.

Many of the games mentioned in the previous section are disguises for Nim, often with the addition of *reversible moves*, which we mentioned in the preceding article. As we saw in the Introduction, every impartial game is equivalent to a *bogus nim-heap*, i.e., a heap of m beans ("m" for "mex"), together with some (reversible) options that *increase* the size of the heap.

To summarize the Sprague-Grundy theory: the nim-value of the sum of two impartial games is the nim-sum of their separate nim-values. Impartial games belong to one of only two outcome classes: all positions are either

$\mathcal{P}$-positions   previous-player-winning   nim-value zero, or
$\mathcal{N}$-positions   next-player-winning   nonzero nim-value.

In the literature, $\mathcal{P}$-positions are sometimes called "safe" or "good" or "winning" without indicating which player enjoys this happy situation.

4. Subtraction Games

(See WW, pp. 83–86, 487–498.) Subtraction games are very simple examples of impartial games, played, like Nim, with heaps of beans. A move in the game $S(s_1, s_2, s_3, \ldots)$ is to take a number of beans from a heap, provided that number is a member of the *subtraction-set*, $\{s_1, s_2, s_3, \ldots\}$. Analysis of such a game and of many other heap games is conveniently recorded by a *nim-sequence*,

$$n_0 n_1 n_2 n_3 \ldots,$$

meaning that the nim-value of a heap of h beans is n_h, $h = 0, 1, 2, \ldots$, i.e., that the value of a heap of h beans in this particular game is the nimber $*n_h$. In this section, and often later, to avoid printing stars, we say that the nim-value of a position is n, meaning that its value is the nimber $*n$.

Table 2 shows some examples: the first is a manifestation of She-Loves-Me-She-Loves-Me-Not; the last is Nim. If the subtraction-set is finite, the nim-sequence is (ultimately) periodic. But little is known about the length of the period *vis à vis* the membership of the subtraction set.

In subtraction games the nim-values 0 and 1 are remarkably related by *Ferguson's Pairing Property* [Ferguson 1974; WW, pp. 86, 422]: if s_1 is the least member of the subtraction-set, then

$$\mathcal{G}(n) = 1 \qquad \text{just if} \qquad \mathcal{G}(n - s_1) = 0.$$

Subtraction game	Nim-sequence	(ultimate) Period
$S(1)$	0̇1̇01010101...	2
$S(2)$	0̇01̇10011001100...	4
$S(3)$	0̇0011̇10001110001110...	6
$S(1,2)$	0̇12̇0120120120...	3
$S(1,2,3)$	0̇123̇0123012301230...	4
$S(1,2,3,4)$	0̇1234̇0123401234012340...	5
$S(2,4,7)$	00112203̇10̇210210210...	3
$S(2,5,6)$	0̇0110213021̇0011021302100...	11
$S(4,10,12)$	0̇00011111002211330022110̇000...	22
$S(1,2,3,4,...)$	0123456789...	(saltus 1 and) 1

Table 2. Nim-sequences and periods for subtraction games.

Here and later "$\mathcal{G}(n) = v$" means that the nim-value of a heap of n beans is v.

5. Take-and-Break Games

(See WW, pp. 81–106.) Guy and Smith [1956] devised a code classifying a broad range of impartial games played with heaps or rows. Suppose a game has code

$$\mathbf{d_0 \cdot d_1 d_2 d_3} \ldots,$$

where the *code digits* $\mathbf{d}_k$ are nonnegative integers. If the binary expansion of $\mathbf{d}_k$ is

$$\mathbf{d}_k = 2^{a_k} + 2^{b_k} + 2^{c_k} + \cdots,$$

where $0 \le a_k < b_k < c_k < \cdots$, then it is legal to remove k beans from a heap, provided that the rest of the heap is left in exactly a_k or b_k or c_k or ... nonempty heaps.

In order that the game should satisfy the ending condition, $\mathbf{d}_0$ must be divisible by 4, i.e., $a_0 \ge 2$.

Subtraction games are the special case $\mathbf{d}_s = 3$ when s is in the subtraction-set, and $\mathbf{d}_k = 0$ otherwise.

Octal games are those with code digits $\mathbf{d}_k \le 7$ for all k. Guy and Smith showed that an octal game is *ultimately periodic* with period p, i.e.,

$$\mathcal{G}(n+p) = \mathcal{G}(n) \quad \text{for all} \quad n > n_0 = 2e + p + t,$$

provided that $\mathcal{G}(n+p) = \mathcal{G}(n)$ for $n \le n_0$ apart from some exceptional values of n, of which e is the largest, and $\mathbf{d}_k = 0$ for $k > t$, i.e., the maximum number of beans that may be taken from a heap in a single move is t. Whether all such finite octal games are ultimately periodic remains a difficult open question. They cannot be *arithmetically periodic*: that is, there is no period p and *saltus* $s > 0$, such that $\mathcal{G}(n+p) = \mathcal{G}(n) + s$ for all large enough n (WW, p. 114).

Table 3 exhibits some specimen games, of which the last three are *hexadecimal games* with $\mathbf{d}_k \leq \mathbf{15} = \mathbf{F}$. Such games may be arithmetically periodic. Anil Gangolli and Thane Plambeck established the ultimate periodicity of four octal games that were previously unknown:

The game **.16** has period 149459 (a prime!), the last exceptional value being $\mathcal{G}(105350) = 16$. The game **.56** has period 144 and last exceptional value $\mathcal{G}(326639) = 26$. The games **.127** and **.376** each have period 4 (with cycles of values 4, 7, 2, 1 and 17, 33, 16, 32 respectively) and last exceptional values $\mathcal{G}(46577) = 11$ and $\mathcal{G}(2268247) = 42$.

Grundy's Game [Grundy 1939; WW, p. 111], in which the move is to split a heap into two *unequal* heaps, continues to defy complete analysis, despite Mike Guy's calculation of the first ten million nim-values. Among these values,

$$0, 1, 6, 7, 10, 11, 12, 13, 18, 19, 20, 21, 24, \ldots$$

occur quite rarely. When written in binary, these values contain an even number of ones *if you ignore the last digit*. These *rare* values form a closed space (the *sparse space*) under nim-addition, while the complement forms the *common coset*:

$$\begin{array}{ccccccc} \text{rare} & \overset{*}{+} & \text{rare} & = & \text{rare} & = & \text{common} & \overset{*}{+} & \text{common} \\ \text{rare} & \overset{*}{+} & \text{common} & = & \text{common} & = & \text{common} & \overset{*}{+} & \text{rare} \end{array}$$

If the nim-values in a sequence begin to cluster in a suitable common coset, this clustering is likely to persist. In Kayles the rare and common values are *evil* and *odious* numbers respectively, with an even and odd number of ones in their binary expansions. On the other hand, Dawson's Kayles doesn't exhibit this sparse space phenomenon. In Grundy's Game only 1273 rare values have appeared; the only one in the range $36184 < n \leq 10^7$ is $\mathcal{G}(82860) = 108$. If the rare values have indeed died out, then Grundy's Game will ultimately be periodic, but the period may be astronomical.

Amongst the comparative chaos, John Conway and Mike Guy have noted a remarkable structure in the nim-values for Grundy's Game, related to the number 59. The probability that $\mathcal{G}(n + d) = \mathcal{G}(n)$ is often as high as $\frac{1}{4}$ if

$$d \text{ is } near \text{ } 59k \quad \text{and} \quad d \equiv k \bmod 3.$$

Examples of these pseudo-periods are 58, 61, 116, 119, 122, 290, 293, 296, 360, 412, 580, 583, 586, 589, 647, 650, 882, 952, 1172, where the last four correspond to $k = 11, 15, 16, 20$.

6. Green Hackenbush

(See ONAG, pp. 165–172; WW, pp. 183–190.) This is played on a picture, as in Blue-Red Hackenbush, but now all the edges are grEen, and may be chopped by Either player, making it an impartial game. Every Green Hackenbush picture has a nim-value: for example (Figure 6, right) the value of a string of 6 green

Code	Game	Nim-sequence
·77	*Kayles*. Knock down one skittle, or two contiguous skittles, from a row. [Dudeney 1908, Loyd 1914].	Ultimate period $p = 12$, $41281472182\dot{7}$ except for $n = 0, 3, 6, 9, 11, 15, 18, 21, 22, 28,$ 34,39,57,70, nim-value is resp. 0,3,3,4, 6,7,3,4,6,5,6,3,4,6.
·137	*Dawson's Chess*. $3 \times n$ board. White and Black pawns on ranks 1 and 3. Capturing obligatory. Looks partizan but isn't. [Dawson 1934; 1935].	$\dot{8}11203110332244559330113021104537\dot{4}$ except 0 for $n = 0, 14, 34$ and 2 for $n = 16, 17, 51$. $p = 34$.
·07	*Dawson's Kayles*. Knock down 2 skittles, but only if they're contiguous.	As for ·137, but shifted one term: 0011203... in place of 011203...
·6	*Officers*. Take 1 counter from any *longer* row. [Descartes 1953].	No period found. $\mathcal{G}(10342) = 256$.
·007	*Treblecross*. One-dimensional tic-tac-toe. (WW, pp. 93–94).	No pattern yet found.
·077	*Duplicate Kayles*. Knock down 2 or 3 contiguous skittles [Guy and Smith 1956].	$p = 24$. Kayles with each nim-value repeated, 00112233114433...
·7777	*Double Kayles*. Take up to 4 beans from a heap; leave rest in at most 2 heaps. [Guy and Smith 1956; WW, p. 98].	$p = 24$. Kayles with each nim-value g replaced by the pair $2g, 2g + 1$ or $2g + 1, 2g$ (according to a certain rule), 0123456732897654328945...
·156	See [Kenyon 1967].	$p = 349$.
·165	See [Austin 1976].	$p = 1550$.
4.3	*Lasker's Nim*	0124356879... $p = s = 4$.
·8	(first cousin of) *Triplicate Nim*. Take 1 from heap, rest left in exactly 3 nonempty heaps.	Arithmetically periodic, $p = 3$, saltus $= 1$. 0000111222333444...
·3F	(F=15) *Kenyon's Game*. Take 1 from heap or take 2 and leave rest in any number of heaps up to 3. [Kenyon 1967].	$p = 6$, $s = 3$ 0120123453456786789...
·E	(E =14) Take 1, leave rest in just 1, 2 or 3 heaps.	001234153215826514... $\mathcal{G}(246) = 128$. No known pattern.

Table 3. Some sample take-and-break games.

edges is *6. It is clear that the six possible moves exactly parallel the six possible moves that you can make from a heap of six beans.

We will see how to evaluate Green Hackenbush trees by the Colon Principle and how to reduce any picture to a forest by the Fusion Principle.

Green Hackenbush trees are examples of the *ordinal sum* $G : H$, which can be defined [WW, p. 214] for any two games G and H

$$G : H = \{G^L, G : H^L \mid G^R, G : H^R\}$$

where any move in G annihilates H, while a move in H leaves G unaffected. The *Colon Principle* [WW, pp. 184–185] states that $H \geq K$ implies $G : H \geq G : K$, and, in particular, that $H = K$ implies $G : H = G : K$. That is, $G : H$ depends only on the *value* of H and not on its *form*. It *may* depend on the *form* of G, because there are games $G_1 = G_2$ for which $G_1 : H \neq G_2 : H$.

The Colon Principle applies at branch points of Green Hackenbush trees, allowing us to do nim-addition "up in the air." For example, at a in Figure 6, left, we have $*3 + *2 = *$; at b, $* + * = 0$; and at c, $* + *2 = *3$, so the value is the same as that of Figure 6, middle, where, at d, $*2 + *2 + * + *4 = *5$, and the tree is worth $*6$. Notice the interplay of ordinary addition along strings, with nim-addition at branch points.

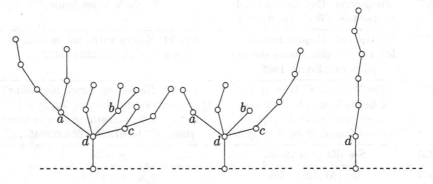

Figure 6. Transforming a tree into a stalk.

Green Hackenbush pictures involving circuits can be evaluated by the *Fusion Principle* (WW, pp. 186–188):

> The value of a picture is unaltered
> if you identify the nodes of a circuit.

The edges of the circuit then become loops, which may be replaced by twigs: compare Figure 7, middle and right. Check that the value of Figure 7, left, is $*8$. In this way, every component of a Green Hackenbush picture can be reduced to a tree, and hence to a string, and the strings are combined by nim-addition.

7. Welter's Game

(See [Welter 1952; 1954; Berlekamp 1972]; ONAG, pp. 153–165; WW, pp. 472–481.) This is another game whose analysis involves the interplay of nim-addition and ordinary addition. It is a form of Nim with unequal heaps, but in order to

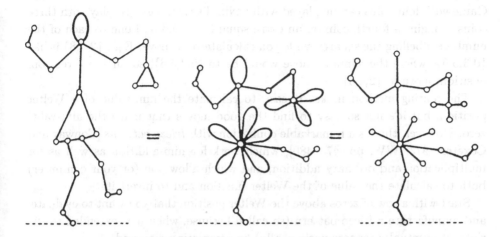

Figure 7. Girl becomes tree.

keep track of empty heaps, only one of which is allowed, it's better to play it with coins on a strip of squares, numbered 0, 1, 2, ..., with at most one coin on a square. A move is to shift a coin leftwards to any unoccupied square, possibly passing over other coins. The game ends when the k coins are on the leftmost squares 0, 1, ..., $k - 1$. Figure 8 shows a position with $k = 7$.

| 0 | ● | ● | ● | 4 | ● | 6 | 7 | ● | 9 | 10 | 11 | 12 | ● | 14 | 15 | 16 | 17 | 18 | 19 | 20 | ● | 22 | 23 | 24 |

Figure 8. The position {1,2,3,5,8,13,21} in Welter's Game.

To calculate the nim-value, or *Welter function*, $[a|b|c|\ldots]_k$ of the position with k coins on squares a, b, c, ..., first note that for $k = 1$, $[a] = a$, and that for $k = 2$, $[a|b]$ is one less than the nim-sum of a and b: e.g., $[1|3] = 1$, $[5|6] = 2$. For more than two coins, *mate* the pair that is congruent modulo the highest power of two (it doesn't matter that this pair may not be unique). Remove the mated pair and find the best mated pair among the remaining $k - 2$ coins. Continue until all coins are mated, except, when k is odd, for one coin, the *spinster*, s. Then, if (a,b), (c,d), ... are the mates, $[a|b|c|d|\ldots]$ may be calculated as the *nim-sum*

$$[a|b] \overset{*}{+} [c|d] \overset{*}{+} \ldots (\overset{*}{+} [s])$$

where the last term is included just if k is odd.

In Figure 8 the best mates are (5,21), then (1,13), then (2,8), and 3 is the spinster, so the nim-value is

$$[1|2|3|5|8|13|21] = [5|21] \overset{*}{+} [1|13] \overset{*}{+} [2|8] \overset{*}{+} 3$$
$$= 15 \overset{*}{+} 11 \overset{*}{+} 9 \overset{*}{+} 3 = 14.$$

It turns out that $[a|b|c|d] = 0$ just if the nim-sum $a \overset{*}{+} b \overset{*}{+} c \overset{*}{+} d = 0$, so Welter's Game with four coins can be played with a Nim-like strategy. To play with three coins, imagine a fourth coin on an extra square -1, and add one to each of the numbers labelling the squares while you calculate your move. E.g., $\{2,5,8\}$ is like $\{0,3,6,9\}$, where the winning move would be to $\{0,3,5,6\}$, so, in the three-coin position, move to $\{2,4,5\}$.

The mating method makes it easy to calculate the nim-value of a Welter position, but it's not so easy to find the good moves that make the nim-value zero. However, there's a remarkable connexion with *frieze patterns* [Conway and Coxeter 1973; WW, pp. 475–480], which work for nim-addition as well as for multiplication and ordinary addition, and which allow you (or your computer) both to calculate the value of the Welter function and to invert it.

Start with a row of zeros above the Welter position that you want to evaluate, and manufacture a frieze pattern (so called because, when it is extended to the right, it eventually repeats periodically) by completing diamonds

$$
\begin{array}{ccc}
 & b & \\
a & & d \\
 & c &
\end{array}
\quad \text{using the rule} \quad a \overset{*}{+} d = (b \overset{*}{+} c) + 1,
$$

so that $c = [a|d] \overset{*}{+} b$, where the sums are still nim-sums. Lo and behold (Figure 9) the value of the Welter function appears at the foot of the pattern, as follows from a formula on page 159 on ONAG.

$$
\begin{array}{cccccccc}
0 & 0 & 0 & 0 & 0 & 0 & 0 & 0 \\
 1 & 2 & 3 & 5 & 8 & 13 & 21 \\
 & 2 & 0 & 5 & 12 & 4 & 23 \\
 & & 3 & 7 & 13 & 15 & 31 \\
 & & & 3 & 12 & 13 & 11 \\
 & & & & 9 & 13 & 10 \\
 & & & & & 15 & 11 \\
 & & & & & & 14
\end{array}
$$

Figure 9. Calculating the Welter function from a frieze pattern.

If you want to change the value $n = [a|b|c| \ldots]$ to some $n' \neq n$, then there are unique $a' \neq a$, $b' \neq b$, $c' \neq c$, ... such that

$$[a'|b|c| \ldots] = n' = [a|b'|c| \ldots] = [a|b|c'| \ldots] = \ldots$$

and $[a|b|c| \ldots] = n$ remains true if we replace any *even* number of a, b, c, ..., n by the corresponding primed letters. This *Even Alteration Theorem* [ONAG, pp. 160–162; WW, p. 477] may be written

$$\begin{bmatrix} a|b|c| \\ a'|b'|c'| \end{bmatrix} \cdots \end{bmatrix} = \frac{n}{n'}$$

To find $a', b', c', \ldots$ corresponding to a given n', continue the bottom row of the frieze pattern, $n, n', n, n', n, \ldots$ alternately, and then extend the pattern to the right, using the same diamond rule. You will discover that the defining row, a, b, c, $\ldots$ continues with the answers, $a', b', c', \ldots$!

```
0    0    0    0    0    0    0    0    0    0    0    0    0    0    0    0
  1    2    3    5    8   13   21   15    0   37   35   10   11   19
    2    0    5   12    4   23   25   14   36    5   40    0   23
      3    7   13   15   31   24   25   41    5   15   45   29
        3   12   13   11   17   25   33   15   12    9   47
          9   13   10    6   31   46    4    7   11    8
           15   11    0    9   41    8   13    7   11
             14    0   14    0   14    0   14    0
```

Figure 10. Inverting the Welter function using a frieze pattern.

In Figure 10 we find the good moves in the position $\{1,2,3,5,8,13,21\}$ by choosing $n' = 0$ and extending the pattern of Figure 9. If you extend it even further to the right, you'll see why it's called a frieze pattern. If you believe the algorithm, and read the second row of Figure 10,

$$\begin{bmatrix} 1\,|2|\,3\,|\,5\,|\,8\,|13|21 \\ 15|0|37|35|10|11|19 \end{bmatrix} = \frac{14}{0}$$

Check that each move leads to a $\mathcal{P}$-position. Some of the suggested moves, e.g., 1 to 15, 3 to 37, are not legal, but, provided $n' < n$, you'll always find one that is legal, in fact there is always an odd number of legal good moves. Here there are three good moves: 2 to 0, 13 to 11, and 21 to 19.

We can even give you a strategy for the misère form (last player losing) of Welter's Game, if you're willing to learn about *Abacus Positions* [WW, pp. 478–481].

8. Coin-Turning Games

(See WW, pp. 429–456.) Several of the impartial games we've already mentioned, and a wide range of new games, can be realized by an idea of Hendrik Lenstra. The $\mathcal{P}$-positions in several of these turn out to correspond to the codewords in some well-known and some not-so-well-known error-correcting codes; see [Pless 1991] and [Fraenkel 1996b] in this volume.

Turning Turtles was originally played with turtles, but it's less cruel to play it with a row of coins (Figure 11). A move is to turn a head to a tail, with the

Figure 11. A Turning Turtles position, with coins 3, 4, 6, 7 showing heads.

additional option of turning at most one other coin, to the left of it. This second coin may go from head to tail, or from tail to head. The game is over when all coins show tails, and the last player wins.

We leave you to verify that this is a disguise for Nim: if you number the coins 1, 2, 3, ... from the left, then the nim-value of coin n is $*n$ if it's a head, and 0 if it's a tail. The nim-value of a general position is the nim-sum of the nim-values, i.e., the nim-sum of the nim-values of the heads. For example, the good moves in Figure 11 are to turn coin 6 to a tail; or to turn 7 to a tail and 1 to a head; or to turn 4 to a tail and 2 to a head.

Mock Turtles is played in the same way, but a move may turn one, two or three coins, provided the rightmost turned goes from head to tail (this is to make the game satisfy the ending condition). We now number the coins from *zero* (the Mock Turtle) and find the nim-value (or Grundy function), $\mathcal{G}(n)$, of the n-th coin, when head up, to be:

$$n \;\;= 0 \;\; 1 \;\; 2 \;\; 3 \;\; 4 \;\; 5 \;\; 6 \;\; 7 \;\; 8 \;\; 9 \;\; 10 \;\; 11 \;\; 12 \;\; 13 \;\; 14 \;\; 15 \;\; 16 \;\; 17 \;\; 18 \; \ldots$$
$$\mathcal{G}(n) = 1 \;\; 2 \;\; 4 \;\; 7 \;\; 8 \;\; 11 \;\; 13 \;\; 14 \;\; 16 \;\; 19 \;\; 21 \;\; 22 \;\; 25 \;\; 26 \;\; 28 \;\; 31 \;\; 32 \;\; 35 \;\; 37 \; \ldots$$

These are the *odious numbers* that we met as common values in Kayles.

$$\mathcal{G}(n) = 2n \text{ or } 2n + 1.$$

To find which, write n in binary and append a check digit, 0 or 1, to make an *odd* number of digits 1.

Moebius, Mogul and *Moidores* are the corresponding games in which a move turns up to t coins, where $t = 5$, 7 and 9. We consider only odd values of t, because the *Mock Turtle Theorem* gives us the results for even values of t:

> Every nim-value for the $t = 2m + 1$ game is an odious number.
> The corresponding value for the $t = 2m$ game
> is got by dropping the final binary digit.

The nim-values for coins 0 to 17 (when head-up) in Moebius are shown in Table 4. The structure of the $\mathcal{P}$-positions in 18-coin Moebius is revealed on replacing the coin numbers by the labels in the third row.

coin #	0	1	2	3	4	5	6	7	8	9	10	11	12	13	14	15	16	17
nim-value	1	2	4	8	16	31	32	64	103	128	171	213	256	301	342	439	475	494
label	∞	1	4	0	−4	−1	5	6	−8	2	−3	−5	8	3	−7	7	−6	−2

Table 4. Eighteen-coin Moebius gives the game its name.

Coins 0 to 5, with labels ∞, 0, ± 1, ± 4, clearly form a $\mathcal{P}$-position (whichever ones you turn over, I'll turn over the rest). Starting from this, or indeed from any $\mathcal{P}$-position, we can find others by operating on the labels with any *Möbius transformation* (modulo 17):

$$x \to \frac{ax + b}{cx + d} \quad \text{with} \quad ad - bc = 1.$$

There are   1 + 102 + 153 + 153 + 102 +   1 $\mathcal{P}$-positions
with respectively 0   6   8   10   12   18   heads.

If we drop the Mock Turtle (at ∞) we have the $t = 4$ game on 17 coins. The $\mathcal{P}$-positions in these two games correspond to the codewords in the [18,9,6] extended quadratic residue code and the [17,9,5] quadratic residue code.

Similarly if we play 24-coin Mogul ($t = 7$, turn up to 7 coins) we find

1 + 759 + 2576 + 759 + 1 $\mathcal{P}$-positions
with 0   8   12   16   24   heads

coinciding with the 2^{12} codewords of the extended [24,12,8] Golay code. With $t = 6$ and 23 coins the $\mathcal{P}$-positions correspond to the codewords in the perfect [23,12,7] Golay code. Robert Curtis [1976; 1977] has given a pictorial representation of this, the Miracle Octad Generator or "MOG", which also shows the connexion with the Steiner system $S(5, 8, 24)$.

In the *Ruler Game* any number of *contiguous* coins may be turned (with the rightmost always going from head to tail). If the coins are numbered from 1, then the nim-value, $\mathcal{G}(n)$, is the highest power of 2 that divides n.

In *Turnips* (or *Ternups*) a move turns three equally spaced coins. Number the coins from 0 and write n in *ternary* (base 3). Then $\mathcal{G}(n)$ is the k-th odious number if the last digit 2 in the ternary expansion is in the k-th place from the right, or $\mathcal{G}(n) = 0$ if there is no digit 2 in the ternary expansion of n.

There is a plethora of such coin-turning games. They can also be played on a two-dimensional array of coins. For example, we can play the Cartesian product, $A \times B$, of two one-dimensional games A and B, in which a move is to turn all coins with coordinates (a_i, b_j), where $\{a_i\}$ and $\{b_j\}$ are sets of coins constituting legal moves in games A and B respectively. To satisfy the ending condition, the "most northeasterly" coin turned must go from head to tail (Figure 12).

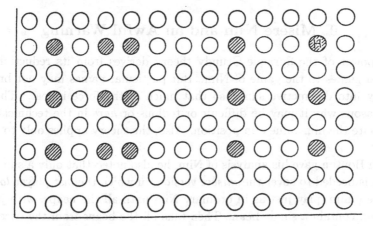

Figure 12. A typical move in Moebius × Turnips.

The nim-value of a (head-up) coin in such a game is given by the *Tartan Theorem*:

> The nim-values for the game $A \times B$ are
> the nim-products of those for A and B:
> $$\mathcal{G}_{A \times B}(a, b) = \mathcal{G}_A(a) \overset{*}{\times} \mathcal{G}_B(b)$$

where $\mathcal{G}_A(a)$ is the nim-value of coin number a in game A, etc., and $\overset{*}{\times}$ denotes *nim-multiplication*. Nim-multiplication [ONAG, pp. 52–53; Lenstra 1977a; 1977b] may be defined from the field laws (e.g., associativity and distributivity over nim-addition), together with the rule

> $$n \overset{*}{\times} N = n \times N \quad \text{for} \quad n < N$$
> $$N \overset{*}{\times} N = 3N/2$$

where N is any *Fermat power* of 2 $(2, 4, 16, \ldots, 2^{2^h}, \ldots)$. For example, $2 \overset{*}{\times} 2 = 3$, because 2 is a Fermat power, while $2 \overset{*}{\times} 3 = 2 \overset{*}{\times} (2 \overset{+}{} 1) = 3 \overset{+}{} 2 = 1$. A more complicated example is

$$13 \overset{*}{\times} 7 = (4 \overset{*}{\times} 3 \overset{+}{} 1) \overset{*}{\times} 7 = (4 \overset{*}{\times} (2 \overset{+}{} 1)) \overset{*}{\times} (4 \overset{+}{} 2 \overset{+}{} 1) \overset{+}{} 7$$
$$= 4 \overset{*}{\times} 4 \overset{*}{\times} (2 \overset{+}{} 1) \overset{+}{} 4 \overset{*}{\times} 2 \overset{*}{\times} (2 \overset{+}{} 1) \overset{+}{} 4 \overset{*}{\times} (2 \overset{+}{} 1) \overset{+}{} 7$$
$$= 6 \overset{*}{\times} (2 \overset{+}{} 1) \overset{+}{} 4 \overset{*}{\times} (3 \overset{+}{} 2) \overset{+}{} 4 \overset{*}{\times} 3 \overset{+}{} 7$$
$$= (4 \overset{+}{} 2) \overset{*}{\times} (2 \overset{+}{} 1) \overset{+}{} 4 \overset{*}{\times} 1 \overset{+}{} 12 \overset{+}{} 7$$
$$= 4 \overset{*}{\times} 2 \overset{+}{} 2 \overset{*}{\times} 2 \overset{+}{} 4 \overset{+}{} 2 \overset{+}{} 4 \overset{+}{} 11$$
$$= 8 \overset{+}{} 3 \overset{+}{} 9$$
$$= 2.$$

The assiduous reader will verify that the nim-cube-roots of 1 are 1, 2 and 3, and the nim-fifth-roots are 1, 8, 10, 13, 14.

9. Misère Nim and an Awful Warning

The power of the Sprague-Grundy theory derives from its reduction of all impartial games to the game of Nim. But particular games may not break up naturally into disjunctive sums, so that much of the force is lost. There are other reasons why it may be difficult or tedious or *hard* in the technical sense, to calculate the nim-value of a position. Also, the theory applies only to normal play.

When Bouton gave his analysis of Nim, he also noted that only a small modification is needed to cover the *misère* version, in which the last player *loses*. To win misère nim, play just as in ordinary Nim until all the heaps, with just one exception, contain a single bean. Then take all the beans from the exceptional heap, or all the beans but one, so as to leave an odd number of heaps of size one.

It's tempting to think (and several people have been tempted to write) that you can play misère impartial games just like normal impartial games until very near the end, when you ...

BUT THAT'S NOT TRUE!

The situation is very complicated. The little that is known in general is given in WW, Chapter 13. However, an intriguing breakthrough was made recently by William Sibert and John Conway [1992], who have given an analysis of *Misère Kayles* and Thane Plambeck [1992] has used their method to analyze a small number of other games.

References

[Austin 1976] R. Austin, "Impartial and partisan games", M.Sc. Thesis, Univ. of Calgary, 1976.

[Berlekamp 1972] E. R. Berlekamp, "Some recent results on the combinatorial game called Welter's Nim", pp. 203–204 in Proc. 6th Ann. Princeton Conf. Information Science and Systems, 1972.

[Berlekamp et al. 1982] E. R. Berlekamp, J. H. Conway, and R. K. Guy, *Winning Ways For Your Mathematical Plays*, Academic Press, London, 1982.

[Bouton 1902] C. L. Bouton, "Nim, a game with a complete mathematical theory", *Ann. of Math.* **3**(2) (1902), 35–39.

[Conway 1976] J. H. Conway, *On Numbers And Games*, Academic Press, London, 1976.

[Conway and Coxeter 1973] J. H. Conway and H. S. M. Coxeter, "Triangulated polygons and frieze patterns, *Math. Gaz.*" **57** (1973), 87–94, 175–183.

[Curtis 1976] R. T. Curtis, "A new combinatorial approach to M_{24}", *Math. Proc. Cambridge Philos. Soc.* **79** (1976), 25–42.

[Curtis 1977] R. T. Curtis, "The maximal subgroups of M_{24}", *Math. Proc. Cambridge Philos. Soc.* **81** (1977), 185–192.

[Dawson 1934] T. R. Dawson, Problem 1603, *Fairy Chess Review*, December 1934, p. 94.

[Dawson 1935] T. R. Dawson, "Caissa's Wild Roses", Reprinted in *Five Classics of Fairy Chess*, Dover, 1973.

[Descartes 1953] B. Descartes, "Why are series musical?", *Eureka* **16** (1953), 18–20. Reprinted *ibid.* **27** (1964), 29–31.

[Dudeney 1908] H. E. Dudeney, *The Canterbury Puzzles, and Other Curious Problems*, Dutton, New York, 1908. Fourth edition reprinted by Dover, New York, 1958.

[Ferguson 1974] T. S. Ferguson, "On sums of graph games with last player losing", *Internat. J. Game Theory* **3** (1974), 159–167.

[Fraenkel 1996a] A. S. Fraenkel, "Scenic trails ascending from sea-level Nim to alpine chess", pp. 13–42 in this volume.

[Fraenkel 1996b] A. S. Fraenkel, "Error-Correcting Codes Derived from Combinatorial Games", pp. 417–431 in this volume.

[Guy 1989] R. K. Guy, *Fair Game: How to Play Impartial Combinatorial Games*, COMAP, Arlington, MA, 1989.

[Guy and Smith 1956] R. K. Guy and C. A. B. Smith, "The G-values of various games", *Proc. Camb. Phil. Soc.* **52** (1956), 514–526.

[Grundy 1939] P. M. Grundy, "Mathematics and Games", *Eureka* **2** (1939), 6–8. Reprinted in *Eureka* **27** (1964), 9–11.

[Kenyon 1967] J. C. Kenyon, "A Nim-like game with period 349", Univ. of Calgary, Math. Dept. Res. Paper **13**.

[Lenstra 1977a] H. W. Lenstra, Jr., "Nim multiplication", Séminaire de Théorie des Nombres, exposé No. 11, Université de Bordeaux, 1977.

[Lenstra 1977b] H. W. Lenstra, Jr., "On the algebraic closure of two", *Proc. Kon. Nederl. Akad. Wetensch.* **A80** (1977), 389–396.

[Loyd 1914] S. Loyd, *Cyclopedia of Puzzles and Tricks*, Franklin Bigelow Corporation, Morningside Press, NY, 1914. A selection, edited by M. Gardner is available as *The Mathematical Puzzles of Sam Loyd* (two volumes), Dover, New York, 1959.

[Plambeck 1992] T. E. Plambeck, "Daisies, Kayles, and the Sibert–Conway decomposition in misère octal games", *Theoret. Comput. Sci. (Math Games)* **96** (1992), 361–388.

[Pless 1991] Vera Pless, "Games and codes", pp. 101–110 *Combinatorial Games* (edited by R. K. Guy), Proc. Symp. Appl. Math. **43**, Amer. Math. Soc., Providence, RI, 1991.

[Sprague 1935–36] R. Sprague, "Über mathematische Kampfspiele", *Tôhoku Math. J.* **41** (1935–36), 438–444.

[Sibert and Conway 1992] W. L. Sibert and J. H. Conway, "Mathematical Kayles", *Internat. J. Game Theory* **20** (1992), 237–246.

[Welter 1952] C. P. Welter, "The advancing operation in a special abelian group", *Nederl. Akad. Wetensch. Proc.* **A55** = *Indag. Math.* **14** (1952), 304–314.

[Welter 1954] C. P. Welter, "The theory of a class of games on a sequence of squares, in terms of the advancing operation in a special group", *Nederl. Akad. Wetensch. Proc.* **A57** = *Indag. Math.* **16** (1954), 194–200.

RICHARD K. GUY
MATHEMATICS AND STATISTICS DEPARTMENT
UNIVERSITY OF CALGARY
2500 UNIVERSITY AVENUE
ALBERTA, T2N 1N4
CANADA

Games of No Chance
MSRI Publications
Volume 29, 1996

Championship-Level Play of Dots-and-Boxes

JULIAN WEST

ABSTRACT. A single-elimination Dots-and-Boxes tournament was held during the MSRI meeting, with a $500 purse. This is an analysis of the finals, in which Daniel Allcock defeated Martin Weber, playing both first and second player. A systematic notation is developed for the analysis.

Dots-and-Boxes, described in Chapter 16 of *Winning Ways* [Berlekamp et al. 1982], is a game played on a finite rectangular unit lattice of dots (or, in dualized form, on an arbitrary graph). A move consists of joining two adjacent dots, that is, dots at distance one; if this completes one or more squares, a point ("box") is awarded for each and the player retains her turn. A move after which turn does not pass to the opponent is known as a *complimenting move* and, under so-called normal win conditions—last player to move wins—leads to so-called *loony* values (explained below). However, Dots-and-Boxes does not have a normal win condition; indeed, analysis is complicated considerably by the unusual who-dies-with-the-most-wins condition.

Under various names, this game is popular with children in many countries. As played by most practitioners, it is a fairly uninteresting game, consisting of a phase of randomly segmenting the board followed by a phase of greedily dividing up the spoils. So it would seem an odd choice for a tournament between serious game theory researchers. But Dots-and-Boxes is a classic example of a game that is Harder Than You Think. *Winning Ways* (p. 535) gives an account of all the stages you will go through in becoming a Dots-and-Boxes expert:

- abandon the greedy approach in favour of double-dealing moves;
- learn the parity rule for long chains;
- become an expert at Strings-and-Coins;
- (and so) become an expert at Nimstrings;
- apply Twopins theory; become an expert at Kayles;
- recognize the exceptional cases where Nimstrings does not suffice to analyse Dots-and-Boxes.

Finally, it goes on to demonstrate that Dots-and-Boxes, played on general graphs, is NP-hard (p. 543).

This is rather daunting, but for our present purposes we shall only really need to know the parity rule and one observation drawn from Nimstrings. We'll be able to illustrate each by using the games to hand, so we exhibit these games now: see Figure 1. The games were played on a grid of six dots by six, and are shown here near completion, well beyond the point where the outcome has become obvious. (Since future play by a sensible player can now be predicted, these diagrams give a complete account of each game. Nevertheless, we'll also present a diagram of one game at an earlier stage, in order to illustrate crucial stages in the play.)

In each of these games, the moves of the first player (Dodie) are shown with Dotted lines; therefore the first edge drawn is the dotted one numbered 1, and the second edge drawn is the solid one numbered 1. Sometimes multiple edges are played in one turn; when this happens we have labelled them as, say, 15a, 15b, 15c. When either player claims a square, a D (for Dodie) or an E (for Evie) is entered; we have placed the letter slightly off centre in the direction of the edge that completed the square. Every edge approached by a D or an E bears a letter, since it was a complimenting move, though not every edge bearing a letter completed a square, since it may have been the last edge of a turn. Note that (by coincidence) Dodie's move 16a completed two boxes in each game. Moves that claim two boxes are called *double-crossing moves*; they are fundamental in the analysis of Dots-and-Boxes.

Consider the situation in Figure 1, top. Evie has just claimed the box at upper right (18a) and with her move at 18b opens up two boxes for Dodie. Such a move is called a *hard-hearted handout*; Dodie has no reason not to take the two free boxes, and therefore no real choice. But after this she is stuck: she must either move in the four-box *chain* or the six-box *loop*, and any move here allows Evie to capture boxes. We say that Evie *has control*. In either region, Evie will be able to *retain control* by "politely declining" the last squares, making a *double-dealing* move—that is, one that invites a double-crossing one in reply. If you don't know how to decline the last two boxes in a chain, you should work it out as an exercise: you will then be able to defeat most casual players of Dots-and-Boxes!

So if Dodie moves in the chain, Evie has a choice between taking all four boxes and surrendering the remaining six, or taking only two boxes and getting the remaining six on her following turn. Alternatively, if Dodie moves in the loop, Evie can either claim six boxes and surrender four, or claim only two boxes (since it requires four boxes—in other words, two double-dealing moves—to decline a loop) and get four on her next turn. Evie thus has a chance to get eight boxes if Dodie moves in the chain, and at best six if Dodie moves in the loop, so Dodie should move in the loop. Nevertheless, Evie will win the game, with 13 boxes to Dodie's 12.

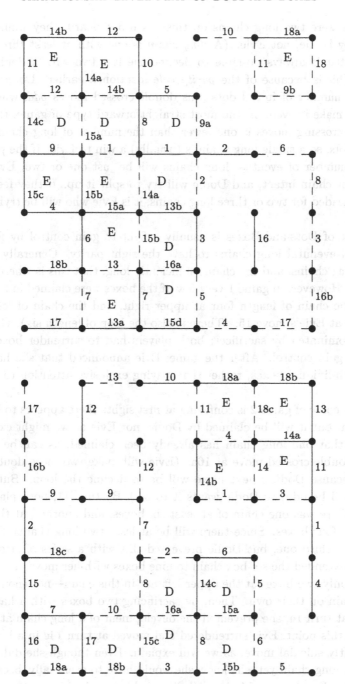

Figure 1. Near-ending position in the two games of the MSRI Dots-and-Boxes tournament finals. Dodie (D) to move in each case; dashed lines indicate her moves.

If there were two long chains on this 6×6 dot board, they would both be claimed by Dodie, not Evie. (A *long chain* is one with at least three boxes in it, so that it is always possible to decline the last two with a double-dealing move.) This is because of the *parity rule* mentioned earlier: Dodie wishes to make the number of initial dots plus double-crossed moves *odd*, whereas Evie wishes to make it even. In the most straightforward type of game, the number of double-crossing moves is one fewer than the number of long chains, and we have 36 dots, so a single long chain is (usually) a win to Evie. If the players feel that the number of eventual long chains will be just one or two, Evie will try to keep the chain intact, and Dodie will try to split it up. If they feel that the board is headed for two or three long chains, it is Evie who will be trying to split things up.

A game of Dots-and-Boxes is usually a fight to gain control by forcing the number of evenutal long chains to have the right parity. Generally, there are enough long chains, and the chains sufficiently long, that this is where the game is settled. However, in game 1 very few of the boxes were claimed in long chains: there is the chain of length four at upper right, and the chain of length three that went at Evie's move 15. (There is also the loop of length six). Game 1 was a game dominated by sacrifices; both players had to surrender boxes in their efforts to gain control. After the game, Evie announced that she had won by counting individual boxes, rather than paying exclusive attention to the parity rule.

The outcome of game 2 is confusing at first sight: there appears to be a single long chain, but it will be claimed by Dodie, not Evie as we might expect. The reason is that one long chain has already been claimed, as can be seen from Dodie's double-crossed move at 16a. (Evie will make two more double-crossed moves, because Dodie's next move will be to decline the loop. But the total number will be odd, as Dodie wants it to be.) By turn 11, both players could see that there was one chain of at least six boxes, and another, at the bottom, of at least four boxes. Since there will be at least two long chains, Evie hoped to create a third one, but Dodie prevented this with a series of strong moves. First she extended the six-box chain to nine boxes with her moves 11, 12 and 13. This left only two boxes at the upper left *not* in this chain—not enough to form a long chain on their own. Then, by sacrificing two boxes with a hard-hearted handout at turn 14, she prevented the development of a long chain at the upper right. At this point, Evie surrendered and moved at turn 14c in a long chain— an explicitly suicidal move, as we will explain. Even though she didn't have to move in a long chain yet, she knew she would have to eventually, because of the parity rule. So her turn 14c is much like a player resigning in chess as soon as the checkmate becomes evident. Both players then played out the game with intentionally cavalier moves—the two boxes that Dodie takes on turn 16 could have been taken by either player on turn 15, but both declined as a gesture of futility. For clarity, we redraw in Figure 2 the situation just after this move.

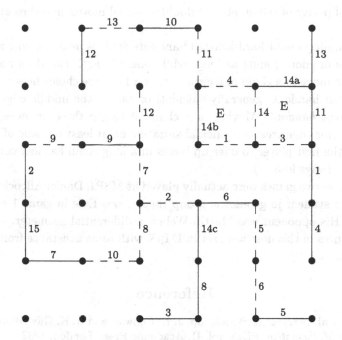

Figure 2. Position in game 2 just after Evie (full lines) plays her move 14c, at which point both players knew Dodie would win.

The reason Evie's move 14c is suicidal comes from a simple strategy-stealing argument in the theory of Nimstrings. Nimstrings is just Dots-and-Boxes with the normal win condition of last-to-move-wins. Move 14c allows the choice of two replies (those edges labelled 15a and 15b in our example). Consider the entire game position with the exception of these two boxes: it must be a win for either the first player or the second player (though it may in general be very hard to know which). Arrange to be the appropriate player by either: moving at 15a, then 15b, then going on to be player 1 on the remaining position; *or* moving at 15b, then stopping so as to become player 2 on the remaining position. (One player or another will eventually claim the two free boxes left by this second option, but in Nimstrings we don't care who.)

In Nimstrings, 14c is therefore an example of a *loony* move. The value loony is an extension to the field of nimbers from Sprague–Grundy theory. It has the peculiar property that its sum with any nimber is again loony. This can be seen in action here, because leaving a loony position is a losing move, whatever is happening elsewhere on the board. (Note that also the sum of two loonies is loony.)

In the language of Dots-and-Boxes, a move such as Evie's 14c is called a *half-hearted handout*. Since it is fatal (i.e., loony) in Nimstrings, it is nearly always fatal in Dots-and-Boxes, amounting as it does to offering your opponent the choice of parity of the number of double-crossed moves, and thus of eventual long chains.

Understanding about hard-hearted handouts enables us to see why a chain of length three or more counts as long, while one of length two does not. If you are forced to move in a chain of length two, you have the choice between making a half-hearted handout (generally suicidal) or taking the middle edge to make a *hard-hearted handout*. Moving in a chain of length three or more, drawing even an interior edge creates a suicidal situation on at least one side of the edge. Therefore, the first player to offer up boxes in a long chain has lost control, and will almost always lose.

When these two games were actually played at MSRI, Daniel Allcock, a Berkeley graduate student in geometric group theory, was Evie in game 1 and Dodie in game 2. His opponent was Martin Weber, a differential geometer.

The pictures in this note were set in LaTeX with some assistance from Jennifer Overington.

Reference

[Berlekamp et al. 1982] E. R. Berlekamp, J. H. Conway, and R. K. Guy, *Winning Ways For Your Mathematical Plays*, vol. II, Academic Press, London, 1982.

JULIAN WEST
MALASPINA UNIVERSITY-COLLEGE
and
DEPARTMENT OF MATHEMATICS AND STATISTICS
UNIVERSITY OF VICTORIA
VICTORIA, B.C.
CANADA V8W 3P4
    julian@math.uvic.ca

Games of No Chance
MSRI Publications
Volume **29**, 1996

Championship-Level Play of Domineering

JULIAN WEST

ABSTRACT. A single-elimination Domineering tournament was held at the
MSRI meeting, with a $500 purse. This is an analysis of the finals of that
tournament, in which Dan Calistrate defeated David Wolfe by three games
to one. An algebraic notation for commenting games is introduced.

Domineering is a game played on subsets of the square lattice by two players,
who alternately remove connected two-square regions (dominoes) from play. Left
may only place dominoes vertically; Right must play horizontally. The normal
win-condition applies, so that the first player unable to move loses.

It is difficult to analyse general Domineering positions, even for quite small
boards. One way to gain insight into the nature of the problem is to watch
actual games between expert players. To determine who were the strongest
players available, an open-registration tournament was held. To insure that the
players gave proper consideration to their play, a prize of $500 was awarded the
winner.

The finalists were Dan Calistrate of Calgary and David Wolfe of Berkeley.
The format for the final was to play two games, each player taking the first turn
once; if the series was split, two more games would be played, and so on until one
player won both games of a round. As in chess, it would be expected that one
set of pieces would provide an advantage, therefore winning with the favoured
set would be like holding serve in tennis. In the event, the first round produced
two first-player wins, and Calistrate won both second-round games.

An 8 × 8 board was selected as large enough to be beyond the range of current
analysis, and therefore apt to provide an interesting game. It seems quite likely
that analysis of the general position is genuinely hard, so that the solution of
the 8 × 8 board would simply necessitate a move to 9 × 9 or 10 × 10 to retain
interest.

Because of symmetry, the square board must be either a first-player or a
second-player win (as opposed to a win for Left or a win for Right). Indeed,
because of this symmetry, we can always orient the board so that the player who

starts is considered to be Left; we have adopted this convention for clarity. The
5×5 board is known to be a second-player win (that is, it has value 0; see *On
Numbers and Games* [Conway 1976]). Conway observes that for larger boards
"0 is an infinitely unlikely value", because this value would only obtain if Left's
best move is to a position of net value to Right. This becomes less and less likely
the more options Left is given.

Furthermore, the actual opening move generally agreed
on as strongest appears to be quite strong. It leaves the
board in the position shown at the right, which is at least
as good for Left as $1-$ (the 8×8 board with a 2×2 bite out
of the corner). That board is also likely to be of net value
to the first player, but presumably has value less than 1.

A word about notation: we have numbered the dominoes in the order in which
they were placed on the board. Thus Left (vertical) dominoes have odd numbers.
When it is possible to compute the game-value of a contiguous region, we will
write it in.

In actual play, the game appears to go through three stages: an early game,
a middle game, and an end game. Nearly all of the interest appears to be in
the middle game. The early game lasts about eight moves (half-turns), and
consists of corner moves similar to the opening explained above. It appears that
experts cannot distinguish between the relative values of these openings, seeing
them rather as variants. The choice of which variant to play may be as much
psychological as anything, trying to guess a terrain that may be less familiar
for the opponent. Unfortunately, we do not have as rich a vocabulary for our
textbook openings as does the chess world, but the actual positions reached in
the championship games were as shown in Figure 1.

In the early game, every move reserves a future move for the same player by
creating a space in which only that player can place a domino. This suggests
that Domineering on a spacious board is very hot, explaining the difficulty of a
thorough analysis.

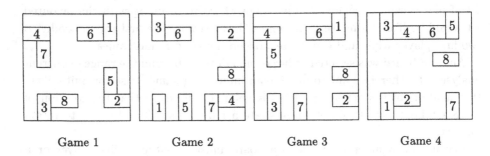

Game 1   Game 2   Game 3   Game 4

Figure 1. Position at end of opening for each of the four games in the finals. In
games 1 and 3, Wolfe is Left (vertical) and Calistrate is Right (horizontal). In
games 2 and 4, the roles are reversed.

In the middle game, in the space of about six moves, the board is converted from the relatively large open space of the "textbook" opening into a fractured board consisting entirely of numbers and switches. Since such a position is easy to analyse and to play flawlessly, the end game need not be played out. As the large, difficult to analyse region of the board is sectioned, it is rarely broken into two hot regions; therefore it is usually apparent in which component a player should play. Finding the correct move can be subtle, and a mistake is often fatal, since the game is really decided in about six moves.

Since the game is presumably a win for Left, and since it is decided in the space of six moves, it is very unlikely that control will be exchanged even twice. Therefore, Left wins exactly those games in which Left makes no errors. Right wins games in which Left makes an error, and Right successfully exploits it. These games can be viewed as "service breaks". Since any winning move is a good move, wins for Left consist entirely of good moves. Wins for Right contain exactly one bad move. So there is only one bad move to be pointed out in these four games.

In one sense, the onus is on the first player to initiate the transition to the middle-game. Imagine the board divided into 2×2 squares. If no dominoes ever cross the borders between these squares, the second player will be able to win by claiming half of the squares. The early game consists of moves that respect these borders, so the transition to the middle game is the breaking of this respect, and it is the first player who benefits. Note that the opening of game 4 is an odditiy in this respect, in that it is the second player who breaks symmetry (at move 8).

Game 3, the only one that led to a second-player win, breaks this expectation in the other sense: the first player *failed* to break symmetry. Since this was the "service break" game, it is the only one in which we can definitively say that a mistake was made, and it thus proves interesting to analyse. After twelve moves, the situation shown at the top in Figure 2 was reached.

In actual play, Left played as in the bottom left diagram of Figure 2. It looks like a reasonable move, since as it continues the paradigm of reserving a free move for Left; but it leaves a devastating future move for Right in the six-square region. The correct move, the only winner, is not easy to spot; it is shown on in the bottom right diagram of Figure 2.

Notice that this fulfils our expectation that this is the moment at which the first player must break out of the division into 2×2 cells.

The other three games simply demonstrate the considerable latitude that the first player has for securing the (assumed) win. Figure 3 shows the positions reached just as they become susceptible to analysis. Figure 4 shows the games again as they cool down to sums of numbers and switches. In game 2, Left's move 13 is indeed the optimal move (to $7 \mid \frac{11}{2} \parallel 5, \{5, 4\}$). Right's reply selects the option $+5$.

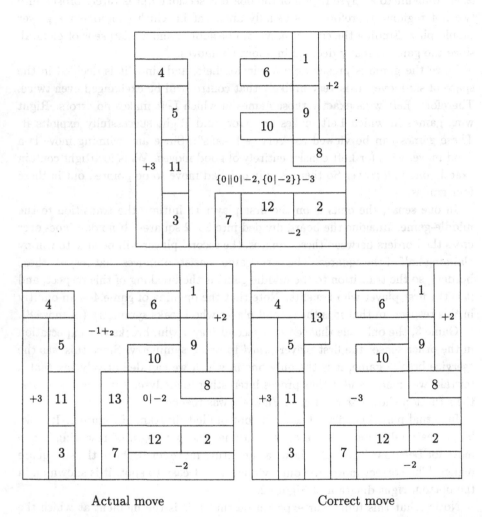

Actual move   Correct move

Figure 2. Left's fatal mistake in Game 3: Wolfe is Left (vertical) and Calistrate is Right (horizontal).

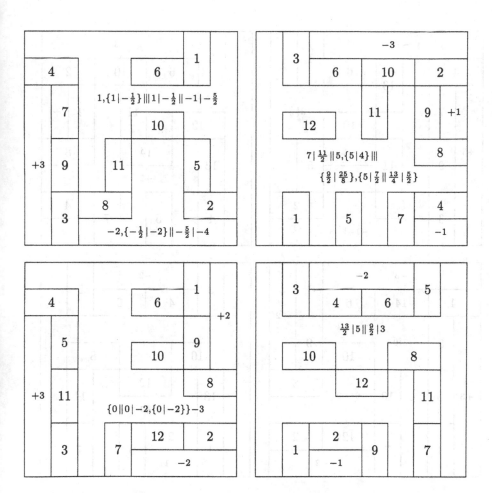

Figure 3. Positions reached in games 1–4 just as they become susceptible to analysis.

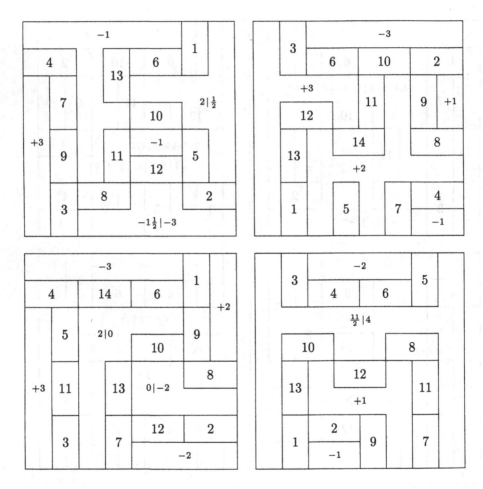

Figure 4. Positions reached in games 1–4 just as they cool down to sums of numbers and switches.

Values were computed using David Wolfe's *games* software. Boards were type-set using Jennifer Turney's awk scripts for domineering, which produce LaTeX code.

Reference

[Conway 1976] J. H. Conway, *On Numbers And Games*, Academic Press, London, 1976.

JULIAN WEST
MALASPINA UNIVERSITY-COLLEGE
and
DEPARTMENT OF MATHEMATICS AND STATISTICS
UNIVERSITY OF VICTORIA
VICTORIA, B.C.
CANADA V8W 3P4
julian@math.uvic.ca

Games of No Chance
MSRI Publications
Volume **29**, 1996

The Gamesman's Toolkit

DAVID WOLFE

ABSTRACT. Tired of all those hand calculations? Of converting to canonical
forms? Of calculating means and temperatures? Of wondering if you've
goofed? Wanted to play with the overheating operator, but don't have any
patience?

The Gamesman's Toolkit, written in C under UNIX, implements virtually all of the material in finite combinatorial game theory. It is invaluable
in analyzing games, and in generating and testing hypotheses. Several of
the results presented at the 1994 Combinatorial Games Workshop were
discovered using this program.

The Gamesman's Toolkit is useful and fun. This paper is an overview
of its features.

1. Overview

The Gamesman's Toolkit, written in C and running under UNIX, implements
virtually all of the material in finite combinatorial game theory. The toolkit can
be used in one of two ways: as a "games calculator", by the gamesman who wishes
to do standard algebraic manipulations on games; and as a programming toolkit,
to analyze a particular game, for instance. A broad base of game-theoretic
functions is provided, along with a parser and output routines. Either way, the
program has proved to be a versatile and powerful tool for the student and the
researcher.

2. Use As a Games Calculator

Table 1 shows a sample run of the Toolkit, and gives an idea of its functionality
as a games calculator. Here is the input notation, in a nutshell:

- The caret, ^, represents ↑; the letter v is reserved for ↓. Triple-up can be
  typed as either ^^^ or ^3.
- You can assign variables to games and later reuse them. For example, you
  can set g = 2|1 and later reuse g in another expression.

unix% games	The program is executed from the UNIX prompt.
Type 'help' and...	You are told how to get help; the prompt > appears.
> * ? ^	How does $*$ compare with $\uparrow$?
<>	Answer: they are incomparable
> * ? ^^	How does $*$ compare with $\uparrow\uparrow$?
<	Answer: $*$ is less than $\uparrow\uparrow$
> -2 + 3/8 + vvv + *6	Compute $-2+\frac{3}{8}+\Downarrow+*6$
-13/8v3*6	Answer: $-\frac{13}{8}+\Downarrow+*6$ (v3 is short for vvv)
> +[2],*|-2	Find the canonical form of $\{+_2, *\|-2\}$;
0,*|-2	brackets indicate subscripts, so +[2] is tiny-two
> $[1/2]	Cool the last result ($) by $\frac{1}{2}$
-1/2|-3/2	$\{0, *\|-2\}$ cooled by $\frac{1}{2}$ is $\{-\frac{1}{2}\|-\frac{3}{2}\}$
> g = $	Assign a variable g to the last result
g = -1/2|-3/2	
> mean g	Calculate g's mean and temperature
-1	
> temp g	
1/2	
> (1|||0||-1|-3)[1/2]	Cool $1\|\|\|0\|\|-1\|-3$ by $\frac{1}{2}$.
1/2||0|g	The program uses the value of g to shorten the answer
> no ups	Please do not print out $\uparrow$ in canonical forms
> ^*	What's the canonical form of $\uparrow*$?
0,*|0	It's $\{0, *\|0\}$
> domino	Enter a domineering position
Enter domin...	
xxxx	
xx	
xxxx	
	The input is ended with an extra carriage return
-3/2|-4	The ten-box position has value $\{-\frac{3}{2}\|-4\}$

Table 1. Sample run of the Gamesman's Toolkit. The first column is the computer dialogue, and the second contains comments. What the computer prints is in typewriter font, and what the user types is underlined. Most commonly, the user enters a game expression, and the computer converts the expression to canonical form.

- Subscripts in games notation are typically enclosed by brackets for the toolkit. Thus, the input format for $+_2$ (tiny-2) is +[2], and that for g_3 (i.e., g cooled by 3) is g[3].
- Similarly, superscripts are enclosed by angle brackets < and >.
- The symbol % represents the heating (integral) symbol. Thus, %[1*]<1>g denotes $\int_{1*}^{1} g$, that is, g overheated from 1* to 1 [Berlekamp 1988; Berlekamp and Wolfe 1994, p. 214].

3. Use As a Programmer's Toolkit

The Gamesman's Toolkit can be extended to analyze a particular game: the user implements the rules to that game, the program can then evaluate positions of that game. If you are an experienced (or patient) programmer you should find the Toolkit quite versatile. (Unfortunately, the extensibility mechanism was not designed for users without programming experience.)

Several people have written such extensions. Konane (programmed by Michael Ernst), Toads and Frogs (Jeff Erickson) and Domineering (David Wolfe) come with the Toolkit. Dan Garcia has written a wonderful interface using Tcl for playing domineering on an X-window system [Garcia 1996].

To get you started, Table 2 summarizes the most useful files and functions you'll need. The listing on page 97 contains a program for evaluating positions in Wyt Queens. (Wyt Queens, or Wythoff's game [Berlekamp et al. 1982, p. 61], is played on a quarter-infinite chess board. One or more queens occupy a square each, and move independently. A move consists of moving a queen north, west, or northwest any number of squares. The last player to move wins.)

4. Availability

The Gamesman's toolkit is free, and is available by request: send e-mail to me at wolfe@cs.berkeley.edu. I've chosen to distribute it in this way, rather than by posting it publicly, so that I can maintain a mailing list for notifying you of significant improvements in the program. You can also download it from http://http.cs.berkeley.edu/~wolfe/games.tar.gz. Enjoy!

Acknowledgement

Elwyn Berlekamp first encouraged me to write the toolkit in a course at UC Berkeley in Spring of 1989. Many people have since contributed ideas and code to the program, including Dan Calistrate, Raymond Chen, Jeff Erickson, Michael Ernst, Dan Garcia, Yonghoan Kim, David Moews, and Martin Mueller.

gameops.h: Performs operations on games (game_type)	
init()	must be called once to initialize the game routines
make(list_type, list_type)	constructs a game from two lists of options and converts it to canonical form; destroys its arguments
zero	the zero game
num(Q_type), up(int), star(int)	construct games consisting numbers, ups and star
plus(game_type, game_type)	add two games
minus(game_type, game_type)	subtract two games
negative(game_type)	negate a game
eq(game_type, game_type), ge(...), etc.	compare games (equal, $\geq$, etc.)
is_int(game_type), is_num(game_type)	true if game is an integer, number
list_type left_options(game_type)	game's list of left options
list_type right_options(game_type)	game's right options
output.h: Output routines for games and lists	
game_printf(game_type)	output a game plus a newline
game_print(game_type)	output a game without the newline
game_sprintn(game_type)	print a game to a string without newline, maximum of n characters
list.h: Manipulates lists of integers or games (list_type)	
list_make()	returns an empty list
list_insert(list_type, int or game_type)	insert an element into a sorted list
list_prepend(list_type, int or game_type)	prepend to an unsorted list
int or game_type list_nth(list_type)	find the nth element of a list
list_copy(list_type)	return a copy of a list
list_free(list_type)	destroy a list and free its space
rational.h: Manipulates low precision rational numbers (Q_type)	
Q(int, int)	construct a rational p/q
int Q_p(Q_type), Q_q(Q_type)	numerator or denominator of a rational
hash.h: Maintains a hash table keyed by an (int, list_type) pair	
boolean hash_test(int, list_type)	true if the hash table contains entry
game_type hash_get_last()	get the last entry tested positive
hash_put(int, list_type, game_type)	put a value into the hash table; destroys its list argument

Table 2. Most common functions used by a programmer of the toolkit.

```
/* A programming example: The game of Wyt Queens */
#include "games.c"        /* Includes all the needed .h files */
#define QUEENS_KEY 1001   /* Hash table keys below 1000 are reserved. */

game_type queens (int x, int y) {
    list_type posn_as_list, left, right;
    game_type g;
    int n, min;

    /* Encode the position as a list of integers, in order to use the hash
       table to store computed positions and avoid recomputing positions */
    posn_as_list = list_make();
    list_prepend (posn_as_list, x);
    list_prepend (posn_as_list, y);
    if (hash_test (QUEENS_KEY, posn_as_list)) { /* If pos'n previously computed */
        list_free (posn_as_list);              /* Free space used by the list */
        return hash_get_last();                /* Return computed value */
    }

    /* Position wasn't already computed, so evaluate the position recursively. */
    left = list_make();
    for (n=0; n<x; n++) list_insert (left, queens (n, y)); /* Horizontal moves */
    for (n=0; n<y; n++) list_insert (left, queens (x, n)); /* Vertical */
    min = ( x<y ? x : y);
    for (n=1; n<=min; n++) list_insert (left, queens (x-n,y-n)); /* Diagonal */
    right = list_copy (left);       /* Right's options are the same as Left's. */
    g = make (left, right);         /* Construct the game's canonical form.
                                       Lists left and right are destroyed. */

    /* Store the value of the position in hash table.  The posn_as_list is
       destroyed and freed by hash_put() by, so no need to list_free() it. */
    hash_put (QUEENS_KEY, posn_as_list, g);

    return g;
}

#define MAXBUFF 10
void main() {
    char s[MAX_BUFF];

    init();                                    /* initialize the toolkit! */
    game_sprintn (s, queens(2,6), MAXBUFF-1); /* Store position in string s */
    printf ("A queen at location (2,6) has value %s\n", s);
}
```

Listing 1. Sample extension program for the Toolkit. Lines marked with a vertical bar on the margin relate to the hash table, and may be ignored on a first reading, or omitted if efficiency is not an issue. The hash table is used to avoid the exponential cost of reevaluating the same positions over and over during recursion.

References

[Berlekamp 1988] E. R. Berlekamp, "Blockbusting and Domineering", *J. Combin. Theory* (Ser. A) **49** (1988), 67–116.

[Berlekamp and Wolfe 1994] E. Berlekamp and D. Wolfe, *Mathematical Go: Chilling Gets the Last Point*, A. K. Peters, Wellesley, MA, 1994.

[Berlekamp et al. 1982] E. R. Berlekamp, J. H. Conway and R. K. Guy, *Winning Ways for Your Mathematical Plays*, Academic Press, London, 1982.

[Garcia 1996] D. Garcia, "Xdom: A graphical, X-based front-end for Domineering", pp. 311–313 in this volume.

DAVID WOLFE
COMPUTER SCIENCE DIVISION
UC BERKELEY, CA 94720
wolfe@cs.berkeley.edu

Strides on Classical Ground

The origins of chess and related games are lost in time; yet, in spite of hundreds of years of analysis, they remain as interesting as ever, because of their fantastically large configuration space. The articles presented here are steps in the continuing endeavor to master these games, an endeavor in which the computer nowadays is often a valuable tool. In fact, the "simpler" board game Nine Men's Morris has succumbed to computer analysis, as reported by R. Gasser. Checkers may well be on its way: J. Schaeffer tells of the development of the program Chinook, and pays a tribute to the extraordinary (human!) player M. Tinsley. N. Elkies and L. Stiller write articles about chess, computerless in one case and computer-heavy in the other. Shogi, also called Japanese chess, is Y. Kawano's subject.

The last four articles of this section deal with Go, a game that has come under intense scrutiny recently. Although it is a territorial game and not, strictly speaking, a combinatorial game according to the definition on page 1, the board breaks up toward the end into a sum of smaller independent games, a situation that the theory of combinatorial games handles well. Other aspects of Go, such as ko, require extensions of the traditional theory, as explained in two of these articles.

Games of No Chance
MSRI Publications
Volume **29**, 1996

Solving Nine Men's Morris

RALPH GASSER

ABSTRACT. We describe the combination of two search methods used to solve Nine Men's Morris. An improved retrograde analysis algorithm computed endgame databases comprising about 10^{10} states. An 18-ply alpha-beta search then used these databases to prove that the value of the initial position is a draw. Nine Men's Morris is the first non-trivial game to be solved that does not seem to benefit from knowledge-based methods.

1. Introduction

In recent years, a number of games have been solved using computers, including Qubic [Patashnik 1980], Connect-4 [Allen 1989; Allis 1988] and Go-Moku [Allis et al. 1993]. All these games were solved using knowledge-based methods. These methods are successful because all these games have a low decision complexity [Allis et al. 1991], that is, the right move is often easy to find. Not all games profit to the same extent. For instance, in checkers, chess and go a multitude of moves often seem fairly equal. Brute-force search is often the most viable means of playing or solving this type of game. This observation is supported by the fact that the best programs for both chess [Hsu 1990] and checkers [Schaeffer 1992] rely heavily on search.

Search methods are not only useful for playing games: they are ubiquitous in many aspects of problem solving. Some of these algorithms are among the best understood in computer science. However, not all search algorithms are equally well studied; in particular, exhaustive search in large state spaces is still in its infancy. Partly this is because the hardware has only recently progressed to a point where interesting problems are within reach. Due to the continual improvement in hardware and software, search methods must be constantly re-evaluated and adapted to new system designs [Stiller 1991; Lake 1992]. Games are ideal for gaining expertise in this area, because they are played in a restricted domain, where the problem difficulty can be controlled and the effectiveness of the approach measured.

We describe the combination of two brute-force search methods used to solve Nine Men's Morris. An improved retrograde analysis algorithm computed endgame databases comprising about 10^{10} states. An 18-ply alpha-beta search then used these databases to prove that the value of the initial position is a draw. Nine Men's Morris is the first non-trivial game to be solved that does not seem to benefit from knowledge-based methods.

2. Nine Men's Morris

Nine Men's Morris is one of the oldest games still played today [Bell 69]. Boards have been found on many historic buildings throughout the world. The oldest (about 1400 BC) was found carved into a roofing slate on a temple in Egypt. Others have been found as widely strewn as Ceylon, Troy and Ireland.

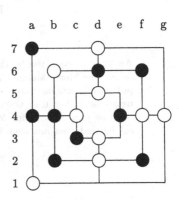

Figure 1. After the opening.

The game is played on a board with 24 points where stones may be placed. Initially the board is empty and each of the two players hold nine stones. The player with the white stones starts.

During the opening, players alternately place their stones on any vacant point (Figure 1).

After all stones have been placed, play proceeds to the midgame. Here a player may slide one of her stones to an adjacent vacant point. If at any time during the game a player succeeds in arranging three of her stones in a row—this is known as *closing a mill*—she may remove any opponent's stone that is not part of a mill. In Figure 2, if White closes a mill in the opening by playing to b6, she can now remove Black's stone on a1, but not the one on d2.

As soon as a player has only three stones left, the endgame commences. When it is her turn, the player with three stones may jump one of her stones to any vacant point on the board.

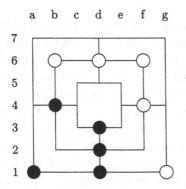

Figure 2. After b6, White removes a1 or b4.

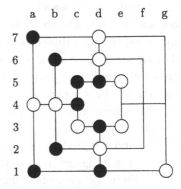

Figure 3. Black has no moves.

The game ends in the following ways:

- A player who has less than three stones loses.
- A player who cannot make a legal move loses (Figure 3).
- If a midgame or endgame position is repeated, the game is a draw.

Two points are subject to debate among Nine Men's Morris enthusiasts. The first hinges on the observation that in the opening it is possible to close two mills simultaneously. Should the player then be allowed to remove one or two opponent's stones? Our implementation only allows one stone to be removed. The second point concerns positions where the player to move has just closed a mill, but all the opponent's stones are also in mills. May she then remove a stone or not? In our implementation she may remove any stone. It seems unlikely that either of these rule variations affect the value of the game.

3. Solving Nine Men's Morris

Nine Men's Morris can be divided into two distinct phases. The opening, where stones are placed on the board and the mid- and endgame, where stones are moved. These two phases lead to a state space graph with specific characteristics. Most importantly, the opening phase induces an acyclic graph, whereas in the mid- and endgame phase move cycles can occur. Another difference is the search depth. The opening is clearly defined to be 18 plies deep for every path; by contrast, the time spent in the mid- and endgame depends on the chosen moves.

These different properties suggest the use of different search methods. We decided to use retrograde analysis to construct databases containing all mid- and endgame positions. This seems advantageous because retrograde analysis handles cycles more efficiently than forward search. In addition the opening search will require the values of many positions. Due to the interdependencies, this probably means computing the value for all or nearly all mid- and endgame positions, an ideal task for retrograde analysis.

Computing further databases for the opening is not reasonable, because their size would be even larger than for the mid- and endgame. In addition, because the value of only a single position (the empty board) is of interest in the opening phase, no intermediate position values must be stored. Alpha-beta search [Knuth and Moore 1975] is ideal for this type of problem.

4. Mid- and Endgame Databases

4.1. State space. To see if applying retrograde analysis is feasible, we must construct an appropriate state space. Each of the 24 board points can either be unoccupied or occupied by a black or white stone, so an upper bound for the state space size is 3^{24}, or approximately 2.8×10^{11}, states. This rather loose bound can be improved by taking the following game constraints into account:

- players have three to nine stones on the board,
- there are unreachable positions, and
- the board is symmetrical.

The first observation allows us to split the state space into 28 subspaces. Figure 4 shows these subspaces and their dependencies; thus the 7-5 positions (seven stones against five) can only be computed after the values for all 7-4 and 6-5 positions have been determined.

Not all states in these subspaces are reachable in the course of a legal game. For example, if one player has a closed mill, her opponent cannot have all nine stones on the board (Figure 5). Similar considerations can be made if two or three closed mills are on the board. The 9-9, 9-8, 9-7, 9-6, 9-5, 9-4, 9-3, 8-8, 8-7 and 8-6 subspaces contain such unreachable states.

The subspaces also contain many symmetrical positions. In general there are five symmetry axes (Figure 6). In the special case of the 3-3 subspace, there are additional symmetries,

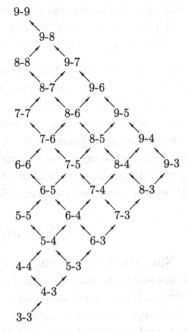

Figure 4. Database dependencies.

because all rings are interchangeable. Since one of these five axis is redundant—for example, $\pi_4 = \pi_1 \circ \pi_2 \circ \pi_3$, we can expect about a 16-fold reduction in the state space size from the symmetries.

Taking all three reductions into account leaves a total of 7,673,759,269 states in the mid- and endgame phase.

For such large state spaces, a memory-efficient storage method is needed. The standard method used in retrograde analysis is to construct a perfect hash function, which, given a state, returns a unique index. This makes it unnecessary to store the state description along with the state value, because the state de-

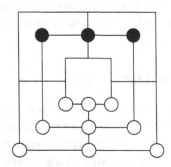

Figure 5. An unreachable state.

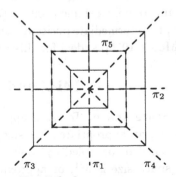

Figure 6. Symmetry axes.

scription is encoded in the index. Ideally the perfect hash function ranges from 0 to the number of states minus 1, and can be computed rapidly. Constructing such a hash function can be difficult. Therefore, as rapid computation is critical, it seems reasonable to loosen the requirements and allow hash functions with slightly larger ranges. This means that some unused entries are included in the file, but this will not impede the computation as long as the file size does not become too large. The hash function we decided to use maps the 7,673,759,269 states into a range of 9,074,932,579 indices.

4.2. Calculation. The first step in retrograde analysis is to initialize all easily recognizable won and lost positions. In Nine Men's Morris there are two types of terminal positions, both of which are losses for the player to move: positions where the player cannot move, and positions where the player has less than three stones. Therefore, if the player to move is blocked, we mark that state as a loss. We would also like to mark positions where the player to move has less than three stones as losses, but because we eliminated these positions from the state space, we mark positions where the player to move can close a mill and reduce the opponent to two stones as wins. We also mark positions where the player with three stones loses in two plies. Such positions are easy to recognize and belong to one of the following categories:

- the opponent has two open mills with no stones in common;
- the opponent has two open mills and the player to move cannot close a mill; or
- the player to move must remove a stone blocking an opponent's mill.

After the initialization has set these wins and losses, it sets the value of all remaining positions to a draw. An iterative process then determines the real values of these "drawn" states. As a first step, consider how wins and losses are propagated in two-player games. If a position is lost for the player to move, all the position's predecessors can be marked as wins for the opponent. Similarly, if a position is won for the player to move, all predecessors are potential losses for the opponent. They are only true losses if all of their successors are also won for the player (Figure 7).

It would be inefficient to check all successors to determine if a position is a loss or not. This can be avoided by using a Count field for each state, to store the number of successors that have not yet been proved a win for the opponent. Then, instead of checking all successors every time a position occurs, we simply decrement Count. When it reaches zero, all successors have been proved wins for the opponent, and the position can be marked a loss. The cost of this improvement is the additional memory used for the Count array.

Since we compute the depth of the wins and losses, one byte is used to store the value of each state, in a Val array. In practice the Count and Val array can use the same memory, since the two pieces of information are never needed

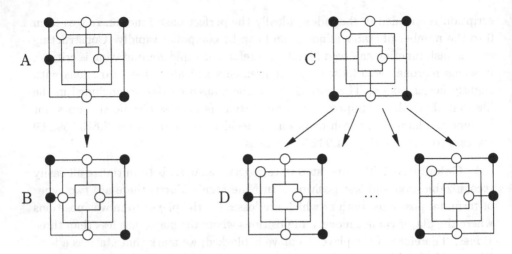

Figure 7. Determining predecessor values. Position A is won for White (to play) since *some* successor, say B, is lost for Black. Position C is lost for Black since *all* of it successors D are won for White.

simultaneously: the Count array is only needed while a position's value may still be drawn. In other words, the Count and Val values are stored as a "union"; see Table 1.

Table 2 shows the algorithm by which we determine the value of all states.

The first database was computed in 1989 on a Macintosh IIx, the last on a Macintosh Quadra 800 in 1993. In between, the algorithm was improved and ported to various machines. These include a Cray X-MP/28, a DEC VAX 9000-420, an IBM RS/6000, a 16-Transputer (T805) system, a 30-Processor MUSIC system (Motorola DSP 96000) and a DEC 3000 (Alpha). Only the DEC 3000 showed better performance than the Apple Macintosh. There are various reason for this surprising result. Many machines executed our code at low priority because of their multi-user environments. Also, many of the more powerful machines are optimized to deal with floating-point and vector operations, whereas their integer performance is not as highly tuned. The parallel machines additionally suffered from insufficient main memory and no direct disk access. This made the Macintosh front-end an I/O bottleneck.

Val entry	Count entry
0 = loss in 0	....
1 = win in 1	252 = "draw" (3 successors unknown)
2 = loss in 2	253 = "draw" (2 successors unknown)
3 = win in 3	254 = "draw" (1 successor unknown)
......	255 = "draw" (all successors unknown)

Table 1. All known information for a position fits in one byte.

```
        Backupable ← [ ];                          {initialization}
        for all states do
            Val[state] ← Initialize(state);
            Count[state] ← 0;
            if Val[state] ≠draw then
                Put(state,Backupable);
            end; {if}
        end; {for}
        while not Empty(Backupable) do             {iteration}
            state ← GetState(Backupable);
            for all Predecessors(state) do
                if Val[pred] = draw then
                    if Val[state] = loss then
                        Val[pred] ← win;
                        Put(pred,Backupable);
                    else   {Val(state) = win}
                        if Count[pred] = 0 then
                            for all Successors(pred) do    {count successors}
                                Count[pred] ← Count[pred]+1;
                            end; {for}
                        end; {if}
                        Count[pred] ← Count[pred]−1;
                        if Count[pred] = 0 then
                            Val[pred] ← loss;
                            Put(pred,Backupable);
                        end; {if}
                    end; {if}
                end; {if}
            end; {for}
        end; {while}
```

Table 2. This algorithm shows how the initialization and iteration determine the value of all states. For simplicity, we don't show the merging of the Count and Val arrays, nor the computation of the depth of the wins and losses.

4.3. Verification. Computations of this size almost inevitably contain some errors. The most common cause are software bugs in the retrograde analysis code, the system software, or the compiler. Hardware problems, such as disk errors or memory glitches, occur as well. For these reasons, the databases were verified on a cluster of 30 Sun Sparcstations [Balz 1994]. This took approximately three months. The verification found half a dozen hardware errors. Additionally, an error was found in the Nine Men's Morris code, which allowed some positions to be classified as draws instead of losses. These errors induced further inconsistencies, so that thousands of positions had to be corrected.

The verification only checked that the wins, losses and draws were consistent. The depth information was not checked, because the additional memory needed to store the databases increases the disk I/O and slows the verification dramatically. Presently, we are running a depth verification algorithm on a new parallel machine which may enable the complete verification. Even if depth verification finds no additional errors, how can we ascertain that none exist? A basic idea behind the verification is to use a different algorithm (forward search) and independent code to verify the data. But there are still some procedures common to both algorithms, such as the file indexing function or the initial state values. Ideally, our results should be independently verified.

5. Opening Search

Now that all mid- and endgame position values are computed and stored in databases, the value of the initial position can be found with a simple 18-ply alpha-beta search. In principle, the 18-ply opening can lead to any of the databases except 8-3, 9-3 and 9-4. This comprises approximately 9 GBytes of data. Since our machine only had 72 MBytes, accessing database values becomes an I/O bottleneck. The following methods were applied to alleviate this problem:

- reduce the size of the databases,
- reduce the number of used databases, and
- reduce the number of disk accesses.

Because no cycles occur in the opening, it is not necessary to compute the depth of the win or loss. Therefore a first idea is to reduce the size of the databases by packing five positions into one byte ($3^5 = 243$). However, in order to further minimize storage requirements, we decided to pack eight states into a byte. Although we then have only one bit per state, this is sufficient to compute the game-theoretic value if we do two alpha-beta searches. In the first search we determine if the initial position is won or not. If it is not won, we do a second search (with different databases) to determine if the initial position is lost or not.

While this reduces the size of the databases by a factor of eight, it still leaves files totaling about 1 GByte. To Nine Men's Morris players, it is clear that most

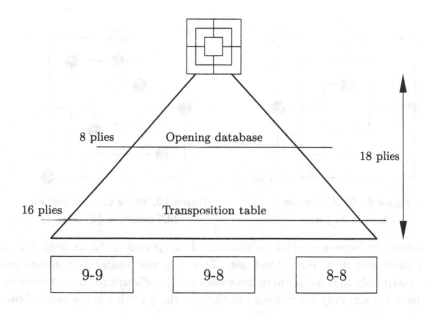

Figure 8. The opening search.

reasonable games will result in at most one or two mills being closed during the opening. For this reason we decided to try to prove the draw using only the 9-9, 9-8 and 8-8 databases, a total of 115 MBytes. It is then no longer possible to compute the correct value for every position. But in a position with White to move, we can compute an upper bound on the correct value by assuming all positions not in the databases are won for white and a lower bound if we assume all such positions are won for black. To prove that the game-theoretic value is a draw, we must show that White has both an upper and lower bound of a draw.

We inserted a transposition table at the 16-ply level to further reduce the number of database accesses. Normally, a transposition table includes positions from any ply, possibly even giving priority to those higher up in the tree. We chose to include only the 16-ply positions because it is far more important to reduce disk I/O than the total number of nodes in the tree. See Figure 8.

Finally, all positions that were visited by the alpha-beta searches at the 8-ply level were stored in an intermediate database. This allows games to be played in real time, because for the first 8 plies only a search to the intermediate database must be performed. Here is the number of 8-ply positions alpha-beta visited in each of the two searches:

	proved	not proved
White at least draws	15,501	12
White at most draws	4,364	29

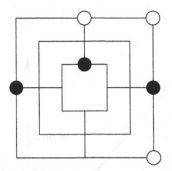

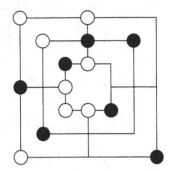

Figure 9. Black to move.
Loss in 26 plies.

Figure 10. White to move and win.
Mill closure in 187 plies.

We differentiate between positions that could be proved to be at least (or at most) a draw and those that could not. This does not mean the positions are lost (or won); only that using three databases was insufficient to show otherwise.

Of the approximately 3.5 million positions at the 8-ply level, the table shows that only 19,906 were evaluated. The time for a node evaluation ranged from 2 seconds to 15 minutes. The total run-time was about three weeks on a Macintosh Quadra 800. Examining the table, one might be inclined to assume that most positions are drawn. This must not be the case; perhaps the move-ordering heuristic was simply successful in selecting drawish moves.

6. Results

6.1. Database Positions. Sifting through the final databases in search of "interesting" positions is a daunting task. Since we have no real intuition about what makes a position interesting, we decided to gather results in a more statistical fashion. The next two figures show examples of positions with long winning sequences. The 3-3 position in Figure 9 is an example of the longest sequence in that database. The 8-7 position in Figure 10 shows a won position with the longest move sequence until a mill is closed.

Figure 11 shows the percentage of positions that are won for the player to move. They are grouped according to the number of stones on the board, and normalized accordingly. The figure shows that the win probability strongly correlates with the stone difference: more stones are better. There seems to be a cut-off at about seven stones, so that having less makes winning difficult. The statistics also show that the 3-3 database seems to be special. This is probably because both players are allowed to jump, essentially making this a different game. One must keep in mind that these statistics can be misleading. For instance, in the 4-3 database, if the player with four stones is to move, it seems she has a small chance of winning. Upon closer examination of the database, we see that all wins are trivial positions where the player can close a mill immediately.

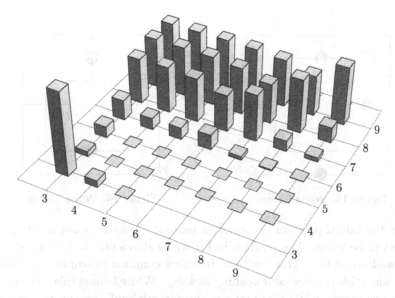

Figure 11. Win percentages for the player to move, as a function of the initial number of stones. The highest percentage, for 3-3 games, is about 83%. The absence of a column (as for 3-9 games) indicates a zero percentage.

These positions will not occur in a real game, so actually the player with three stones has a small advantage. Similar reservations apply to all these databases. It is not clear how the realistic positions can be filtered from the rest.

6.2. Opening Positions. It is now also possible to examine some common opening positions. Between equally matched players, games tend to be very drawish during the opening, that is, most possible moves result in a draw, as for the example in Figure 12. However, occasional blunders do occur. Figure 13 shows a position that occurred in a match our Nine Men's Morris program played against Mike Sunley (British Champion) at the Second Computer Olympiad in London, 1990. The program, which was then still based on heuristics, won four games and drew two.

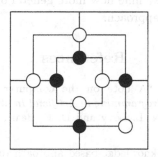

Figure 12. Opening position with black to move. Any move leads to a drawn position.

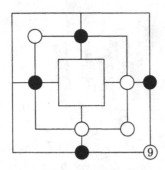
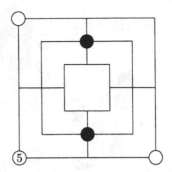

Figure 13. White 9 loses.   **Figure 14.** White 5 loses.

Since the initial position is a draw, a natural question to ask is how soon a mistake can be made. The position in Figure 14 shows one of the earliest losing moves we have found. This positions refutes a common misconception, namely that closing mills is a desirable opening strategy. White follows this strategy with her first two moves, while Black ignores the potential mill threats. Subsequently, White's third move loses even though she will now easily be able to close two mills during the opening.

7. Conclusion

Nine Men's Morris is a draw. This result was achieved using a combination of alpha-beta search and endgame databases. Our new efficient retrograde analysis algorithm allowed the 10^{10} database states to be solved on a desktop computer. This makes Nine Men's Morris the first classic board game to be solved that does not profit from knowledge-based methods.

The expertise gained in the management and use of large, pre-computed databases can also be applied to other problems. Many games and puzzles profit from combining databases with search, for instance Awari, Checkers, the 15-Puzzle or Rubik's Cube. Since games and puzzles are a restricted problem domain, future work will examine how more general optimization problems can be made amenable to this approach.

References

[Allen 1989]   J. D. Allen, "A note on the computer solution of connect-four", pp. 134–135 in *Heuristic Programming in Artificial Intelligence: The First Computer Olympiad* (edited by D. N. L. Levy and D. F. Beal), Ellis Horwood, Chichester, England, 1989.

[Allis 1988]   L. V. Allis, "A knowledge-based approach to connect-four. The game is solved: White wins", M.Sc. thesis, Faculty of Mathematics and Computer Science, Vrije Universiteit, Amsterdam, 1988.

[Allis et al. 1991] L. V. Allis, H. J. van den Herik, and I. S. Herschberg, "Which games will survive", pp. 232–243 in *Heuristic Programming in Artificial Intelligence: The Second Computer Olympiad* (edited by D. N. L. Levy and D. F. Beal), Ellis Horwood, Chichester, England, 1991.

[Allis et al. 1993] L. V. Allis, H. J. van den Herik, and M. P. H. Huntjens, "Go-Moku solved by new search techniques", pp. 1–9 in *Proc. AAAI Fall Symposium on Games: Planning and Learning*, AAAI Press Tech. Report FS93-02, Menlo Park, CA.

[Balz 1994] G. Balz, "Verification of state databases", Diploma thesis, ETH Zürich, 1994.

[Bell 1969] R. C. Bell, *Board and Table Games from Many Civilisations*, Oxford University Press, Oxford, 1969.

[Hsu 1990] F. H. Hsu, "Large-scale parallelization of alpha-beta search: an algorithmic study with computer chess", Ph.D. thesis, Carnegie-Mellon University, Pittsburgh, 1990.

[Knuth and Moore 1975] D. E. Knuth and R. W. Moore, "An analysis of alpha-beta pruning", *Artificial Intelligence* **6 1975**, 293–326.

[Lake et al. 1992] R. Lake, P. Lu, and J. Schaeffer, "Using retrograde analysis to solve large combinatorial search spaces", pp. 181–188 in *The 1992 MPCI Yearly Report: Harnessing the Killer Micros* (edited by E. D. Brooks et al.), Lawrence Livermore National Laboratory (UCRL-ID-107022-92), 1992.

[Patashnik 1980] O. Patashnik, "Qubic: $4 \times 4 \times 4$ tic-tac-toe", *Math. Mag.* **53** (1980), 202–216.

[Schaeffer et al. 1992] J. Schaeffer et al., "A World Championship caliber checkers program", *Artificial Intelligence* **53** (1992), 273–290.

[Stiller 1991] L. Stiller, "Group graphs and computational symmetry on massively parallel architecture", J. Supercomputing **5** (1991), 99–117.

RALPH GASSER
INSTITUT FÜR THEORETISCHE INFORMATIK
EIDGENÖSSISCHE TECHNISCHE HOCHSCHULE (ETH)
8092 ZÜRICH
SWITZERLAND

Games of No Chance
MSRI Publications
Volume 29, 1996

Marion Tinsley:
Human Perfection at Checkers?

JONATHAN SCHAEFFER

Marion Tinsley died on April 3, 1995, at the age of 68. Why does the death of this checkers (8 × 8 draughts) player attract our attention? His record speaks for itself:

Since an accidental loss in the 1950 U.S. Championship, Tinsley finished in undivided first place in every tournament that he played in, except the last (in the 1994 U.S. Championship, he tied for first place with the computer program *Chinook* and Don Lafferty). He contested nine World Championship matches, winning each usually by an embarrassingly large margin. Over the last forty-five years of his life, comprising thousands of tournament, World Championship, match, exhibition and casual games, Tinsley lost the unbelievable number of seven games. Seven games!? In forty-five years? This is as close to perfection as is humanly possible.

Tinsley once remarked that he had become bored playing humans; there wasn't any challenge left. When he was young, he began to acquire the reputation of being unbeatable. For forty-five years, most of his opponents would play for the draw; going for a win was unthinkable. Tinsley's enjoyment of checkers waned, and at one point he retired from the game for twelve years because of a lack of competition.

When the program *Chinook* came on the scene, Tinsley relished the opportunity to play it. *Chinook* had no respect for Tinsley's abilities, willingly taking risks: anything to increase the chances of winning. Tinsley said that playing *Chinook* made him feel like a young man again. In 1990, *Chinook* earned the right to play Tinsley for the (human) World Checkers Championship. The American Checker Federation and English Draughts Association (A.C.F. and E.D.A., the governing bodies for international checkers) refused to sanction the *Chinook*–Tinsley match, retroactively deciding that computers were ineligible for the World Championship. He was unable to change their decision, and he took what he perceived to be his only option. In 1992 he resigned as World Champion, then immediately turned around and signed a contract to play the match with *Chinook*. Although this match was not the "official" World Championship, Tinsley was the champion and *Chinook* the official challenger. In their embarrassment, the A.C.F. and E.D.A. hastily created a new Man-Machine World Championship title so they could be part of the *Chinook*–Tinsley match. The two governing bodies were now faced with the prospect of crowning a new human World Champion, which was meaningless as long as Tinsley was alive. They decided to award him the title World Champion Emeritus.

The match was held in London in August 1992, and was sponsored by Silicon Graphics. The final result was four wins and two losses for Tinsley, with 33 draws. These were only the sixth and and seventh losses by Tinsley since 1950.

In August 1994, Tinsley battled *Chinook* again for the Man-Machine title. After six games, all draws, Tinsley resigned the match and the title to *Chinook*, citing health reasons. A week later he was diagnosed with cancer, and after seven months he succumbed to the disease.

In the domain of competitive mind games, such as chess or bridge, Tinsley's record is unparalleled. Strong players come to the fore frequently, but their tenure at the top is usually short-lived. Emmanuel Lasker, chess World Champion for over twenty years, is an isolated exception. Being the best over a 45-year period is unprecedented.

Was the Tinsley of 1994 as good a player as the Tinsley of, say, 1950 or 1970? Knowledgeable checkers players say the answer is yes. However, in old age he became lazy, winning the minimum number of games needed to carry a tournament or match, and then effortlessly drawing the rest (even not bothering to follow through in won positions because it required too much effort).

What made Tinsley such a strong player? I asked him that question several times, without getting an adequate answer. Like many strong game players, he would say that when playing a game he just "knew" the right move to make. The solution to a problem was immediately obvious to him, even without having analyzed all the ramifications of the move. He was rarely wrong.

Having observed him for several years, I can make a few comments on his strength. First, there is no doubt that Tinsley had an incredible memory. Before I first met him, I was acquainted with the stories—apocryphal, I assumed—concerning his memory. One man reported that, when Tinsley was young, he studied checkers eight hours a day, five days a week. As he got older, this dropped to one day a week. This person claimed that Tinsley could recall every one of those sessions. Needless to say, I found this difficult to believe.

After *Chinook*'s first game against Tinsley in 1990, we started analyzing the game. He began recounting the history of the opening line we played, recalling games he had played in the 1940's! The move sequences flowed easily from him, without hesitation, sometimes annotated with the name of the opponent, date or place where the game was played! Games played in 1947 were as vivid in his memory as if it they had taken place only yesterday.

The second facet to his play was his incredible sixth sense. A glance at a position was sufficient to tell him everything he needed to know. For example, in 1990 *Chinook* was playing Tinsley the tenth game of a fourteen-game match (won by Tinsley 1–0 with thirteen draws). I reached out to play *Chinook*'s tenth move. I no sooner released the piece than he looked up in surprise and said "You're going to regret that". Being inexperienced in the ways of the great Tinsley, I sat there silent, thinking "What do you know? My program is searching twenty moves deep and says it has an advantage". Several moves later, *Chinook*'s assessment dropped to equality. A few moves later, it said Tinsley was better. Later *Chinook* said it was in trouble. Finally, things became so bad we resigned. In his notes to the game, Tinsley revealed that he had seen to the end of the game and knew he was going to win on move 11, one move after our mistake. *Chinook* needed to look ahead sixty moves to know that its tenth move was a loser.

My observations of tournament chess and checkers players have led me to conclude that the sixth sense is experience. It is well known how intensely Tinsley studied checkers, analyzing anything from a grandmaster game to a game between novices. His uncanny ability to know good from bad and safe from dangerous is the direct result of all his hard work. Strong chess players have the same ability, but perhaps it is not quite as evident as it was with Tinsley.

What made Tinsley special was his play away from the board. He was universally liked. He was kind and gentle, eager to talk equally with checkers master and checkers novice. The first time I met him is indelibly imprinted on my mind. I was attending *Chinook*'s first human tournament, the 1990 Mississippi Open. I had never attended such an event before, and knew no one. I walked in the door, tried to get my bearings, but must have looked lost. A tall slim man walked up to me and said "You look like a checkers player. Can I help you?" He proceeded to introduce me to the tournament organizer and director, and asked whether I needed any help with the hotel. This kind stranger turned out to be Marion Tinsley. I was a stranger walking in off the street and was treated royally by the World Champion. Even though Tinsley was our adversary over the checkerboard, away from it he was our friend.

Professionally, Tinsley was a professor of mathematics at the University of Florida at Tallahassee. Away from work and the checkers board, he was a Baptist minister.

Tinsley could have said no when faced with the prospect of defending his title against a computer. Instead he bravely said yes, setting a precedent that we hope will serve as an example to champions in other competitive games.

It was Tinsley's fervent wish to remain an undefeated champion. As long as his health remained sound, he claimed he would never lose a match to *Chinook* or anyone else. It was the *Chinook* team's wish to defeat Tinsley and become the first computer World Champion (in any game). Ironically, in the end both sides saw their wishes realized, although under less happy circumstances than both would have liked. *Chinook* reluctantly became the first computer World Champion by winning on forfeit from Tinsley. Tinsley, however, remained undefeated over-the-board by resigning a drawn match.

JONATHAN SCHAEFFER
DEPARTMENT OF COMPUTING SCIENCE
UNIVERSITY OF ALBERTA
EDMONTON, ALBERTA
CANADA T6G 2H1

Games of No Chance
MSRI Publications
Volume **29**, 1996

Solving the Game of Checkers

JONATHAN SCHAEFFER AND ROBERT LAKE

ABSTRACT. In 1962, a checkers-playing program written by Arthur Samuel
defeated a self-proclaimed master player, creating a sensation at the time
for the fledgling field of computer science called artificial intelligence. The
historical record refers to this event as having solved the game of checkers.
This paper discusses achieving three different levels of solving the game:
publicly (as evidenced by Samuel's results), practically (by the checkers
program *Chinook*, the best player in the world) and provably (by considering the 5×10^{20} positions in the search space). The latter definition may
be attainable in the near future.

1. Introduction

Checkers is a popular game around the world, with over 150 documented
variations. Only two versions, however, have a large international playing community. What is commonly known as checkers (or draughts) in North America is
widely played in the United States and the British Commonwealth. It is played
on an 8×8 board, with checkers moving one square forward and kings moving one
square in any direction. Captures take place by jumping over an opposing piece,
and a player is allowed to jump multiple men in one move. Checkers promote to
kings when they advance to the last rank of the board. So-called international
checkers is popular in the Netherlands and the former Soviet Union. This variant uses a 10×10 board, with checkers allowed to capture backwards, and kings
allowed to move many squares in one direction, much like bishops in chess. This
paper is restricted to the 8×8 variant, but many of the ideas presented here also
apply to the 10×10 game.

People enjoy playing games of skill because of the intellectual challenge and
the satisfaction derived from playing well. Many board games, such as chess
and checkers, have too many possibilities for a human to understand them all.
Hence they use knowledge and search to make their decisions at the board. The
person with the best "algorithm" for playing the game wins in the long run.
Without perfect knowledge, mistakes are made, and even World Champions will

lose occasionally (for example, the world chess champion, Gary Kasparov, may lose three or four games a year).

This gives rise to an intriguing question: is it possible to program a computer to play a game perfectly? Can the game-theoretic value of checkers be determined? In other words, is it possible to solve the game? In recent years, some games have been solved, including Connect-Four [Allen 1989; Allis 1988], Qubic [Patashnik 1980], Nine Men's Morris [Gasser 1994] and Go-Moku [Allis et al. 1993].

This paper describes three ways in which it is possible to solve the game of checkers. Section 2 deals with *publicly* solving the game, creating the impression in the media that checkers has been solved. Section 3 deals with *practically* solving the game, creating a computer player that is better than all humans but is not perfect. Section 4 deals with *provably* solving the game, determining the game-theoretic value and a strategy for always achieving that value. For the game of checkers, publicly solving the game is a thing of the past, practically solving it is the present, and provably solving it is the near future.

2. Publicly Solving Checkers

In the late 1950's and early 1960's, Arthur Samuel did pioneering work in artificial intelligence using the game of checkers as his experimental testbed [Samuel 1959; Samuel 1967]. Thirty years later, his work is still remembered, both for the significance of his research contributions, and for the legacy of his checkers-playing program. In 1962, Robert Nealy, blind checkers champion of Stamford, Connecticut, lost a single game to Dr. Samuel's program. For the fledgling field of artificial intelligence, this event was viewed as a milestone and its significance was misrepresented in the media. Reporting of this event resulted in the game of checkers being labeled as "solved": computers were better than all humans. Thus, a 1965 article by Richard Bellin in the *Proceedings of the National Academy of Sciences* declared that "... it seems safe to predict that, within ten years, checkers will be a completely decidable game" (vol. 53, p. 246), while Richard Restak, in the influential book *The Brain: The Last Frontier* (1979), stated that "... an improved model of Samuel's checker-playing computer is virtually unbeatable, even defeating checkers champions foolhardy enough to 'challenge' it to a game" (p. 336).

How good was Samuel's program? Although Nealy advertised himself as a master, the highest level he achieved was the class below master [Fortman 1978]. Analysis of the fateful game showed that Nealy made a trivial mistake, possibly indicating that he was not taking his electronic opponent seriously. A match played a year later resulted in a decisive victory for Nealy, without loss of a single game.

As a result of the one win against Nealy, checkers was classified as an uninteresting problem domain and all the artificial intelligence research that might have used checkers as an experimental testbed switched to using chess. The perception that checkers is a solved game persists to the present time, and has been a major obstacle to anyone conducting research using this game. Modern artificial intelligence books now treat the subject of Samuel's program's performance more realistically, for example indicating that it "came close to expert play" [Barr and Feigenbaum 1981].

Dr. Samuel had no illusions about the strength of his program. In personal correspondence, he apologized for the problems created by the misconception that checkers was solved [Tinsley 1994] and admitted he had no idea how he could make his program good enough to compete with the world champion [Smith 1994].

3. Practically Solving Checkers

The first checkers-playing program is credited to Strachey [1952]. Samuel began his program in the early 1950's and continued working on it on and off for over two decades [Samuel 1959; 1967]. In 1965, the program played four games each against Walter Hellman and Derek Oldbury (then playing a match for the World Championship), and lost all eight games. In the late 1970's a team at Duke University headed by Eric Jansen developed the first program capable of defeating masters [Truscott 1978]. In May 1977, their program *PAASLOW* had the distinction of defeating Grandmaster Elbert Lowder in a nontournament game (the program recovered from a lost position), but lost their friendly match with one win, two losses and two draws [Fortman 1978]. The Duke team originally accepted a $5,000 challenge match to play the World Champion, Dr. Marion Tinsley, but eventually decided not to play [Anonymous 1979].

In 1989 and 1990, the First and Second Computer Olympiads were held in London, and three strong programs emerged [Levy and Beal 1989; 1991]. The programs *Colossus* (Martin Bryant) and *Checkers* (Gil Dodgen) were aimed at the PC market, while *Chinook* (Jonathan Schaeffer *et al.*) was a research project. *Colossus* and *Checkers* are still commercially available and are continually being updated. Both are ranked among the top twenty players in the world.

Chinook is a checkers program developed at the University of Alberta beginning in 1989. In 1990, it earned the right to play for the human World Checkers Championship, by coming an undefeated second to Tinsley, the World Champion, at the U.S. Open. This was the first time a program had earned the right to play for a human World Championship. Computers have played World Champions before—for example *BKG9* at backgammon [Berliner 1980] and *DeepThought* at chess [Kopec 1990]—but these were exhibition events with no title at stake.

The American Checker Federation and English Draughts Association (the governing bodies for checkers) refused to sanction the *Chinook*–Tinsley match, arguing that the World Championship was for humans, not machines. Eventually, they created the Man-Machine World Championship title to allow the match to proceed, but only after Tinsley resigned his "human" World Championship title in protest. In 1992, Tinsley defeated *Chinook* in their title match by a score of four wins to two with 33 draws [Schaeffer 1992; Schaeffer et al. 1993a].

In 1994, *Chinook* became World Champion when, after six games in their rematch (all draws), Tinsley resigned the match and the title citing health concerns. *Chinook* is the first computer program to become World Champion in any nontrivial game of skill. Obviously this was not a satisfactory way to win a World Championship. Subsequently, *Chinook* successfully defended its title twice against Grandmaster Don Lafferty. Unfortunately, with the recent passing of Marion Tinsley [Schaeffer 1995], mankind has lost its best chance of wresting the title away from the computer.

Chinook has not solved the game, but it is playing at a level that makes it almost unbeatable. In its last 242 games against the strongest players in the world (over 220 of which were against Grandmasters), *Chinook* only lost one game (to be fair, it also drew one game from a lost position). Since the program continues to automatically increase its checkers knowledge every day, *Chinook* will eventually be almost unbeatable. In effect, checkers will be solved in practice, if not in theory.

Chinook's strength comes from deep searches (deciding which positions to examine), a good evaluation function (deciding how favorable a position is), endgame databases (containing perfect information on all positions with eight pieces or less) and a library of opening moves [Schaeffer et al. 1992].

SEARCH: *Chinook* uses an iterative, alpha-beta search with transposition tables and the history heuristic [Schaeffer 1989]. Under tournament conditions (thirty moves an hour), the program searches to an average minimum depth of nineteen ply (one ply is one move by one player). The search uses selective deepening to extend lines that are tactically or positionally interesting. Consequently, major lines of play are often searched many plies deeper. It is not uncommon for the program to produce analyses that are thirty-ply deep or more.

KNOWLEDGE: The evaluation function has 25 heuristic components, each of which is weighted and summed to give a position evaluation. The game is divided into four phases, each with their own set of weights. The definition of the heuristics and their weights was arrived at only after long discussions with a checkers expert. The knowledge acquisition process goes on. The knowledge we add to the program nowadays tends to be subtle: it is harder to define and often covers exceptions to the general knowledge that is already programmed into *Chinook*.

DATABASES: All checkers positions with eight or fewer pieces (checkers or kings) on the board have had their game-theoretic value (win, loss or draw) computed [Lake et al. 1994]. The knowledge contained in the eight-piece databases exceeds human capabilities. Databases are discussed in more detail in the next section.

OPENING BOOK: The program has a library of 60,000 opening positions, each of which has been computer verified. Most of the positions originate from the checkers literature. However, computer verification has resulted in hundreds of corrections to the literature and some refutations of major lines of play (not published, until we have a chance to use them in a game!).

It is difficult for humans to compete with a program searching as deep as *Chinook* does. Only the best players in the world can maintain a level of vigilance sufficient to compensate for this. Nevertheless, it is clear that there is an important class of positions where even fifty-ply searches are insufficient to uncover the subtleties of the position [Schaeffer et al. 1993b]. The rare times that the program loses all have the same pattern. The opponent gets the program out of its opening library early in the game where the strategic aspects of the position require a search of roughly more than 25 ply. If the program does not have the proper knowledge to play this position, it may make a mistake and, against only the very best players, possibly lose a game. With a deeper search, an improving evaluation function, more endgame databases and a growing opening library, the probability of this happening is rapidly decreasing.

Although *Chinook* is not perfect and can still lose a game, in practice checkers can be regarded as "solved".

4. Provably Solving Checkers

What does it mean to solve a game? Allis [1994] defines three levels:

ULTRA-WEAKLY SOLVED: The game-theoretic value for the initial position has been determined.

WEAKLY SOLVED: The game is ultra-weakly solved and a strategy exists for achieving the game-theoretic value from the opening position, assuming reasonable computing resources.

STRONGLY SOLVED: For all possible positions, a strategy is known for determining the game-theoretic value for both players, assuming reasonable computing resources.

Ultra-weakly solving a game is of theoretical interest but means little in practice. For example, Allis cites the game of Hex, which is known to be a win for the first player on all diamond-shaped boards, but no one knows a strategy for achieving this result. Strongly solved games include simple domains such as tic-tac-toe, and more complex domains such as chess and checkers endgames (for which databases enumerating all possibilities have been constructed).

Note the qualification on resources in the above definitions. In some sense, a game such as chess is solved since we have an algorithm (alpha-beta searching) that could return the game-theoretic value given an unlimited amount of time. The resource constraints preclude such solutions.

To these definitions, we can add a fourth:

ULTRA-STRONGLY SOLVED: For all positions in a strongly solved game, a strategy is known that improves the chances of achieving more than the game-theoretic value against a fallible opponent.

This point is very important in practice. Knowing that a position is, for example, a draw only means that this is the best one can achieve against a perfect opponent. Many opponents (particularly humans) are fallible, and selecting a move based on the probability of opponent error increases the expected outcome of a game.

How difficult is it to solve a game? Allis, van den Herik and Herschberg argue that there are two dimensions [Allis et al. 1991]:

SPACE COMPLEXITY: the number of positions in the search space.

DECISION COMPLEXITY: the difficulty required to make correct decisions.

A first impression suggests that, the larger the search space, the harder the problem is to solve. This is not true, as illustrated by a trivial variant of the game of Go on a 19×19 board (361 squares). In this game, the two players alternate making moves, where a move consists of placing a stone on an empty square. The first person to fill in 181 squares wins [Allis et al. 1991]. This game has a search complexity of $3^{361} \approx 10^{170}$, but is trivial to solve.

All the games solved thus far have either low decision complexity (10^{30} positions for Qubic, 10^{105} positions for Go-Moku) or low space complexity (10^{11} positions for Nine Men's Morris) or both (10^{14} positions for Connect-Four). Checkers is rated as having high decision complexity (more complex than 9×9 Go and the play of bridge hands, on a par with backgammon and 10×10 draughts, and less complex than chess) and moderate space complexity (10^{20} positions versus unsolved games such as backgammon with 10^{19} positions, chess with 10^{44} positions, and Scrabble with 10^{150} positions).

A bound on the space complexity of checkers can be derived. The naive calculation of 5^{32} (each of 32 squares can be occupied by a checker or king of either color, or be empty) can be improved upon. Let b be the number of black checkers, w the number of white checkers, B the number of black kings, W the number of white kings and f the number of black checkers on the first rank. In the following, 12 is the maximum number of pieces per side, 32 is the number of squares on the board that can be occupied, 28 is the number of squares that it is legal to place a checker on (checkers become kings on the last rank) and 4 is the number of squares on black's first rank (these squares cannot hold any white

pieces	positions
1	120
2	6,972
3	261,224
4	7,092,774
5	148,688,232
6	2,503,611,964
7	34,779,531,480
8	406,309,208,481
9	4,048,627,642,976
10	34,778,882,769,216
11	259,669,578,902,016
12	1,695,618,078,654,976
13	9,726,900,031,328,256
14	49,134,911,067,979,776
15	218,511,510,918,189,056
16	852,888,183,557,922,816
17	2,905,162,728,973,680,640
18	8,568,043,414,939,516,928
19	21,661,954,506,100,113,408
20	46,352,957,062,510,379,008
21	82,459,728,874,435,248,128
22	118,435,747,136,817,856,512
23	129,406,908,049,181,900,800
24	90,072,726,844,888,186,880
total	500,995,484,682,338,672,639

Table 1. Search space complexity of checkers.

checkers). The number of positions having n or fewer pieces on the board is

$$\sum_{b=0}^{\min(n,12)} \sum_{B=0}^{\min(n,12)-b} \sum_{w=0}^{\min(n-b-B,12)} \sum_{W=0}^{\min(n-b-B,12)-w} \sum_{f=0}^{\min(b,4)} \text{Num}(b,B,w,W,f)-1,$$

where

$$\text{Num}(b,B,w,W,f)=\binom{4}{f}\binom{24}{b-f}\binom{28-(b-f)}{w}\binom{32-b-w}{B}\binom{32-b-w-B}{W}.$$

The subtraction of one position in the above formula handles the impossible case of zero pieces on the board. This yields a search complexity of roughly 5×10^{20} positions (Table 1).

This number is misleading, since it includes positions that are not legally reachable from the start of the game. For example, although there are 9×10^{19}

plausible positions with 24 pieces on the board, only a small fraction of this number are legal. It is possible to place 24 pieces on the board and have 10 of them be kings. However, there is no legal move sequence from the start of the game that can reach such a position, since the forced capture rule precludes this. It is possible to have 24 pieces on the board with two of them being kings, but this move sequence is contorted and cooperative and has no relationship to a reasonable playing strategy.

A bound on the number of reasonable positions one might encounter in trying to prove checkers is obtained using the following assumptions. First, consider only those positions where the difference between the number of black and white pieces is less than three, that is, $|(b + B) - (w + W)| < 3$. More lopsided positions will always be a win for the strong side unless (1) there is an immediate capture present to restore the material balance (which reduces the position to one satisfying the above condition), or (2) the pieces of the strong side are constricted in a manner that is impossible to reach either by the rules of the game, or by considering a reasonable playing strategy.

Within the subset of positions created by the above assumption, a number of positions can be further discarded because they are prevented from occurring by the rules of the game or by reasonable playing strategy. For example, the rules condition eliminates all positions with 24 kings on the board. The reasonable playing strategy assumption enables us to consider only those positions that will occur in normal play. One observation is that as the number of pieces on the board decreases, the probability of there being multiple kings increases. The number of kings can be restricted so that scenarios such as 12 kings versus 12 kings are eliminated. Define a subset of the search space that limits the number of kings for positions with 12 or more pieces as follows:

$$(B \leq \lceil 24 - \tfrac{1}{2}n \rceil, W \leq \lfloor 24 - \tfrac{1}{2}n \rfloor) \cup (B \leq \lfloor 24 - \tfrac{1}{2}n \rfloor, W \leq \lceil 24 - \tfrac{1}{2}n \rceil)$$

subject to the conditions $12 \leq n \leq 24$ and $|(b + B) - (w + W)| < 3$. Even this assumption is generous: it includes unlikely positions such as 6 black kings against 6 white checkers.

Summing the positions satisfying these conditions reduces the search space to slightly less than 10^{18} positions. Still included in this number are some positions not legally reachable from the start of the game. We have found no way to mathematically define this set. It is possible that the exclusion of these positions may reduce the plausible search space from 10^{18} to 10^{17} or less.

A search space of 10^{18} still seems impossibly large. If 10^6 positions were solved per second, $100,000$ years of computing would be required to enumerate all the positions. What makes the problem of finding the game-theoretic value of checkers feasible is the idea of a proof tree. When analyzing a position that is a win, one need only consider one winning move; the alternatives moves are either inferior or redundant. When considering a position that is a loss or a draw, all alternatives must be considered (since you have to prove you cannot do better). In the best case—a game that is provably winning—the search only has to explore roughly the square root of the search space: a single move at winning positions and all moves at losing positions. This means that a proof of the game-theoretic value of checkers might require a search of as few as 10^9 positions. Of course, knowing which 10^9 positions out of 10^{18} is the hard part.

To weakly solve checkers (determine the game-theoretic value), we have devised a three-pronged approach:

BOTTOM-UP: The only positions with known game-theoretic values, as defined by the rules of the game, are those positions where one side has no pieces or no moves (a loss). Using retrograde analysis, the value of these terminal positions can be backed up to positions many moves from the end of the game [Thompson 1986]. We have built endgame databases that solve all 443,748,401,247 positions with eight or fewer pieces on the board [Lake et al. 1994].

Endgame databases essentially reduce the number of moves required to be played from the start of the game before one reaches a position with a known game-theoretic value.

MIDDLE: We have collected thousands of positions with more than eight pieces on the board for which we have determined the game-theoretic value. They come from games played by *Chinook*, where the search was deep enough to demonstrate that an endgame database value could be backed up to the root position of the search. For example, we have proved that a position with as many as 22 pieces on the board is a win. The proof is simpler than one might expect, because forced capture moves in the analysis quickly reduce the number of pieces to eight.

In effect, we are building a middlegame database of solved positions with more than eight pieces on the board.

TOP-DOWN: A proof consists of demonstrating a sequence of moves from the start of the game to the endgame databases that achieves the maximum possible value (of win, draw or loss). Once the start of the game is assigned an accurate value, the game theoretic value of checkers has been determined. In tournament checkers, the first three half-moves are randomly chosen (or balloted) among a set of 144 opening sequences. In other words, the proof can be treated as 144 subproblems.

Starting at the beginning of the game and doing a search is counterproductive: the search will be too large. Instead, we use published checkers analysis to identify the "best" lines of play for each side in each opening. By playing down these lines, we try to solve positions further down the line of play (thereby creating new entries for the middlegame database) and then use those results to try and solve positions closer to the start of the line. In effect, the opening is used to guide the search to the parts of the search space most likely to be profitable.

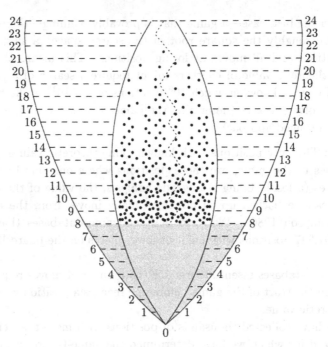

Figure 1. Checkers search space.

Figure 1 illustrates this three-pronged approach. Despite the deceptive appearance, the diagram is drawn to scale based on the numbers in Table 1. It shows the number of pieces on the board (vertically) versus the logarithm of the number of positions (base 10) with that many pieces on the board (horizontally). Many of the positions are not legally reachable by the rules of the game, and the logarithm of the estimated illegal number is shown in the region of dashed lines. The endgame phase of the proof is the shaded area: all positions with eight or fewer pieces. The middlegame database contains positions with more than eight pieces for which a value has been proved (shown as dots in the diagram). Note that a dot is misleading because each represents the root of a search tree which has been solved (and, by implication, all positions in that proof tree must also have been solved). None of these lower-level solved positions is presented in the diagram. The dotted line represents an opening line of play (as suggested by the checkers literature). It goes from the start of the opening to the endgame databases. By repeatedly searching that line, more middlegame database positions can be solved, thereby simplifying the task of proving the start position.

Critical to the success of this approach is the search algorithm chosen. Our initial experience is with two search approaches: proof number search (PNS) and alpha-beta. Allis has been successful with his PNS approach [Allis et al. 1994], a variation on McAllester's conspiracy numbers algorithm [McAllester 1988]. PNS builds a tree to variable depth trying to answer a binary question (for example, is this position a win?). Each node in a search tree whose value has not been proven is given a *proof* and *disproof* number. The proof number indicates the number of nodes in the tree whose value has to be determined to prove the assertion (the position is a win), while the disproof number indicates the number of nodes in the tree whose value has to be shown false to prove the assertion untrue (the position is not a win). The search algorithm examines the path of least resistance, the line of play with the lowest proof/disproof number. These numbers exploit the nonuniform branching factor in the search tree. For example, in chess, a forced sequence of checking moves leaves the opponent with few responses and, hence, few positions that need to be proven. The algorithm is intuitively appealing and has been successfully used to solve Connect-Four and Qubic.

There is a simple alternative search method to consider: alpha-beta. It can be used to prove the value of a position by modifying the evaluation function to return ∞ for a win, 0 for a draw and $-\infty$ for a loss or unproven position. Under the pessimistic assumption that all unproven positions are losses, a nonloss value returned by a search is a useful result and can only be improved by finding the correct value for the unproven positions. Thus, if a search returns a win value we know the position is a win, and if it returns a draw value, the position is at least a draw (it might still be a win).

Initial experience with the two algorithms has had some surprising results. Generally alpha-beta significantly outperforms PNS, but occasionally PNS wins by an enormous amount. Experiments performed by Victor Allis and Jonathan Schaeffer on the games of the 1994 *Chinook*–Oldbury match showed that many positions could be solved with 25 ply of search and were quickly found by alpha-beta. PNS, on the other hand, would explore some lines to excessive search depths looking for a more complicated solution than was needed. PNS excels when there is a nonuniform branching factor in the search tree. Ignoring capture moves (which are forced), the number of legal moves in a position changes little from move to move. Thus PNS has little information to guide its search. Research needs to be done on finding a more suitable heuristic for indicating to PNS where success is likely to be found.

Figure 2 illustrates the potential for deep alpha-beta searching. There are eighteen pieces on the board, so the endgame databases seem quite distant. It is white's tenth move (or ninth if you consider that white's first move was balloted). *Chinook* (white) played its checker from g1 to h2, whereupon Tinsley immediately remarked "You're going to regret that". On move 36, *Chinook* resigned. Since this game was played in 1990, the endgame databases have expanded from six to eight pieces and the search depths of the program have significantly increased. Today, the program can prove that the move g1 to h2 is a loss and b4 to a5 is a draw. This proof was done using alpha-beta by starting at the end of the game (move 36), searching that position and proving it to be lost, adding that result to the middlegame database, and then backing up to move 35 and trying to prove that position, etc. Thus this position, only ten

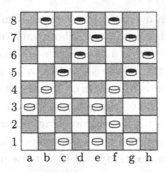

Figure 2. *Chinook*–Tinsley, 1990: White to play

moves from the start of the game (nine if you exclude balloted moves) has been solved. Subsequent searches of other positions are now simplified, since if they encounter this complicated position in their search, they need look no further; the middle game database contains the result.

The above discussion has concentrated on weakly solving the game. Ultra-strongly solving the game requires additional information; the program must choose the move to maximize its chances of improving its result. For example, given a drawn (lost) position, the program must select the move that maintains the draw (loss) and maximizes the chances of the opponent making a mistake that changes the game's outcome. In general this is a difficult problem, since maximizing difficulty is usually a function of the strengths and weaknesses of the opponent.

The biggest software advance in *Chinook* in the past two years has been an improvement in its ability to differentiate between drawing moves. In most game-playing programs, the value of a draw is 0, equality. *Chinook* differentiates between drawn positions by assigning them a value in the range of $+\frac{1}{3}$ of a checker to $-\frac{1}{12}$ of a checker. Note the asymmetric range of values. A very weak draw is still a draw and must have a score close to zero. If not, a weak nondrawing move might lead to a position with a higher score and be selected as the move to play. In tournaments, this scheme has resulted in several drawn games becoming wins when the program made it difficult for the opponent to find the drawing line.

The current status of the project is:

- Our endgame database consists of 4.4×10^{11} positions, that is, all positions with eight or fewer pieces. Completing the nine-piece databases will require computational and storage requirements beyond our current capabilities. We have begun the computing but expect to have only a small percent of the 10^{12} positions completed in 1995.
- The opening lines have been determined. This was done in consultation with our checkers expert, Martin Bryant, and his extensive literature on the game.
- The middlegame database contains only positions that occurred in *Chinook*'s tournament games. We are currently starting a systematic search of one

particular opening where it appears likely we can prove a draw. Success in one opening is all that is necessary to validate our approach.

5. Conclusions

With current technology, we believe it is possible to weakly solve the game of checkers. A database containing a proof tree for the game may contain as few as 10^9 nodes—a small number by today's standards. Strongly solving the game will be much harder, since it requires knowledge of all 10^{18} legally reachable positions in the game. The combination of weakly solving the game with the deep search, endgame databases and draw differentiation facilities in *Chinook* means that it will be possible to build a program that is close to ultra-strongly solving the game in practice.

Acknowledgements

Many people have contributed to *Chinook* over the years, including Martin Bryant, Joseph Culberson, Randal Kornelson, Brent Knight, Paul Lu, Steve Sutphen, Duane Szafron and Norman Treloar. Aske Plaat provided detailed feedback on the final draft of this paper. The support from Silicon Graphics International and the MPCI project at Lawrence Livermore Laboratories is greatfully appreciated. This research was supported by the Natural Sciences and Engineering Research Council of Canada (grant number OGP-8173) and the Netherlands Organization for Scientific Research (NWO).

References

[Allen 1989] J. D. Allen, "A note on the computer solution of connect-four", pp. 134–135 in *Heuristic Programming in Artificial Intelligence: The First Computer Olympiad* (edited by D. N. L. Levy and D. F. Beal), Ellis Horwood, Chichester, England, 1989.

[Allis 1988] L. V. Allis, "A knowledge-based approach to connect-four. The game is solved: White wins", M.Sc. thesis, Faculty of Mathematics and Computer Science, Vrije Universiteit, Amsterdam, 1988.

[Allis 1994] L. V. Allis, "Searching for solutions in games and artificial intelligence", Ph.D. thesis, Dept. of Computer Science, University of Limburg, 1994.

[Allis et al. 1991] L. V. Allis, H. J. van den Herik, and I. S. Herschberg, "Which games will survive", pp. 232–243 in *Heuristic Programming in Artificial Intelligence: The Second Computer Olympiad* (edited by D. N. L. Levy and D. F. Beal), Ellis Horwood, Chichester, England, 1991.

[Allis et al. 1993] L. V. Allis, H. J. van den Herik, and M. P. H. Huntjens, "Go-Moku solved by new search techniques", pp. 1–9 in *Proc. AAAI Fall Symposium on Games: Planning and Learning*, AAAI Press, Report FS9302, Menlo Park, California, 1993.

[Allis et al. 1994] L. V. Allis, M. van der Meulen, and H. J. van den Herik, "Proof-number search", *Artificial Intelligence* **66** (1994), 91–124.

[Anonymous 1979] Anonymous, "Computer Checkers: The Tinsley Challenge", *Personal Computing* (July 1979), 88–89.

[Barr and Feigenbaum 1981] A. Barr and E. Feigenbaum, *The Handbook of Artificial Intelligence*, Kaufmann, Los Altos, CA, 1981.

[Berliner 1980] H. Berliner, "Backgammon computer program beats World Champion", *Artificial Intelligence* **14** (1980), 205–220.

[Fortman 1978] R. L. Fortman, "Duke vs. Lowder", *Personal Computing* (November 1978), 26–28.

[Gasser 1994] R. Gasser, "Harnessing computational resources for efficient exhaustive search", Ph.D. thesis, Swiss Federal Institute of Technology, Zurich, 1994.

[Kopec 1990] D. Kopec, "Advances in Man-Machine Play", pp. 9–32 in *Computers, Chess and Cognition* (edited by T. A. Marsland and J. Schaeffer), Springer, New York, 1990.

[Lake et al. 1994] R. Lake, J. Schaeffer, and P. Lu, "Solving large retrograde analysis problems using a network of workstations", pp. 135–162 in *Advances in Computer Chess* 7 (edited by H. J. van den Herik et al.), University of Limburg, Maastricht, Netherlands, 1994.

[Levy and Beal 1989] *Heuristic Programming in Artificial Intelligence: The First Computer Olympiad* (edited by D. N. L. Levy and D. F. Beal), Ellis Horwood, Chichester, England, 1989.

[Levy and Beal 1991] *Heuristic Programming in Artificial Intelligence: The Second Computer Olympiad* (edited by D. N. L. Levy and D. F. Beal), Ellis Horwood, Chichester, England, 1991.

[McAllester 1988] D. A. McAllester, "Conspiracy numbers for min-max search", *Artificial Intelligence* **35** (1988), 287–310.

[Patashnik 1980] O. Patashnik, "Qubic: $4 \times 4 \times 4$ tic-tac-toe", *Math. Mag.* **53** (1980), 202–216.

[Samuel 1959] A. L. Samuel, "Some studies in machine learning using the game of checkers", *IBM J. Res. Devel.* **3** (1959), 210–229.

[Samuel 1967] A. L. Samuel, "Some studies in machine learning using the game of checkers: recent progress", *IBM J. Res. Devel.* **11** (1967), 601–617.

[Schaeffer 1989] J. Schaeffer, "The history heuristic and alpha-beta search enhancements in practice", *IEEE Trans. on Pattern Anal. and Machine Intelligence* **11** (1989), 1203–1212.

[Schaeffer 1992] J. Schaeffer, "Man versus machine: The Silicon Graphics World Checkers Championship", Technical Report 92-20, Dept. of Computing Science, University of Alberta, 1992.

[Schaeffer 1995] J. Schaeffer, "Marion Tinsley: Human Perfection at Checkers?", pp. 115–118 in this volume.

[Schaeffer et al. 1992] J. Schaeffer et al., "A World Championship caliber checkers program", *Artificial Intelligence* **53**(2–3) (1992), 273–290.

[Schaeffer et al. 1993a] J. Schaeffer, N. Treloar, P. Lu, and R. Lake, "Man versus machine for the World Checkers Championship", *AI Magazine* **14**(2) (1993), 28–35.

[Schaeffer et al. 1993b] J. Schaeffer, P. Lu, D. Szafron, and R. Lake, "A re-examination of brute-force search", pp. 51–58 in *Proc. AAAI Fall Symposium on Games: Planning and Learning*, AAAI Press, Report FS9302, Menlo Park, California, 1993.

[Smith 1994] H. Smith, personal communication citing letter sent to Smith by A. L. Samuel (1994).

[Strachey 1952] C. S. Strachey, "Logical or nonmathematical programs", *Proceedings of the ACM Conference, Toronto*, 1952.

[Thompson 1986] K. Thompson, "Retrograde analysis of certain endgames", *J. International Computer Chess Assoc.* **9**(3) (1986), 131–139.

[Tinsley 1994] M. Tinsley, personal communication citing letter sent to Tinsley by A. L. Samuel (1994).

[Truscott 1978] T. Truscott, "The *Duke* checkers program", preprint, Duke University, 1978.

JONATHAN SCHAEFFER
DEPARTMENT OF COMPUTING SCIENCE
UNIVERSITY OF ALBERTA
EDMONTON, ALBERTA
CANADA T6G 2H1

ROBERT LAKE
DEPARTMENT OF COMPUTING SCIENCE
UNIVERSITY OF ALBERTA
EDMONTON, ALBERTA
CANADA T6G 2H1

Games of No Chance
MSRI Publications
Volume **29**, 1996

On Numbers and Endgames:
Combinatorial Game Theory in Chess Endgames

NOAM D. ELKIES

> Neither side appears to have any positional advantage in the normal sense....
> the player with the move is able to arrange the pawn-moves to his own advantage
> [and win] in each case. It is difficult to say why this should be so, although the
> option of moving a pawn one or two squares at its first leap is a significant factor.
>
> — *Euwe and Hooper [1960], trying to explain why the position below*
> *(Example 87 in their book) should be a first-player win.*

> What shall we do with an Up?
>
> — *[Parker and Shaw 1962, 159–160, bass 2]*

ABSTRACT. In an investigation of the applications of CGT to chess, we construct novel mutual Zugzwang positions, explain the pawn endgame above,
show positions containing non-integer values (fractions, switches, tinies,
and loopy games), and pose open problems concerning the values that may
be realized by positions on either standard or nonstandard chessboards.

1. Introduction

It was already noted in *Winning Ways* [Berlekamp et al. 1982, p. 16] that combinatorial game theory (CGT) does not apply directly to chess, because the winner of a chess game is in general not determined by who makes the last move, and indeed a game may be neither won nor lost at all but drawn by infinite play.[1]

Still, CGT has been effectively applied to other games such as Dots-and-Boxes and Go, which are not combinatorial games in the sense of *Winning Ways*. The main difficulty with doing the same for chess is that the 8×8 chessboard is too small to decompose into many independent subgames, or rather that some of the chess pieces are so powerful and influence such a large fraction of the board's area that even a decomposition into two weakly interacting subgames (say a kingside attack and a queenside counteroffensive) generally breaks down in a few moves.

Another problem is that CGT works best with "cold" games, where having the move is a liability or at most an infinitesimal boon, whereas the vast majority of chess positions are "hot": Zugzwang[2] positions (where one side loses or draws but would have done better if allowed to pass the move) are already unusual, and positions of mutual Zugzwang, where neither side has a good or even neutral move, are much rarer.[3]

This too is true of Go, but there the value of being on move, while positive, can be nearly enough constant to be managed by "chilling operators" [Berlekamp and Wolfe 1994], whereas the construction of chess positions with similar behavior seems very difficult, and to date no such position is known.

To find interesting CGT aspects of chess we look to the endgame. With most or all of the long-range pieces no longer on the board, there is enough room for a decomposition into several independent subgames. Also, it is easier to construct mutual Zugzwang positions in the endgame because there are fewer pieces that could make neutral moves; indeed, it is only in the endgame that mutual Zugzwang ever occurs in actual play. If a mutual Zugzwang is sufficiently localized on the chessboard, one may add a configuration of opposing pawns that must eventually block each other, and the first player who has no move on that configuration loses the Zugzwang. Furthermore this configuration may split up as a sum of independent subgames. The possible values of these games are sufficiently varied that one may construct positions that, though perhaps

---

[1] Of course infinite play does not occur in actual games. Instead the game is drawn when it is apparent that neither side will be able to checkmate against reasonable play. Such a draw is either agreed between both opponents or claimed by one of them using the triple-repetition or the fifty-move rule. These mechanisms together approximate, albeit imperfectly, the principle of draw by infinite play.

[2] This word, literally meaning "compulsion to move" in German, has long been part of the international chess lexicon.

[3] Rare, that is, in practical play; this together with their paradoxical nature is precisely why Zugzwang and mutual Zugzwang are such popular themes in composed chess problems and endgame studies.

not otherwise intractable, illustrate some of the surprising identities from *On Numbers and Games* [Conway 1976]. Occasionally one can even use this theory to illuminate the analysis of a chess endgame occurring in actual play.

We begin by evaluating simple pawn subgames on one file or two adjacent files; this allows us to construct some novel mutual Zugzwang positions and explain the pawn endgame that baffled Euwe. We then show positions containing more exotic values: fractions, switches and tinies, and loopy games. We conclude with specific open problems concerning the values that may be realized by positions either on the 8 × 8 chessboard or on boards of other sizes.

NOTES. (1) In the vast majority of mutual Zugzwangs occurring in actual play only a half point is at stake: one side to move draws, the other loses. (In tournament chess a win, draw or loss is worth 1, $\frac{1}{2}$, or 0 points, respectively.) We chose to illustrate this article with the more extreme kind of mutual Zugzwang involving the full point: whoever is to move loses. This is mainly because it is easier for the casual player to verify a win than a draw, though as it happens the best example we found in actual play is also a full-point mutual Zugzwang. The CGT part of our analysis applies equally to similar endgames where only half a point hinges on the mutual Zugzwang.

(2) We use "algebraic notation" for chess moves and positions: the ranks of the chessboard are numbered 1 through 8 from bottom to top; the columns ("files") labeled a through h from left to right; and each square is labeled by the file and rank it is on. Thus in the diagram on page 135 the White king is on f3. Pawns, which stay on the same file when not capturing, are named by file alone when this can cause no confusion. Pawn moves are described by the destination square, and other moves by the piece (capital letter, N = knight) followed by the destination square. A capture is indicated with a colon, and the file of departure is used instead of a capital letter in captures by pawns: thus b:a3 means that the b-pawn captures what's on a3. A + indicates a check, and a capital letter after a pawn move indicates a promotion.

(3) All game and subgame values are stated from White's perspective, that is, White is "Left", Black is "Right".

2. Simple Subgames with Simple Values

Integers. Integer values, indicating an advantage in spare tempo moves, are easy to find. (A *tempo move* or *waiting move* is one whose only effect is to give the opponent the turn.) An elementary example is shown in Diagram 1.

The kingside is an instance of the mutual Zugzwang known in the chess literature as the "trébuchet": once either White or Black runs out of pawn moves, he must move his king, losing the g-pawn and the game. Clearly White has one free pawn move on the e-file, and Black has two on the a-file, provided he does

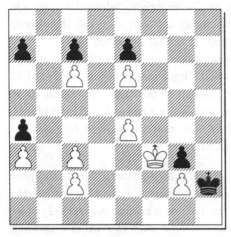

Diagram 1   Diagram 2

not rashly push his pawn two squares on the first move. Finally the c-file gives White four free moves (the maximum on a single file), again provided Pc2 moves only one square at a time. Thus the value of Diagram 1 is $1 - 2 + 4 = 3$, and White wins with at least two free moves to spare regardless of who moves first.

Infinitesimals. Simple subgames can also have values that are not numbers, as witness the b- and h-files in Diagram 2. The b-file has value $\{0 \mid 0\} = *$; the same value (indeed an isomorphic game tree) arises if the White pawn is placed on a3 instead of b4. The e-file has value zero, since it is a mutual Zugzwang; this is the identity $\{* \mid *\} = 0$. The h-file, on the other hand, has positive value: White's double-move option gives him the advantage regardless of who has the move. Indeed, since the only Black move produces a $*$ position, while White to move may choose between 0 and $*$, the h-file's value is $\{0, * \mid *\} = \uparrow$. (While $\uparrow$ is usually defined as $\{0 \mid *\}$, White's extra option of moving to $*$ gives him no further advantage; in *Winning Ways* parlance (p. 64), it is *reversible*, and bypassing it gives White no new options. This may be seen on the chessboard by noting that in the position where White has pawns on h2 and f4 and Black has pawns on h5 and f7, Black to move loses even if White is forbidden to play h3 until Black has played h4: $1. \ldots h4\ 2. f5$, or $1. \ldots f6\ 2. f5$.)

This accounts for all two-pawn positions with the pawns separated by at most two squares on the same file, or at most three on adjacent files. Putting both pawns on their initial squares of either the same or adjacent files produces a mutual Zugzwang (value zero). This leaves only one two-pawn position to evaluate, represented by the a-file of Diagram 3.

From our analysis thus far we know that the a-file has value $\{0, * \mid \uparrow\}$. Again we bypass the reversible $*$ option, and simplify this value to $\Uparrow * = \{0 \mid \uparrow\}$. Equivalently, Diagram 3 (in which the c-, d- and e-files are $\Downarrow *$, and the kingside is mutual Zugzwang for chess reasons) is a mutual Zugzwang, and remains so if

Diagram 3

Diagram 4

White is forbidden to play a2-a4 before Black has moved his a6-pawn. This is easily verified: WTM (White to move) 1. c5 a5 and wins by symmetry, or 1. a4 d3 2. a5(c5) e5 or 1. a3 d3 etc.; BTM 1. ... a5 2. c5 wins symmetrically, 1. ... d3 2. c5 e5 (else 3. e5 and 4. a3) 3. c6 a5 4. a4, 1. ... c6 2. e5 and 3. a3, 1. ... c5 2. d3 e5 (e6 3. e5 a5 4. a4) 3. a3 a5 4. a4.

On a longer chessboard we could separate the pawns further. Assuming that a pawn on such a chessboard still advances one square at a time except for an initial option of a double move on its first move, we evaluate such positions thus: a White pawn on a2 against a Black one on a7 has value $\{0, * \mid \Uparrow *\} = \Uparrow\Uparrow\Uparrow$; against a Black Pa8, $\{0, * \mid \Uparrow\Uparrow\Uparrow\} = \Uparrow\Uparrow\Uparrow\Uparrow *$; and by induction on n a Black pawn on the $(n + 4)$-th rank yields n ups or n ups and a star according to whether n is odd or even, provided the board is at least $n + 6$ squares wide so the Black pawn is not on its initial square. With both pawns on their initial squares the file has value zero unless the board has width 5 or 6, when the value is $*$ or $* 2$, respectively. Of course if neither pawn is on its starting square the value is 0 or $*$, depending on the parity of the distance between them, as in the b- and e-files of Diagram 2.

Diagram 4 illustrates another family of infinitesimally valued positions. The analysis of such positions is complicated by the possibility of pawn trades that involve entailing moves: an attacked pawn must in general be immediately defended, and a pawn capture parried at once with a recapture. Still we can assign standard CGT values to many positions, including all that we exhibit in diagrams in this article, in which each entailing line of play is dominated by a non-entailing one (*Winning Ways*, p. 64).

Consider first the queenside position in Diagram 4. White to move can choose between 1. b3 (value 0) and 1. a4, which brings about $*$ whether or not Black interpolates the *en passant* trade 1. ... b:a3 2. b:a3. White's remaining choice 1. a3

would produce an inescapably entailing position, but since Black can answer
1. a3 with b:a3 2. b:a3 this choice is dominated by 1. a4 so we may safely ignore
it. Black's move a4 produces a mutual Zugzwang, so we have $\{0,* \mid 0\} = \uparrow*$
(*Winning Ways*, p. 68). Our analysis ignored the Black move 1....b3?, but
2. a:b3 then produces a position of value 1 (White has the tempo move 3. b4 a:b4
4. b3), so we may disregard this option since $1 > \uparrow*$.

We now know that in the central position of Diagram 4 Black need only
consider the move d5, which yields the queenside position of value $\uparrow*$, since after
1....e3? 2. d:e3 White has at least a spare tempo. WTM need only consider
1. d4, producing a mutual Zugzwang whether or not Black trades *en passant*,
since 1. d3? gives Black the same option and 1. e3?? throws away a spare tempo.
Therefore the center position is $\{0 \mid \uparrow*\} = \Uparrow$ (*Winning Ways*, p. 73). Both this
and the queenside position turn out to have the same value as they would had
the pawns on different files not interacted. This is no longer true if Black's rear
pawn is on its starting square: if in the center of Diagram 4 Pd6 is placed on
d7, the resulting position is no longer $*$, but mutual Zugzwang (WTM 1. d4 e:d3
2. e:d3 d6; BTM 1....d5 2. e3 or d6 2. d4). Shifting the Diagram 4 queenside
up one or two squares produces a position (such as the Diagram 4 kingside) of
value $\{0 \mid 0\} = *$: either opponent may move to a mutual Zugzwang (1. h5 or
1....g6), and neither can do any better, even with Black's double-move option:
1. g5 h5 is again $*$, and 1....g5 2. h:g5 h:g5 is equivalent to 1....g6, whereas
1....h5? is even worse. With the h4-pawn on h3, though, the double-move
option becomes crucial, giving Black an advantage (value $\{* \mid 0, *\} = \downarrow$, using
the previous analysis to evaluate 1. h4 and 1....g6 as $*$).

3. Schweda versus Sika, Brno, 1929

We are now ready to tackle a nontrivial example from actual play—specifically,
the subject of our opening quote from [Euwe and Hooper 1960]. We repeat it in
Diagram 5 for convenience.

On the e- and f-files the kings and two pawns are locked in a vertical trébuchet;
whoever is forced to move there first will lose a pawn, which is known to be
decisive in such an endgame. Thus we can ignore the central chunk and regard
the rest as a last-mover-wins pawn game.

As noted in the introduction, the 8×8 chessboard is small enough that a
competent player can play such positions correctly even without knowing the
mathematical theory. Indeed the White player correctly evaluated this as a win
when deciding earlier to play for this position, and proceeded to demonstrate
this win over the board. Euwe and Hooper [1960, p. 56] also show that Black
would win if he had the move from the diagram, but they have a hard time
explaining why such a position should a first-player win—this even though Euwe
held both the world chess championship from 1935 to 1937 and a doctorate in

Diagram 5

mathematics [Hooper and Whyld 1992]. (Of course he did not have the benefit of CGT, which had yet to be developed.)

Combinatorial game theory tells us what to do: decompose the position into subgames, compute the value of each subgame, and compare the sum of the values with zero. The central chunk has value zero, being a mutual Zugzwang. The h-file we recognize as $\Downarrow*$. The queenside is more complicated, with a game tree containing hot positions (1. a4 would produce $\{2 \mid 0\}$) and entailing moves (such as after 1. a4 b5); but again it turns out that these are all dominated, and we compute that the queenside simplifies to $\uparrow$. Thus the total value of the position is $\uparrow + \Downarrow* = \downarrow*$. Since this is confused with zero, the diagram is indeed a first-player win. To identify the queenside value as $\uparrow$ we show that White's move 1. h4, converting the kingside to $\downarrow$, produces a mutual Zugzwang, using values obtained in the discussion of Diagram 4 to simplify the queenside computations. For instance, 1. h4 a5 2. h5 a4 3. h6 b6 4. b4 wins, but with WTM again after 1. h4, Black wins after 2. a4 a5, 2. a3 h5, 2. b4 h5 (mutual Zugzwang) 3. a3/a4 b5/b6, 2. b3 a5 (mutual Zugzwang $+ \downarrow$), or 2. h5 a5 and 3. h6 a4 or 3. a4(b3) h6. Black to move from Diagram 7 wins with 1. ... a5, reaching mutual Zugzwang after 2. h4 a4 3. h5 h6 or 2. a4(b3) h6, even without using the ... h5 double-move option (since without it the h-file is $*$ and $\uparrow + * = \uparrow*$ is still confused with zero).

4. More Complicated Values: Fractions, Switches and Tinies

Fractions. Fractional values are harder to come by; Diagram 6 shows two components with value $\frac{1}{2}$. In the queenside component the c2 pawn is needed to assure that Black can never safely play b4; a pawn on d2 would serve the same

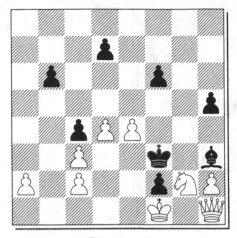

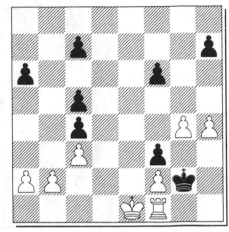

Diagram 6   Diagram 7

purpose. In the configuration d4, e4/d7, f6 it is essential that White's e5 forces a pawn trade, i.e., that in the position resulting from 1. e5 f5? 2. d5 White wins, either because the position after mutual promotions favors White or because (as in Diagram 6) the f-pawn is blocked further down the board. Each of these components has the form $\{0, * \mid 1\}$, but (as happened in Diagrams 2 and 3) White's $*$ option gives him no further advantage, and so each component's value simplifies to $\{0 \mid 1\} = \frac{1}{2}$. Since the seven-piece tangle occupying the bottom right corner of Diagram 6 not only blocks the f6-pawn but also constitutes a (rather ostentatious) mutual Zugzwang, and Black's h-pawn provides him a free move, the entire Diagram is itself a mutual Zugzwang illustrating the identity $\frac{1}{2} + \frac{1}{2} - 1 = 0$.

What do we make of Diagram 7, then? Chess theory recognizes the five-man configuration around f2 as a mutual Zugzwang (the critical variation is WTM 1. Rh1 K:h1 2. Kd2 Kg1! 3. Ke3 Kg2, and Black wins the trébuchet; this mutual Zugzwang is akin to the kingside mutual Zugzwang of Diagram 3, but there the d-pawns simplified the analysis). Thus we need to evaluate the three pure-pawn subgames, of which two are familiar: the spare tempo-move of Pc7, and the equivalent of half a spare tempo-move White gets from the upper kingside.

To analyze the queenside position (excluding Pc7), we first consider that position after Black's only move a5. From that position Black can only play a4 (value 1), while White can choose between a4 and a3 (values 0 and $*$), but not 1. b3? c:b3 2. a:b3 c4! and the a-pawn promotes. Thus we find once more the value $\{0, * \mid 1\} = \frac{1}{2}$. Returning to the Diagram 7 queenside, we now know the value $\frac{1}{2}$ after Black's only move a5. White's moves a3 and a4 produce 0 and $*$, and 1. b3 can be ignored because the reply c:b3 2. a:b3 c4 3. b4 shows that this is no better than 1. a3. So we evaluate the queenside of Diagram 7 as $\{0, * \mid \frac{1}{2}\} = \frac{1}{4}$, our first quarter.

Diagram 8

Thus the whole of Diagram 7 has the negative value $\frac{1}{2} - 1 + \frac{1}{4} = -\frac{1}{4}$, indicating a Black win regardless of who has the move, though with BTM the only play is 1.... a5! producing a $\frac{1}{2} + \frac{1}{2} - 1$ mutual Zugzwang.

On a longer chessboard we could obtain yet smaller dyadic fractions by moving the b-pawn of Diagram 6 or the Black a-pawn of Diagram 7 further back, as long as this does not put the pawn on its initial square. Each step back halves the value. These constructions yield fractions as small as $1/2^{N-7}$ and $1/2^{N-6}$, respectively, on a board with columns of length $N \geq 8$.

Switches and tinies. We have seen some switches (games $\{m \mid n\}$ with $m > n$) already in our analysis of four-pawn subgames on two files such as occur in Diagrams 4 and 5. We next illustrate a simpler family of switches.

In the a-file of Diagram 8, each side has only the move a6. If Black plays a6 the pawns are blocked, while White gains a tempo move with a6 (compare with the e-file of Diagram 1), so the a-file has value $\{1 \mid 0\}$. On the c-file whoever plays c4 gets a tempo move, so that file gives $\{1 \mid -1\} = \pm 1$. Adding a Black pawn on c7 would produce $\{1 \mid -2\}$; in general, on a board with files of length N, we could get temperatures as high as $(N - 5)/2$ by packing as many as $N - 3$ pawns on a single file in such a configuration.[1]

The f-file is somewhat more complicated: White's f5 produces the switch $\{2 \mid 1\}$, while Black has a choice between f6 and f5, which yield $\{1 \mid 0\}$ and 0. Bypassing the former option we find that the f-file shows the three-stop game $\{2 \mid 1 \parallel 0\}$. Likewise $\{4 \mid 2 \parallel 0\}$ can be obtained by adding a White pawn on f2, and on a longer board $n + 1$ pawns would produce $\{2n \mid n \parallel 0\}$. The h-file shows

---

[1] For large enough N, it will be impossible to pack that many pawns on a file starting from an initial position such as that of 8×8 chess, because it takes at least $\frac{1}{4}n^2 + O(n)$ captures to get n pawns of the same color on a single file. At any rate one can attain temperatures growing as some multiple of $\sqrt{N}$.

Diagram 9

the same position shifted down one square, with Black no longer able to reach 0 in one step. That file thus has value $\{2 \mid 1 \parallel 1 \mid 0\}$, which simplifies to the number 1, as may be seen either from the CGT formalism or by calculating directly that the addition of a subgame of value -1 to the h-file produces mutual Zugzwang.

Building on this we may construct a few tinies and minies, albeit in more contrived-looking positions than we have seen thus far (though surely no less natural than the positions used in [Fraenkel and Lichtenstein 1981]). Consider Diagram 9. In the queenside (apart from the pawns on a5 and b7, which I put there only to forestall a White defense based on stalemate) the Black pawn on c2 and both knights cannot or dare not move; they serve only to block Black from promoting after ...d:c3. That is Black's only move, and it produces the switch $\{0 \mid -1\}$ as in the a-file of Diagram 8. White's only move is 1. c:d4 (1. c4? d:c4 2. d:c4 d3 3. N:d3 Nb3, or even 2. ... Nb3 3. N:b3 d3), which yields mutual Zugzwang. Thus the queenside evaluates to $\{0 \parallel 0 \mid -1\}$, or tiny-one. Adding a fourth Black d-pawn on d7 would produce tiny-two, and on larger boards we could add more pawns to get tiny-n for arbitrarily large n. In the kingside of Diagram 9 the same pawn-capture mechanism relies on a different configuration of mutually paralyzing pieces, including both kings. With a White pawn on g3 the kingside would thus be essentially the same as the queenside with colors reversed, with value miny-one; but since White lacks that pawn the kingside value is miny-zero, or $\downarrow$ (*Winning Ways*, p. 124). Black therefore wins Diagram 9 regardless of whose turn it is, since his kingside advantage outweighs White's queenside edge.

Some loopy chunks. Since pawns only move in one direction, any subgame in which only pawns are mobile must terminate in a bounded number of moves. Subgames with other mobile pieces may be unbounded, or *loopy* in *Winning Ways*

Diagram 10 Diagram 10'

terminology (p. 314); indeed, unbounded games must have closed cycles (loops) of legal moves because there are only finitely many distinct chess positions.

Consider for instance the queenside of Diagram 10, where only the kings may move. Black has no reasonable options since any move loses at once to Kb5 or Kd5; thus the queenside's value is at least zero. White could play Kc3, but Black responds Kc5 at once, producing Diagram 10' with value ≤ 0 because then Black penetrates decisively at b4 or d4 if the White king budges. Thus Kc3 can never be a good move from Diagram 10. White can also play Kb3 or Kd3, though. Black can then respond Kc5, forcing Kc3 producing Diagram 10'. In fact Black might as well do this at once: any other move lets White at least repeat the position with 2. Kc4 Kc6 3. Kb3(d3), and White has no reasonable moves at all from Diagram 10' so we need not worry about White moves after Kb3(d3). We may thus regard 1. Kb3(d3) Kc5 2. Kc3 as a single move that is White's only option from the Diagram 10 queenside.

By the same argument we regard the Diagram 10' queenside as a game where White has no moves and Black has only the "move" 1. ... Kb6(d6) 2. Kc4 Kc6, recovering the queenside of Diagram 10. We thus see that these queenside positions are equivalent to the loopy games called **tis** and **tisn** in *Winning Ways*, p. 322 (**istoo** and **isnot** in American English). Since the kingside has value 1, Diagram 10 (1 + **tis**) is won for White, as is Diagram 10' (1 + **tisn** = **tis**) with BTM, but with WTM Diagram 10' is drawn after 1. h5 Kb6(d6) 2. Kc4 Kc6 3. Kb3(d3) Kc5 4. Kc3 etc.

We draw our final examples from the Three Pawns Problem (Diagram 11). See [Hooper and Whyld 1992] for the long history of this position, which was finally solved by Szén around 1836; Staunton [1847] devoted twelve pages (487–500) to its analysis. (Thanks to Jurg Nievergelt for bringing this Staunton reference to my attention.) Each king battles the opposing three pawns. Three pawns on

Diagram 11   Diagram 12

adjacent files can contain a king but (unless very far advanced) not defeat it. Eventually Zugzwang ensues, and one player must either let the opposing pawns through, or push his own pawns when they can be captured. As with our earlier analysis, we allow only moves that do not lose a pawn or unleash the opposing pawns; thus the last player to make such a move wins. The Three Pawns Problem then in effect splits into two equal and opposite subgames. One might think that this must be a mutual Zugzwang, but in fact Diagram 11 is a first-player win. Diagram 12 shows a crucial point in the analysis, which again is a first-player win despite the symmetry. The reason is that each player has a check (White's a5 or c5, Black's f4 or h4) that entails an immediate king move: Black is not allowed to answer White's 1. a5+ with the Tweedledum move (*Winning Ways*, p. 4) 1.... h4+, and so must commit his king before White must answer the pawn check. This turns out to be sufficient to make the difference between a win and a loss in Diagrams 11 and 12.

Diagram 13 is a classic endgame study by J. Behting using this material ([Sutherland and Lommer 1938, #61]; originally published in *Deutsche Schachzeitung* 1929). After 1. Kg1! the kingside shows an important mutual Zugzwang: BTM loses all three pawns after 1.... g3 2. Kg2 or 1.... f3/h3 2. Kf2/h2 h3/f3 3. Kg3, while WTM loses after 1. Kh2(f1) h3, 1. Kf2(h1) f3, or 1. Kg2 g3, when at least one Black pawn safely promotes to a queen. On other king moves from Diagram 13 Black wins: 1. Kf2(f1) h3 or 1. Kh2(h1) f3 followed by ... g3 and White can no longer hold the pawns, e.g., 1. Kh1 f3 2. Kh2 g3+ 3. Kh3 f2 4. Kg2 h3+ 5. Kf1 h2. Thus we may regard Kg1 as White's only kingside option in Diagram 13. Black can play either ... g3 reaching mutual Zugzwang, or ... f3/h3+ entailing Kf2/h2 and again mutual Zugzwang but BTM; in effect Black can interpret the kingside as either 0 or ∗. In the queenside, White to move can only play a6 reaching mutual Zugzwang. Black to move plays 1.... Ka7 or Kc7, when

Diagram 13

Diagram 14

2. a6 Kb8 is mutual Zugzwang; but the position after 1.... Ka7(c7) is not it-self a mutual Zugzwang because White to move can improve on 2. a6 with the sacrifice 2. b8Q+! K:b8 3. a6, reaching mutual Zugzwang with BTM. The posi-tions with the Black king on b8 or a7(c7) are then seen to be equivalent: Black can move from one to the other and White can move from either to mutual Zugzwang. Thus the Diagram 13 queenside is tantamount to the loopy game whose infinitesimal but positive value is called **over** = $1/$**on** in *Winning Ways*, p. 317. White wins Behting's study with 1. Kg1! Ka7(c7) 2. b8Q+! K:b8 3. a6 reaching mutual Zugzwang; all other alternatives (except 2. Kg2 Kb8 repeating the position) lose: 1. a6? g3!, or 2. a6? Kb8. Since **over** exceeds $*$ as well as 0, White wins Diagram 13 even if Black moves first: 1.... Ka7 2. Kg1! Kb8 3. a6, 1.... g3 2. a6, or 1.... f3/h3+ 2. Kf2/h2 Ka7 3. b8Q+ K:b8 4. a6 etc.

The mutual Zugzwang in the analysis of the Diagram 13 queenside after 2. b8Q+ K:b8 3. a6 is the only mutual Zugzwang involving a king and only two pawns. In other positions with a king in front of two pawns either on adjacent files or separated by one file, the king may not be able to capture the pawns, but will at least have an infinite supply of tempo moves. Thus such a position will have value **on** or **off** (*Winning Ways*, p. 317 ff.) according to whether White or Black has the king. For instance, in the kingside of Diagram 14, White must not capture on h4 because then the f-pawn promotes, but the king can shuttle endlessly between h2 and h3 while Black may not move (1.... f2? 2. Kg2 h3+ 3. K:f2! and the h-pawn falls next). If White didn't have the pawn on b2, the queenside would likewise provide Black infinitely many tempo moves and the entire Diagram would be a draw with value **on** + **off** = **dud** (*Winning Ways*, p. 318). As it is, White naturally wins Diagram 14, since Black will soon run out of queenside moves. We can still ask for the value of the queenside; it turns out to give another realization of **over**. Indeed, the Black king can only shuttle between b7 and b8, since moving to the a- or c-file loses to c6 or a6, respectively; and until the b-pawn reaches b4 White may not move his other pawns since c6/a6 drops a pawn to Kc7/a7. We know from Diagram 13 that if the b-pawn were on b5 the Diagram 14 queenside would be mutual Zugzwang. The same is true with that pawn on b4 and the Black king on b7: WTM 1. b5 Kb8, BTM 1.... Kb8 2. b5 or 2. c6/a6 Kc7/Ka7 3. b5. Thus pawn on b4 and king on b8 give *, as do Pb3/Kb7, while Pb3/Kb8 is again mutual Zugzwang. From b2 the pawn can move to mutual Zugzwang against either Kb7 or Kb8 (moving to * is always worse, as in the Diagram 13 queenside), yielding a value of **over** as claimed. Positions such as this one, which show an advantage of **over** thanks to the double-move option, are again known to chess theory; see for instance [Sutherland and Lommer 1938, endgame #55] by H. Rinck (originally published in *Deutsche Schachzeitung*, 1913), which uses a different pawn trio. Usually, as in that Rinck endgame, the position is designed so that White can only win by moving a pawn to the fourth rank in two steps instead of one.

5. Open Problems

We have seen that pawn endgames can illustrate some of the fundamental ideas of combinatorial game theory in the familiar framework of chess. How much of CGT can be found in such endgames, either on the 8×8 or on larger boards? Of course one could ask for each game value in *Winning Ways* whether it can be shown on a chessboard. But it appears more fruitful to focus on attaining specific values in endgame positions. I offer the following challenges:

Nimbers. Do $*2$, $*4$ and higher nimbers occur on the 8×8 or larger boards? We have seen already that on a file of length 6 the position Pa2 vs. Pa5 gives $*2$, and Pa2 vs. Pb6 does the same for files of length 7. But these constructions extend neither to longer boards nor to nimbers beyond $*2$ and $*3 = * + *2$.

Positive infinitesimals. We have seen how to construct tiny-x for integers $x \geq 0$ (Diagram 9). How about other x such as $\frac{1}{2}$ or $1\uparrow$? Also, do the higher ups $\uparrow^2$, $\uparrow^3$, etc. of *Winning Ways* (pp. 277 and 321) occur?

Fractions. We can construct arbitrary dyadic fractions on a sufficiently large chessboard. Does $\frac{1}{8}$ exist on the standard 8×8 board? Can positions with value $\frac{1}{3}$ or other non-dyadic rationals arise in loopy chess positions? (Thirds do arise as mean values in Go [Berlekamp and Wolfe 1994; Gale 1994], thanks to the Ko rule.)

Chilled chess. Is there a class of chess positions that naturally yields to chilling operators as do the Go endgames of [Berlekamp and Wolfe 1994]?

In other directions, one might also hope for a more systematic CGT-style treatment of *en passant* captures and entailing chess moves such as checks, captures entailing recapture, and threats to capture; and ask for a class of positions on an $N \times N$ board that bears on the computational complexity of pawn endgames as [Fraenkel and Lichtenstein 1981] does for unrestricted $N \times N$ chess positions.

Acknowledgements

This paper would never have been written without the prodding, assistance and encouragement of Elwyn Berlekamp. In the Fall of 1992 he gave a memorable expository talk on combinatorial game theory, during which he mentioned that at the time CGT was not known to have anything to do with chess. I wrote a precursor of this paper shortly thereafter in response to that implied (or at least perceived) challenge. The rest of the material was mostly developed in preparation for or during the MSRI workshop on combinatorial games, which was largely Berlekamp's creation. I am also grateful for his comments on the first draft of this paper, particularly concerning [Berlekamp and Wolfe 1994].

This paper was typeset in LaTeX, using Piet Tutelaers' chess fonts for the diagrams. The research was made possible in part by funding from the National Science Foundation, the Packard Foundation, and the Mathematical Sciences Research Institute.

References

[Berlekamp and Wolfe 1994] E. R. Berlekamp and D. Wolfe, *Mathematical Go: Chilling Gets the Last Point*, A K Peters, Wellesley (MA), 1994.

[Berlekamp et al. 1982] E. R. Berlekamp, J. H. Conway, and R. K. Guy, *Winning Ways For Your Mathematical Plays, I: Games In General*, Academic Press, London, 1982.

[Conway 1976] J. H. Conway, *On Numbers And Games*, Academic Press, London, 1976.

[Euwe and Hopper 1960] M. Euwe and D. Hooper, *A Guide to Chess Endings*, David McKay, New York, 1960. Reprinted by Dover, New York, 1976.

[Fraenkel and Lichtenstein 1981] A. S. Fraenkel and D. Lichtenstein, "Computing a perfect strategy for $n \times n$ chess requires time exponential in n", *J. Comb. Theory* **A31** (1981), 199–214.

[Gale 1994] D. Gale, "Mathematical Entertainments: Go", *Math. Intelligencer* **16**(2) (Spring 1994), 25–31.

[Hooper and Whyld 1992] D. Hooper, K. Whyld, *The Oxford Companion to Chess* (second edition), Oxford Univ. Press, 1992.

[Parker and Shaw 1962] A. Parker and R. Shaw (arrangers), *What Shall We Do With the Drunken Sailor?*, Lawson-Gould, New York, 1962.

[Staunton 1847] H. Staunton, *The Chess-Player's Handbook*, H. G. Bonh, London, 1847. Reprinted by Senate, London, 1994.

[Sutherland and Lohmer 1938] M. A. Sutherland and H. M. Lommer: 1234 *Modern End-Game Studies*, Printing-Craft Ltd., London, 1938. Reprinted by Dover, New York, 1968.

NOAM D. ELKIES
DEPARTMENT OF MATHEMATICS
HARVARD UNIVERSITY
CAMBRIDGE, MA 02138
elkies@math.harvard.edu

Games of No Chance
MSRI Publications
Volume 29, 1996

Multilinear Algebra and Chess Endgames

LEWIS STILLER

ABSTRACT. This article has three chief aims: (1) To show the wide utility
of multilinear algebraic formalism for high-performance computing. (2) To
describe an application of this formalism in the analysis of chess endgames,
and results obtained thereby that would have been impossible to compute
using earlier techniques, including a win requiring a record 243 moves.
(3) To contribute to the study of the history of chess endgames, by focusing
on the work of Friedrich Amelung (in particular his apparently lost analysis
of certain six-piece endgames) and that of Theodor Molien, one of the
founders of modern group representation theory and the first person to
have systematically numerically analyzed a pawnless endgame.

1. Introduction

Parallel and vector architectures can achieve high peak bandwidth, but it can
be difficult for the programmer to design algorithms that exploit this bandwidth
efficiently. Application performance can depend heavily on unique architecture
features that complicate the design of portable code [Szymanski et al. 1994; Stone
1993].

The work reported here is part of a project to explore the extent to which
the techniques of multilinear algebra can be used to simplify the design of high-
performance parallel and vector algorithms [Johnson et al. 1991]. The approach
is this:

- Define a set of fixed, structured matrices that encode architectural primitives
  of the machine, in the sense that left-multiplication of a vector by this matrix
  is efficient on the target architecture.
- Formulate the application problem as a matrix multiplication.
- Factor the matrix corresponding to the application in terms of the fixed matrices using addition, tensor product, and matrix multiplication as combining
  operators.
- Generate code from the matrix factorization.

Supported by U.S. Army Grant DAAL03–92–G–0345.

This approach has been used in signal processing algorithms [Granata and Tolimieri 1991; Granata et al. 1991; Johnson et al. 1990; Tolimieri et al. 1989; Tolimieri et al. 1993]. The success of that work motivates the present attempt to generalize the domain of application of the multilinear-algebraic approach to parallel programming [Stiller 1992b].

I have used the methodology presented in this paper in several domains, including parallel N-body codes [Stiller 1994], Fortran 90 communication intrinsic functions [Stiller 1992a], and statistical computations [García and Stiller 1993]. In each case, significant speedup over the best previous known algorithms was attained. On the other hand, it is clear that this methodology is intended to be applicable only to a narrow class of domains: those characterized by regular and oblivious memory access patterns. For example, parallel alpha-beta algorithms cannot be formulated within this paradigm.

This paper describes the application of the multilinear algebraic methodology to the domain of chess endgames. Dynamic programming was used to embed the state space in the architecture. By successively unmoving pieces from the set of mating positions, the set of positions from which White could win can be generated. This application presents an interesting challenge to the multilinear-algebraic parallel-program design methodology:

- The formalism for the existing multilinear algebra approach had been developed to exploit parallelization of linear transformations over a module, and had to be generalized to work over Boolean algebras.
- The symmetry under a noncommutative crystallographic group had to be exploited without sacrificing parallelizability.
- The state-space size of 7.7 giganodes was near the maximum that the target architecture could store in RAM.

Two main results are reported here:

- Table 1 on page 168 gives equations defining the dynamic programming solution to chess endgames. Using the techniques described in this paper, the factorizations can be modified to produce efficient code for most current parallel and vector architectures.
- Table 2 on page 175 presents a statistical summary of the state space of several six-piece chess endgames. This table could not have been generated in a practicable amount of time using previous techniques.

Section 2 below provides the background of the chess endgame problem. A survey of some human analysis of chess endgames is given, followed by a survey of previous work in the area of computer endgame analysis. Readers interested only in chess and not in the mathematical and programming aspects of this work can skip from this section to Section 8.

Section 3 introduces some basic concepts of parallel processing.

Section 4 describes previous work in the area of tensor product formalism for signal-processing computations.

Section 5 develops a generalized version of the formalism of Section 4, and describes the chess endgame algorithm in terms of this formalism.

Section 6 presents equations defining the dynamic programming solution to chess endgames (Table 1). Section 6.2 describes how these equations are modified to exploit symmetry. The derivation of crystallographic FFTs is used as a motivating example in the derivation of symmetry-invariant equations.

Section 7 discusses very briefly some possible refinements to the algorithm.

Section 8 presents some of the chess results discovered by the program, including the best play from a position that requires 243 moves to win.

Section 9 gives some implementation details and discusses run times.

2. Endgame Analysis by Humans and Computers

We start with a brief historical survey of the analysis of endgames, particularly those containing at most six pieces.

In listing the pieces of an endgame, the order will be White King, other White pieces, Black King, other Black pieces. Thus, ♔♖♚♘ is the same as ♔♖♚♞, and comprises the endgame of White King and White Rook against Black King and Black Knight.

2.1. Human analysis. Endgame analysis appears to date from at least the ninth century, with al-'Adlī's analysis of positions from ♔♖♚♘ and ♔♖♘♚♖ ['Adlī, plates 105 and 112]. However, the rules were slightly different in those days, as stalemate was not necessarily considered a draw. The oldest extant collection of compositions, including endgames, is the Alfonso manuscript, ca. 1250, which seems to indicate some interest during that time in endgame study [Perez 1929, pp. 111–112].

Modern chess is generally considered to have begun roughly with the publication of Luis Ramirez de Lucena's *Repetición de amores y arte de ajedrez* [Lucena 1497?]. Ruy Lopez de Sigura's book [Lopez 1561] briefly discusses endgame theory, but its main impact on this work would be the introduction of the controversial fifty-move rule [pp. 55–56], under which a game that contains fifty consecutive moves for each side without the move of a pawn or a capture could be declared drawn [Roycroft 1984].

Pietro Carrera's *Il gioco de gli scacchi* [Carrera 1617] discussed a number of fundamental endgames such as ♔♕♚♙♙, and certain six-piece endgames such as ♔♖♖♚♖♘ and ♔♖♖♚♖♙ [Book 3, pp. 176–178]. A number of other authors began developing the modern theory of endgames [Greco 1624; Stamma 1737; Philidor 1749?]. Giovanni Lolli's monumental *Osservazioni teorico-pratiche sopra il giuoco degli scacchi* [Lolli 1763] would be one of the most significant advances in endgame theory for the next ninety years [Roycroft 1972]. Lolli analyzed the endgame ♔♕♔♙♙, and he agreed with the earlier conclusion of Salvio [1634] that the endgame was a general draw. This assessment would stand substantially unchanged until Kenneth Thompson's computer analysis demonstrated a surprising 71-move win [Thompson 1986]. Notwithstanding this error, Lolli did discover the unique ♔♕♔♙♙ position in which White to play draws but Black to play loses [pp. 431–432].

Bernhard Horwitz and Josef Kling's *Chess Studies* [Kling and Horwitz 1851, pp. 62–66] contained a number of influential endgame studies, although their analysis of ♔♙♙♔♘ was questioned in [Roycroft 1972, p. 207]. The Horwitz and Kling assessment was definitively shown to be incorrect by two independent computer analyses [Thompson 1983; Roycroft 1984].

Alfred Crosskill [1864] gave an analysis of ♔♖♙♔♖ in which he claimed that more than fifty moves were required for a win; this was confirmed by computer analysis of Thompson. The Crosskill analysis was the culmination of a tradition of analysis of ♔♖♙♔♖ beginning at least from the time of Philidor [Philidor 1749?, pp. 165–169].

Henri Rinck and Aleksei Troitzky were among the most influential endgame composers of the next two generations. Troitzky is well-known for his analysis of ♔♘♘♔♙, in which he demonstrated that more than fifty moves are at times required for a win [Troitskiĭ 1934]. Rinck was a specialist in pawnless endgames, composing more than 500 such studies [Rinck 1950; Rinck and Malpas 1947], including some with six pieces. Troitzky summarized previous work in the area of ♔♘♘♔♙, beginning with a problem in ♔♘♔♙ from the thirteenth-century Latin manuscript *Bonus Socius* [Anonymous], and reserved particular praise for the systematic analysis of this endgame in an eighteenthth-century manuscript [Chapais 1780]. An early version of the program reported in the present paper resulted in the first published solution for this entire endgame [Stiller 1989].

The twentieth century saw the formal codification of endgame theory in works such as [Berger 1890; Chéron 1952; Euwe 1940; Fine 1941; Averbakh 1982], and many others. Some work focusing particularly on pawnless six-piece endings has also appeared, for example, [Roycroft 1967; Kopnin 1983; Berger 1922].

Currently the Informator *Encyclopedia of Chess Endings* series [Matanović 1985], which now uses some of Thompson's computer analysis, is a standard reference. The books [Nunn 1992; Nunn 1994] are based on that work.

Additional historical information can be found in [Hooper and Whyld 1992; Golombek 1976; Murray 1913; Roycroft 1972].

2.2. Friedrich Amelung and Theodor Molien. Friedrich Ludwig Amelung (born March 11, 1842; died March 9, 1909) was a Latvian chess player and author who edited the chess column of the Riga newspaper *Düna-Zeitung*. He studied philosophy and chemistry at the University of Dorpat from 1862 to 1879, and later became a private teacher and director of a mirror factory [Lenz 1970, p. 11; Düna-Zeitung 1909a; 1909b]. He published a number of endgame studies and analyses of endgames, and began a systematic study of pawnless endgames. For example, he explored the endgame ♔♕♔♖♘ in detail [Amelung 1901]; this endgame was shown to have unexpected depth, requiring up to 46 moves to win, in [Stiller 1989]. He also published an article [Amelung 1893] on ♔♗♘♔♘ and ♔♗♗♔♘, combinations that were not exhaustively analyzed until the 1980s [Thompson 1986; Stiller 1989].

However, his main interest to our work actually lies in two major projects: an analysis of the four-piece endgame ♔♖♔♘ [Amelung 1900], and his studies of certain pawnless six-piece endgames [Amelung 1902; 1908a; 1908b; 1908c].

Amelung's 1900 analysis of ♔♖♔♘ was significant because it contained the first histogram, to my knowledge, of a pawnless endgame or, for that matter, of any endgame [Amelung 1900, pp. 265–266]. This table listed the approximate number of positions in ♔♖♔♘ from which White could win and draw in 2–5 moves, 5–10 moves, 10–20 moves, and 20–30 moves. Such tables have been a mainstay of computer-age endgame analysis, of course. The existence of this early analysis does not appear to have been known to contemporary workers, although it appeared in a widely read and influential publication, the *Deutsche Schachzeitung*.

Even more intriguing, however, is Amelung's comment that an even earlier, exact, numerical analysis containing the number of win-in-*k* moves for each *k* of a four-piece chess endgame was known, and was due to "Dr. Th. Mollien, der Mathematiker von Fach ist"; that is, to the professor Th. Mollien.

Theodor Molien was born on September 10, 1861 in Riga, and died on December 25, 1941. (The biographical information here has been taken from [Kanunov 1983], which was translated for me by Boris Statnikov; see also [Bashmakova 1991]. Kanunov gives his name Фёдор Эдуардович Молин, or Fyodor Eduardovich Molin; we write Theodor Molien, in conformity with his publications. Amelung variously used Mollin, Mollien and Molien.)

His father, Eduard, was a philologist and teacher, and Theodor eventually became fluent in a number of languages, including Hebrew, Greek, Latin, French, Italian, Spanish, Portuguese, English, Dutch, Swedish, and Norwegian, as well as German and Russian, of course. "Read a hundred novels in a language," Molien liked to say, "and you will know that language."[1] He studied celestial mechanics at Dorpat University (1880–1883) and also took courses from Felix Klein in Leipzig (1883–1885).

He studied celestial mechanics at Dorpat University (1880–1883) and took courses from Felix Klein in Leipzig (1883–1885). His doctoral dissertation, published in *Mathematische Annalen* [Molien 1893], proved a number of the fundamental structure theorems of group representation theory, including the decomposability of group algebras into direct sums of matrix algebras.

---

[1] «Прочитайте сто романов на каком-либо языке,—любил говорить он позднее,—и Вы будете знать этот язык.» [Kanunov 1983, p. 9].

Molien's early papers on group representation theory [1893; 1897a; 1897b; 1898], despite their importance, were obscure and difficult to understand. They anticipated Frobenius' classic paper on the determinant of a group-circulant matrix [Frobenius 1896], a fact that Frobenius readily admitted [Hawkins 1974], although he had tremendous difficulty understanding Molien's work (letter to Alfred Knezer, May 6, 1898). Referring to [Molien 1893], Frobenius wrote to Dedekind:

> You will have noticed that a young mathematician, Theodor Molien in Dorpat, has independently of me considered the group determinant. He has published ... a very beautiful but hard-to-read work "On systems of higher complex numbers", in which he investigated non-commutative multiplication and obtained important general results of which the properties of the group determinant are special cases.[2]

(This letter of February 24, 1898 was kindly supplied by Thomas Hawkins, in a transcription made by Walter Kaufmann-Bühler; it is quoted here with the permission of Springer-Verlag. I have benefited from Hawkins' translation in supplying mine.)

Despite these results, and despite Frobenius' support, Molien was rejected from a number of Russian academic positions, partly because of the Czarist politics of the time (according to Kanunov) and, at least in one case, because the committee considered his work too theoretical and without practical applications [Kanunov 1983, pp. 35–36]. After studying medieval mathematical manuscripts at the Vatican Library in 1899, he accepted a post at the Tomsk Technological Institute in Siberia, where he was cut off from the mathematical mainstream and became embroiled in obscure administrative struggles (he was, in fact, briefly fired). His remaining mathematical work had little influence and he spent most of his time teaching.

As a consequence, Molien's work was unknown or underestimated in the West for a long while; for example, the classic text [Wussing 1969] barely mentions him. With the publication of Hawkins' series of articles [1971; 1972; 1974] on the history of group representation theory, the significance of his work became better-known, and more recent historical appraisals gives him due credit [van der Waerden 1985, pp. 206–209, 237–238].

Not mentioned in Kanunov's biography is that, before Molien moved to Tomsk, he was one of the strongest chess players in Dorpat and was particularly known for his blindfold play (Ken Whyld, personal communication, 1995). He was president of the Dorpat chess club, and several of his games were published in

---

[2] "Sie werden bemerkt haben, daß sich ein jungerer Mathematiker Theodor Molien in Dorpat unabhängig von mir mit der Gruppendeterminante beschäftigt hat. Er hat im 41. Bande der Mathematischen Annalen eine sehr schöne, aber schwer zu lesende Arbeit 'Ueber Systeme höherer complexer Zahlen' veröffentlicht, worin er die nicht commutative Multiplication untersucht hat und wichtige allgemeine Resultate erhalten hat, von denen die Eigenschaften der Gruppendeterminant specielle Fälle sind."

a Latvian chess journal, *Baltische Schachblätter*, edited, for a time, by Amelung [Balt. Schachbl. 1893; 1902, p. 8]; one of the games he lost fetched a "best-game" prize in the main tournament of the Jurjewer chess club in 1894 [Molien 1900].

In 1898 Molien published four chess studies [Molien 1898a] based on his research into the endgame ♔♖♔♗ [Amelung 1900, p. 5]. These numerical studies are alluded to several times in the chess journals of the time [Amelung 1898; 1909, pp. 5, 265; Molien 1901], but I have not been able to locate a publication of his complete results, despite the historical significance of such a document.

It is an interesting coincidence that within a span of a few years Molien performed groundbreaking work in the two apparently unrelated areas of group representation theory and quantitative chess endgame analysis, although his work in both areas was mostly ignored for a long time. There is, perhaps, some mathematical affinity between these areas as well, since, as we shall see, the chess move operators can be encoded by a group-equivariant matrix; rapid multiplication of a group-equivariant matrix by a vector, in general, relies on the algebra-isomorphism between a group algebra and a direct sum of matrix algebras first noted by Molien [Clausen and Baum 1993; Diaconis 1988; Diaconis and Rockmore 1990; Karpovsky 1977]; massively parallel implementations are described in [Stiller 1995, Chapter 7].

We now continue with our discussion of chess endgame history proper, particularly Amelung's work on pawnless endgames, of which his work on ♔♖♗♔♘♘ deserves special mention.

Partly in response to the first edition of the influential manual of endings [Berger 1890, pp. 167–169], in 1902 Amelung published a three-part series in *Deutsche Schachzeitung*, perhaps the premier chess journal of its time, analyzing the endings of King, Rook and minor piece (♘ or ♗) against King and two minor pieces [Amelung 1902], and representing a continuation of Amelung's earlier work with Molien on the endgame ♔♖♔♘ [Amelung 1900]. Amelung indicated that the case ♔♖♗♔♘♘ was particularly interesting, and in 1908 he published a short article on the topic in *Für Haus und Familie*, a biweekly supplement to the Riga newspaper *Düna-Zeitung*, of which he was the chess editor [Amelung 1908a].

Amelung's interest in this endgame was so great that he held a contest in *Düna-Zeitung* for the best solution to a particular example of this endgame [Amelung 1908c]. A solution was published the next year, but Amelung died that year and was unable to continue or popularize his research. Consequently, succeeding commentators dismissed many of his more extreme claims, and his work seemed to have been largely forgotten. It is discussed in [Berger 1922, p. 223–233], but it was criticized by the mathematician and chess champion Machgielis (Max) Euwe in his titanic study of pawnless endgames [Euwe 1940, Volume 5, pp. 50–53]:

"This endgame [♔♖♗♔♘♘] offers the stronger side excellent winning chances. F. Amelung went so far as to say that the defense was hopeless, but this assessment seems to be untrue."[3]

The *Düna-Zeitung* has turned out to be an elusive newspaper; I was not able to locate any references to it in domestic catalogues and indices. The only copy I was able to find was archived at the National Library of Latvia. In addition to the remark about Molien, the research reported here argues for a renewed appreciation of the accuracy and importance of Amelung's work.

2.3. Computer endgame analysis. Although some have dated computer chess from Charles Babbage's brief discussion of automated game-playing in 1864, his conclusion suggests that he did not appreciate the complexities involved:

> In consequence of this the whole question of making an automaton play any game depended upon the possibility of the machine being able to represent all the myriads of combinations relating to it. Allowing one hundred moves on each side for the longest game at chess, I found that the combinations involved in the Analytical Engine enormously surpassed any required, even by the game of chess. [Babbage 1864, p. 467]

Automated endgame play appears to date from the construction by Leonardo Torres-Quevedo of an automaton to play ♔♖♔ endings. Although some sources give 1890 as the date in which this machine was designed, it was exhibited at about 1915 [Bell 1978, pp. 8–11; Simons 1986]. According to [Scientific American 1915, p. 298], "Torres believes that the limit has by no means been reached of what automatic machinery can do, and in substantiation of his opinions presents his automatic chess-playing machine."

Unlike most later work, Torres-Quevedo's automaton could move its own pieces. It used a rule-based approach [Scientific American 1915; Torres-Quevedo 1951], like that of [Huberman 1968]. By contrast, we are concerned with exhaustive analysis of endgames, in which the value of each node of the state-space is computed by backing up the game-theoretic values of the leaves.

The mathematical justification for the retrograde analysis chess algorithm was already implicit in [Zermelo 1913]. Additional theoretical work was done by John von Neumann and Oskar Morgenstern [1944, pp. 124–125].

The contemporary dynamic programming methodology, which defines the field of retrograde endgame analysis, was discovered by Richard Bellman [1965]. (Strangely enough, this article is not generally known to the computer game community, and is not included in the comprehensive bibliography by van den

---

[3] "Dit eindspel biedt de sterkste partij zeer goede winstkansen. F. Amelung ging zelfs zoo ver, dat hij de verdediging als kansloos beschouwde, maar deze opvatting schijnt ojuist te zijn." (Translation from the Dutch by Peter Jansen.)

Herik and I. S. Herschberg [1986]. Bellman's work was the culmination of work going back several years:

Checkers and Chess. Interesting examples of processes in which the set of all possible states of the system is indescribably huge, but where the deviations are reasonably small in number, are checkers and chess. In checkers, the number of possible moves in any given situation is so small that we can confidently expect a complete digital computer solution to the problem of optimal play in this game. In chess, the general situation is still rather complex, but we can use the method described above to analyze completely all pawn-king endings, and probably all endings involving a minor piece and pawns. Whether or not this is desirable is another matter [Bellman 1961, p. 3].

Bellman [1954; 1957] had considered game theory from a classical perspective as well, but his work came to fruition with [Bellman 1965], where he observed that the entire state-space could be stored and that dynamic programming techniques could then be used to compute whether either side could win any position. Bellman also sketched how a combination of forward search, dynamic programming, and heuristic evaluation could be used to solve much larger state spaces than could be tackled by either technique alone. Bellman predicted that checkers could be solved by his techniques, and the utility of his algorithms for solving very large state spaces has been validated by Jonathan Schaeffer et al. for checkers [Lake et al. 1994; Schaeffer et al. 1992; Schaeffer and Lake 1996] and Ralph Gasser for Nine Men's Morris [Gasser 1991; 1996]. On the other hand, $4 \times 4 \times 4$ tic-tac-toe has been solved by Patashnik [1980] using forward search and a variant of isomorph-rejection based on the automorphism group computation of Silver [1967].

E. A. Komissarchik and A. L. Futer [1974] studied certain special cases of ♔♕♙♚♛, although they were not able to solve the general instance of such endgames. J. Ross Quinlan [1979; 1983] analyzed ♔♖♚♞ from the point of view of a machine learning testbed. Hans Berliner and Murray S. Campbell [1984] studied the Szén position of three connected passed pawns against three connected passed pawns by simplifying the promotion subgames. Campbell [1988] has begun to extend this idea to wider classes of endgames. Jansen [1992a; 1992b; 1992c] has studied endgame play when the opponent is presumed to be fallible. H. Jaap van den Herik and coworkers have produced a number of retrograde analysis studies of various four-piece endgames, or of endgames with more than 4 pieces whose special structure allows the state-space size to be reduced to about the size of the general four-piece endgame [Dekker et al. 1987; van den Herik and Herschberg 1985a; 1988]. Danny Kopec has written several papers in the area as well [Kopec et al. 1988].

The first retrograde analysis of general five-piece endgames with up to one pawn was published in [Thompson 1986]. The significance of this work was twofold. First, many more moves were required to win certain endgames than had previously been thought. Second, Thompson's work invalidated generally accepted theory concerning certain five-piece endgames by demonstrating that certain classes of positions that had been thought to be drawn were, in fact, won. The winning procedure proved to be quite difficult for humans to understand [Michie and Bratko 1987]. The pawnless five-piece work of Thompson was extended to all pawnless five-piece endgames and many five-piece endgames with one pawn by an early version of the program discussed in this paper.

3. Parallel Processing

The motivation for using parallel processing is to achieve increased computation bandwidth by using large numbers of inexpensive processors [van de Velde 1994; Stone 1993]. In particular, it is hoped that the so-called "von Neumann bottleneck" between the CPU and the memory of a standard serial computer could be alleviated by massive parallelism [Hwang and Briggs 1985]. There are, of course, many trade-offs that can be made in the architecture of a computer, such the type of interconnection network, the granularity of the processors, and whether each processor executes the same instruction (SIMD) or different instructions (MIMD) [Hillis 1985; Valiant 1990a; 1990b; Blelloch 1990].

One common form of interconnection network is the hypercube [Leighton 1992]. Consider a parallel computer with 2^n processors, each with some local memory. Place each processor at a distinct vertex of the unit cube in Euclidean n-space, and imagine each edge of the cube as a wire that directly connects the processors at its endpoints. Two processors connected by an edge can communicate directly in a single timestep. Each processor in an n-dimensional hypercube can be viewed as having a length n binary address, with processors connected if and only if their addresses differ in exactly one bit location.

The hypercube can compute the effect of the *end-off shift* matrix E_8. Applied to an array, E_8 shifts each element of the array down, filling the initial element with 0:

$$E_8 \begin{pmatrix} v_1 \\ v_2 \\ \vdots \\ v_8 \end{pmatrix} = \begin{pmatrix} 0 \\ v_1 \\ \vdots \\ v_7 \end{pmatrix} \qquad (3.1)$$

The computation is performed by Gray coding the coordinates of the elements of the array v so that v_i and v_{i+1} are physically adjacent in the hypercube [Gilbert 1958]; see Figure 1. Higher-dimensional shift patterns are also easy to compute on a hypercube [Ho and Johnsson 1990].

It is also possible to perform arbitary permutations on a hypercube, although we shall not discuss the techniques required for that here. In practice general

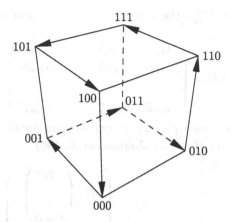

Figure 1. Left multiplication by an end-off shift matrix can be computed by Gray coding the coordinates of each element of the array. This figure illustrates a three-bit Gray code, which can be thought of as an embedding of a cycle of length eight into a three-cube.

processor permutation is typically performed with hardware assistance [Nassimi and Sahni 1982; Leiserson et al. 1992].

4. Parallel Processing and Tensor Products

This section briefly summarizes some previous work on the application of tensor products to parallel processing, particularly to the parallel and vectorized computation of fast Fourier transform. The chess algorithm will be developed in the next section by generalizing this approach.

4.1. Mathematical preliminaries. Let $\mathfrak{V}_n$ be the space of length-n vectors with entries in a field $\mathfrak{F}$. We let $\{e_i^n\}_{i=1}^n$ be the standard basis consisting of vectors (having one component 1 and all others 0). An element of $\mathfrak{V}_n$ may be thought of as a length-n array whose elements are in $\mathfrak{F}$, or as an $n \times 1$ matrix over $\mathfrak{F}$ [Marcus 1973].

The mn basis elements of $\mathfrak{V}_n \otimes \mathfrak{V}_m$, $\{e_i^n \otimes e_j^m\}_{i=0,j=0}^{n-1,m-1}$, are ordered by

$$e_i^n \otimes e_j^m \mapsto e_{mi+j}^{mn}.$$

In this manner an element of $\mathfrak{V}_n \otimes \mathfrak{V}_m$ may be considered to be a vector of length mn with elements drawn from $\mathfrak{F}$. Let $\mathfrak{M}_m^n$ be the space of $n \times m$ matrices over $\mathfrak{F}$. In the following, a linear transformation will be identified with its matrix representation in the standard basis. Let $\mathfrak{M}_n = \mathfrak{M}_n^n$. Let I_n be the $n \times n$ identity transformation of $\mathfrak{V}_n$.

Write $\mathrm{diag}(v_0, v_1, \ldots, v_{n-1}) \equiv \mathrm{diag}(v)$ for the diagonal matrix in $\mathfrak{M}_n$ whose diagonal elements are taken from the coordinates of v.

If $A \in \mathfrak{M}_m^n$ and $B \in \mathfrak{M}_{m'}^{n'}$, the matrix of the tensor product $A \otimes B \in \mathfrak{M}_{mm'}^{nn'}$ is given by

$$A \otimes B = \begin{pmatrix} A_{11}B & A_{12}B & \cdots & A_{1m}B \\ A_{21}B & A_{22}B & \cdots & A_{2m}B \\ \vdots & \vdots & \ddots & \vdots \\ A_{n1}B & A_{n2}B & \cdots & A_{nm}B \end{pmatrix}$$

The importance of the tensor-product to our work in parallel processing lies in the following identity, for $B \in \mathfrak{M}_m$ [Johnson et al. 1991]:

$$(I_n \otimes B) \begin{pmatrix} v_0 \\ v_1 \\ \vdots \\ v_{nm-1} \end{pmatrix} = \begin{pmatrix} B \cdot \begin{pmatrix} v_0 \\ \vdots \\ v_{m-1} \end{pmatrix} \\ B \cdot \begin{pmatrix} v_m \\ \vdots \\ v_{2m-1} \end{pmatrix} \\ \vdots \\ B \cdot \begin{pmatrix} v_{(n-1)m} \\ \vdots \\ v_{nm-1} \end{pmatrix} \end{pmatrix} \quad (4.1)$$

Suppose $n = ml$. The n-point stride l permutation matrix P_l^n is the $n \times n$ matrix defined by $P_l^n (v \otimes w) = w \otimes v$, where $v \in \mathfrak{V}_m$ and $w \in \mathfrak{V}_l$. The effect of P_l^n on a vector is to stride through the vector, taking m steps of size l. For example, taking $m = 3$, $l = 2$, and $n = 6$, we have:

$$P_2^6 \begin{pmatrix} v_0 \\ v_1 \\ v_2 \\ v_3 \\ v_4 \\ v_5 \end{pmatrix} = \begin{pmatrix} v_0 \\ v_2 \\ v_4 \\ v_1 \\ v_3 \\ v_5 \end{pmatrix}$$

Stride permutations are important due to the following *Commutation Theorem* [Tolimieri et al. 1989]:

THEOREM 4.1. *If $A \in \mathfrak{M}_m$, $B \in \mathfrak{M}_l$, and $n = ml$, we have*

$$P_l^n (A \otimes B) P_m^n = B \otimes A.$$

This theorem, which is easy to prove even when the entries are from a semiring, allows the order of evaluation in a tensor product to be varied. We shall see in Section 4.2 that some evaluation orders naturally correspond to vectorization, and some to parallelizations; the Commutation Theorem will be the method by which one type of execution is traded off for another.

4.2. Code generation: Conversion from factorization to code. We now describe the relationship between the matrix representation of a formula and the denoted machine code. Because many of the algorithms to be presented will be presented in the tensorial manner, with the code-generation phase only represented implicitly, this section is fundamental to this work.

The matrix notation we use is nothing more than an informal notation for describing algorithms. It differs from standard notations primarily in its explicit denotation of data distribution, communication, and operation scheduling. Whereas most high-level languages, and even special-purpose parallel languages, leave the distribution of data over the processors and the scheduling of operations within processors to the discretion of the compiler, the notation we use, at least potentially, encodes all such scheduling. This has both advantages and disadvantages: although it gives the programmer a finer level of control, which can be important for time-critical applications, it requires some conscious decision-making over data-distribution that is unnecessary in some other languages. On the other hand, the functional nature of the notation does make it potentially amenable to compiler reordering. The most serious disadvantage is its narrowness of application. Originally developed for signal processing codes, this work demonstrates its wider application, but there are many applications which would not easily fall under its rubric.

The target architecture of the language is a machine comprising m parallel processors, each with shared memory. However, it is easy to see that the results go through also, with an extra communication step or two, on local-memory machines. Each processor may also have vector capabilities, so that computations within the processors should be vectorized. We do not assume restrictions on the vector length capability of the processors.

User data is always stored conceptually in the form of a vector

$$\begin{pmatrix} v_0 \\ v_1 \\ \vdots \\ v_{n-1} \end{pmatrix}.$$

Assuming that m divides n, elements $v_0, \ldots, v_{(n/m)-1}$ are stored in processor 0, elements $v_{n/m}, \ldots, v_{(2n/m)-1}$ in processor 1, and so on. Matrices are stored in column-major order. It is assumed that certain general classes of specific matrices are already implemented on the architecture, in particular, the stride permutations and any specific permutations corresponding to the interconnection network.

Suppose $B \in \mathfrak{M}_l$, and let $\mathbf{code}(B)$ be any sequence of machine instructions that computes the result of left-multiplication by B. That is, $\mathbf{code}(B)$ is a program that takes as input an array v of l elements of $\mathfrak{F}$, and returns as output the array $B \cdot v$ of l elements of $\mathfrak{F}$, where vectors are identified with their coordinates in the standard basis.

Given $\mathbf{code}(B)$ and $\mathbf{code}(B')$ for two matrices B and B', it is easy to compute some $\mathbf{code}(B + B')$. Simply let $\mathbf{code}(B + B')$ be the program that, given its input array $\boldsymbol{v}$, first runs as a subroutine $\mathbf{code}(B)$ on $\boldsymbol{v}$ (saving the result), then runs $\mathbf{code}(B')$ on $\boldsymbol{v}$, and then returns the coordinate-wise sum of the arrays that are returned by these two subroutine calls.

Similarly, given $\mathbf{code}(M)$ and $\mathbf{code}(M')$, it is easy to find $\mathbf{code}(M \cdot M')$, assuming the dimensions of M and M' are compatible: run $\mathbf{code}(M)$ on the result of running $\mathbf{code}(M')$ on the argument $\boldsymbol{v}$.

Of course, $\mathbf{code}(I_l)$ is the code that returns its argument, an l-vector.

Consider a parallel processor with m processors, $p_1, \ldots, p_m$, each with some local memory. We make the convention that a length ml array will be stored with its first l elements in processor p_1, its second l elements in processor p_2, and so on.

Given this convention, one can interpret $\mathbf{code}(I_m \otimes B)$ as code that runs on this m-processor architecture. To construct $\mathbf{code}(I_m \otimes B)$, load $\mathbf{code}(B)$ in each p_i. When called on a length ml array $\boldsymbol{v}$, p_i runs $\mathbf{code}(B)$ on the l elements of $\boldsymbol{v}$ that are stored in its local memory, and outputs the result to its local memory. Equation (4.1) shows that this will compute the tensor product. Similar rules can be derived when the number of processors is different from m.

The code corresponding to $A \otimes I_l$, for $A \in \mathfrak{M}_m$, is a bit more subtle. The interpretation of $\mathbf{code}(A \otimes I_l)$ is as the code corresponding to A, except that it operates on l-vectors rather than on scalars. This code can be constructed (loosely speaking) from $\mathbf{code}(A)$ by interpreting the length ml argument array $\boldsymbol{v}$ as being an element of the m-module over the ring $\mathfrak{F}^l$. This corresponds closely to hardware primitives on certain vector architectures.

The relation $A \otimes B = (A \otimes I_l)(I_m \otimes B)$ can be used to compute general tensor products.

By combining a fixed set of transformations reflecting the hardware primitives of the underlying architecture with combining rules like $+$, $\cdot$ and $\otimes$, and some simple tensor product identities, one can define concise expressions that can be translated into efficient code for certain classes of functions [Granata et al. 1992].

4.3. Fast Fourier transforms. We discuss briefly the formulation of the FFT in the tensor product framework presented above [Tolimieri et al. 1993, pp. 16–20]. The presentation is intended to illustrate the parallel code development methodology used to describe parallelization of the chess endgame algorithm, in Section 6.1 and Table 1. The exposition of the chess material, however, does not depend on any of the results here.

Let F_n be the n-dimensional Fourier transform matrix $(\omega^{ij})_{i,j=0}^{n-1}$, where ω is a primitive n-th root of unity.

Let $n = ml$. The Singleton [1967] mixed-radix version of the Cooley–Tukey fast Fourier transform [Cooley and Tukey 1965] can be expressed recursively by

$$F_n = (F_m \otimes I_l)T_l(I_m \otimes F_l)P_m^n,$$

where T_l is a diagonal matrix encoding the twiddle factors:

$$T_l = \bigoplus_{j=0}^{m-1} \left(\mathrm{diag}(1, \omega, \dots, \omega^{l-1}) \right)^j.$$

This can be interpreted as a mixed parallel/vector algorithm. Given an input vector v, $P_m^n v$ forms a list of m segments, each of length l. The $I_m \otimes F_l$ term performs m l-point FFTs in parallel on each segment. T_l just multiplies each element by a twiddle factor. Finally, the $F_m \otimes I_l$ term performs an m-point FFT on vectors of size l.

The commutation theorem can be used to derive a parallel form

$$F_n = P_m^n \left(I_l \otimes F_m \right) P_l^n T_l \left(I_m \otimes F_l \right) P_m^n, \tag{4.2}$$

and a vector form

$$F_n = \left(F_m \otimes I_l \right) T_l P_m^n \left(F_l \otimes I_m \right). \tag{4.3}$$

The parallel Pease FFT [Pease 1968] can be derived by unrolling the recursion in (4.2), and the vectorized Korn–Lambiotte FFT [Korn and Lambiotte 1979] can be derived by unrolling (4.3).

By using the commutation theorem and varying the factorization, many different FFT algorithms have been derived, with different tradeoffs between parallelization and vectorization [Chamberlain 1988; Averbuch et al. 1990; Johnson 1989; Granata et al. 1991; Auslander and Tolimieri 1985].

5. Application to Chess

This section describes the chess endgame algorithm in a generalization of the tensor product formalism described in Section 4.

Imagine, for the moment, that a matrix over $\mathfrak{GF}_2$ is actually a Boolean matrix whose entries are taken from the Boolean algebra $\{0, 1, \vee, \wedge\}$. We write $+$ and $\cdot$ for $\vee$ and $\wedge$. The notion of linear transformations then changes, as does, therefore, $\mathfrak{M}_m^n$, in the natural way.

This generalization has been used for expressing graph algorithms [Backhouse and Carré 1975; Lehmann 1977; Abdali and Saunders 1985; Tarjan 1981a; 1981b]. The definitions of $\otimes$, matrix product, and matrix sum remain essentially unchanged.

In particular, the commutation theorem, the notion of code(M), and the relation between $\otimes$ and parallelization still holds.

These ideas could, of course, be presented categorically using the approach of [Skillicorn 1993; Bird et al. 1989], or using the mathematics-of-arrays formalism of [Mullin 1988].

5.1. Definitions. For simplicity of exposition, captures, pawns, stalemates, castling, and the fifty-move rule will be disregarded unless otherwise stated.

Let S be an ordered set of k chess pieces. For example, if $k = 6$ one could choose $S = \langle \text{♟}, \text{♔}, \text{♕}, \text{♖}, \text{♛}, \text{♜} \rangle$.

An S-*position* is a chess position that contains exactly the k pieces in S. We write $S = \langle S_1, S_2, \dots, S_k \rangle$. An S-position can be viewed as an assignment of each piece $S_i \in S$ to a *distinct* square of the chessboard (note that captures are not allowed).

We denote by $\mathfrak{V}_n$ the space of length-n Boolean vectors. The space of 8×8 Boolean matrices is thus

$$\mathfrak{C} \equiv \mathfrak{V}_8 \otimes \mathfrak{V}_8.$$

Let $\{e_i\}_{i=1}^8$ be the standard basis for $\mathfrak{V}_8$, and $\bigotimes^j \mathfrak{V}$ the j-th tensor power of $\mathfrak{V}$.

Let $\mathfrak{B} \equiv \bigotimes^k \mathfrak{C}$ be the *hyperboard* corresponding to S. It can be thought of as a cube of side-length 8 in $\mathbb{R}^{2k}$. Each of the 64^k basis elements corresponds to a point with integer coordinates between 1 and 8.

Each basis element of $\mathfrak{C}$ is of the form $e_i \otimes e_j$ for $1 \leq i, j \leq 8$. Any such basis element, therefore, denotes a unique square on the 8×8 chessboard. Any element of $\mathfrak{C}$ is a sum of distinct basis elements, and therefore corresponds to a set of squares [White 1975].

Each basis element of $\mathfrak{B}$ is of the form $c_1 \otimes c_2 \otimes \cdots \otimes c_k$, where each c_s is some basis element of $\mathfrak{C}$. Since each c_s is a square on the chessboard, each basis element of $\mathfrak{B}$ can be thought of as a sequence of k squares of the chessboard. Each position that is formed from the pieces of S is thereby associated with a unique basis element of $\mathfrak{B}$. Any set of positions, each of which is formed from pieces of S, is associated with a unique element of $\mathfrak{B}$: the sum of the basis elements corresponding to each of the positions from the set.

This correspondence between sets of chess positions and elements of $\mathfrak{B}$ forms the link between the chess algorithms and the tensor product formulation. In the following, the distinction between sets of chess positions formed from the pieces in S and elements of the hyperboard $\mathfrak{B}$ will be omitted when the context makes the meaning clear.

If $p \in \{\text{♔}, \text{♕}, \text{♖}, \text{♗}, \text{♘}\}$ is a piece, the *unmove operator* $\mathrm{III}_{p,s}$ is the function that, given an S-position P returns the set of S-positions that could be formed by unmoving S_s in P as if S_s were a p.

$\mathrm{III}_{p,s}$ can be extended to a linear function from elements of $\mathfrak{B}$ to itself, and thereby becomes an element of $\mathfrak{M}_{64^k}$. (Technically, the unmove operators are only quasilinear, since the Boolean algebra is not a ring, and thus $\mathfrak{B}$ is not a module. However, we need not worry about this distinction.)

The core of the chess endgame algorithm is the efficient computation of the $\mathrm{III}_{p,s}$. We now describe a factorization of $\mathrm{III}_{p,s}$ in terms of primitive operators. The ideas of Section 4.2 may then be used to derive efficient parallel code from this factorization.

5.2. Group actions. We introduce a few group actions [Fulton and Harris 1991]. We will use the group-theoretic terminology both to give concise descriptions of certain move operators and to describe the exploitation of symmetry. There is a close correspondence between multilinear algebra, combinatorial enumeration, and group actions which motivates much of this section [Merris 1980; 1981; 1992; Merris and Watkins 1983].

The symmetric group on k elements $\mathfrak{S}_k$ acts on $\mathfrak{B}$ by permuting the order of the factors: $\mathfrak{s} \bigotimes_{s=1}^{k} c_s = \bigotimes_{s=1}^{k} c_{\mathfrak{s}s}$, for $\mathfrak{s} \in \mathfrak{S}_k$ and $c_s \in \mathfrak{C}$.

The dihedral group of order 8, $\mathfrak{D}_4$, is the group of symmetries of the square. It is generated by two elements $\mathfrak{r}$ and $\mathfrak{f}$ with relations $\mathfrak{r}^4 = \mathfrak{f}^2 = \mathfrak{e}$ and $\mathfrak{r}^3\mathfrak{f} = \mathfrak{f}\mathfrak{r}$. It acts on $\mathfrak{C}$ by

$$\mathfrak{r}(e_i \otimes e_j) = e_{8-j+1} \otimes e_i,$$
$$\mathfrak{f}(e_i \otimes e_j) = e_i \otimes e_{8-j+1}.$$

Thus, $\mathfrak{r}$ rotates the chessboard counterclockwise $90°$ and $\mathfrak{f}$ flips the chessboard about the horizontal bisector. $\mathfrak{D}_4$ acts diagonally on $\mathfrak{B}$ by

$$\mathfrak{d}\bigotimes_{s=1}^{k} c_s = \bigotimes_{s=1}^{k} \mathfrak{d}c_s$$

Let $\mathfrak{C}_4$ be the cyclic group generated by $\mathfrak{r}$.

A group $\mathfrak{G}$ acting on $\mathfrak{V}_n$ and $\mathfrak{V}_m$ acts on $\mathfrak{M}_n^m$ by conjugation: $(\mathfrak{g}M)v = \mathfrak{g}(M\mathfrak{g}^{-1}(v))$. We let

$$\int_{\mathfrak{G}} x = \sum_{\mathfrak{g} \in \mathfrak{G}} \mathfrak{g}x.$$

The notation $\int_{\mathfrak{G}} x$ is intended to represent the group average of x with respect to $\mathfrak{G}$ [Fulton and Harris 1991, p. 6]. It is a fixed point of the $\mathfrak{G}$ action: $\mathfrak{g}\int_{\mathfrak{G}} x = \int_{\mathfrak{G}} x$ for all $\mathfrak{g} \in \mathfrak{G}$.

6. Endgame Algorithm

We now present the endgame algorithm using the notation developed in Section 5. Section 6.1 gives the fundamental factorization. Section 6.2 describes the modification of the equations of Table 1 to exploit symmetry. Section 6.3 describes the control structure of the algorithm.

6.1. Factoring the unmove operator. Recall from (3.1) that E_8 was defined to be the unit one-dimensional 8×8 end-off shift matrix. The unit multidimensional shift along dimension s is defined by

$$U_s \in \mathfrak{M}_{64^k} \equiv I_{64^{s-1}} \otimes (E_8 \otimes I_8) \otimes I_{64^{k-s}}.$$

Such multidimensional shifts are commonly used in scientific computation.

$$\text{III}_{\text{♖},s} = \int_{\mathfrak{C}_4} LU_s(I_{64^k} + LU_s)^6$$

$$\text{III}_{\text{♘},s} = L\int_{\mathfrak{D}_4} U_s \cdot (\mathfrak{r}(U_s^2))$$

$$\text{III}_{\text{♗},s} = \int_{\mathfrak{D}_4} LU_s(I_{64^k} + LU_s\mathfrak{r}U_s)^6$$

$$\text{III}_{\text{♔},s} = L\int_{\mathfrak{C}_4} U_s + U_s\mathfrak{r}U_s$$

$$\text{III}_{\text{♕},s} = \text{III}_{\text{♖},s} + \text{III}_{\text{♗},s}$$

Table 1. These equations define the core of a portable endgame algorithm. By modifying the factorizations, code suitable for execution on a wide range of high-performance architectures can be derived.

Fix a basis $\{c_i\}_{i=1}^{64}$ of $\mathfrak{C}$, and define

$$L \in \mathfrak{M}_{64^k} \equiv \text{diag}\left(\int_{\mathfrak{S}_k} \sum_{i_1 < \cdots < i_k} c_{i_1} \otimes \cdots \otimes c_{i_k}\right)$$

Certain basis elements of $\mathfrak{B}$ do not correspond to legal S-positions. These "holes" are elements of the form $\bigotimes_{s=1}^{k} c_s$ such that there exist distinct s, s' for which $c_s = c_{s'}$. If $v \in \mathfrak{B}$ then Lv is the projection of v onto the subspace of $\mathfrak{B}$ generated by basis elements that are not holes.

Table 1 defines the piece-unmove operators. Figure 2 illustrates the computation of the integrand in the expression for $\text{III}_{\text{♖},1}$ in Table 1.

This corresponds to moving the ♖ to the right. The average over $\mathfrak{C}_4$ means that the ♖ must be moved in 4 directions. For example, conjugation by $\mathfrak{r}$ of the operation of moving the ♖ right corresponds to moving the ♖ up: if one rotates the chessboard clockwise 90°, moves the ♖ right, and then rotates the chessboard counterclockwise 90°, the result will be the same as if the ♖ had been moved up to begin with.

As in the case of fast Fourier transforms (4.2) and (4.3), by varying the factorization, one can derive code suitable for different architectures. For example, if the interconnection architecture is a two-dimensional grid, only U_s for $s = 1$ can be directly computed. By using the relations $U_s = (1\,s)U_1$ and $\text{III}_{\mathbf{p},s} = (1\,s)\text{III}_{\mathbf{p},1}$, where $(1\,s) \in \mathfrak{S}_k$ interchanges 1 and s, equations appropriate for a grid architecture can be derived.

These equations are vectorizable as well [Smitley 1991]. The vectorized implementation of Table 1 by Burton Wendroff et al. has supported this claim [Wendroff et al. 1993].

Other factorizations appropriate for combined vector and parallel architectures, such as a parallel network of vector processors, can also be derived [Kaushik et al. 1993].

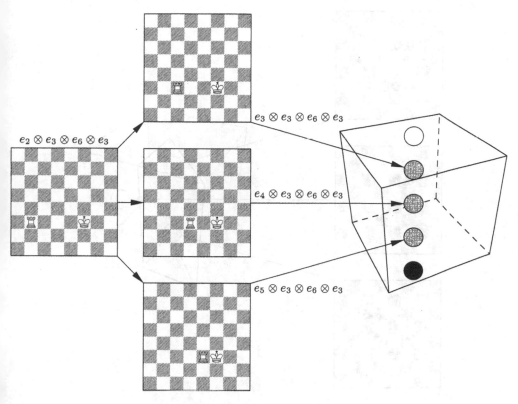

$e_2 \otimes e_3 \otimes e_6 \otimes e_3$

$e_3 \otimes e_3 \otimes e_6 \otimes e_3$

$e_4 \otimes e_3 \otimes e_6 \otimes e_3$

$e_5 \otimes e_3 \otimes e_6 \otimes e_3$

Figure 2. Unmoving the ♖ to the right from the position on the left results in the three positions shown in the center. Here, $S = \langle$♖, ♔$\rangle$. Each position corresponds to a point in the hyperboard, shown on the right. The position $e_6 \otimes e_3 \otimes e_6 \otimes e_3$ is illegal and is zeroed out by L.

6.2. Exploiting symmetry.

The game-theoretic value of a chess position without pawns is invariant under rotation and reflection of the chessboard. Therefore, the class of positions considered can be restricted to those in which the ♚ is in the lower left-hand octant, or fundamental region, of the chessboard, as shown in Figure 3. These positions correspond to points in a triangular wedge in the hyperboard.

Algebraically, because each $\text{III}_{\mathbf{p},s}$ is a fixed point of the $\mathfrak{D}_4$ action, we need only consider the $10 \cdot 64^{k-1}$-space $\mathfrak{B}' \equiv \mathfrak{C}/\mathfrak{D}_4 \otimes \bigotimes^{k-1}\mathfrak{C}$, rather than the bigger

Figure 3. The chessboard may be rotated $90°$ or reflected about any of its bisectors without altering the value of a pawnless position. Therefore, the ♚ may be restricted to lie in one of the ten squares shown.

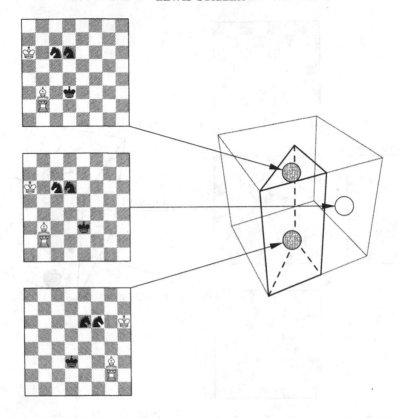

Figure 4. Only an eighth of the hyperboard is physically stored. When the ♚ is moved outside the squares marked in Figure 3, as in going from the top to the middle positions, we apply a symmetry transformation that puts the ♚ back in the allowed area; this is represented by the position at the bottom. These three positions correspond to three points in the hyperboard, only the first and third of which are physically stored. The Black-to-move position at the top requires 222 moves against best play for White to win (see Table 2).

64^k-space $\mathfrak{B}$. By convention, the first piece of S, corresponding to the first factor in the expression for $\mathfrak{B}'$, is the ♚.

When pieces other than the ♚ are moved, the induced motion in the hyperboard remains within the wedge. Thus, the induced functions $\mathrm{III}'_{\mathbf{p},s} \colon \mathfrak{B}' \mapsto \mathfrak{B}'$ have the same form as Table 1 when $s \geq 1$.

However, when the ♚ is moved outside its fundamental region, the resulting position must be transformed so that the ♚ is in its fundamental region. This transformation of the chessboard induces a transformation on the hyperboard (Figure 4).

Algebraically,

$$\mathrm{III}'_{\text{♚},1} = \sum_{\mathfrak{d} \in \mathfrak{D}_4} \mathrm{III}'_{\text{♚},1_{\mathfrak{d}}} \otimes \bigotimes^{k-1} \mathfrak{d},$$

where $\mathrm{III}'_{\text{♚},1_{\mathfrak{d}}} \in \mathfrak{M}_{10}$.

The sum over $\mathfrak{d} \in \mathfrak{D}_4$ corresponds to routing along the pattern of the Cayley graph of $\mathfrak{D}_4$ (see Figure 5).

This is a graph whose elements are the eight transformations in $\mathfrak{D}_4$, and whose edges are labeled by one of the generators $\mathfrak{r}$ or $\mathfrak{f}$. An edge labeled $\mathfrak{h}$ connects node $\mathfrak{g}$ to node $\mathfrak{g}'$ if $\mathfrak{h}\mathfrak{g} = \mathfrak{g}'$. The communication complexity of the routing can be reduced by exploiting the Cayley graph structure [Stiller 1991a]. The actual

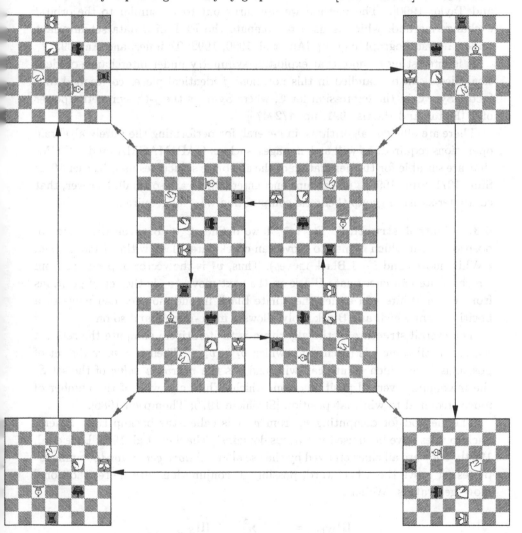

Figure 5. The Cayley graph for $\mathfrak{D}_4$. Each node is pictured by showing the effect of its corresponding transformation on a position in ♔♗♘♘♚♖; thus, the chess value of each of these nodes is the same. Black arrows correspond to $\mathfrak{r}$, and rotate the board counterclockwise 90°. Gray (diagonal) arrows correspond to $\mathfrak{f}$, and flip the board horizontally. The position shown arose during a game between Anatoly Karpov and Gary Kasparov in Tilburg, October 1991.

communication pattern used is that of a group action graph, which looks like a number of disjoint copies of the Cayley graph, plus some cycles [White 1984].

The problem of parallel application of a structured matrix to a data set invariant under a permutation group has been studied in the context of finite-element methods by Danny Hillis and Washington Taylor as well. Although their terminology is different from our terminology, their general ideas are similar [Hillis and Taylor 1990]. The method we use turns out to be similar to the orbital exchange method, which is used to compute the FFT of a data set invariant under a crystallographic group [An et al. 1990; 1992; Tolimieri and An 1990].

It is interesting to note that exploiting symmetry under interchange of identical pieces can be handled in this notation: j identical pieces correspond to a factor $\mathrm{Sym}^j \mathfrak{C}$ in the expression for $\mathfrak{C}$, where Sym^j is the j-th symmetric power of $\mathfrak{C}$ [Fulton and Harris 1991, pp. 472–475].

There are efficient algorithms, in general, for performing the purely algebraic operations required, as well as languages, such as GAP, MAGMA, and AXIOM, that are suitable for the denotation of the algebraic structures used [Butler 1992; Sims 1971; Sims 1994]. The groups encountered here are so small, however, that computer-assisted group-theoretic computation was not required.

6.3. Control structure. For $i \geq 1$ we define $v_i \in \mathfrak{B}$ to be the vector of positions from which White to move can checkmate Black within i moves (i.e., i White moves and $i - 1$ Black moves). Thus, v_1 is the vector of positions from which White can checkmate Black on the next move. v_2 is the set of positions from which White can either checkmate Black in one move or can move to a position from which any Black reply allows a mate-in-one, and so on.

The overall structure of the algorithm is to iteratively compute the sets v_1, v_2, ..., until some i is reached for which $v_i = v_{i+1}$. Then $v = v_i$ is the set of positions from which White can win, and i is the *maximin value* of the set S: the maximum, over all positions from which White can win, of the number of moves required to win that position [Ströhlein 1970; Thompson 1986].

The method for computing v_i from v_{i-1} is called the backup rule. Several backup rules have been used in various domains [Schaeffer et al. 1992; Lake et al. 1994]. They are all characterized by the use of an *unmove generator* to "unmove" pieces, or move them backward, possibly in conjunction with more traditional move generators. We let

$$\mathrm{III}_{\mathrm{White}} = \sum_{\{s \,:\, S_s \text{ is White}\}} \mathrm{III}_{S_s, s},$$

$$\mathrm{III}_{\mathrm{Black}} = \sum_{\{s \,:\, S_s \text{ is Black}\}} \mathrm{III}_{S_s, s}.$$

The backup rule used is $v_{i+1} = \mathrm{III}_{\mathrm{White}}(\overline{\mathrm{III}_{\mathrm{Black}}(\overline{v_i})})$, where a vertical bar denotes complement.

7. Refinements

7.1. Captures and pawns. The algorithms developed so far must be modified to account for captures and pawns.

Each subset of the original set of pieces S induces a *subgame*, and each subgame has its own hyperboard [Bellman 1965]. Without captures, moving and unmoving are the same, but when captures are considered they are slightly different. The equations for $\text{III}_{p,s}$ developed in the preceding section refer to *unmoving* pieces, not to moving them [Thompson 1986]. Unmoving pieces cannot capture, but they can uncapture, leaving a piece in their wake. This is simulated via interhyperboard communication.

The uncapture operation can be computed by using outer products, corresponding to the parallel broadcast, or **SPREAD** primitive [Adams et al. 1992; Stiller 1992a]. An uncapture is the product from left to right of an unmove operator in the parent game, a diagonal matrix, a sequence of stride matrices, and a broadcast. The broadcast is a tensor product of copies of an identity matrix with the 1×64 matrix of 1's.

Each pawn position induces a separate hyperboard. Pawn unpromotion induces communication between a quotient hyperboard and a full hyperboard, again implemented by multiplication by $\mathfrak{D}_4$.

7.2. Database. There are two values that can be associated with a position: *distance-to-mate* and *distance-to-win*.

The distance-to-mate is the number of moves by White required for White to checkmate Black, when White plays so as to checkmate Black as soon as possible, and Black tries to avoid checkmate as long as possible [Zermelo 1913]. Although the distance-to-mate might seem like the natural metric to use, it can produce misleadingly high distance values because the number of moves to mate in trivial subgames, like ♔♖♚, would be included in the count of something like ♔♖♚♘. In fact, in ♔♖♚♘, it does not matter for most purposes how many moves are required to win the subgame ♔♖♚, once White captures the ♘, as long as the ♘ is captured safely [Rasmussen 1988].

The more usual distance-to-win metric is simply the number of moves required by White to force conversion into a winning subgame. In practice, this metric is more useful when the position has no pawns. It also is the metric of relevance to the fifty-move rule. If a particular position has a distance-to-win of m, then against perfect play the win value would be altered by an m' move rule for $m' < m$. Although our program has implemented distance-to-mate metric for five-piece endgames, the results presented here use the more conservative distance-to-win metric.

The *max-to-win* for a set of pieces is the maximum, over all positions using those pieces from which White can win, of the distance-to-win of that position.

The distance-to-win of each point in the hyperboard can be stored so that a two-ply search permits optimum play. By Gray coding this distance, the increment of the value can be done by modifying only one bit.

Curiously, the motif of embedding Cayley graphs into Cayley graphs arises several times in this work. Gray codes, which can be viewed as embedding the Cayley graph for $\mathbb{Z}_{2^n}$ into that of $\mathbb{Z}_2^n$, are used both for implementing U (and, therefore, $\mathrm{III}_{\mathbf{p},s}$) and for maintaining the database. Embedding the Cayley graph for $\mathfrak{D}_4$ in that of $\mathbb{Z}_2^n$ arises during unpromotion and moving the ♔. Because many interconnection networks are Cayley graphs or group action graphs [Annexstein 1990; Rosenberg 1988; Draper 1990; 1991] this motif will reappear on other implementations.

8. Chess Results

The combinatorially possible pawnless five-piece games and many five-piece games with a single pawn were solved using an early version of the current program. This work resulted in the first publication of the 77-move ♔♗♘♚♘ max-to-win, which at the time was the longest known pawnless max-to-win [Stiller 1989]. Some endgames were solved under the distance-to-mate metric as well. The distance-to-mate results were not very illuminating.

Later, several pawnless six-piece endgames were also solved. Table 2 presents statistical information about these six-piece endgames. For most endgames, we considered $6, 185, 385, 360 = 462 \cdot 62 \cdot 61 \cdot 60 \cdot 59$ positions, a number explained as follows: there are 462 arrangements of two nonadjacent kings, modulo the dihedral symmetry; for each such arrangement, the remaining four pieces are placed in any available square. Thus the state space per endgame has about $6.2 \cdot 10^9$ nodes, although the size of each hyperboard is $462 \cdot 64^4 \approx 7.7 \cdot 10^9$ nodes.

Note that this inflates the statistics for wins, because of the advantage of the first move in a random position: White is already won if the ♚ is in check, or it may be able to capture a piece in one move. When there are repeated pieces, the aggregate statistics (though not the percentages) are also inflated, because games arising from permutations of the identical pieces are counted as different. Thus, for example, there is really only a single mutual zugzwang in ♔♖♖♖♚♛, but it is counted six times.

The max-to-win values were significantly higher than previously known endgames. No five-piece endgame had a max-to-win over 100, and most of the nontrivial ones had max-to-wins of approximately 50. ♔♖♘♚♘♘ has the longest known max-to-win of 243, although it is not a general win.

(The phrase "general win" is not susceptible to precise definition. It applies to any class of games, like ♔♕♘♚♘, where all "normal" starting configurations lead to a win, although some special configurations may not be wins for obvious reasons, such as the immediate loss of a piece by White due to a fork.)

We remark that ♔♖♗♚♘♘ is a general win, with 223 moves required to win in the worst case. Roycroft, a leading endgame expert, said in 1972 that this

pieces	# wins	%	D	Z	pieces	# wins	%	D	Z
(chess pieces)	4821592102	78	243	18176	(chess pieces)	5257968414	85	63	6670
(chess pieces)	5948237948	96	223	456	(chess pieces)	4529409548	73	54	1030
(chess pieces)	4433968114	72	190	8030	(chess pieces)	1015903231	65	52	256
(chess pieces)	5338803302	86	153	1858	(chess pieces)	5058432960	82	51	2820
(chess pieces)	4734636964	77	140	1634	(chess pieces)	3302327120	53	49	1270
(chess pieces)	5843483696	94	101	1520	(chess pieces)	5689213742	92	48	32
(chess pieces)	4242312073	69	99	1010	(chess pieces)	4079610404	66	46	22
(chess pieces)	5359504406	87	98	1478	(chess pieces)	5122186896	83	44	32
(chess pieces)	5324294232	86	92	6300	(chess pieces)	1185941301	75	44	396
(chess pieces)	5125056553	83	92	243	(chess pieces)	981954704	63	38	1662
(chess pieces)	5834381682	94	86	12918	(chess pieces)	1483910528	94	37	26
(chess pieces)	5707052904	92	85	342	(chess pieces)	4213729734	68	36	78
(chess pieces)	5935067734	96	82	388	(chess pieces)	4626525594	75	35	17688
(chess pieces)	1123471668	72	75	95	(chess pieces)	3825698576	62	32	6
(chess pieces)	5365200098	87	73	1410	(chess pieces)	3789897570	61	32	35
(chess pieces)	5023789102	81	73	1410	(chess pieces)	6130532196	99	31	58
(chess pieces)	5808880660	94	72	2228	(chess pieces)	3920922433	63	29	152
(chess pieces)	5553239408	90	71	1780	(chess pieces)	3533291870	57	27	3
(chess pieces)	4944693522	80	69	48	(chess pieces)	4136045492	67	18	16
(chess pieces)	1497242834	95	68	83	(chess pieces)	970557572	62	12	18
(chess pieces)	6054654948	98	65	6					

Table 2. Statistics gathered for six-piece endgames: description of endgame, number and percentage of positions that are wins for White, maximum D of distance-to-win over all positions that are wins for White, and number Z of mutual zugzwangs. Bishops linked with a bar are constrained to lie on squares of opposite colors; this reduces the state space roughly by a factor of four, from 6.19×10^9 to 1.67×10^9 nodes.

configuration was "known to be a draw", while (chess pieces), which was considered a draw by most players, he called only "controversial or unknown". Most of the standard works concurred with the opinion that (chess pieces) was not a general win [Euwe 1940, Vol. 5, pp. 50–53; Chéron 1952, p. 417; Berger 1890, pp. 167–169; Fine 1941, p. 521]. Chéron, however, seems to reserve judgment.

The fifty-move rule would affect the value of each endgame listed with max-to-win of fifty or more. The 92-move win in (chess pieces) is somewhat surprising too.

A mutual zugzwang is closely related to a game whose Conway value is zero: it is a position in which White to move can only draw, but Black to move loses. Such positions seem amusing because, particularly when no pawns are involved, chess is a very hot game in the sense of *Winning Ways* [Berlekamp et al. 1982].

Unlike the "maximin" positions such as the one in Table 3, whose analysis is fairly impenetrable, mutual zugzwangs can sometimes be understood by humans.

Figure 6. Mutual zugzwang: White to play draws, but Black to play loses.

Figure 6 shows an example, discovered by the program. The ♜ is trapped on h8, since g8 is guarded by the ♝ on a2, and the ♘'s guard each other. If the Black ♜ were to capture a ♘, it would be captured, and the resulting subgame of ♔♝♘♚ would be winning for White. The position seems to be a race between Kings to see who will reach the upper right corner area first. If the Black ♚ reaches g7 or e8 first, the Black ♜ can sacrifice itself for a White ♘, and then the Black ♚ captures the other White ♘, leaving the drawn endgame ♔♝♚. On the other hand, if the White ♔ reaches g7 first, it simply captures the Black ♜h8. Note also that neither ♘ can move, as the ♜ would immediately capture the other ♘.

It is not difficult to see that Black to play loses: White gets in first. For example, 1. ... ♚c3 2.♔b1 ♚d4 3.♔c2 ♚c5 (If 3. ... ♚e5?, 4.♘g6+ wins the ♜) 4♔d3 ♚d6 5.♔e4 ♚c7 (If ♚e7? 6.♘g6+ wins) 6. ♔f5 ♚d8 7.♔g6 ♚e8 8. ♔g7 and White wins.

However, White to move from the position in Figure 6 must move the ♝. 1.♝b1+ ♚c3 forces 2.♝a2 ♚c2, since other moves by White on the second move allow the ♜h8 to escape via g7. Chess theory, confirmed by the program, shows that this general position in ♔♝♘♘♚♜ is drawn. Any other move of the ♝ on move 1 allows Black to win the race. For example, 1. ♝f7 ♚c3 2.♔b1 ♚d4 3.♔c2 ♚c5 4.♔d3 ♚d6 5.♔e4 ♚e7! draws.

Now consider Figure 7, left. If Black moves the ♛a7 then ♛hg1 or ♛a2 mates. If the ♛f6 moves then ♛b2 or ♛f1 will mate. If ♚b1 then ♛c2 mates. Thus, any Black move loses. On the other hand, if White moves first then Black can force the draw. This mutual zugzwang, discovered by the computer, is somewhat inelegant in that includes promoted pieces. Noam Elkies used it as the basis for a much more elegant composition, one that follows accepted aesthetic practice by avoiding the use of promoted pieces in the original position and by appearing more natural. This composition [Elkies 1993; Rusinek 1994, #546] is shown in Figure 7, right. We quote from Elkies' analysis: "1. ♛g7+ Not 1.♛d6+? ♚xg2 2.f8/♛ (interpolating further checks does not help) when 2. ... ♛h3+ 3.♔g5 ♛e3+ forces either perpetual check or a queen trade, drawing. 1. ... ♚h2 2.f8/♛. If 2.♛e5+ ♚xg2 3.f8/♛ ♛h3+ 4.♔g5 b1/♛ with ♚h1 and ♝e4 draws, but now 2. ... b1/♛ loses to 3.♛f4+ ♚g1 4.♝e4+ and mate. Thus, Black tries for perpetual check, and not with 2 ... ♛d1+? 3.♝f3. 2 ... ♛b5+

Figure 7. Left: mutual zugzwang found by the program. Right: Position constructed by Noam Elkies [1993], leading to a version of the position on the left. White to play and win.

3. ♔h6 ♛b6+ 4. ♗c6! Not yet 4. ♔xh7 b1/♛+ 5. ♔h8 ♛b8! drawing. Now Black must take the bishop because 4. ... ♛e3+ 5. ♛g5 ♛xg5+ 6. ♔xg5 b1/♛ 7. ♛f2+ mates. 4. ... ♛xc6+ 5. ♔xh7 b1/♛+. So Black does manage to give the first check in the four-queen endgame, but he is still in mortal danger. 6. ♔h8 ♔h1! Black not only cannot continue checking, but must play this modest move to avoid being himself checked to death! For instance, 6. ... ♛g2 7. ♛c7+ ♔g1 8. ♛fc5+ ♔h1 9. ♛h5+ and the Black king soon perishes from exposure. But against the quiet 6. ... ♔h1 White wins only with 7. ♛fg8!!, a second quiet move in this most tactical of endgames, bringing about [a rotated version of the ♔♛♛♔♛♛ mutual zugzwang]."

Other analyses of mutual zugzwang can be found in [Elkies 1996] in this book.

Pawnless six-piece endgames are extremely rare in tournament play, but they do arise sometimes. During an elite tournament in Tilburg, a game between Anatoly Karpov and Gary Kasparov reached the position shown in Figure 5. After another fifty moves a draw was reached, but it was far from clear that a win could not be achieved given unlimited play. An exhaustive analysis by the six-piece program [Stiller 1991b] proved that it couldn't.

We conclude this section displaying the best play line for the ♔♖♘♔♘♘ position with maximal distance-to-win, namely 243. Recall that this means it takes 243 moves against optimal Black play in order for White to be able to safely capture a piece, thereby ensuring an easy win. The initial position is shown in Figure 8, and the line of play in Table 3.

Figure 8. Starting point for the longest ♔♖♘♔♘♘ fight. See Table 3.

	moves 1–40	moves 41–80	moves 81–120	moves 121–160
1	Kf7-e6 Nc6-b4	Ke5-d4 Nb4-a6	Kd6-e5 Nf2-g4	Nd6-e4 Ne7-g6
2	Ke6-e5 Nb4-d3	Rc3-c2 Kc6-d7	Ke5-d4 Nh3-f4	Ke5-f5 Ng6-f8
3	Ke5-e4 Nd3-f2	Nc8-b6 Kd7-d6	Nc5-e4 Kg5-g6	Rh7-h6 Kc6-c7
4	Ke4-f3 Nf2-d3	Nb6-c4 Kd6-c6	Ra8-a6 Kg6-f5	Rh6-h1 Nf8-d7
5	Kf3-e2 Nc2-b4	Nc4-e3 Kc6-d6	Ra6-a5 Kf5-e6	Rh1-b1 Nd7-b8
6	Ke2-e3 Rb1-b2	Ne3-f5 Kd6-e6	Ne4-c5 Ke6-e7	Kf5-e5 Nd5-e3
7	Ke3-d4 Nd3-f4	Nf5-g7 Ke6-f7	Ra5-a7 Ke7-f6	Ke5-d4 Ne3-f5
8	Kd4-c4 Nb4-d5	Ng7-h5 Nc5-e6	Kd4-e4 Kf6-g5	Kd4-d5 Nf5-e3
9	Rg7-h7 Nd5-e3	Kd4-e5 Na6-b4	Ra7-a5 Nf4-h5	Kd5-c5 Nb8-d7
10	Kc4-d4 Ne3-c2	Rc2-e2 Nb4-d3	Nc5-e6 Kg5-g6	Kc5-d4 Ne3-g4
11	Kd4-e4 Nf4-e6	Ke5-e4 Nd3-b4	Ra5-b5 Kg6-f7	Rb1-c1 Kc7-d8
12	Ke4-e5 Ne6-g5	Re2-b2 Kf7-g6	Ne6-c5 Kf7-e7	Rc1-e1 Ng4-f6
13	Rh7-h5 Nc2-e1	Nh5-g3 Ne6-g5	Rb5-b2 Ke7-d6	Ne4-g5 Kd8-c7
14	Ke5-f5 Ng5-f3	Ke4-d4 Ng5-e6	Nc5-b7 Kd6-e7	Ng5-f7 Nd7-f8
15	Kf5-e4 Nf3-d2	Kd4-c4 Nb4-a6	Rb2-a2 Nh5-g7	Re1-f1 Nf6-g4
16	Ke4-e3 Nd2-b3	Rb2-f2 Ne6-g5	Ra2-e2 Ke7-d7	Rf1-g1 Ng4-f6
17	Rh5-h1 Ne1-c2	Rf2-f1 Na6-c7	Re2-g2 Ng7-e8	Rg1-e1 Kc7-d7
18	Ke3-d3 Nb3-c1	Ng3-e2 Ng5-f7	Ke4-f4 Ng4-f6	Kd4-e5 Nf6-e8
19	Kd3-e4 Nc1-b3	Ne2-f4 Kg6-g5	Kf4-e5 Nd7-e7	Nf7-h8 Kd7-e7
20	Rh1-h3 Nb3-c5	Kc4-d4 Nc7-b5	Rg2-e2 Ke7-d7	Ke5-d5 Ke7-d7
21	Ke4-e5 Nc2-e1	Kd4-c5 Nb5-d6	Nb7-a5 Nf6-g4	Re1-f1 Ne8-c7
22	Ng8-f6 Ne1-d3	Nf4-e6 Kg5-g6	Ke5-f5 Ng4-h6	Kd5-e5 Nf8-e6
23	Ke5-d6 Nc5-b7	Ne6-f8 Kg6-g5	Kf5-g6 Nh6-g8	Nh8-g6 Ne6-c5
24	Kd6-c7 Nb7-c5	Kc5-d5 Nd6-f5	Na5-c4 Ne8-c7	Rf1-b1 Kd7-c6
25	Kc7-c6 Kb2-c2	Rf1-b1 Nf5-g3	Kg6-f7 Ng8-h6	Ng6-e7 Kc6-d7
26	Rh3-h2 Kc2-b3	Rb1-b7 Nf7-h6	Kf7-f6 Nh6-g8	Ne7-f5 Kd7-c6
27	Kc6-d5 Kb3-b4	Rb7-g7 Ng5-f4	Kf6-e5 Ng8-e7	Nf5-d4 Kc6-d7
28	Kd5-d4 Nd3-f4	Nf8-e6 Kf4-f3	Re2-d2 Kd7-c6	Rb1-d1 Nc7-a6
29	Rh2-h4 Kb4-b5	Rg7-b7 Ng3-h5	Rd2-c2 Nc7-a6	Nd4-f5 Kd7-c6
30	Nf6-e8 Nc5-b3	Rb7-b4 Nh5-f6	Nc4-e3 Kc6-d7	Rd1-h1 Na6-b4
31	Kd4-e4 Nf4-g6	Kd5-d4 Nf6-h5	Rc2-d2 Kd7-c6	Rh1-h6 Kc6-d7
32	Rh4-h7 Nb3-c5	Kd4-d3 Nh6-g4	Rd2-d6 Kc6-b5	Ke5-d4 Nc5-e6
33	Ke4-d4 Ng6-f4	Ne6-g5 Kf3-g3	Rd6-h6 Ne7-c8	Kd4-c4 Nb4-a6
34	Ne8-d6 Kb5-c6	Ng5-e4 Kg3-h4	Ke5-d4 Na6-b4	Rh6-h7 Kd7-c6
35	Rh7-h6 Nc5-b3	Rb4-a4 Nh5-f4	Rh6-h5 Kb5-c6	Rh7-h1 Na6-c7
36	Kd4-e4 Nf4-e6	Kd3-d4 Nf4-e6	Ne3-c4 Nc8-e7	Rh1-d1 Nc7-e8
37	Ke4-e5 Ne6-d4	Kd4-d5 Ne6-f4	Rh5-h6 Kc6-c7	Nf5-e7 Kc6-c7
38	Rh6-h3 Nb3-c5	Kd5-d6 Nf4-h3	Rh6-h7 Kc7-d7	Kc4-d5 Ne6-f8
39	Nd6-c8 Nd4-c2	Ra4-a8 Ng4-f2	Kd4-e5 Nb4-d5	Ne7-g8 Kc7-d7
40	Rh3-c3 Nc2-b4	Ne4-c5 Kh4-g5	Nc4-d6 Kd7-c6	Kd5-c5 Kd7-e6

Table 3. An optimal line of play for the position of Figure 8. Occasional local variations are possible (for instance, 15. f5-f4), but are not shown.

	moves 161–180	moves 181–200	moves 201–220	moves 221–243
161	♖d1-e1 ♔e6-d7	♖d5-a5 ♘h6-g4	♘h6-f5 ♔f8-f7	♔d5-d6 ♘c7-e8
162	♖e1-e7 ♔d7-d8	♔e5-d4 ♔f8-f7	♖e1-e2 ♘d5-b6	♔d6-e7 ♘e8-c7
163	♖e7-a7 ♘f8-d7	♖a5-a7 ♔f7-f6	♖e2-e7 ♔f7-f8	♖h7-h6 ♘a6-b8
164	♔c5-c6 ♘d7-e5	♔d4-e4 ♘g7-e8	♖e7-e1 ♘b6-d5	♘a4-b6 ♔c8-b7
165	♔c6-d5 ♘e5-g6	♖a7-a6 ♔f6-g7	♖e1-e5 ♘d5-b6	♘b6-c4 ♘b8-c6
166	♖a7-h7 ♘e8-c7	♖a6-b6 ♘g4-f6	♕g5-g6 ♘e8-c7	♔e7-d6 ♘c6-b4
167	♔d5-c6 ♘g6-e5	♔e4-f5 ♘f6-d7	♘f5-d6 ♘b6-d5	♖h6-h8 ♘b4-a6
168	♔c6-d6 ♘e5-c4	♘f4-e6 ♔g7-f7	♖e5-e1 ♘c7-e6	♖h8-h7 ♔b7-c8
169	♔d6-c5 ♘c4-e5	♘e6-g5 ♔f7-f8	♕g6-f5 ♘e6-c7	♘c4-a5 ♔c8-d8
170	♖h7-h5 ♘e5-f7	♖b6-a6 ♘e8-g7	♕f5-e5 ♘d5-b4	♘a5-c6 ♔d8-c8
171	♔c5-c6 ♘c7-e6	♔f5-g6 ♘d7-e5	♖e1-f1 ♕f8-e7	♘c6-e7 ♔c8-d8
172	♖h5-a5 ♔d8-e8	♔g6-h7 ♘g7-e8	♖f1-f7 ♔e7-d8	♘e7-d5 ♘c7-e8
173	♘g8-f6 ♔e8-e7	♖a6-e6 ♘e5-f7	♘d6-b7 ♔d8-c8	♔d6-c6 ♘a6-b8
174	♘f6-d5 ♔e7-f8	♘g5-f3 ♘f7-d6	♘b7-c5 ♘c7-b5	♔c6-b5 ♘e8-d6
175	♔c6-d7 ♘e6-d4	♔h7-g6 ♘d6-f5	♖f7-g7 ♔c8-d8	♔b5-c5 ♘d6-c8
176	♘d5-f4 ♘f7-h6	♖e6-e1 ♘f5-e7	♖g7-b7 ♘b4-c6	♖h7-h8 ♔d8-d7
177	♖a5-d5 ♘d4-f5	♔g6-g5 ♔f8-f7	♔e5-e6 ♔d8-c8	♘d5-f6 ♔d7-c7
178	♔d7-e6 ♘f5-g7	♘f3-e5 ♔f7-g7	♖b7-h7 ♘c6-b4	♖h8-h7 ♔c7-d8
179	♔e6-f6 ♘h6-g8	♘e5-g4 ♔g7-f8	♘c5-a4 ♘b4-a6	♖h7-b7 ♘b8-a6
180	♔f6-e5 ♘g8-h6	♘g4-h6 ♘e7-d5	♔e6-d5 ♘b5-c7	♔c5-c6 ♘c8-e7
				♔c6-b6 ♘a6-b4
				♖b7-d7 ♔d8-c8
				♖d7×♘e7

Table 3 (continuation)

9. Implementation Notes and Run Times

The implementation was on a 64K processor CM-2/200 with 8 GBytes RAM. The processors were interconnected in a hypercube and clocked at 7MHz (10 MHz for the CM-200). The CM-2 six-piece code required approximately 1200 seconds for initialization and between 111 and 172 seconds to compute K_{i+1} from K_i. Exact timings depend on S (for instance, as is clear from Table 1, $III_{♕,s}$ is slower than either $III_{♖,s}$ or $III_{♗,s}$) as well as run-time settable factorization choices and load on the front end.

Per-node time per endgame (time to solve the endgame divided by number of nodes in the state-space) is faster by a factor of approximately 6000 than timings *of different endgames* reported using classical techniques [van den Herik and Herschberg 1985b; Thompson 1986; Nefkens 1985; van den Herik et al. 1988] based on the five-piece timings of the code reported here.

In unpublished personal communication Thompson has indicated that the per-node time of the fastest serial endgame code is currently only a factor of approximately 700 times slower than that of the code reported in this paper (depending on the endgame) [Thompson 1990].

Unfortunately, direct comparison of six-piece timing against other work is, of course, not currently possible since six-piece endgames could not have been solved in a practicable amount of time using classical techniques on previous architectures. However, with larger and faster serial machines, and with enough spare cycles, six-piece endgames are in fact coming within reach of classical solution techniques. This would permit a more informative timing comparison.

Thus, although per-node timing comparisons based on radically differently sized state-spaces are not very meaningful, the large per-node timing differential of the current program compared to classical programs does tend to support the hypothesis that the techniques reported here lend themselves to efficient parallel implementation.

The only program with per-node time of comparable speed to the author's CM-200 implementation is the vectorized implementation of Table 1 by Burton Wendroff et al. [1993], although this implementation currently solves only a single four-piece endgame.

The CM-200 source code implementing Table 1 is currently available from ftp://ftp.cs.jhu.edu/pub/stiller/snark.

10. Future Work

The main historical open question is to find out Molien's exact contribution to the history of numerical chess endgame analysis, and to locate and check his analysis of ♔♖♔♗. Kanunov [1983, p. 6] refers to private papers held by Molien's daughter; currently we are trying to locate these papers in the hope that they might shed light on the questions raised in Section 2.2. Amelung himself is also a figure about whom little is known, and the remarks here would seem to suggest that a detailed reassessment of his contribution to the endgame study would be desirable.

The question of Molien and Amelung's contributions to quantitative endgame analysis is part of the larger historical question of pre-digital precursors to computer chess algorithms. In addition to the work of Babbage, Molien, Amelung, Zermelo, and Quevedo, we remark that K. Schwarz, in a little-known 1925 article in *Deutsche Schachzeitung*, argued for a postional evaluation function similar to the squares-attacked heuristic used in some full-chess programs [Schwarz 1925].

From a computational point of view, it might seem that the next logical step in the evolution of the current program should be the exhaustive solution of pawnless seven-piece endgames. In fact, in my opinion a more promising approach would be to follow up on the suggestions first made by Bellman [Bellman 1961; Bellman 1965; Berliner and Campbell 1984] and solve endgames with multiple pawns and minor pieces. Such an approach would combine heuristic evaluation of node values corresponding to promotions with the exhaustive search techniques described here. Although the use of heuristics would introduce some errors, the results of such a search would, in my opinion, have considerable impact on the evaluation of many endgames arising in practical play.

Even more speculatively, it is also possible to search for certain classes of endgames considered artistic by endgame composers; such endgames typically depend on a key mutual-zugzwang or domination position some moves deep in the tree.

Acknowledgments

Thanks go to my advisor, Simon Kasif, for his help in the formulation and presentation of these ideas.

Throughout the six-piece project, Noam Elkies patiently provided invaluable chess advice and suggestions, including the mutual zugzwang recognition algorithm. Burton Wendroff made possible much of the work and developed and implemented the vectorized Cray algorithm.

A. John Roycroft provided much of the historical information on human endgame analysis.

Harold van der Heijden graciously provided all studies in his database of approximately 36000 endgame studies in which certain pawnless six-piece endgames arose.

I am also grateful to the following people for useful conversations and suggestions: Alan Adler, Hans Berliner, Murray Campbell, Richard Draper, Charles Fiduccia, Ralph Gasser, Thomas Hawkins, Feng-Hsiung Hsu, Steven Salzberg, Jonathan Schaeffer, Ken Thompson, and David Waltz.

Silvio Levy skillfully and scrupulously edited the manuscript.

Assistance in translating many of the non-English sources was provided by Peter Jansen, Michael Klepikov, George Krotkoff, A. John Roycroft, Claudia Salzberg, Boris Statnikov, and the staff of the Cleveland Public Library.

Access to most of the manuscripts, rare books and chess journals cited in the paper was obtained during my visit to the John Griswold White Collection of Chess, Orientalia and Fine Arts, located at the Cleveland Public Library, 325 Superior Avenue, Cleveland, OH 44414. Thanks to Alice N. Loranth and Motoko B. Reece of the Cleveland Public Library for their assistance during his visits to the collection.

The National Library of Latvia in Riga (Latvijas Nacionālā Bibliotēka) assisted in providing copies of a number of articles by Friedrich Amelung; Kenneth Whyld helped narrow down the search for them to the city of Riga.

Computing facilities were graciously provided by the Advanced Computing Laboratory of the Los Alamos National Laboratory, Los Alamos, NM 87545.

References

[Abdali and Saunders 1985]   Syed Kamal Abdali and Benjamin David Saunders, "Transitive closure and related semiring properties via elimination", *Theoretical Computer Science* **40** (1985), 257–274.

[Adams et al. 1992]   Jeanne C. Adams, Walter S. Brainerd, Jeanne T. Martin, Brian T. Smith, and Jerrold L. Wagener, *Fortran 90 Handbook: Complete ANSI/ISO Reference*, McGraw-Hill, New York, 1992.

['Adlī]   أَلْعَدلِي, كِتَاب الشَّطرَنج [Al-'Adlī, *Book of Chess*], ninth century, photographic copy of Arabic manuscript (available at the John G. White Collection; see Acknowledgements).

[Amelung 1893]   Friedrich Ludwig Amelung, "Das Endspiel von zwei Offizieren gegen einen Springer" [The endgame of two pieces against a rook], *Baltische Schachblätter* **4** (1893), 290–297.

[Amelung 1898]   Friedrich Ludwig Amelung, "Auszüge aus den Briefen von A. Ascharin an F. Amelung" [Excerpts from the letters of A. Ascharin to F. Amelung], *Baltische Schacblätter* **5** (1898), 12–38.

[Amelung 1900]   Friedrich Ludwig Amelung, "Das Endspiel von Thurm gegen Springer" [The endgame of rook against knight], *Deutsche Schachzeitung*, **55** (1900) 1–5 (January), 37–41 (February), 101–105 (April), 134–138 (May), 198–202 (July), 261–266 (September).

[Amelung 1901]   Friedrich Ludwig Amelung, "Das Endspiel von Thurm und Springer gegen die Dame" [The endgame of rook and knight against queen], *Deutsche Schachzeitung*, **56** (1901) 193–197 (July), 225–229 (August).

[Amelung 1902]   Friedrich Ludwig Amelung, "Die Endspiele mit Qualitätsvortheil, insbesondere das Endspiel von Thurm und Läufer gegen Läufer und Springer" [The endgames with exchange advantage, especially the endgame of rook and bishop against bishop and knight], *Deutsche Schachzeitung*, **57** (1902) 265–268 (September), 297–300 (October), 330–332 (November).

[Amelung 1908a]   Friedrich Ludwig Amelung, "Das Endspiel von Turm und Läufer gegen zwei Springer" [The endgame of rook and bishop against two knights], *Für Haus und Familie* (supplement to *Düna-Zeitung*) **40** (16–29 February 1908), 52–53.

[Amelung 1908b]   Friedrich Ludwig Amelung, Endspiel 1028. *Deutsches Wochenschach*, **24**(14) (5 April 1908), 130.

[Amelung 1908c]   Friedrich Ludwig Amelung, Lösungspreis-endspiel Nr. 178 [Prize-endgame 178], *Für Haus und Familie* (supplement to *Düna-Zeitung*) **63** (15–28 March 1908), 87. Solution in **13** (17–30 January 1909), 20–21.

[An et al. 1990]   Myoung An, James William Cooley, and Richard Tolimieri, "Factorization method for crystallographic Fourier transforms", *Advances in Applied Mathematics* **11** (1990), 358–371.

[An et al. 1992]   Myoung An, Chao Lu, E. Prince, and Richard Tolimieri, "Fast Fourier transforms for space groups containing rotation axes of order three and higher", *Acta Crystallographica* A**48** (1992), 346–349.

[Annexstein et al. 1990] Fred S. Annexstein, Marc Baumslag, and Arnold Leonard Rosenberg, "Group action graphs and parallel architectures", *SIAM Journal on Computing*, **19** (1990), 544–569.

[Anonymous] [Nicholas de St Nicholai?], "*Bonus Socius* [Good companion], 13th century.

[Auslander and Tolimieri 1985] Louis Auslander and Richard Tolimieri, "Ring structure and the Fourier transform", *Mathematical Intelligencer*, **7**(3) (1985), 49–52, 54.

[Averbakh 1982] Yuri Lvovich Averbakh, Шахматные окончания; ферзевые [*Chess endings: queen*], Fizkul'tura i sport, Moscow (second edition), 1982.

[Averbuch et al. 1990] Amir Averbuch, Eran Gabber, Boaz Gordissky, and Yoav Medan, "A parallel FFT on a MIMD machine", *Parallel Computing*, **15** (1990), 61–74.

[Babbage 1864] Charles Babbage, "Games of skill", pp. 465–471 in *Passages From the Life of a Philosopher*, Longman and Green, London, 1864.

[Backhouse and Carré 1975] Roland C. Backhouse and B. A. Carré, "Regular algebra applied to path-finding problems", *Journal of the Institute of Mathematics and its Applications*, **15**(2) (1975), 161–186.

[Balt. Schachbl. 1893] "Baltische Schachpartien aus den Jahren 1858 bis 1892", Partie 114 [Baltic chess games from the years 1892 to 1858, Game 114], *Baltische Schachblätter* **4** (1893), 266–267. Score of consultation Theodor Molien and A. Hasselblatt v. Friedrich Amelung.

[Balt. Schachbl. 1902] [Carl Behting and Paul Kerkovius?] "Das zweite Baltische Schachturnier" [The second Baltic chess tournament], *Baltische Schachblätter* **9** (1902), 1–24.

[Bashmakova 1991] J. G. Bashmakova, "Fedor Eduardovich Molin", pp. 1739–1740 in *Biographical Dictionary of Mathematicians*, Vol. 3, Scribner's, New York, 1991.

[Bell 1978] Alex G. Bell, *The Machine Plays Chess?*, Pergamon Chess Series, Pergamon Press, Oxford, 1978.

[Bellman 1954] Richard Ernest Bellman, "On a new iterative algorithm for finding the solutions of games and linear programming problems", Technical Report P-473, The RAND Corporation, U. S. Air Force Project RAND, Santa Monica, CA, 1 June 1954.

[Bellman 1957] Richard Ernest Bellman, "The theory of games", Technical Report P-1062, The RAND Corporation, Santa Monica, CA, 15 April 1957.

[Bellman 1961] Richard Ernest Bellman, "On the reduction of dimensionality for classes of dynamic programming processes", Technical report, The RAND Corporation, Santa Monica, CA, 7 March 1961.

[Bellman 1965] Richard Ernest Bellman, "On the application of dynamic programming to the determination of optimal play in chess and checkers", *Proceedings of the National Academy of Sciences of the United States of America*, **53** (1965), 244–246.

[Berger 1890] Johann Berger, *Theorie und Praxis der Endspiele: Ein Handbuch für Schachfreunde* [*Theory and practice of the endgame: a handbook for chessplayers*], Veit, Leipzig, 1890.

[Berger 1922] Johann Berger, *Theorie und Praxis der Endspiele: Ein Handbuch für Schachfreunde* [*Theory and practice of the endgame: a handbook for chessplayers*], de Gruyter, Berlin and Leipzig (second edition), 1922.

[Berlekamp et al. 1982] Elwyn Ralph Berlekamp, John Horton Conway, and Richard K. Guy, *Winning Ways for Your Mathematical Plays* I, Academic Press, London, 1982.

[Berliner and Campbell 1984] Hans Jack Berliner and Murray S. Campbell, "Using chunking to solve chess pawn endgames, *Artificial Intelligence* **23** (1984), 97–120.

[Bird et al. 1989] Richard S. Bird, Jeremy Gibbons, and Geraint Jones, "Formal derivation of a pattern matching algorithm, *Science of Computer Programming*, **12** (1989), 93–104.

[Blelloch 1990] Guy E. Blelloch, *Vector Models for Data-Parallel Computing*, Artificial Intelligence [series], MIT Press, Cambridge, MA, 1990.

[Butler 1992] Gregory Butler, *Fundamental Algorithms For Permutation Groups*, Lecture Notes in Computer Science **559**, Springer, Berlin/New York, 1992.

[Campbell 1988] Murray S. Campbell, "Chunking as an abstraction mechanism", Technical Report CMU-CS-88-116, Carnegie-Mellon University, 22 February 1988.

[Carrera 1617] Pietro Carrera, *Del gioco de gli scacchi* [*T he game of chess*], volume 3, de Rossi[?], Trento[?], 1617.

[Chamberlain 1988] R. M. Chamberlain, "Gray codes, fast Fourier transforms, and hypercubes", *Parallel Computing* **6** (1988), 225–233.

[Chapais 1780] Chapais, *Essais analytiques sur les échecs, avec figures* [*Analytical essay on chess, with illustrations*], [ca. 1780].

[Chéron 1952] André Chéron, *Nouveau traité complet d'échecs: La fin de partie* [*New complete treatise of chess: The endgame*], Yves Demailly, Lille, France, 1952.

[Clausen and Baum 1993] Michael Clausen and Ulrich Baum, *Fast Fourier Transforms*, "Bibliographisches Institut Wissenschaftsverlag, Mannheim, 1993.

[Cooley and Tukey 1965] James William Cooley and John Wilder Tukey, "An algorithm for the machine calculation of complex Fourier series", *Mathematics of Computation* **19** (1965), 297–301.

[Crosskill 1864] [Alfred Crosskill], "The rook and bishop against rook", *The Chess-Player's Magazine* **2** (1864), 305–311.

[Dekker et al. 1987] Sito T. Dekker, H. Jaap van den Herik, and I. S. Herschberg, "Complexity starts at five", *International Computer Chess Association Journal* **10**(3) (September 1987), 125–138.

[Diaconis 1988] Persi Diaconis, *Group Representations in Probability and Statistics*, Lecture Notes—Monograph Series **11**, Institute of Mathematical Statistics, Hayward, CA, 1988.

[Diaconis and Rockmore 1990] Persi Diaconis and Daniel Nahum Rockmore, "Efficient computation of the Fourier transform on finite groups", *Journal of the American Mathematical Society* **3** (1990), 297–332.

[Draper 1990] Richard Noel Draper, "A fast distributed routing algorithm for supertoroidal networks", Technical report, Supercomputing Research Center, Bowie, Maryland, July 1990.

[Draper 1991] Richard Noel Draper, "An overview of supertoroidal networks", Technical Report SRC-TR-91-035, Supercomputing Research Center, Bowie, Maryland, 17 January 1991.

[Düna-Zeitung 1909a] Announcement of Amelung's death in *Für Haus und Familie* (supplement to *Düna-Zeitung*) (10–23 March 1909), 52.

[Düna-Zeitung 1909b] "Friedrich Ludwig Amelung", *Für Haus und Familie* (supplement to *Düna-Zeitung*) **70** (28 March–10 April 1909), 76–78.

[Elkies 1993] Noam David Elkies, "Chess art in the computer age", *American Chess Journal*(2) **1** (September 1993), 48–52.

[Elkies 1996] Noam David Elkies, "On numbers and endgames: Combinatorial game theory in chess endgames", pp. 135–150 in this volume.

[Euwe 1940] Machgielis Euwe, *Het Eindspel* [*The endgame*], G. B. van Goor Zonen, 's-Gravenhage, 1940. German translation: *Das Endspiel*, Das Schach-Archiv, F. L. Rattman, Hamburg, 1957.

[Fine 1941] Reuben Fine, *Basic Chess Endings*, David McKay, New York, 1941.

[Frobenius 1896] Ferdinand Georg Frobenius, "Über die Primfactoren der Gruppendeterminante" [On prime factors of group determinants], *Sitzungsberichte der Königlich Preußischen Akademie der Wissenschaften zu Berlin* (1896), 1343–1382.

[Fulton and Harris 1991] William Fulton and Joe Harris, *Representation Theory: A First Course*, Graduate Texts in Mathematics **129**, Springer, New York, 1991.

[García and Stiller 1993] Angel E. García and Lewis Benjamin Stiller, "Computation of the mean residence time of water in the hydration shells of biomolecules", *Journal of Computational Chemistry* **14** (1993), 1396–1406.

[Gasser 1991] Ralph Gasser, "Applying retrograde analysis to Nine Men's Morris", pp. 161–173 in *Heuristic Programming in Artificial Intelligence: The Second Computer Olympiad* (edited by D. N. L. Levy and D. F. Beal), Ellis Horwood, Chichester, England, 1991.

[Gasser 1996] Ralph Gasser, "Solving Nine Men's Morris", pp. 101–113 in this volume.

[Gilbert 1958] E. N. Gilbert, "Gray codes and paths on the n-cube", *Bell Systems Technical Journal* **37** (1958), 815–826.

[Golombek 1976] Harry Golombek, *Chess: A History*, G. P. Putnam's Sons, New York, 1976.

[Granata and Tolimieri 1991] John A. Granata and Richard Tolimieri, "Matrix representations of the multidimensional overlap and add technique", *IEEE Transactions on Circuits and Systems for Video Technology* **1** (1991), 289–90.

[Granata et al. 1991] John A. Granata, Michael Conner, and Richard Tolimieri, "A tensor product factorization of the linear convolution matrix", *IEEE Transactions on Circuits and Systems* **38** (1991), 1364–1366.

[Granata et al. 1992] John A. Granata, Michael Conner, and Richard Tolimieri, "The tensor product: a mathematical programming language for FFTs and other fast DSP operations", *IEEE Signal Processing Magazine* **9** (1992), 40–48.

[Greco 1624] Gioacchino Greco, *Trattato del nobilissimo et militare essercitio de scacchi nel qvale si contengono molti bellissimi tratti et la vera scienza di esso gioco* [*Treatise on the very noble and military exercise of chess: the science of that game*], 1624.

[Hawkins 1971] Thomas William Hawkins, "The origins of the theory of group characters", *Archive for History of Exact Sciences* **7** (1971), 142–170.

[Hawkins 1972] Thomas William Hawkins, "Hypercomplex numbers, Lie groups, and the creation of group representation theory", *Archive for History of Exact Sciences* **8** (1972), 243–287.

[Hawkins 1974] Thomas William Hawkins, "New light on Frobenius' creation of the theory of group characters", *Archive for History of Exact Sciences* **12** (1974), 217–143.

[van den Herik and Herschberg 1985a] H. Jaap van den Herik and I. S. Herschberg, "The construction of an omniscient endgame database", *International Computer Chess Association Journal* **8**(2) (June 1985), 66–87.

[van den Herik and Herschberg 1985b] H. Jaap van den Herik and I. S. Herschberg, "Elementary theory improved, a conjecture refuted", *International Computer Chess Association Journal* **8**(3) (September 1985), 141–149.

[van den Herik and Herschberg 1986] H. Jaap van den Herik and I. S. Herschberg, "A data base on data bases", *International Computer Chess Association Journal* **9**(1) (March 1986), 29–34.

[van den Herik et al. 1988] H. Jaap van den Herik, I. S. Herschberg, T. R. Hendriks, and J. P. Wit, "Computer checks on human analyses of the KRKB endgame", *"International Computer Chess Association Journal* **11**(1) (March 1988), 26–31.

[Hillis 1985] William Daniel Hillis, *The Connection Machine*, ACM Distinguished Dissertations, MIT Press, Cambridge, MA, 1985.

[Hillis and Taylor 1990] William Daniel Hillis and Washington Taylor IV, "Exploiting symmetry in high-dimensional finite-difference calculations", *Journal of Parallel and Distributed Computing* **8** (1990), 77–79.

[Ho and Johnsson 1990] Ching-Tien Ho and S. Lennart Johnsson, "Embedding meshes in Boolean cubes by graph decomposition", *Journal of Parallel and Distributed Computing* **8** (1990), 325–339.

[Hooper and Whyld 1992] D. Hooper, K. Whyld, *The Oxford Companion to Chess* (second edition), Oxford Univ. Press, 1992.

[Huberman 1968] Barbara Jane Huberman, "A program to play chess end games", Technical Report CS 106, Stanford Artificial Intelligence Project Memo AI-65, Stanford University Department of Computer Science, 1968.

[Hwang and Briggs 1985] Kai Hwang and Faye Alaye Briggs, *Computer Architecture and Parallel Processing*, McGraw-Hill Series in Computer Organization and Architecture, McGraw-Hill, New York, 1985.

[Jansen 1992a] Peter Jozef Jansen, "KQKR: Assessing the utility of heuristics", *International Computer Chess Association Journal* **15**(4) (December 1992), 179–191.

[Jansen 1992b] Peter Jozef Jansen, "KQKR: Awareness of a fallible opponent", *International Computer Chess Association Journal* **15**(3) (September 1992), 111–131.

[Jansen 1992c] Peter Jozef Jansen, "Using Knowledge about the Opponent in Game-Tree Search", Ph.D. thesis, Carnegie Mellon University Department of Computer Science, Pittsburgh, PA, September 1992. Also published as Tech. Report CMU-CS-92-192.

[Johnson et al. 1990] Jeremy R. Johnson, Robert W. Johnson, Domingo Rodriguez, and Richard Tolimieri, "A methodology for designing, modifying and implementing Fourier transform algorithms on various architectures", *Circuits, Systems, and Signal Processing* **9** (1990), 449–500.

[Johnson 1989] Robert W. Johnson, "Automatic implementation of tensor products", manuscript, 24 April 1989.

[Johnson et al. 1991] Robert W. Johnson, Chua-Huang Huang, and Jeremey R. Johnson, "Multilinear algebra and parallel programming", *Journal of Supercomputing* **5** (1991), 189–217.

[Kanunov 1983] Nikolai Fedorovich Kanunov, Федор Эдуардович Молин 1861–1941 [*Fyodor Eduardovich Molin* 1861–1941], Nauka, Moscow, 1983.

[Karpovsky 1977] Mark Girshevich Karpovsky, "Fast Fourier transforms on finite non-Abelian groups", *IEEE Transactions on Computers* C-**26** (1977), 1028–1030.

[Kaushik et al. 1993] Shivnandan D. Kaushik, Sanjay Sharma, and Chua-Huang Huang, "An algebraic theory for modeling multistage interconnection networks", *Journal of Information Science and Engineering* **9** (1993), 1–26.

[Kling and Horwitz 1851] Josef Kling and Bernhard Horwitz, "*Chess Studies, or Endings of Games. Containing Upwards of Two Hundred Scientific Examples of Chess Strategy*, C. J. Skeet, Charing Cross, England, 1851.

[Komissarchik and Futer 1974] E. A. Komissarchik and Aaron L. Futer, "Об анализе ферзевого эндшпиля при помощи ЭВМ [Analysis of a queen endgame using an IBM computer], Problemy Kibernetiki **29** (1974), 211–220.

[Kopec et al. 1988] Danny Kopec, Brent Libby, and Chris Cook, "The endgame king, rook and bishop vs. king and rook (KRBKR)", pp. 60–61 in *Proceedings: 1988 Spring Symposium Series: Computer Game Playing* (Stanford, 1988), American Association for Artificial Intelligence.

[Kopnin 1983] Aleksey Grigoryevich Kopnin, "Some special features of the endgame struggle Rook and Knight against 2 Knights (GBR class 0107)", *EG* (ARVES, Amsterdam) **5**(#70) (January 1983), 89–92.

[Korn and Lambiotte 1979] David G. Korn and Jules J. Lambiotte, Jr. "Computing the fast Fourier transform on a vector computer", *Mathematics of Computation* **33** (1979), 977–992.

[Lake et al. 1994] Robert Lake, Jonathan Schaeffer, and Paul Lu, "Solving large retrograde analysis problems using a network of workstations", pp. 135–162 in *Advances in Computer Chess* 7 (edited by H. J. van den Herik et al.), University of Limburg, Maastricht, Netherlands, 1994.

[Lehmann 1977] Daniel J. Lehmann, "Algebraic structures for transitive closure", *Theoretical Computer Science* **4** (1977), 59–76.

[Leighton 1992] Frank Thomson Leighton, *Introduction to Parallel Algorithms and Architectures: Arrays, Trees, Hypercubes*, M. Kaufmann, San Mateo, CA, 1992.

[Leiserson et al. 1992] Charles Eric Leiserson et al., "The network architecture of the CM-5", pp. 272–285 in *Proceedings of the 4th Annual ACM Symposium on Parallel Algorithms and Architectures* (San Diego, 1992), ACM Press, New York, 1992.

[Lenz 1970] Wilhem Lenz, editor, *Deutschbaltisches biographisches Lexikon 1710–1960* [*German-Baltic biographical dictionary 1710–1960*], Böhlau, Köln, 1970.

[Lolli 1763] Giovanni Battista Lolli, *Osservazione teorico-pratiche sopra il giuoco degli scacchi* [*Theoretical-practical observations on the game of chess*], Bologna, 1763.

[Lopez 1561] Ruy Lopez de Sigura, *Libro de la invencion liberal y arte del juego del axedrez* [*The book of liberal invention and the art of the game of chess*], Alcala, 1561.

[Lucena 1497?] Luis Ramirez de Lucena, *Repetición de Amores y Arte de Ajedrez* [*Treatise on love and the game of chess*], Salamanca, ca. 1497. Facsimile reprint Colección Joyas Bibliográficas, Madrid, 1953.

[Marcus 1973] Marvin Marcus, *Finite Dimensional Multilinear Algebra*, Part 1, Marcel Dekker, New York, 1973.

[Matanović 1985] Aleksandar Matanović, *Enciklopedija Šahovskih Završnica* [*Encyclopedia of Chess Endings*], Šahovski informator, Beograd, 1985.

[Merris 1980] Russell Merris, "Pattern inventories associated with symmetry classes of tensors", *Linear Algebra and Applications* **29** (1980), 225–230.

[Merris 1981] Russell Merris, "Pólya's counting theorem via tensors", *American Mathematical Monthly* **88** (March 1981), 179–185.

[Merris 1992] Russell Merris, "Applications of multilinear algebra", *Linear and Multilinear Algebra* **32** (1992), 211–224.

[Merris and Watkins 1983] Russell Merris and William Watkins, "Tensors and graphs", *SIAM Journal on Algebraic and Discrete Methods* **4** (1983), 534–547.

[Michie and Bratko 1987] Donald Michie and Ivan Bratko, "Ideas on knowledge synthesis stemming from the KBBKN endgame", *International Computer Chess Association Journal* **10**(1) (March 1987), 3–10.

[Molien 1893] Theodor Molien, "Ueber Systeme höherer complexer Zahlen" [On systems of higher complex numbers], *Mathematische Annalen*, **41** (1893), 83–156; errata in **42** (1893), 308–312.

[Molien 1897a] Theodor Molien, "Eine Bemerkung zur Theorie der homogenen Substitutionsgruppen" [A remark on the theory of homogeneous substitution groups], *Sitzungberichte Naturforscher-Gesellschaft Dorpat* (Yurev, Estonia) **11** (1897), 259–274.

[Molien 1897b] Theodor Molien, "Über die Anzahl der Variabeln einer irreductibelen Substitutionsgruppen" [On the number of variables of irreducible substitution groups], *Sitzungberichte der ber Naturforscher-Gesellschaft Dorpat* (Yurev, Estonia) **11** (1897), 277–288.

[Molien 1898a] Theodor Molien, pp. 208–209 in "Sammlung baltischer Schachprobleme und Endspiele aus den Jahren 1890 bis 1897", *Baltische Schachblätter* **6** (1898), 179–212.

[Molien 1898b] Theodor Molien, "Über die Invarianten der linearen Substitutionsgruppen" [On the invariants of linear substitution groups], *Sitzungsberichte Akademie der Wissenschaft. Berlin* (1898), 1152–1156.

[Molien 1900] Theodor Molien, Partie 73 [score of game against W. Sohn], "*Baltische Schachblätter* **7** (1900), 346.

[Molien 1901] Theodor Molien, "Ziffermässig genaue Ausrechnung aller 12 millionen Gewinne und Remisen im Endspiel „Thurm gegen Läufer" [Numerical exact computation of all 12 million wins and draws in the endgame "rook against bishop"], April 1897. Cited in the index of *Baltische Schacblätter* **8** (1901), page 72.

[Mullin 1988] Lenore M. Restifo Mullin, "A mathematics of arrays", Technical Report 8814, CASE Center, Syracuse University, December 1988.

[Murray 1913] Harold James Ruthven Murray, *A History of Chess*, Oxford University Press, London, 1913.

[Nassimi and Sahni 1982] David Nassimi and Sartaj Kumar Sahni, "Parallel permutation and sorting algorithms and a new generalized connection network", *Journal of the Association for Computing Machinery* **29** (1982), 642–667.

[Nefkens 1985] Harry J. Nefkens, "Constructing data bases to fit a microcomputer", *International Computer Chess Association Journal* **8**(4) (December 1985), 217–224.

[von Neumann and Morgenstern 1944] John von Neumann and Oskar Morgenstern, *Theory of Games and Economic Behavior*, Wiley, New York, 1944.

[Nunn 1992] John Nunn, *Secrets of Rook Endings*, Batsford Chess Library, H. Holt, New York, 1992.

[Nunn 1994] John Nunn, *Secrets of Pawnless Endings*, Batsford Chess Library, H. Holt, New York, 1994.

[Patashnik 1980] Oren Patashnik, "Qubic: $4 \times 4 \times 4$ tic-tac-toe", *Math. Mag.* **53**(4) (September 1980), 202–216.

[Pease 1968] Marshall Carleton Pease, "An adaptation of the fast Fourier transform for parallel processing", *Journal of the Association for Computing Machinery* **15** (1968), 252–264.

[Perez 1929] Juan Bautista Sanchez Perez, *El Ajedrez de D. Alfonso el Sabio* [*The chess of D. Alfonso the Wise*], Franco-Española, Madrid, 1929.

[Philidor 1749?] François-André Danican Philidor, *L'analyze des échecs: contenant une nouvelle methode pour apprendre en peu temps à se perfectionner dans ce noble jeu* [*The analysis of chess: containing a new method for perfecting oneself in a short time in this noble game*], London, [1749?].

[Quinlan 1979] John Ross Quinlan, "Discovering rules by induction from large collections of examples", pp. 168–201 in *Expert systems in the micro-electronic age* (edited by Donald Michie), Edinburgh University Press, 1979.

[Quinlan 1983] John Ross Quinlan, "Learning efficient classification procedures and their application to chess end games", pp. 463–482 in *Machine Learning: An Artificial Intelligence Approach* (edited by Ryszard Stanislaw Michalski et al.), Tioga, Palo Alto, CA, 1983.

[Rasmussen 1988] Lars Rasmussen, "Ultimates in KQKR and KQKN", *International Computer Chess Association Journal* **11**(1) (March 1988), 21–25.

[Rinck 1950] Henri Rinck, *1414 Fins de partie* [*1414 endgames*], La Academica, Barcelona, 1950.

[Rinck and Malpas 1947] Henri Rinck and Louis Malpas, *Dame contre tour et cavalier* [*Queen against rook and knight*], L'Echiquier, Bruxelles, 1947.

[Rosenberg 1988] Arnold Leonard Rosenberg, "Shuffle-oriented interconnection networks", Technical Report COINS 88-84, Computer and Information Science Department, University of Massachusetts at Amherst, 11 October 1988.

[Roycroft 1967] Arthur John Roycroft, "A note on 2S's v R + B", *EG* (ARVES, Amsterdam) **1** (April 1967), 197–198.

[Roycroft 1972] Arthur John Roycroft, *Test Tube Chess: A Comprehensive Introduction to the Chess Endgame Study*, Faber and Faber, London, 1972.

[Roycroft 1984] Arthur John Roycroft, "Two bishops against knight", *EG* (ARVES, Amsterdam) **5** (#75, April 1984), 249.

[Roycroft 1984] Arthur John Roycroft, "A proposed revision of the '50-move rule': Article 12.4 of the Laws of Chess", *International Computer Chess Association Journal* **7**(3) (September 1984), 164–170.

[Rusinek 1994] Jan Rusinek, *Selected Endgame Studies*, vol. 3: *Almost Miniatures: 555 Studies With Eight Chessmen*, University of Limburg, 1994.

[Salvio 1634] Alessandro Salvio, *Trattato dell'inventione et art liberale del gioco di scacchi* [*Treatise of the invention and liberal art of the game of chess*], Napoli, 1634.

[Schaeffer et al. 1992] Jonathan Schaeffer, Joseph Culberson, Norman Treloar, Brent Knight, Paul Lu, and Duane Szafron, "A World Championship caliber checkers program", *Artificial Intelligence* **53** (1992), 273–290.

[Schaeffer and Lake 1996] Jonathan Schaeffer and Robert Lake, "Solving the Game of Checkers", pp. 119–133 in this volume.

[Schwarz 1925] K. H. Schwarz, "Versuch eines mathematischen Schachprinzips" [An attempt toward a mathematical principle for chess], *Deutsche Schachzeitung* **53**(11) (November 1925), 321–324.

[Silver 1967] Roland Silver, "The group of automorphisms of the game of 3-dimensional Ticktacktoe", "*American Mathematical Monthly* **74** (March 1967), 247–254.

[Simons 1986] Geoff Leslie Simons, *Is Man a Robot?*, Wiley, Chichester, England, 1986.

[Sims 1971] Charles Coffin Sims, "Computations with permutation groups", pp. 23–28 in *Proceedings of the Second Symposium on Symbolic and Algebraic Manipulation, Los Angeles, 1971* (edited by Stanley Roy Petrick), ACM Press, New York 1971.

[Sims 1994] Charles Coffin Sims, *Computation With Finitely Presented Groups*, Encyclopedia of Mathematics and its Applications **48**, Cambridge University Press, Cambridge, England, 1994.

[Singleton 1967] Richard C. Singleton, "On computing the fast Fourier transform", *Communications of the ACM* **10** (1967), 647–654.

[Skillicorn 1993] David Benson Skillicorn, "Structuring data parallelism using categorical data types", pp. 110–115 in *Proceedings of the 1993 Workshop on Programming Models for Massively Parallel Computers*, Berlin, 1993, IEEE Computer Society Press, Los Alamitos, CA, 1993.

[Smitley 1991] David Smitley and Kent Iobst, "Bit-serial SIMD on the CM-2 and the Cray-2", *Journal of Parallel and Distributed Computing* **11** (1991), 135–145.

[Stamma 1737] Philip Stamma, *Essai sur le jeu des échecs* [*Essay on the game of chess*], Paris, 1737.

[Stiller 1989] Lewis Benjamin Stiller, "Parallel analysis of certain endgames", *International Computer Chess Association Journal* **12**(2) (June 1989), 55–64.

[Stiller 1991a] Lewis Benjamin Stiller, "Group graphs and computational symmetry on massively parallel architecture", *Journal of Supercomputing* **5** (1991), 99–117.

[Stiller 1991b] Lewis Benjamin Stiller, "Karpov and Kasparov: the end is perfection", *"International Computer Chess Association Journal* 14(4) (December 1991), 198–201.

[Stiller 1992a] Lewis Benjamin Stiller, "An algebraic foundation for Fortran 90 communication intrinsics", Technical Report LA-UR-92-5211, Los Alamos National Laboratory, August 1992.

[Stiller 1992b] Lewis Benjamin Stiller, "An algebraic paradigm for the design of efficient parallel programs", Technical Report JHU-92/26, Department of Computer Science, The Johns Hopkins University, Baltimore, November 1992.

[Stiller 1994] Lewis Benjamin Stiller, Luke L. Daemen, and James E. Gubernatis, *"n*-body simulations on massively parallel architectures", *Journal of Computational Physics* 115 (1994), 550–552.

[Stiller 1995] Lewis Benjamin Stiller, "Exploiting Symmetry on Parallel Architectures", Ph.D. thesis, Department of Computer Science, The Johns Hopkins University, Baltimore, 1995.

[Stone 1993] Harold Stuart Stone, *High-Performance Computer Architecture*, Addison-Wesley, Reading, MA, 1993.

[Ströhlein 1970] Thomas Ströhlein, "Untersuchungen über Kombinatorische Spiele" [Investigations of combinatorial games], Ph.D. thesis, Fakultät für Allgemeine Wissenschaften der Technischen Hochshule München, February 1970.

[Scientific American 1915] "Torres and his remarkable automatic devices, *"Scientific American Supplement*, 53 (November 1915), 296–298.

[Szymanski et al. 1994] Boleslaw K. Szymanski, James Hicks, R. Jagannathan, Vivek Sarkar, David Benson Skillicorn, and Robert K. Yates, "Is there a future for functional languages in parallel programming?" pp. 299–304 in *Proceedings of the 1994 International Conference on Computer Languages*, Toulouse, IEEE Computer Society Press, Los Alamitos, CA, 1994.

[Tarjan 1981a] Robert Endre Tarjan, "Fast algorithms for solving path problems", *Journal of the Association for Computing Machinery* 28 (1981), 594–614.

[Tarjan 1981b] Robert Endre Tarjan, "A unified approach to path problems", *Journal of the Association for Computing Machinery* 28 (1981), 577–593.

[Thompson 1983] [Kenneth Lane Thompson], *EG* (ARVES, Amsterdam) 5(74) (November 1983).

[Thompson 1986] Kenneth Lane Thompson, "Retrograde analysis of certain endgames", *International Computer Chess Association Journal* 9 (1986), 131–139.

[Thompson 1990] Kenneth Lane Thompson, personal communication, April 1990.

[Tolimieri and An 1990] Richard Tolimieri and Myoung An, "Computations in X-ray crystallography", pp. 237–250 of *Recent advances in Fourier analysis and its applications* (edited by J. S. Byrnes and Jennifer L. Byrnes) NATO ASI Series C: Mathematical and Physical Sciences 315, Kluwer, Dordrecht and Boston, 1990.

[Tolimieri et al. 1989] Richard Tolimieri, Myoung An, and Chao Lu, *Algorithms for Discrete Fourier Transform and Convolution*, Springer, New York, 1989.

[Tolimieri et al. 1993] Richard Tolimieri, Myoung An, and Chao Lu, *Mathematics of Multidimensional Fourier Transform Algorithms*, Springer, New York, 1993.

[Torres-Quevedo 1951] Gonzalo Torres-Quevedo, "Présentation des appareils de Leonardo Torres-Quevedo" [Presentation of the machines of Leonardo Torres-Quevedo], pp. 383–406 in *Les machines à calculer et la pensée humaine*, Paris, 1951, Colloques internationaux du Centre National de la Recherche Scientifique **37**, Publications du CNRS, Paris, 1953.

[Troitskiĭ 1934] Alekseii Alekseevich Troitskiĭ, "Два коня против пешек (теоретический очерк)" [Two knights against pawn (theoretical essay)], pp. 248–288 in Сборник шахматных этюдов [*Collection of chess studies*], Fizkul'tura i turizm, Moscow, 1934.

[Valiant 1990a] Leslie G. Valiant, "A bridging model for parallel computation", *Communications of the ACM* **33**(8) (1990), 103–111.

[Valiant 1990b] Leslie G. Valiant, "General purpose parallel architectures", pp. 945–971 in *Handbook of Theoretical Computer Science: Algorithms and Complexity*, volume A (edited by Jan van Leeuwen), Elsevier, Amsterdam, 1990.

[van, von...] See under last name.

[van de Velde 1994] Eric F. van de Velde, *Concurrent Scientific Computing*, Texts in Applied Mathematics **16**, Springer, New York, 1994.

[van der Waerden 1985] Bartel Leendert van der Waerden, *A History of Algebra: From Al-Khwārizmī to Emmy Noether*, Springer, Berlin/New York, 1985.

[Wendroff et al. 1993] Burton Wendroff, Tony Warnock, Lewis Benjamin Stiller, Dean Mayer, and Ralph Brickner, "Bits and pieces: constructing chess endgame databases on parallel and vector architectures", *Applied Numerical Mathematics* **12** (1993), 285–295.

[White 1984] Arthur Thomas White, *Graphs, Groups, and Surfaces*, North-Holland/ Elsevier, New York, 1984.

[White 1975] Dennis Edward White, "Multilinear enumerative techniques, "*Linear and Multilinear Algebra* **2** (1975), 341–352.

[Wussing 1969] Hans Wussing, *Die Genesis des abstrakten Gruppenbegriffes: ein Beitrag zur Entstehungsgeschichte der abstrakten Gruppentheorie*, Deutscher Verlag der Wissenschaften, Berlin, 1969. English translation by Abe Shenitzer: *The Genesis of the Abstract Group Concept: A Contribution to the History of the Origin of Abstract Group Theory*, MIT Press, Cambridge, MA, 1984.

[Zermelo 1913] Ernst Zermelo, "Über eine Anwendung der Mengenlehre auf die Theorie des Schachspiels" [On an application of set theory to the theory of playing chess], pp. 501–504 in *Proceedings of the Fifth International Congress of Mathematicians*, Cambridge, 1912, vol. 2 (edited by Ernest William Hobson and Augustus Edward Hough Love), Cambridge University Press, England, 1913.

LEWIS STILLER
DEPARTMENT OF COMPUTER SCIENCE
THE JOHNS HOPKINS UNIVERSITY
BALTIMORE, MD 21218
    stiller@symmetry.xo.com

Games of No Chance
MSRI Publications
Volume **29**, 1996

Using Similar Positions to Search Game Trees

YASUHITO KAWANO

ABSTRACT. We propose a method that uses information on similar positions to solve Tsume-shogi problems (mating problems in Japanese chess). Two notions, priority and simulation, are defined for this method. A given problem is solved step-by-step according to the priority. Simulation often allows us to omit searching on each step. A program made by the author and based on this method solved a celebrated problem requiring over 600 plies (half-moves) to mate.

1. Introduction

It would be useful in board games if the experience learned from one position could be applied to other, similar, positions. However, positions that seem to only slightly differ on the board can actually differ greatly in terms of strategy. This paper proposes a method that uses information on similar positions to solve *Tsume-shogi* problems (mating problems in Japanese chess, the rules of which are explained in Section 2).

The method is based on the notions of priority and simulation. Priority means that we look at certain moves of our opponent earlier than others. Intuitively, we say that a position **P** simulates another position **Q** if we can mate on **Q** according to the mating sequence of **P**. Roughly speaking, our problem-solving strategy has two steps. First, we restrict ourselves to our opponent's moves of high priority. Second, we check whether the position **P** where the solution has already been obtained simulates a position **Q** having a lower priority. If **P** simulates **Q**, we can dispense with the search for the latter position.

We discuss a Tsume-shogi program by the author, based on this method and considered to be one of the strongest Tsume-shogi programs available today. It was the first program to solve a celebrated and long-unsolved (by computer) problem, requiring over 600 plies (half-moves) to mate.

2. Shogi and Tsume-shogi

Shogi is a board game, similar in origin and feel to chess [Fairbairn 1989; Wilkes 1950; Leggett 1966]. It was established around the fourteenth century in Japan. The biggest difference between Shogi and chess is that captured pieces in Shogi can be put back into play, on the capturer's side, so the number of pieces in play tends to remain high.

Shogi is played on a 9×9 board, with, initially, twenty pieces for each player. All pieces are the same color (so they can be reused by the opponent), ownership being distinguished by the way they face. In diagrams, pieces are represented by Kanji characters, but here we use the initial letters of their English translations (with a circumflex on top of S and N, which resemble their upside-down counterparts). The three rows nearest a player are than player's domain. A piece is *promoted* when it moves into, inside, or out of the opponent's domain, and it remains promoted until it is taken. Here are the names and movements of the pieces:

Unpromoted		Promoted	
↑	P: Pawn	⬚	PP: Promoted Pawn
↑	L: Lance	⬚	PL: Promoted Lance
V	Ñ: Knight	⬚	PN: Promoted Knight
⬚	Ŝ: Silver	⬚	PS: Promoted Silver
⬚	G: Gold	—	
✕	B: Bishop	✳	PB: Promoted Bishop
✛	R: Rook	⬚	PR: Promoted Rook
⬚	K: King	—	

The Pawn can move only one square straight ahead. It does not have the chess pawn's initial move or diagonal capturing move. The Lance moves straight ahead. The Knight moves two squares forward and one sideways (a subset of the moves of the chess knight); it is the only piece that can jump over others. The Silver can move one square forward in any direction or one square diagonally backward. The Gold can move one square forward in any direction, one square to the side, or one square straight back. When the Pawn, Lance, Knight, and Silver are promoted, their allowed moves become the same as the Gold's. The Rook and Bishop have the same moves as their chess counterparts. When they are promoted, they have the power to move one square in the other four directions. The King can move one square in any direction. The Gold and the King, however, can not be promoted.

Instead of a regular move, a player can *drop* a piece previously captured by him into a vacant square, provided this does not break one of the following rules:

DOUBLED PAWN: When a player's Pawn is already on a column, he is not allowed to drop another Pawn on the same column.

DROPPED-PAWN MATE: Mating by dropping a Pawn is prohibited. (It is legal to mate by moving a Pawn that is already on the board.)

DEADLOCK: It is illegal to move or drop a piece onto a square if it cannot move from there.

Here is the initial Shogi set-up:

L	Ñ	Ŝ	G	K	G	Ŝ	Ñ	L
	R						B	
P	P	P	P	P	P	P	P	P
P	P	P	P	P	P	P	P	P
	B					R		
L	Ñ	Ŝ	G	K	G	Ŝ	Ñ	L

Tsume-shogi is a Shogi mating problem. The data is a Shogi position; a solution consists in a sequence of legal moves satisfying several conditions:

- Each move of the mating player must be optimal, in the sense of leading to mate as soon as posbible.
- The position after each move of the mating player must be a check.
- Each response of the opponent must be optimal, in the sense of delaying mate as long as possible. (Interpositions by moving or dropping a piece must not be *muda*: this term is explained below.)

The chain of moves is called *Hontejun* (main sequence); its length, the number of plies, is called *Tsume-tesu*. The main sequence should be unique; more specifically, in the positions of the main sequence, each mating player's move except for the last one must be unique, otherwise the problem is regarded as defective.

It is said that over 100,000 complete Tsume-shogi problems have already been published. For most of them, mating is completed in less than twenty moves, but there are also hundreds of problems requiring more than 100 plies. At present, the 1519-ply "Microcosmos" problem is believed to be the one requiring the largest number of moves to complete.

We call the solution of Tsume-shogi a *mating sequence* in this paper.

3. Tsume-shogi Programs

Tsume-shogi programs use one of two approaches: iterative deep search, or best-first search. At present, the best programs using iterative deep search can solve problems with 25 or less plies without fail and in a practical time. They do, however, tend to fail on problems with 27 or more plies since a lot of time is needed. On the contrary, programs using the idea of best first search often solve problems requiring large numbers of moves in a short time; but they cannot be guaranteed to solve any given problem [Ito and Noshita 1994; Matsubara 1993]. Besides this, a hardness result for generalized Tsume-shogi problems has been shown in [Adachi et al. 1987].

Because of the possibility of dropping, the number of candidate move is usually much larger than in chess. This rule causes problems not found in computer chess. One big problem is the treatment of interposed pieces. Consider the following easy problem in Tsume-shogi:

9	8	7	6	5	4	3	2	1	
								К	a
								d	b
									c
									d
									e
									f
									g
									h
R				K			L		i

no piece in hand

The solution to this problem starts with the move 9i–9a+. (We will use * to denote a dropped piece and + for promotion.) However, the opponent can drop any piece on any square between the Promoted Rook and the King, because all absent pieces are in the opponent's hand. This can be done seven times, so the mating sequence is 15-ply long. These interposed pieces are called *muda-ai*, and human players usually ignore them, because they just prolong the mating sequence without altering it in any essential way. In contrast, a computer will search through hundreds of thousands of positions if it does not understand muda-ai.

Muda-ai has no formal definition. Informally, it means any interposed piece incapable of essentially changing the mating sequence. Yoshikazu Kakinoki [1990] proposed an algorithm to determine muda-ai. He defines muda-ai recursively as follows:

(i) An interposed piece is muda if it is captured at the moment the King is mated.

(ii) Suppose the King is mated after n moves in the absence of a given interposition. The interposed piece is muda if it is immediately captured and not reused, and the King is mated in n moves, not counting the interposing move and the capture.

Although there exists no exact algorithmical expression of muda-ai, it is known that Kakinoki's algorithm determines muda-ai precisely in most cases, so his criterion is regarded as a standard. In practice, a definition that "localizes" the concept of muda-ai by basing it on "imaginary pieces" is adopted by many programmers. More precisely, suppose the opponent has an infinite supply of pieces to drop. We judge whether the interposition of such an imaginary piece at a given square is muda by Kakinoki's algorithm, and we decree that the mudaness of a real dropped interposition at that square is the same. This criterion for muda-ai sometimes leads to different results from Kakinoki's original criterion (which takes into account whether or not the opponent actually has additional pieces to drop), but it is computationally much more manageable, since it avoids an explosion of the search space.

4. Using Similar Positions

Since muda-ai reflects the essential equality of mating sequences, Kakinoki's algorithm should be recognized as a method capable of determining this essential equality and its applications. We consider the extension of this decision of muda-ai.

Priority. We introduce the notion of priority before we define the notion of simulation. Priority is the total searching order of the opponent's next moves at a position: moves of low priority will not be searched for until moves of high priority are searched for. The same priority is used with moves of corresponding positions. In Kakinoki's algorithm, moves other than interpositions have the highest priority, and interpositions near the King have higher priorities than those farther away. In addition, for example, captures by many opponent's pieces except for the King are noted as low priorities. The priority is set in the following discussion.

Though priority is defined locally, we can extend it to a partial order on a set of moves (nodes) in a game tree for a given problem. We set the priority of each mating player's move to be equal to its parent's. We also introduce the following relations:

(a) The problem (the root node) has the highest priority in the game tree,

(b) The priority of a move having the highest priority in the local sense is the same as its parent's, and

(c) The priority of a move not having the highest priority in the local sense is lower than its older siblings' descendents'.

We solve each given problem according to this extended priority. We first restrict ourselves to our opponent's moves of highest priority. Each position Q not having highest priority goes unnoticed until we solve the older sibling of the root node of the subtree consisting of moves having the same priority as Q. If there exists a position P that simulates the position Q, we can dispense with the search for Q. To find such a position P, we appoint candidates beforehand for each position Q, where the candidates satisfy the relation with Q in high probability; otherwise, we use a transposition table to find candidates.

Simulation. We define a relation $P \geq Q$ between positions, read "Q simulates P". Let $\mathrm{pri}(m)$ and $\mathrm{pri}(Q)$ be the priorities of m and Q, where m is a move and Q is a node of the game tree. Let $P^{\mathrm{pri}(Q)}$ be the problem obtained from P by restricting ourselves to moves m of our opponent that satisfy the condition $\mathrm{pri}(m) \not\leq \mathrm{pri}(Q)$. We write $P \geq Q$ when we can mate the King of $P^{\mathrm{pri}(Q)}$ and we can mate the King of Q according to one of the mating sequences for $P^{\mathrm{pri}(Q)}$. Formally, we define $P \geq Q$ recursively as follows:

(i) Q is a position for a mating player's move. Let m be the first move of one of the mating sequences for $P^{\mathrm{pri}(Q)}$. If m is not a checking move for Q, then $P \not\geq Q$. Otherwise, let P', Q' be the positions obtained from P, Q by move m. By definition, $P \geq Q$ if and only if $P' \geq Q'$.

(ii) Q is a position for an opponent's move. If Q is a mate, set $P \geq Q$. If Q is not a mate, generate all of the opponent's next moves m_i, for $i \in \Lambda$. If they are not properly included in the set of next moves of $P^{\mathrm{pri}(Q)}$, then $P \not\geq Q$. Otherwise, let P'_i and Q'_i be the positions obtained from P and Q by move m_i. By definition, $P \geq Q$ if and only if $P'_i \geq Q'_i$ for all $i \in \Lambda$.

Unfortunately, the relation $\geq$ is not suitable for application to searching in Tsume-shogi, even though it is natural. For example, an interposed piece at in the first of these diagrams would be muda, although $\mathbf{P} \not\geq \mathbf{Q}$:

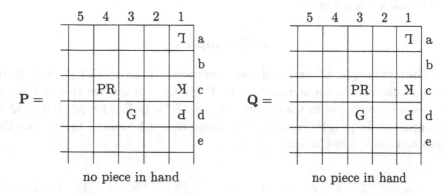

$$\mathbf{P} = \qquad\qquad\qquad \mathbf{Q} =$$

no piece in hand   no piece in hand

The reason why $\mathbf{P} \not\geq \mathbf{Q}$ is this. In $\mathbf{Q}$, we assume that the opponent will drop a Pawn on 2c. The recorded next move of $\mathbf{P}^{\mathrm{pri}(\mathbf{Q})}$ is 4c–2c. But since 4c in $\mathbf{Q}$ is vacant, $\mathbf{Q}$ does not simulate $\mathbf{P}$.

To avoid cases such as this, in which the pieces cannot be moved, we introduce the relation $\geq_f$, which permits the mating player to choose the next move in the above definition. We introduce a choice function for the mating player $f :$ $(m, \delta) \mapsto \Omega_{m\delta}$, where we denote the difference of two positions by δ, and the set of candidates by $\Omega_{m\delta}$. We denote by $\geq_f$ the binary relation defined by replacing case 1 in the definition of $\geq$ with the following sentence:

(i) $\mathbf{Q}$ is a position for a mating player's move. Let m be the first move of one of mating sequences of $\mathbf{P}^{\mathrm{pri}(\mathbf{Q})}$, and let Ω be the set of checking moves in $\Omega_{m\delta}$ for $\mathbf{Q}$. If Ω is empty, then $\mathbf{P} \not\geq_f /\mathbf{Q}$. Otherwise, let $\mathbf{P}'$ and $\mathbf{Q}'$ be the positions obtained from $\mathbf{P}$ and $\mathbf{Q}$ by a move $m' \in \Omega$. By definition, $\mathbf{P} \geq_f \mathbf{Q}$ if and only if there exists $m' \in \Omega$ such that $\mathbf{P}' \geq_f \mathbf{Q}'$.

We now define muda-ai as follows.

DEFINITION. Let $\mathbf{P}$ be a given position in which an interposed piece can forestall a check. Let $\mathbf{Q}$ be a position obtained by capturing the interposed piece immediately, checking, and eliminating the captured piece from the mating player's hand. The interposed piece is muda if and only if $\mathbf{P} \geq_f \mathbf{Q}$.

Under this definition, an interposition at 3c in the preceding example is muda, since $\mathbf{P} \geq_f \mathbf{Q}$ holds, where the difference δ is based on the positions of the PR, and we assume that $f(4c-?, \delta)$ contains 3c–?.

The relation $\geq_f$ depends on the function f. The safety of our method is guaranteed for any function f, since the following theorem holds independently of this choice.

THEOREM. *Suppose* $\mathbf{P} \geq_f \mathbf{Q}$. *Then we can mate in* $\mathbf{Q}$, *and there exists a mating sequence in* $\mathbf{P}^{\mathrm{pri}(\mathbf{Q})}$ *whose length is greater than or equal to the number of moves to mate the King of* $\mathbf{Q}$.

COROLLARY. *An interposed piece that is muda under our definition is muda under Kakinoki's definition.*

5. Examples

Another example of determining muda-ai. An interposed piece at 3c in the next diagram is non-muda, that is, $P \not\geq_f Q$. We can see this as follows. The recorded response to the opponent's move P*2c in P is 5a–2d. But in Q, if the opponent drops a Pawn at 2c, the Bishop cannot be moved to 2d, since the Promoted Rook is in the way.

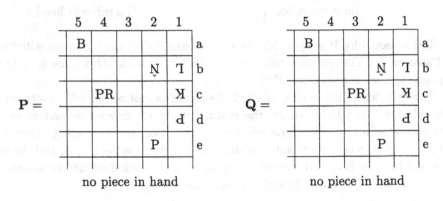

An example of a one-piece difference. The next diagram is an example of $P \geq_f Q$ and $Q \not\geq_f P$. We shall show that $Q \not\geq_f P$.

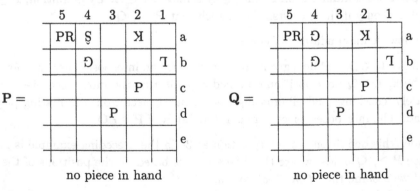

(i) The recorded next move of Q is 5a–4a. Therefore, we move the PR on P.

(ii) In the resulting position, we assume that the opponent captures this PR with his Gold.

(iii) We look at the recorded next move for the moves 5a–4a and 4b–4a in Q. It is G*2b.

(iv) We assume that $f(G*2b, \delta)$ contains S*2b here.

(v) We cannot find the opponent's next move 2a-3b in the mating sequence of $Q^{pri(P)}$.

6. Final Remarks

The proposed best-first search algorithm was implemented by the author. It performs very well. For example, it is said that today's human experts, regardless of the time spent, can only solve about 80% of the problems posed in the well-known "Shogi Zuko" [Kanju 1755], written in the Edo period. This program, however, can solve 85% of them in a short time; other programs can solve at most 70%. The very famous last problem, "Kotobuki" (happiness), is illustrated below. It requires 611 plies, the longest mating sequence of any Tsume-shogi from the Edo period. The author's program solved it in about 8 minutes on an HP9000 workstation (model735/99MHz). Details are given in [Ito et al. 1995].

9	8	7	6	5	4	3	2	1	
Ɔ				ꓒ			ꓤꓒ		a
		ꓕ	P	Ɔ			ꓒ		b
	Ş					ꓒ	ꓒ	ꓒ	c
	ꓕ		B						d
L			ꓭ		ꓘ	P	L		e
		ꓒꓒ					PR	ꓒꓒ	f
		ꓒꓒ				Sꓒ		Ŝ	g
Nꓒ		ꓒꓒ		Ɔ	Ŝ				h
	Ɔ	ꓒꓒ			P		L		i

four pawns in hand

Acknowledgement

The author wishes to thank Professor Kohei Noshita, Dr. Kiyoshi Shirayanagi, Dr. Kenji Koyama, and all the members of the CSA (Computer Shogi Association) for their helpful suggestions.

Note Added in Proof

After this paper had been written, Masahiro Seo pubished new results of his research. His Tsume-shogi program can solve 99% of problems in [Kanju 1755]. See [Seo 1995] for details.

References

[Adachi et al. 1987] H. Adachi, H. Kamekawa and S. Iwata, "Shogi on $n \times n$ board is complete in exponential time" (in Japanese), *Trans. IEICE* **J70-D**, 1843–1852.

[Fairbairn 1989] J. Fairbairn, *Shogi for Beginners*, Ishi Press, Tokyo, 1989.

[Ito and Noshita 1994] T. Ito, K. Noshita, "Two fast programs for solving Tsume-shogi and their evaluation" (in Japanese), *J. Information Processing Soc. Japan*, **35** (1994), 1531–1539.

[Ito et al. 1995] T. Ito, Y. Kawano, K. Noshita, "On the algorithms for solving Tsume-shogi with extremely long solution-steps" (in Japanese), *J. Information Processing Soc. Japan* **36** (1995), 2793–2799.

[Kanju 1755] Kanju Ito, "Shogi Zuko", 1755. Included in Yoshio Kadowaki, *Tsumuya Tsumazaruya* (in Japanese), Toyo bunko 282, Heibon-sha, 1975.

[Kotani et al. 1990] Y. Kotani, T. Yoshikawa, Y. Kakinoki, K. Morita, "Computer Shogi" (in Japanese), Saiensu-sha, 1990.

[Leggett 1966] T. Leggett, *Shogi: Japan's game of strategy*, C. E. Tuttle, Rutland, VT, 1966.

[Matsubara 1993] H. Matsubara, "Shogi (Japanese chess) as the AI research target next to chess", Technical Report ETL-TR-93-23, Electrotechnical Laboratory (Ministry of International Trade and Industry), Tsukuba, Japan, 1993.

[Seo 1995] M. Seo, "Application of conspiracy numbers to a Tsume-shogi program" (in Japanese), *Proceedings of Game Programming Workshop*, Japan 1995, pp. 128–137.

[Wilkes 1950] C. F. Wilkes, *Japanese chess: Shogi* (second edition), Yamato Far Eastern Enterprises, Berkeley, CA, 1950.

YASUHITO KAWANO
NTT COMMUNICATION SCIENCE LABORATORIES
HIKARIDAI 2-2, SEIKA-CHO, SORAKU-GUN KYOTO, JAPAN
   kawano@progn.kecl.ntt.jp

Games of No Chance
MSRI Publications
Volume **29**, 1996

Where Is the "Thousand-Dollar Ko"?

ELWYN BERLEKAMP AND YONGHOAN KIM

ABSTRACT. This paper features a problem that was composed to illustrate
the power of combinatorial game theory applied to Go endgame positions.

The problem is the sum of many subproblems, over a dozen of which
have temperatures significantly greater than one. One of the subproblems
is a conspicuous four-point ko, and there are several overlaps among other
subproblems. Even though the theory of such positions is far from complete, the paper demonstrates that enough mathematics is now known to
obtain provably correct, counterintuitive, solutions to some very difficult
Go endgame problems.

1. Introduction

Consider the boards in Figures 1 and 2, which differ only by one stone (the
White stone a knight's move south-southwest of E in Figure 1 is moved to a
knight's move east-southeast of R in Figure 2). In each case, it is White's turn
to move. Play will proceed according to the Ing rules, as recommended by the
American Go Association in February 1994. There is no komi. Both sides have
the same number of captives when the problem begins.

We will actually solve four separate problems. In Figure 1, Problem 1 is
obtained by removing the two stones marked with triangles, and Problem 2 is
as shown. Does the removal of the two stones matter? In Figure 2, Problem 3
is obtained by removing the two marked stones, and Problem 4 is as shown.
(Problem 3 appeared on the inside cover of *Go World*, issue 70.)

In each case, assume that the winner collects a prize of $1,000. If he wins by
more than one point, he can keep the entire amount, but if he wins by only one
point, he is required to pay the loser $1 per move. How long will each game last?

2. Summary of the Solutions

We think that White wins Problem 1 by one point, but Black wins Problem 2
by one point.

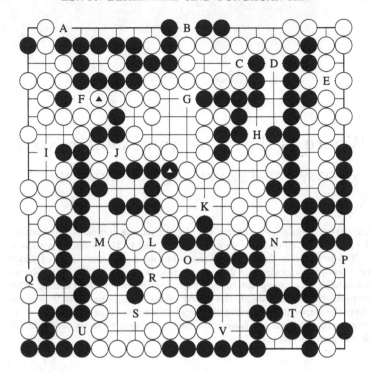

Figure 1. Starting position of Problems 1 and 2.

Despite an enormous number of possible lines of play, the assertions that suggest that conclusion are all proved mathematically, without computers. Surprisingly little read-ahead or searching is required.

For the first 53 moves, the canonical play of both problems is identical. At move 7, White must play a nominally smaller move in order to reduce the number of Black kothreats. Black begins the ko at the point marked E on move 12. But then, even though White has enough kothreats to win the ko, he must decline to fight it! Black fills the ko on move 14.

The ko located near the point marked E in Figure 1 is the only ko that is clearly visible in the problem statements shown in Figure 1. It became mistakenly known as the Thousand-Dollar Ko. It was difficult enough to stump all of the contestants in a recent contest sponsored by Ishi Press, International.

After all moves whose unchilled mathematical temperature exceeds one point (or equivalently, "two points in gote") have been played, there is an interesting and nontrivial contest on the regions that chill to infinitesimals. Using methods described in *Mathematical Go* [Berlekamp and Wolfe 1994], one shows that White has a straightforward win in Japanese scoring. However, under the Ing rules, one of the regions can assume a slightly different value because of a potential small Chinese ko. Black must play under the assumption that he can win this ko. This ko, which is difficult to foresee, is the real Thousand-Dollar Ko. It begins to affect the play soon after the first 20 moves of either problem.

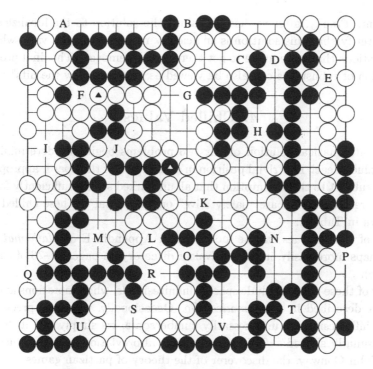

Figure 2. Starting position of Problems 3 and 4.

The players next play the regions that chill to numbers (or, in Go jargon, moves worth less than two points in gote). Because each dame can play a decisive role in the forthcoming Chinese kofight, this subsequence of play requires considerable care.

Under Japanese scoring, the game ends at move 53 with a one-point win for White. This is true in every version of the original problem.

Move 54 marks the beginning of final preparations for the decisive thousand dollar kofight, which is *not* located near E. Beginning at move 56, optimal play of Problems 1 and 2 diverge significantly. In Problem 2, Black attacks White's kothreats and wins most quickly by removing as many of them as she can before the kofight begins. But in the Problem 1, which Black cannot win, she prolongs the game by creating more kothreats of her own. White can then win only by starting the ko surprisingly early, while many nominally "removable" kothreats still remain on the board.

3. Mathematical Modeling

The philosophy that we follow is common in applied mathematics. We begin by making various assumptions that simplify the problem to one that is more easily modelled mathematically. Then we solve the mathematical problem, using whatever simplifying approximations and assumptions seem reasonable and convenient. Finally, after we have obtained the solution to the idealized problem, we verify that this solution is sufficiently robust to remain valid when the simplifications that were made to get there are all removed. The final process of verification of robustness can also be viewed as "picking away the nits".

4. Chilled Values

The mathematical solution of our two problems begins by determining the chilled value of each nontrivial position in each region of the board, appropriately marked, subject to the assumption that all unmarked stones sufficiently far away from the relevant letter are "safely alive" or "immortal". These chilled values are shown in Table 1.

Most of these values can be looked up in Appendix E of *Mathematical Go* (or, perhaps more easily, in the equivalent directory on pages 71–76 of the same reference).

Many of these values include specific infinitesimals, whose notations and properties are described in *Mathematical Go*. Occasionally, when the precise value of some infinitesimal is unnecessarily cumbersome, we simply call it "ish", for "infinitesimally shifted". This now-common abbreviation was introduced long ago by John Conway, the discoverer of the theory of partizan games.

Values must also be computed for those regions that do not appear verbatim in the book. Such calculations are done as described in Chapter 2 of *Mathematical Go*. As an example of such a calculation, we consider the region Q. We have $Q = \{Q^L \mid Q^R\}$, where Q^R is the position after White has played there and Q^L is the position after black has played there. After appropriate changes of markings (which affects only the integer parts of the relative score), we may continue as follows:

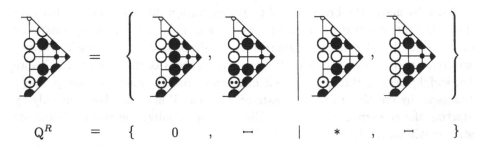

$$Q^R \quad = \quad \{ \quad 0 \quad , \quad \multimap \quad | \quad * \quad , \quad \multimap \quad \}$$

Although 0 and $\multimap$ are both formal Black followers of Q^R, the miny option is dominated and we have

$$Q^R = \{0 \mid *, \multimap\}.$$

The two formal White followers are $*$ and $\multimap$, which are incomparable. However, White's move from Q^R to $\multimap$ reverses through a hot position to 0, and we

region	chilled value	Δ^L	Δ^R
A	$0^3 \mid +\!\!+$	$\leftharpoonup \mid 0^3$	$\{0^3 \mid +\!\!+\}\{\leftharpoonup \mid 0^2\}$
B	$1\frac{3}{4} \mid -2$		$3\frac{3}{4} \mid 0$
C	$-\frac{1}{4}$		$-\frac{1}{4}$
DH	$\frac{1}{8}$		$\frac{1}{8}$
E	$ko[1,0]$		
F	$*$		$*$
G	$\{\frac{5}{8} \parallel \leftharpoonup_{1/2} \mid 0\} \mid -2$		$2\frac{5}{8} \mid 2\{\leftharpoonup_{1/2} \mid 0\} \parallel 0$
G^L	$\frac{5}{8} \parallel \leftharpoonup \mid 0$		$\frac{5}{8}\{0 \mid +\!\!+\} \mid 0$
I	$1 \mid -\frac{1}{2}$		$1\frac{1}{2} \mid 0$
J	$4 \mid -5$		$9 \mid 0$
K	$0 \mid -\frac{1}{2}$		$\frac{1}{2} \mid 0$
M	$\frac{5}{4} \mid -\frac{1}{2}$		$1\frac{3}{4} \mid 0$
$L[S^L]$	θ		
N	$0 \mid +\!\!+ \parallel -\frac{7}{8} \mid -1$		$\{1 \mid \frac{7}{8}\}\{0 \mid +\!\!+\} \parallel 0$
N^R	$-\frac{7}{8} \mid -1$		$\frac{1}{8} \mid 0$
O	$\approx +\!\!+_{13}$	$\approx \leftharpoonup_{13}$	$\approx \{13 \mid 0\}+\!\!+_{13}$
P	$\{12-\theta\} \mid \{-12+\theta\}$		$\approx 24 \mid 0$
Q	$2ish \mid -2{\downarrow}*$		$4ish \mid 0$
$R[S^L]$	θ		
S	$\{\frac{1}{2}+\theta\theta\} \mid -1{\downarrow}{\downarrow}{\downarrow}{\downarrow}*$		$1\frac{1}{2}\theta\theta{\uparrow}{\uparrow}{\uparrow}* \mid 0$
T	$\approx +\!\!+_6$	$\approx \leftharpoonup_6$	$\approx \{6 \mid 0\}+\!\!+_6$
U	$*$		$*$
V	$2 \parallel \mid 0 \mid +\!\!+_1 \parallel -1*$		$3* \mid 2\{\leftharpoonup_1 \mid 0\} \parallel 0$
V^R	$0 \mid +\!\!+ \parallel -1*$		$1*\{0 \mid +\!\!+\} \mid 0$

Table 1. Chilled value and incentives of each region. For most regions, Δ^L and Δ^R are the same. The symbol θ is defined on page 209.

have the canonical form

$$Q^R = \{0 \mid *, 0\} = {\downarrow}*.$$

To compute Q^L, we begin with the standard assumption that "sufficiently distant stones are safely alive", which in this case suggests that we assume White has already played from I twice to reach I^{RR}. In that environment, it is easily seen that, after an appropriate adjustment of markings, $Q^{LLL} = \leftharpoonup$ and $Q^L = \leftharpoonup \mid 0^2$.

We then obtain the value of

$$Q = \{2\{- \mid 0^2\} \mid -2{\downarrow}*\}.$$

Other values shown in Table 1 are computed by similar calculations.

In general, values of games may be conveniently partitioned into three sets: cold, tepid, and hot. Cold values are numbers. In Table 1, the regions with cold values are DH and C. Tepid values are infinitesimally close to numbers. In Table 1, tepid values are A, F, L, O, T, and U. The other values shown in Table 1 are hot. In general, hot moves, tepid moves, and cold moves should be played in that order. So, to determine the first phase of play, we will need to determine the order to play the hot moves in Table 1.

5. Incentives

As explained in Section 5.5 of *Mathematical Go*, a useful way to compare different choices of moves is to compute their incentives. In general, the game

$$Z = \{Z^L \mid Z^R\}$$

has a set of Left incentives of the form

$$\Delta^L(Z) = Z^L - Z,$$

where there are as many Left incentives as there are Left followers Z^L. Similarly, the Right incentives of this game have the form

$$\Delta^R(Z) = Z - Z^R.$$

Calculations of canonical incentives often require considerable attention to mathematical detail. For example, consider a generic three-stop game

$$G = \{x \parallel y \mid z\}$$

where x, y, and z are numbers with $x > y > z$. Then

$$\begin{aligned}
\Delta^L(G) &= G^L - G \\
&= x - \{x \parallel y \mid z\} \\
&= x + \{-z \mid -y \parallel -x\} \\
&= \{x-z \mid x-y \parallel 0\}
\end{aligned}$$

by the *Number Translation Theorem* on page 48 of *Mathematical Go*. But

$$\begin{aligned}
\Delta^R(G) &= G - G^R \\
&= \{x \parallel y \mid z\} + \{-z \mid -y\} \\
&= \{\{x-z \mid x-y\}, \{x-z \parallel y-z \mid 0\} \mid 0, \{x-y \parallel 0 \mid z-y\}\}.
\end{aligned}$$

Deleting Right's dominated option gives

$$\Delta^R(G) = \{\{x-z \mid x-y\}, \{x-z \parallel y-z \mid 0\} \mid 0\}.$$

The second Left option is dominated if $x-y \geq \{y-z \mid 0\}$ or, equivalently, if $x - y \geq y - z$, or if $x + z \geq 2y$. In that case, $\Delta^R(G) = \Delta^L(G)$. However, if $x - y < y - z$, then $\Delta^R(G) > \Delta^L(G)$. Furthermore, if $2(x - y) < y - z$, then

$$\text{temperature}(\Delta^R(G)) \geq \text{temperature}(\Delta^L(G)).$$

Most of the positions shown in Table 1 have a single canonical Left incentive and a single canonical Right incentive. Although these two are often formally different, further calculation often reveals their values to be equal. In Table 1, this happens on all rows except A, E, L, O, R, and T.

If Black plays from S to S^L, he converts each of the regions L and R into a Black kothreat. Although each of these kothreats has a mathematical value of zero, it can, of course, be used to affect the outcome of the ko. We denote each such kothreat by the Greek letter θ, because this letter looks somewhat like 0 and because "theta" and "threat" start with the same sound. Similarly, a White kothreat can be denoted by $-\theta$.

Incentives are games. Just like any pair of games, a pair of incentives may be comparable or incomparable. Fortunately, it happens that most pairs of incentives shown in Table 1 are comparable. Except for a possible infinitesimal error in one case (another "nit" that will be picked away much later), the incentives can be conveniently ordered as shown in Table 2. The top portion of this table shows the positions P, J, Q, ..., N^R, each of which has a unique incentive, which is the same for both Left and Right, and which is greater-ish than the incentives of all positions listed below it and less-ish than incentives of all positions listed above it.

Notice that some positions with nonzero values, such as minies or tinies, may also serve as kothreats for Black or White, respectively. Although their *values* are infinitesimal or zero, tinies, minies, and θ's of either sign, all of them have hot *incentives* for at least one player, and so we will need to include them in our studies of "hot" moves.

Either player will always prefer to play a move with maximum incentive. So, (ignoring a potential infinitesimal nit) each player will prefer to play on P, J, Q, B, ..., in the order listed in the top portion of Table 2.

There are some hot moves that cannot be so easily fit into the ordering in the top portion of Table 2: the ko at E, potential White kothreats at T and O, potential Black kothreats at P^R, $R[S^L]$, and $L[S^L]$, and the position at S, whose play creates or destroys two Black kothreats. These moves are listed in the bottom portion of Table 2, in the order in which they are expected to be played: S will be played before the ko; once the kofight starts, kothreats will alternate; and finally someone will eventually fill the ko.

Good play must somehow intersperse the ordered set of moves shown at the bottom of Table 2 into the ordered set shown at the top. In order to do this, it is very helpful to consider the *pseudo-incentives* of E and S.

P
|
J
|
Q
|
B
|
V
|
G
/ |
S   M
\ |
    I
    |
    V^R
  /
E   N
|
   G^L
E^L   K
|
   N^R
Infinitesimals
|
Numbers

region	Δ	I	II	III	IV	V
P	$\approx 24\,\vert\,0$	1	1	1	1	1
J	$\approx 9\,\vert\,0$	2	2	2	2	2
Q	$4ish\,\vert\,0$	3	3	3	3	3
B	$3\frac{3}{4}\,\vert\,0$	4	4	4	4	4
V	$3*\,\vert\,2\{-_1\,\vert\,0\}\,\Vert\,0$	5	5	5	5	5
G	$2\frac{5}{8}\,\vert\,2\{-_{1/2}\,\vert\,0\}\,\Vert\,0$	6	6	6	6	6
M	$1\frac{3}{4}\,\vert\,0$	8	8	8	7(?)	7(?)
I	$1\frac{1}{2}\,\vert\,0$	9	9	9	9	9
V^R	$1*\{0\,\vert\,+\}\,\vert\,0$	10	10	10	10	10
N	$\{1\,\vert\,\frac{7}{8}\}\{0\,\vert\,+\}\,\Vert\,0$	11	11	11	11	12
G^L	$\frac{5}{8}\{0\,\vert\,+\}\,\vert\,0$	13	22	12(?)	13	13
K	$\frac{1}{2}\,\vert\,0$	15	24	13	14	14
N^R	$\frac{1}{8}\,\vert\,0$	16	25	24	31	
S	$1\frac{1}{2}\theta\theta{\uparrow}{\uparrow}{\uparrow}{\uparrow}*\,\vert\,0$	7	7	7	8	8
E	ko	12	12	14	12	
E^L	$-ko$				15	
$R[S^L]$	$\Delta\theta$				16	
E	ko				18	
T	$-\Delta\theta$		13(?)	15	19	
E^L	$-ko$		15	17	21	
P^R	$\Delta\theta$		16	18	22	
E	ko		18	20	24	
O	$-\Delta\theta$		19	21	25	
E^L	$-ko$		21	23	27	
$L[S^L]$	$\Delta\theta$				28	
E	ko				30	
E^L	$-ko$					
kofill		14	23	25	32	11
totals		0	$0\,\vert\,+$	$-\frac{3}{8}ish$	$\frac{1}{8}ish$	$\frac{1}{8}ish$

Table 2. Sorted incentives (Δ) for the regions, and the dominant lines of hot play. Columns I–V indicate different possibilities for the order in which the moves can be made (see Section 7). Regions of Table 1 that have no hot incentives are omitted.

6. Pseudo-Incentives Related to the Ko at E

Suppose we have a set of games $A, B, C, \ldots, K, L, \ldots$ with ordered incentives $\Delta A > \Delta B > \Delta C > \cdots > \Delta K > \Delta L > \cdots$. Suppose also that there is a kofight in progress and that White has six more kothreats, $-\theta[6], -\theta[5], \ldots, -\theta[1]$. Then a competently played game will continue as follows:

Black: kotake	1	7	13	19	25	31
White: threat	2: $-\theta[6]$	8: $-\theta[5]$	14: $-\theta[4]$	20: $-\theta[3]$	26: $-\theta[2]$	32: $-\theta[1]$
Black: response	3	9	15	21	27	33
White: kotake	4	10	16	22	28	34
Black:	5: A	11: C	17: E	23: G	29: I	35: K
White:	6: B	12: D	18: F	24: H	30: J	36: kofill

Notice that White declines to fill the ko at moves 6, 12, 18, 24 or 30. He eventually fills the ko at move 36, because he has no more kothreats. In the above situation, if X is any canonical value such that $X = \Delta X$ and that $\Delta K > \Delta X > \Delta L$, we may treat the ko as if its value were X. Such treatment gives the correct time at which Right will fill the ko. And both ko and X contribute the same amount to the outcome of the game, since $ko^R = X^R = 0$. For these reasons, we write $ko \equiv X$ and we say that the ko has a *pseudo-incentive* $\Delta^R(ko) = \Delta(X)$.

This viewpoint is especially helpful when we can predict an appropriate value of X from global considerations. Suppose, for example that E, the ko in this problem, appears in an environment containing a few pairs of games with incentives between $\{1 \mid 0\}$ and $\{0 \mid 0\}$, more White kothreats, and even more infinitesimals. Then we claim that $ko^R = 0ish$, $ko^L = \{1 \mid 0\}ish$, $\Delta^R(ko) = 0ish$, $\Delta^L(ko) = \{1 \mid 0\}ish$.

Left begins fighting the ko as soon as she has no more moves of incentive $\{1 \mid 0\}ish$ or greater. Right eventually fills the ko when he runs out of kothreats, which happens during the play of infinitesimals.

In the present problem, there are at least 2 and at most 4 moves with incentives between $\{1 \mid 0\}ish$ and $0ish$: N, G^L, K and N^R. There are many potentially tepid moves, on games such as A, F, G^{LR}, N^L, Q^R, S^R and U, V^{RL}. So, if there are 2 net White kothreats, then $E \equiv 0ish$. Similarly, if there are 2 net Black kothreats, then $E^L \equiv 1ish$ and $E \equiv \{1 \mid 0\}ish$. In either case, we obtain a useful bound on the pseudo-incentives of E, namely,

$$\Delta E \leq \{1 \mid 0\}ish.$$

This allows us to assume that canonical players prefer to play on $V^R ish$ rather than on E.

A continuation of this argument allows us to obtain an upper bound on the incentive of $(S + E)$.

Let $Z = \{S^L + \theta\theta \mid S^R - \theta\theta\}$. Then, if Z is played by Left, there are two net Black kothreats and $E \equiv \{1 \mid 0\}ish$, but if Z is played by Right, there are two net White kothreats and $E \equiv 0ish$. So

$$\Delta(Z + E) = \Delta(\{1\tfrac{1}{2} \mid \tfrac{1}{2} \parallel -1\}ish) = \{2\tfrac{1}{2} \mid 1\tfrac{1}{2} \parallel 0\}.$$

Since Z is always played before E, we have $\Delta(Z+E) = \Delta(Z)$; and since kothreats are at worst harmless, we have the following bounds on pseudo-incentives:

$$\Delta S < \Delta Z < \{2\tfrac{1}{2} \mid 1\tfrac{1}{2} \parallel 0\}.$$

We may now obtain bounds on the incentive of $S = \{\tfrac{1}{2} \mid -1\}ish$. Since Black kothreats can not harm Black, we have

$$\Delta S \geq \Delta(\tfrac{1}{2} \mid -1) = \{1\tfrac{1}{2} \mid 0\}.$$

Thus,

$$\{2\tfrac{1}{2} \mid 1\tfrac{1}{2} \parallel 0\} \geq \Delta S \geq \{1\tfrac{1}{2} \mid 0\}.$$

7. The Search

Using the bounds on the pseudo-incentives of S and E, we obtain the partial ordering shown at the left of Table 2. Using this partial ordering, canonical players face only two choices: whether to play S before or after M, and how to play the ko at E.

In each column of Table 2, the entries show the move numbers on which that row is played.

The choice between S and M occurs at move 7. The dominant continuations following 7S appear in columns I, II and III; the dominant continuations following 7M appear in columns IV and V. But both players may need to decide when to play their first move at E or E^L. After 7M, White later chooses between 11E (column V) and 11N (column IV). After 7S, White clearly does better not to fill a ko that he can win later, so he plays 11N and Black may then choose between 12G^L (column III) and 12E (column I and II). After 12E, White can still choose either to fight the ko with 13T (column II) or not (column I).

We assert that the canonical solution is column I, which corresponds to the game played in Figure 3.

The bottom row of Table 2 shows the total value of the outcome of each line of play, *relative to the line of play shown in column* I. Each of these values can be easily computed, using only the information shown in Table 1. For example, to compare the outcome of column III with the outcome of column I, we tally the rows where the entries have different parities. In row G^L, White (odd numbers) played in column I, but Black (even numbers) played in column III, and so that row's contribution to the difference is

$$(G^L)^L - (G^L)^R = 2\Delta(G^L) = \tfrac{5}{8}ish.$$

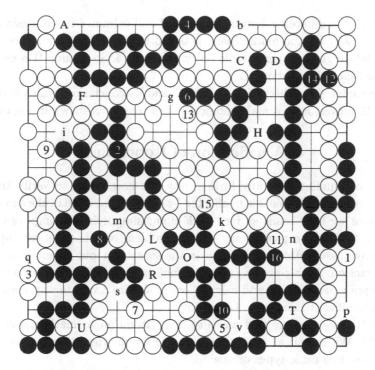

Figure 3. Canonical play of hot regions: moves 1–16.

The only other row that contributes any difference is the *kofill*. The raw (warmed) difference in scores depending on who eventually fills the ko at E is 4 points (one for the stone captured in the ko itself, and three points of white territory along the top right edge of the board). But there is also a difference of three moves, and so the chilled value of this ko is 1 point. Thus, the "kofill" line contributes −1 point to the total. Then the bottom line of column III is obtained as $\frac{5}{8}ish - 1 = -\frac{3}{8}ish$.

The calculation of the bottom line for column V is only slightly more subtle:

$$-2\Delta(M) = -1\tfrac{3}{4}$$
$$+2\Delta(N) = \{1 \mid \tfrac{7}{8}\}ish$$
$$+2\Delta(K) = +\tfrac{1}{2}$$
$$-1\Delta(N^R) = \{0 \mid -\tfrac{1}{8}\}$$
$$+2\Delta(S) = +1\tfrac{1}{2}ish$$
$$\text{kofill} = -1$$
$$\overline{\text{relative total} = +\tfrac{1}{8}ish}$$

The totals of other columns are computed by the same method. From the bottom row, we deduce that the moves followed by question marks in columns III, IV and V are suboptimal.

Although the difference between column II and column I is infinitesimal, $\{0 \mid +\}$ is positive. Furthermore, Black gets the next (tepid) play in column II, whereas White gets the next (tepid) play in column I. So column II is decidedly better for Black then column I.

This completes our discussion of Table 2 and its conclusion, which is that for the first 16 moves, an optimal canonical line of play is as shown in Figure 3.

8. Verification of Robustness

Before continuing with any further play, we now pause to verify that the dominant column of Table 2 is unaffected by the removal of all of the simplifying assumptions we have made so far. Each of our simplifying assumptions is now seen as a "nit" to be picked out. In any problem of this type, we hope that our intuitive assumptions are benign in the sense that our conclusions survive their extraction. If this proves false, the "nit" is no longer a nit, but a cause to redo the prior partitioning and calculations in a more accurate way, even if that entails a significant increase in complexity.

Many Go players may find our nitpicking attention to mathematical detail distasteful. We view it as a small price to pay for the rigor that results.

There are two basic types of nits:

• Some sets of adjacent regions are dependent.

• Some incentives in the top part of table 1 are not quite strictly ordered.

In the present pair of problems, we have thus far included five nits:

8.1. Nit 1: Q and I are dependent. Let Q and I be the independent idealized regions obtained by immortalizing the string of six white stones just below the letter I, and let W be the actual region that is the union of these regions, including the string of six white mortal white stones between them. Then, by playing the difference games, it can be verified that

$$W \geq I + \{2\{- \mid 0^2\} \mid -2\downarrow*\} \quad \text{and} \quad W \leq I + \{2- \mid -2\downarrow*\}.$$

Hence, $W = I + \{2ish \mid -2\downarrow*\}$. Thus no errors result from setting $W = I + Q$ if Q is taken as $2ish \mid -2\downarrow*$.

8.2. Nit 2: N and V are dependent. There is a slight interaction between the regions N and V shown in Figure 1. For example, in the presence of O^{RL}, the chilled value of $N^{RL} + V^{RRR}$ is $\uparrow$, instead of the value implicit in Table 1, which is $\frac{1}{8}$. This nit is closely related to Nit 3.

8.3. Nit 3: $\Delta(N)$ is not strictly less than $\Delta(V^R)$. Let N and V be the independent idealized regions obtained by immortalizing the black string just left of N and just above V; let X be the actual region that is the union of these regions, including the mortal black stones between them. We have seen that

$$N = 0 \mid +_1 \parallel -\tfrac{7}{8} \mid -1,$$
$$V = 2 \parallel \mid 0 \mid +_1 \parallel -1*,$$
$$X \le V + N.$$

But, by playing the difference game, it can be verified that $X \ge V + Y$, where $Y = 0 \mid +_1 \parallel -1{\uparrow} \mid -1$. We notice that

$$\Delta(V^R) > \Delta(Y) > \Delta(N),$$

from which we conclude that all plays on X will be taken in the same order as the corresponding plays on $(V + N)$. This eliminates the concern of Nit 3.

To address Nit 2, we observe that no difference can arise between X and $(V + N)$ unless N is played to N^{RL}, which happens only in columns I and III of Table 2. In both of these cases, V is played to V^{RL}. Under these circumstances, the $(V + N)$ has been played to

$$(V + N)^{RLRL} = V^{RL} + N^{RL} = \{0 \mid +_1\} - \tfrac{7}{8},$$

whereas X has been played to

$$X^{RLRL} = \{0 \mid +_{1\frac{1}{8}\downarrow}\} - \tfrac{7}{8}.$$

Thus, the concern of Nit 2 is eliminated by suppressing the subscript of the tiny that appears in V^{RL}.

8.4. Nit 4: J and M are dependent. Let Z be the union of the dependent regions J and M. By careful calculation, it can be shown that, to within *ish*,

$$\Delta^L Z = \Delta^R Z = \Delta Z = \{9 + \{\tfrac{5}{4} \mid -\tfrac{1}{2}\} + \{1 \mid -1\} \parallel 0\}$$

whence

$$\{10 \mid 0\} > \Delta Z > \{9 \mid 0\}.$$

Hence, ΔZ fits into the total rank ordering of Table 2 in precisely the same position as ΔJ. It follows that, in the dominant line of play, Black move 2 will change Z to $Z^L = J^L + M$, because J^L and M are independent.

8.5. Nit 5: T and P are dependent. Mathematically, this dependence is relatively trivial. It affects only the subscript of the tiny value at T, and thus has no affect on Table 2.

However, this dependence underlies a very interesting opening move: White 1 at T. Although Black can then win, her chilled margin of victory is only infinitesimal, and so great care is required. Black can play 2 at T^R. Then play continues as in column II (delayed by 2 moves), with White 3 at P, Black 4 at J, ..., until White plays move 13 at N. Black then plays $14G^L$. If White neglects to fill the ko on move 15, then Black plays E at move 16. She then fights the ko until she wins and fills it, attaining a bottom-line score on the hot regions of $+\tfrac{1}{2}ish$ or better. If instead White plays 15E, then play continues as in column II ending with a positive value on the bottom line.

Having now extracted all known nits, we now resume our investigation of how play should continue from Figure 3.

9. Tepid Regions

A general theorem in (ko-free) Combinatorial Game Theory states that one need never play a tepid move if one can play a hot move instead. This theorem suggests a natural breakpoint in endgame play, roughly corresponding to the break between the moves shown in Figure 3 and the moves shown in subsequent figures. For convenience of exposition, we assume that White continues play by eliminating the tinies at O and T at moves 17–20. In fact, these moves might occur either earlier or later.

Omitting the integer-valued markings, and excepting the tinies, which have no atomic weight, the tepid regions of Figure 3 are tabulated in Table 3.

region	value	atomic weight
A	$0^4 \mid +\!\!\!+$	$+4$
v	$0 \mid +\!\!\!+$	$+1$
g	$-\!\!\mid 0$	-1
q	$\downarrow *$	-1
s	$\Downarrow\!\!\Downarrow *$	-4
F	$*$	0
u	$*$	0

Table 3. Atomic weights.

In general, if no hot moves are available, Black will strive to increment the atomic weight; White, to decrement it. To this end, Black will typically attack regions of negative atomic weight, while White will attack regions of positive atomic weight.

The total of the atomic weights in Table 3 is -1, and so White, moving first, has a relatively straightforward win.

In the Japanese rules, Black will soon resign. However, under the Ing rules there is a flaw in this analysis, which is shown in the picture of s^{LLLLL} on the right.

Using the Ing rules or any other variation of Chinese rules, an assertive Black may hope to score a point at y. Even after White has played at z, Black may claim one point of Chinese territory at y. To uphold such a claim, Black will need sufficiently many kothreats that she can avoid filling y until after all dame are filled and White is forced to either pass or fill in a point of his own territory while the node y remains empty and surrounded by Black. So we have two different views of chilled values:

Chinese scoring with many θ's	Japanese scoring
$s^{LLLLL} = 1$	$s^{LLLLL} = 0$
$s = 1 \mid 0^5$	$s = 0 \mid 0^5$
$= \frown \mid 0^3$	$= \Downarrow *$

So Black's best hope is to play the tepid regions under the assumption that she can win a terminal Chinese kofight. White must play to win under Japanese rules, because if White fails to win under Japanese rules, the Chinese kofight will not help him. According to a theorem of David Moews, the optimal sequence of play is given by the ordering explained in Appendix E of *Mathematical Go*. This canonical line of play is shown in Figure 4.

The result is that

$$\{0^4 \mid \dotplus\} + \downarrow* + \Downarrow* < 0$$

but

$$\{0^4 \mid \dotplus\} + \downarrow* + \{\frown \mid 0^3\} > 0.$$

Since the subscript on the tiny in game A exceeds the subscripts in games v and g, these inequalities remain valid when applied to the sum of all tepid regions of Figure 3. No matter who plays next, Black can win the game if $s^{LLLLL} = 1$, but White can win if $s^{LLLLL} = 0$.

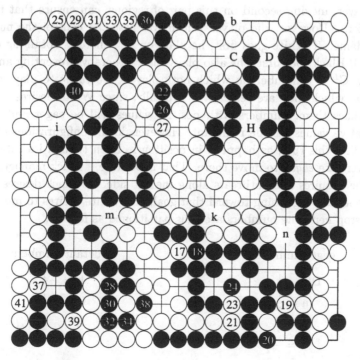

Figure 4. Play of tepid regions: moves 17–41.

The game should therefore continue as shown in Figure 4. Moves 21–36 all have incentive of at least $\uparrow*$. After move 36, White's view of the total value is $\downarrow$, from which White can win only by playing at 37, because it is the only move on the board with incentive $\downarrow$. Black's 38 is hot in her view, but in White's view the incentive of Black's move 38 is only $*$. Moves 39, 40, and 41 are all stars.

White's 41 is the last tepid move. This milestone of play occurs at the conclusion of Figure 4 and the start of Figure 5. The number avoidance theorem mentioned on page 48 of *Mathematical Go* ensures that this same milestone occurs in all well-played combinatorial games.

10. Play of Numbers

The seven regions that remain active at the conclusion of Figure 4 all have values that are numbers. Subsequent play on each of these numbers can effect not only its mathematical value, but also the kothreats available for the Chinese kofight that will occur later. In this kofight, dame will prove helpful to White. The number of such dame can depend on how the numbers are played. For example, consider a marked and chilled position such as

$$\mathrm{n}^{RR} + \mathrm{C}^L = \tfrac{1}{2} - \tfrac{1}{2} = 0$$

Either player, moving second on this pair of regions, can ensure that the total score is zero. However, the player who moves first can control the number of pairs of dame. If White goes first, he plays on C^L, thereby obtaining one pair of dame. If Black goes first, she also plays on C^L, thereby preventing any dame. Evidently, C^L is a number on which both players are *eager* to play; n^{RR} is a number on which they are *reluctant*. Typically, since White desires dame and Black dislikes them, numbers with positive values will be eager and numbers with negative values will be reluctant. However, local configurations can make exceptions to this rule, as happens in position i. Even though its value is $-\tfrac{1}{2}$, both players are *eager* to move there. That is because if Black manages to play to i^{LL}, she acquires two kothreats along the left side. So Black desires to move on i in hopes of progressing toward that goal, and White desires to move on i in order to stop her. The numbers are thus partitioned as follows:

eager numbers	reluctant numbers
$\mathrm{DH} = \tfrac{1}{8}$	$\mathrm{b} = -\tfrac{1}{4}$
$\mathrm{n} = \tfrac{1}{8}$	$\mathrm{C} = -\tfrac{1}{4}$
$\mathrm{m} = \tfrac{1}{4}$	$\mathrm{k} = -\tfrac{1}{2}$
$\mathrm{i} = -\tfrac{1}{2}$	

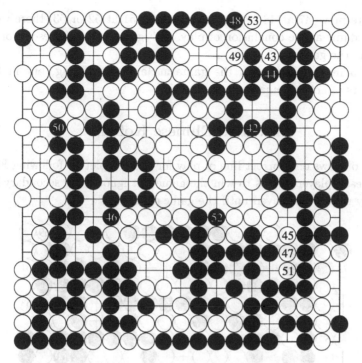

Figure 5. Play of numbers: moves 42–53.

This tabulation provides the rationale for the sequence of play displayed in Figure 5. Notice how the incentives decrease monotonically as play progresses:

formal incentive	moves
eager $-\frac{1}{16}$	42–43
eager $-\frac{1}{8}$	44–45
eager $-\frac{1}{4}$	46–47
reluctant $-\frac{1}{4}$	48–49
eager $-\frac{1}{2}$	50–51
reluctant $-\frac{1}{2}$	52–53

By partitioning the eager and reluctant numbers into separate subsystems, each of which sums to an integer, White can play each integer independently and thereby ensure that he gets at least one pair of dame. However, Black gets everything else. At each turn, she gets to choose among at least two alternatives, and White is then forced to take whichever is left over. Black blocks a potential White kothreat by playing at 46 rather than at 47. Black's move 50 gives her the opportunity to play 58 later in Figure 7.

If White plays 47 at 48 or 49, or if he plays 51 at 52 or 53, there would be two fewer dame at the conclusion of play. In Figure 5, any of these errors would cost White the game.

Move 53 marks the end of the game in Japanese scoring, but in Chinese scoring, lots of action remains.

11. The Chinese Kofight

Figure 6 shows one correct line of play for the completion following Figure 5, in which each player strives to win the game but ignores the secondary issue of how many moves are played until the game ends.

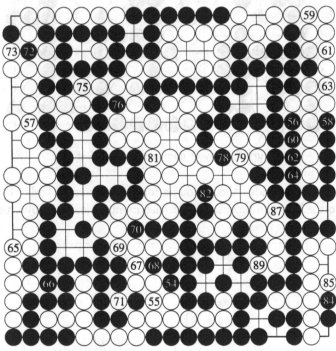

71, 74, 77, 80, 83 and 86: Take the Ko
88: Fill the Ko

Figure 6. Play of Chinese Ko: moves 54–89.

The reader who is familiar only with Japanese scoring should begin by studying the final stage of the fight, which begins after move 70. As White takes the ko at move 71, Black has 3 kothreats, but White has 6 dame. Each pair of dame provides a kothreat for White. After White's move 87, Black has no kothreats. Only one dame remains, but if Black plays it, then White wins the ko that gives him a margin of victory of 3 points. So Black fills the ko at move 88, and White then fills the last dame to win the game by one point.

Suppose, however, that there were no dame at locations 75 and 81, and that the last dame was filled by Black's move 83. In that case, after Black's ko capture at move 85, White could do no better than pass, and Black would win the game by one point.

The unusually wide variety of kothreats and potential kothreats shown in Figures 4 or 5 makes this a surprisingly interesting game. The mathematical tools to handle such games are still being developed, and so the reader may enjoy working out the present problem by the traditional Go-playing method of reading out the various lines of play.

It is somewhat surprising that each player can play this portion of the game in several different ways, all of which lead to equivalent results. From the position shown after move 53 in Figure 5, White can win. However, if this position contains two fewer dame, then Black could win. The variation of the position that contains two fewer dame might arise either because the starting position was Problem 2 rather than Problem 1, or because in Problem 1 White failed to play the numbers on moves 42 to 53 with sufficient care.

Either player can elect to prolong the Chinese kofight.

Figure 7 shows a variation in which Black, unable to win, forces White to

59 at 56
64, 67, 70, 73, 76, 79, 82, 85, 88, 91, 94, 97 and 100: Take the Ko
102: Fill the Ko

Figure 7. Black stalls his loss: moves 54–103.

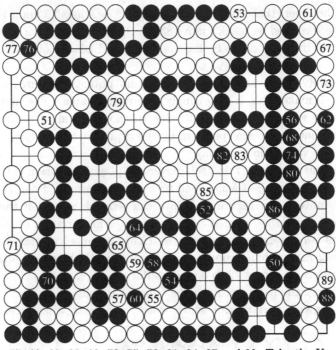

57, 60, 63, 66, 69, 72, 75, 78, 81, 84, 87 and 90: Take the Ko

Figure 8. Black wins Problem 2 quickly: moves 50–90.

play until move 103. This line of play includes Black's surprising move 56, which sacrifices a kothreat prematurely in order to lengthen the game. A slightly less brilliant line of play by Black would eliminate moves 56 and 57, leading to a White win after 101 moves instead of 103.

Figure 8 shows a conclusion to Problem 2, in which Black can prevail, but apparently no sooner than move 90. To achieve this rapid victory, Black deviates from prior lines of play at move 50. There are other ways in which Black can attempt to expedite his victory by deviating from canonical play, but we think that none of them succeed. For example, if Black postpones playing at 22 until after he has played at 38, then White can play at 41 instead of at 37, after which Black cannot win because his early kothreat never materializes.

12. The $1,000 Dame

It is tempting to think that all of our prior analysis of Problems 1 and 2 should also apply to Problems 3 and 4. Canonical play for the first 53 moves is indeed the same for all four problems. Reading further ahead, it is very plausible to foresee the same Chinese kofight. Eventually, the number of Black's threats will exceed the number of pairs of dame and so for several weeks we believed that Problem 4 behaved like Problem 2: Black should win in about ninety moves.

However, Kuo-Yuan Kao then pointed out a fatal flaw in this reasoning, which originates at s^{LLLLL}:

If White plays at z, while other "dame" remain on board, then Black has enough kothreats to fight and win the Chinese ko at y. Eventually, White runs out of other dame to fill and he is forced to pass or fill in his own territory. Only after White has done that can Black afford to fill y and still win the game.

Mathematically, after Black plays y and z remains vacant the unchilled value is $*$. But, after White plays z and y remains vacant, then if Black can win the Chinese Ko, its unchilled Chinese value is 1. So the original position just shown has unchilled value equal to $\{* \mid 1\} = 0$.

This equation reveals how White can win Problem 4. After canonical play through move 53, he should refuse to play at z. Black and White both play dame elsewhere until no other dame remain. White then has no kothreats and Black has many, but they do her no good, because both y and z are still vacant and there are no other dame anywhere else on the board. On move 73, White fills the last dame elsewhere. If Black then fills y, White fills z and wins by one point. If Black passes at move 74, White fills z and still wins by one point in Ing scoring. Black gets her point at y (either filled or unfilled), but she still loses the game.

So, thanks to the $1,000 "false dame" at z, we believe that White wins Problem 4. If he plays canonically for the first 53 moves, he eventually fills the last dame on move 75. But Black then prolongs the game another ten moves by playing out each of her five kothreats.

Problem 3 is an easier version of Problems 1 and 4.

The same argument reveals why, in Problem 1, in Figure 6 Black cannot afford to play Black 54 anywhere else. If instead she played Black 54 at 57, White could play 55 at 54 and then refrain from playing at 55, thereby winning the game by avoiding the Chinese kofight.

13. The Non-Canonical Surprise

Virtually all of the contestants in Ishi Press's Thousand-Dollar Ko competition made fatal mistakes on early moves. But among these weak lines of play, two contestants uncovered a real gem:

$$1P, \ 2J, \ 3Q, \ 4B, \ 5S!$$

From the partial order chart at the left of Table 2, this looks hopelessly wrong. After all, $\Delta S < \Delta G < \Delta V$.

So we played the continuation: 1P, 2J, 3Q, 4B, 5S, 6V, 7G, and reached the same position that might also have arisen from 1P, 2J, 3Q, 4B, 5G, 6V, 7S. This

region	Δ	VI	VII	VIII
B	$3\frac{3}{4} \mid 0$	4	4	4
V	$3* \mid 2\{-\!\!\!-_1 \mid 0\} \parallel 0$	6	6	6
G	$2\frac{5}{8} \mid 2\{-\!\!\!-_{1/2} \mid 0\} \parallel 0$	5	5	5
M	$1\frac{3}{4} \mid 0$	8	8	8
I	$1\frac{1}{2} \mid 0$	9	9	9
V^R	$1*\{0 \mid +\!\!\!+\} \mid 0$			
N	$\{1 \mid \frac{7}{8}\}\{0 \mid +\!\!\!+\} \parallel 0$	11	20	10(?)
G^L	$\frac{5}{8}\{0 \mid +\!\!\!+\} \mid 0$			
K	$\frac{1}{2} \mid 0$	13	22	11
N^R	$\frac{1}{8} \mid 0$	14		
S	$1\frac{1}{2}\theta\theta\!\uparrow\!\uparrow\!\uparrow\!\uparrow* \mid 0$	7	7	7
E	ko	10	10	12
T	$-\Delta\theta$		11(?)	13
kofill		12	21	23
totals		$0ish$	$\frac{3}{8}ish$	$-\frac{1}{8}ish$

Table 4. An extension of Table 2.

less dramatic sequence focuses our attention on an intriguing inequality,

$$\Delta G < \Delta V,$$

whose truth depends on both of the following true assertions:

$$2\tfrac{5}{8} < 3 * \qquad \text{and} \qquad \{\tfrac{1}{2} \mid 0 \parallel 0\} = -\!\!\!-_{1/2} < -\!\!\!-_1 = \{1 \mid 0 \parallel 0\}.$$

So this line is definitely *not* canonical. White's 5G is dominated by the canonical 5V. However, our investigations of continuations of this sequence soon led to the extension of Table 2 shown in Table 4.

The canonical continuation after 5G 6V is evidently as shown in column VI. After the hot games are played, the value is 0-*ish*. The difference between the precise values at the bottom of column VI and column I is

$$\{-\!\!\!-_1 \mid 0\} + \{0 \mid +\!\!\!+_{1/2}\} > 0.$$

It is easy to see that this infinitesimal difference, of atomic weight zero, evaporates near the beginning of the tepid phase of play. The continuations of columns VI and I quickly converge to the same plays of regions A, s, and q, the same careful play of numbers, and the same Chinese kofight. Additional study verifies that there is no difference in Chinese kothreats or dame. The only difference is that White gets there 8 moves sooner via the line of play shown in column VI. So, if White seeks to win as quickly as possible, column VI is his best opening for Problem 1. To stall his defeat in Problem 2, he should open instead as in column I. Black has a superb response to those White contestants who played the surprise opening to Problem 1:

$$1P, \ 2J, \ 3Q, \ 4B, \ 5S!, \ 6G!!$$

To address the question of whether this problem has any other surprise openings, one can re-examine the partial ordering of incentives shown at the left of Table 2. The inequalities shown in this diagram may be partitioned into strong inequalities and weak inequalities, where an inequality of the form $\Delta W < \Delta X$ is *strong* if there exist numbers y and z such that;

$$2\Delta W < y < z < 2\Delta X.$$

An order of play that transposes two consecutive moves related by a strong inequality is not only non-canonical; it loses the game. So the search for more winning non-canonical lines of play must violate only weak inequalities in the partial order to the left of Table 2. The only inequalities that might be weak are these:

$$\Delta V > \Delta G \qquad \Delta V^R > \Delta E \qquad \Delta E > ishes$$
$$\Delta S > \Delta I \qquad \Delta V^R > \Delta N \qquad \Delta E^L > ishes$$

The player who is destined to lose the game anyway is less constrained; he might try very strange moves to prolong the play.

Although most non-canonical moves lead to quick defeats, the number of possibilities is substantial and we have not attempted a complete search or further arguments that might greatly reduce the size of the requisite search.

Mathematical Go includes a collection of powerful techniques for rigorous analysis of intricate endgame positions, including the four problems at the beginning of this paper. The techniques of Mathematical Go provide a "canonical" line of play, which leads to a best-possible score. If any strategy can win, then the canonical strategy will win. In the problems considered here, a search of only five lines of play, summarized in Table 2, was sufficient to determine that White wins Problems 3 and 4. The same analysis shows that the outcome of Problems 1 and 2 depends on a Chinese kofight at $S^{RLLLLLR}$. This location, not E, is the Thousand-Dollar Ko. The mathematics proves that, against a determined opponent, neither player can win Problem 1 or Problem 2 without creating this ko and then winning it.

However, the mathematical techniques used in our analysis of hot and tepid games do not say anything about the number of pairs of dame remaining, nor about the number of moves required to conclude the game. Since the outcome of the Chinese kofight depends on the number of dame, we are not even completely sure who can win! It is conceivable that still other non-canonical lines of play not yet discovered might leave more or fewer pairs of dame on the board when the crucial kofight at $S^{RLLLLLR}$ begins. But based on everything we know, we now think that

- White wins Problem 1 by one point after 95 moves.
- Black wins Problem 2 by one point after 90 moves.
- White wins Problem 3 by one point after 79 moves.
- White wins Problem 4 by one point after 77 moves.

Readers who yearn for another problem of this type will enjoy Figure C.19 of *Mathematical Go*, which White wins in 101 moves. The software companion contains a degenerate version of this problem, just as Problems 3 and 4 above are degenerate versions of Problems 1 and 2.

References

[Berlekamp and Wolfe 1994] E. Berlekamp and D. Wolfe, *Mathematical Go: Chilling Gets the Last Point*, A. K. Peters, Wellesley, MA, 1994. Also published in paperback, with accompanying software, as *Mathematical Go: Nightmares for the Professional Go Player*, by Ishi Press International, San Jose, CA.

ELWYN BERLEKAMP
DEPARTMENT OF MATHEMATICS
UNIVERSITY OF CALIFORNIA AT BERKELEY
BERKELEY, CA 94720
    berlek@math.berkeley.edu

YONGHOAN KIM
DEPARTMENT OF MATHEMATICS
UNIVERSITY OF CALIFORNIA AT BERKELEY
BERKELEY, CA 94720
    yonghoan@math.berkeley.edu

Games of No Chance
MSRI Publications
Volume **29**, 1996

Eyespace Values in Go

HOWARD A. LANDMAN

ABSTRACT. Most of the application of combinatorial game theory to Go has been focussed on late endgame situations and scoring. However, it is also possible to apply it to any other aspect of the game that involves counting. In particular, life-and-death situations often involve counting eyes. Assuming all surrounding groups are alive, a group that has two or more eyes is alive, and a group that has one eye or less is dead.

This naturally raises the question of which game-theoretical values can occur for an eyemaking game. We define games that provide a theoretical framework in which this question can be asked precisely, and then give the known results to date. For the single-group case, eyespace values include 0, 1, 2, $\int \frac{1}{2}$, $\int 1\frac{1}{2}$, $\int \frac{3}{4}$, $\int 1\frac{1}{4}$, $\int 1*$, and several ko-related loopy games, as well as some seki-related values. The $\int \frac{1}{2}$ eye is well-understood in traditional Go theory, and $\int 1\frac{1}{2}$ only a little less so, but $\int \frac{3}{4}$, $\int 1\frac{1}{4}$, and $\int 1*$ may be new discoveries, even though they occur frequently in actual games.

For a battle between two or more opposed groups, the theory gets more complicated.

1. Go

1.1. Rules of Go. Go is played on a square grid with Black and White stones. The players alternate turns placing a stone on an unoccupied intersection. Once placed, a stone does not move, although it may sometimes be captured and removed from the board.

Stones of the same color that are adjacent along lines of the grid are considered to be connected into a single indivisible *unit*. For example, if one takes the subgraph of the board grid whose vertices are the intersections with Black stones and whose edges are the grid lines that connect two such vertices, the Black units are the connected components of that subgraph. (There is little terminological consistency in the English-language literature; a unit has also variously been called a chain [Remus 1963; Zobrist 1969; Harker 1987; Kraszek 1988], a string [Hsu and Liu 1989; Berlekamp and Wolfe 1994], a group [Thorp and Walden 1964; 1972; Fotland 1986; Becker 1987], a connected group [Millen 1981], a block

[Kierulf et al. 1989; Kierulf 1990; Chen 1990], or an army [Good 1962]. This is unfortunate because the terms chain, group and army are frequently used for completely different concepts.)

An empty intersection adjacent to one or more stones of a unit along a grid line is said to be a *liberty* for that unit. When a player's move reduces a unit to zero liberties, that unit is *captured*, and its stones are removed from the board; the intersections formerly occupied by them revert to being empty. A player may play on any empty intersection, except that the move may not recreate a position that existed at the end of any earlier move (the generalized ko rule), and (in some rules) may not end up part of a unit with no liberties (the no suicide rule). Note that capture of enemy stones happens before one's own liberties are counted, and any capture creates at least one liberty, so that any move that captures is guaranteed not to be a suicide.

For a more detailed discussion of the rules of Go, their variants, and the mathematical formalization thereof, see appendices A and B of [Berlekamp and Wolfe 1994]. There have also been earlier attempts to formalize the rules of Go, including [Thorp and Walden 1964; 1972; Zobrist 1969; Ing 1991].

1.2. Fundamentals of life and death in Go. Single-point eyes. When a unit (or a set of units of the same color) completely surrounds a single empty intersection, we call that intersection a *single-point* eye. One of the most fundamental "theorems" of Go is that it is possible for sets of units to become uncapturable even against an arbitrarily large number of consecutive moves by the opponent. Consider a single Black unit that surrounds two or more single-point eyes (Figure 1, left); even if White removes all other liberties first, it is still illegal by the suicide rule for White to play in either single-point eye. Thus the unit is unconditionally alive, with no further need for Black to play to defend it.

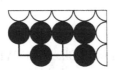

Figure 1. Left: A unit with two single-point eyes cannot be captured. Right: Units may share eyes to make life.

1.3. Static life and topological life. More generally, a set of units can achieve life through shared eyes. We introduce some definitions. A *group* is a set of units of the same color. A group is said to be *alive* if no unit of the group can be captured given optimal defense, even if the opponent moves first. (This definition is more vague and flawed than it might seem at first: to give just one example, it is possible for two disjoint groups to both be alive by this definition, and yet have it be the case that the opponent has a move that threatens both simultaneously so that only one of the two can be saved.) A group is said to be *statically alive*

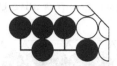

Figure 2. This Black group has one real eye and one false eye, and is dead.

if the opponent cannot capture any unit of the set even given an arbitrarily large number of consecutive moves. Most alive groups are alive because they can achieve static life, although other possibilities (such as seki) exist. A group is said to be *topologically alive* if (1) the units of the group completely surround two or more single-point eyes, and (2) each unit of the group is adjacent to at least two of those eyes. It is straightforward to prove that topologically alive implies statically alive: each unit of the set has at least two liberties from the eyes alone, so an opponent move into any of the eyes captures no units, hence would form a unit with no liberties that dies immediately (such a move is forbidden by the no suicide rule if it applies). The opposite is not true, since a group may be statically alive even if one or more of its eyes are not single-point. However, it appears to be the case (although I will not attempt to prove it here) that any statically alive group can be made into a topologically alive one, and also that any topologically alive group can be made into one with only two eyes. This is the basis for the traditional beginner's guideline that "you need two eyes to live".

For an alternate formulation of life and death fundamentals, which goes beyond that presented here, see [Benson 1976; 1980; Müller 1995, pp. 61–65].

1.4. False eyes. When condition (2) of topological life is violated, the units adjacent to only one eye may be subject to capture (which would destroy the eye) unless they are connected to the group by filling in the eye (which, since the eye is single-point, also destroys it). Such an eye is called a *false eye* in the Go literature, since it does not contribute to life (Figure 2).

The traditional criterion for determining whether an eye is false is local in nature: In addition to occupying all the intersections adjacent to the eye, one must also control enough of the diagonal points (one if the eye is in a corner, two if on an edge, and three if in the middle of the board). Thus the following eyes are all considered false:

This criterion is sometimes inaccurate, since a locally "false" eye may be globally real if the topology of the group includes a loop that contains the eye. This leads to the apparent paradox of "living with two false eyes", which is really not a paradox at all since it clearly falls within the definition of topological life given above. The following Black group is alive even though both its eyes are locally false:

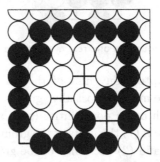

Situations like this are extremely rare, but not unknown, in actual play. Nakayama [1989] comments on one occurrence in a game between Hotta Seiji 4-dan (White) and Nakano Hironari 5-dan (Black) on November 3, 1988. There was also a group that lived with one real and one false eye in a game between Shinohara 9-dan and Ishigure 8-dan some twenty years earlier [Haruyama 1979].

1.5. Eyespaces. An *eyespace* for a group X is a set E of connected intersections such that there exists a sequence of moves that results in at least one intersection of E being an eye of a group X', and the stones of X' include all the stones of X. It sometimes occurs that a group has the potential to make eyes in more than one area, and that moves made in one of these eyespaces do not affect which moves are legal in the other ones. In this case we say that the eyespaces are *independent*. When this happens, the normal theory of sums of games (described in the next section) applies; we can evaluate each area separately, then just add up the results.

To get an approximate measure of the frequency with which multiple separate eyespaces occur in real situations, I examined all the problems in Maeda's tsume-go series [Maeda 1965a; 1965b; 1965c]. This set of books presents 585 life-and-death problems, ranging from elementary to advanced. Sixty-one of them (10.4%) seem to have two independent eyespaces. There are also several problems whose key is that two eyespaces that naively appear to be independent in fact have a subtle interaction that can be exploited (see for example [Maeda 1965b, problems 88, 127, and 163]), and at least one [Maeda 1965b, problem 67] where the key is to make two captures simultaneously so that the opponent can only play in one of the resulting independent eyespaces.

Determining independence can sometimes be quite difficult. However, if we are unable to prove that two eyespaces are independent, we can just lump them together and treat them as a single eyespace. Also, if we merely want to construct an example of a group with multiple independent eyespaces, it is simple to do so by putting solid walls of stones between them. Thus the occasional difficulty of establishing independence does not present a serious impediment to developing a theory of eyespace values.

1.6. Conventions for eyespace diagrams. As in [Berlekamp and Wolfe 1994], we draw diagrams with the convention that White stones cut by the diagram boundary are assumed to be *immortal*, that is, safely connected to a White group with two or more eyes in such a way that no set of moves within the diagram itself can have any effect on their safety. Unlike [Berlekamp and Wolfe 1994], however, Black stones cut by the boundary are not considered to be immortal, but only to be connected out to the remainder of the Black group (which may have other eyespaces) with sufficient liberties that their liberty count does not affect the analysis within the diagram. This condition on liberties is necessary, as can be seen in this example:

As long as the Black unit has at least one liberty outside of the diagram, it is worth one eye for Black; White has no legal move. However, if we maliciously assume that no such liberty exists, then White can capture the Black unit by playing on the empty intersection.

Thus, all the eyespace diagrams should be interpreted as follows: The Black stones cut by the boundary are part of a "backbone" unit that solidly connects all the eyespaces of the group, and that also has enough external liberties that the local analysis need not worry about it being captured.

Normally the backbone will only connect to the diagram at one point. When multiple separate Black units are cut by the boundary, we need to know whether loops exist containing the eyespace (i.e., which of the units is connected to the backbone), and thus whether any eyes that are locally false are also globally false. We assume that all Black units that are contiguous along the cut-line are connected, and only those separated by intervening White stones (as in the false-eye diagrams on page 229) are not.

2. Modeling Life and Death Problems

2.1. The phases of a life and death battle. From a Go player's perspective, life-and-death battles proceed through two or three distinct phases:

- A hot eye-making phase, where Black and White each attempt to make eyes for themselves, and to destroy the other player's eyes. This continues until each group either has at least two eyes (so it lives), or has at most one eye (so it dies), or rarely the situation becomes played out in some other way (seki, bent-four, mannen ko, and so on).
- If both sides live, then comes a warm point-making phase in which the remaining points between the Black and White groups are decided. This phase may include some moves that threaten a group's life, but they will almost always be answered.

- A cool phase (what [Berlekamp and Wolfe 1994, p. 124] calls an *encore*), in which neither player's moves are worth much. In Japanese rules this phase is not even played; in Chinese rules it is part of the final filling in of territory and capturing of dead stones.

Since the difference between a group living and dying is usually large (locally at least fourteen points, and often much more), the first phase can be very hot. Given the current state of game theory, it is difficult (and not very enlightening) to try to analyze a life-and-death battle in terms of points. Our intuition is that the first phase above, the eye-making phase, is the most important. Further, optimal play in an eyespace usually does not depend on the number of points at stake if the group lives or dies. It seems worthwhile to develop a theoretical framework in which only the number of eyes made matters.

2.2. Tsume-go and Bargo. The study of life-and-death problems in Go is called, in Japanese, *tsume-go*; but the term has no direct connotation of life or death. It derives from the transitive verb *tsumeru*, whose main meaning is to stuff, fill, pack, or plug up, and that can also mean to checkmate or to hold one's breath. The sense is that life-and-death problems occur when a group becomes closely surrounded and can no longer run away to safety.

When a Black group is surrounded by White groups that are already alive, and is thus isolated from other Black groups, it must make life on its own. Such a Black group will usually die unless it can make at least two eyes. Although for the purposes of life-and-death we do not care whether Black makes more than two eyes, it is natural to first study the games that may occur without worrying about that limit. To model this, we define the game Bargo, which has the same rules as Go except for scoring; in Bargo, the final score is the number of distinct Black eyes. It doesn't seem to matter much whether we count any kind of eye or only single-point eyes, since any eye should be convertible to a single-point one. The reader demanding mathematical rigor can assume we are counting single-point eyes as defined above.

2.3. Ignoring Infinitesimals. Cooling, chilling, and warming. The $\mathcal{E}$ Operator. This paper presumes that the reader is familiar with the theory of two-person, zero-sum, perfect-information games developed in [Milnor 1953; Hanner 1959; Conway 1976; Berlekamp et al. 1982; Yedwab 1985; Berlekamp 1988; Berlekamp and Wolfe 1994]. In this theory, every move matters, and games are played out until one player has no moves (and thus loses). When considering the eyes of a group, however, we are normally uninterested in whether or not there are moves remaining after the number of eyes has been decided. This leads to some theoretical difficulties: the standard theory treats $\{1\,|\,0\}$ as different from $\{1\,|\,*\}$ or $\{1*\,|\,0\}$, but for our purposes here they are identical.

Roughly, this difference corresponds to the question of allowing or disallowing passes. If passing is allowed, it is impossible to not have a legal move. The precise

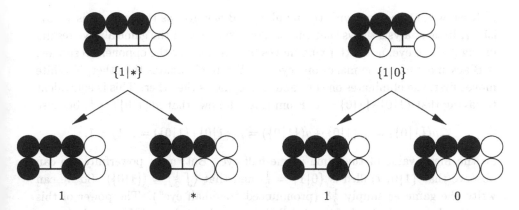

Figure 3. Each of these positions chills to $\{0|1\} = \frac{1}{2}$.

Go endgame theory requires either forbidding passes, or charging one point for them [Berlekamp and Wolfe 1994, Appendices A and B]. By doing neither, we lose some rigor and exactness. It is not even clear (yet) who "wins" Bargo, since we can no longer define winning as getting the last move.

One approach to this dilemma is to define equivalence classes of games modulo *small* [Conway 1976, p. 100–101] or *infinitesimal* games. That is, two games are in the same equivalence class if their difference is infinitesimal. We define $\langle G \rangle$ to be the class of games H such that $G - H$ is infinitesimal. Most standard operations on games have natural mappings into operations on these classes; $\langle G \rangle + \langle H \rangle = \langle G + H \rangle$, $\mu(\langle G \rangle) = \mu(G)$, etc.

Cooling [Conway 1976, p. 103] by any amount greater than zero eliminates these infinitesimal differences. Cooling by 1, or *chilling*, has an important place in both the endgame theory and here. It is easy to find eye-making games that differ only slightly, and chill to the same game: see Figure 3.

Warming is the approximate inverse of chilling. Both games in Figure 3 are infinitesimally close to $\int \frac{1}{2}$, where $\int$ represents the warming operator of mathematical Go endgame theory [Berlekamp and Wolfe 1994, p. 52 ff.]. We can encompass all such games by defining a one-to-many warming $G \mapsto \langle \int G \rangle$, and gain further notational convenience by defining a postfix operator $\mathcal{E}$ with dimension of "eyes" such that $G\mathcal{E} = \langle \int G \rangle$ eyes. Since $\int$ is linear, so is $\mathcal{E}$. As in the endgame theory, integers are unchanged by warming, so that if n is an integer, $n\mathcal{E} = \langle \int n \rangle$ eyes $= \langle n \rangle$ eyes $= n$-ish eyes. The "ish" suffix can be read as "infinitesimally shifted"; for the purposes of this paper it means "plus or minus an infinitesimal". $\langle G \rangle$ is just the set of all games that are G-ish.

2.4. "Half eye". The situation of Figure 3, where a potential eye can be made by Black (in gote) or permanently destroyed by White (also in gote), is fairly common. This is sometimes called a "half eye" [Davies 1975, p. 71]. In what sense is it really one-half?

If we write this game with the number of Black eyes as the score, it is $\{1|0\}$-ish. A little analysis shows that the sum of two copies of the game always results in exactly one eye, no matter who moves first, as long as the opponent responds. If Black moves first, he makes one eye and White eliminates the other; if White moves first, she eliminates one eye and Black makes the other. This is equivalent to saying that $\{1|0\}+\{1|0\} = 1$. From this it follows that $\mu(\{1|0\}) = \frac{1}{2}$, because

$$2\mu(\{1|0\}) = \mu(\{1|0\})+\mu(\{1|0\}) = \mu(\{1|0\}+\{1|0\}) = \mu(1) = 1.$$

So the mean value of the game is one-half eye. But, more powerfully, we can observe that $\{1|0\}$ chilled is $\{0|1\} = \frac{1}{2}$, and that $\langle\int\frac{1}{2}\rangle = \langle\{1|0\}\rangle$. So we can write the game as simply $\frac{1}{2}\mathcal{E}$ (pronounced "one-half eye"). The power of this notation is great; instead of the detailed case analysis above, it is enough to note that

$$\tfrac{1}{2}\mathcal{E} + \tfrac{1}{2}\mathcal{E} = (\tfrac{1}{2} + \tfrac{1}{2})\mathcal{E} = 1\mathcal{E} = 1\text{ish eye}.$$

This simple formulaic reduction constitutes a rigorous proof that "half an eye plus half an eye equals one eye".

2.5. An eye and a half. Another game that appears fairly frequently is $\{2|1\}$ eyes, which can be written $1\frac{1}{2}\mathcal{E}$. The simplest example of this is a three-point eyespace:

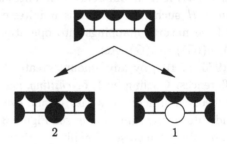

Most Go books refer to this kind of eyespace with terms such as "unsettled shape" [Davies 1975], or merely note that it is "intermediate" between life and death [Segoe 1960]. Such wording is unacceptably vague; as we shall see, there are several different values of eyespaces that could be covered by those descriptions. In addition, a group with an "unsettled" eyespace may in fact be quite settled if it has another eyespace worth half an eye or more. That is, "unsettled" is really best applied to an entire group, and it does not make much sense to apply it to a single eyespace of a group with multiple eyespaces.

Since games can be translated by adding numbers [Conway 1976, p. 112], we have the equality

$$\{2|1\} = 1+\{1|0\},$$

or, in terms of eyes,

$$1\tfrac{1}{2}\mathcal{E} = 1\text{ eye} + \tfrac{1}{2}\mathcal{E}.$$

In Go terms, this means that an "eye and a half" situation like that shown above is exactly equivalent (in terms of eyes made) to a single secure eye plus an eye in gote:

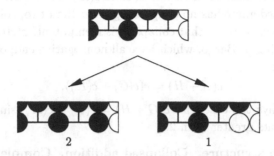

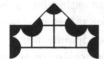

2   1

2.6. Larger "unsettled" shapes. Restriction to two eyes. Bargo$_{[0,2]}$. Collapsing. When we analyze larger "unsettled" shapes in Bargo, we frequently find games that have one or more integer endpoints of greater than two eyes. For example, the common four- and five-point nakade shapes

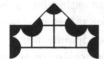   

have Bargo values of $\{3|1\}$-ish eyes and $\{\{3|2\}|1\}$-ish eyes, respectively. These values can be written $2*\mathcal{E}$ and $1\frac{3}{4}\mathcal{E}$. However, for living, there is no value to additional eyes beyond two. To model this effectively, we need to restrict Bargo so that more than two eyes don't count. Since Bargo is also restricted by definition to nonnegative eyes, I call this restricted game Bargo$_{[0,2]}$.

To calculate game values in Bargo$_{[0,2]}$, we first analyze the game as in Bargo (simplifying to canonical form with number-ish stopping points), then restrict the game by changing all stopping points that are greater than 2 to 2 (resimplifying if necessary). This process of restriction I call *collapsing*. We write the game that G collapses into as $c(G)$. Formally:

(i) $c(G) = 2$ if $G \geq 2$, else
(ii) $c(G) = G$ if G is number-ish, else
(iii) $c(\{G^B|G^W\}) = c\{c(G^B)|c(G^W)\}$ if $c(G^B) \neq G^B$ or $c(G^W) \neq G^W$, else
(iv) $c(G) = G$.

Using the above shapes as examples, we see that $\{3|1\}$ collapses to $\{2|1\}$, and $\{\{3|2\}|1\}$ collapses to $\{\{2|2\}|1\} = \{2*|1\}$. Generalizing collapsing to apply to the equivalence classes of games defined earlier, we can write $c(2*\mathcal{E}) = 1\frac{1}{2}\mathcal{E}$, and also $c(1\frac{3}{4}\mathcal{E}) = 1\frac{1}{2}\mathcal{E}$. So in Bargo$_{[0,2]}$, because of the restriction to two eyes, both these shapes have the same eye value as a three-point eye, although the larger eyeshapes ("big eyes") are worth more liberties. This is in good accord with the way Go players view these shapes. Another way of looking at this is that collapsing induces an equivalence relation $\sim$, where $G \sim H$ if and only if $c(G) = c(H)$; we can say $2*\mathcal{E} \sim 1\frac{3}{4}\mathcal{E} \sim 1\frac{1}{2}\mathcal{E}$. In each equivalence class, there is a unique element G for which $c(G) = G$; we say such a G is already *collapsed*. G is the natural representative of the class, since $c(H) = G$ for all H in the class.

Since a collapsed game has no endpoints greater than two, collapsing it again has no effect, which is to say that collapsing is idempotent: $c(c(G)) = c(G)$. For the games occurring in Bargo, which have all nonnegative endpoints, it also has the property that

$$c(G + H) = c(c(G) + c(H)).$$

But it is not always the case that $c(G + H) = c(G) + c(H)$, since $c(G) + c(H)$ may have endpoints greater than 2.

2.7. Semigroup structure. Collapsed addition. Complement. The set $\mathbf{Ug}_{[0,]}$ of games for which all integer-ish endpoints are ≥ 0 is closed under addition, and forms a partially ordered regular abelian semigroup with identity 0. It is different from the set $\mathbf{Ug}^+$ of all games that are nonnegative; neither is a proper subset of the other, since $\downarrow$ is in $\mathbf{Ug}_{[0,]}$ but not $\mathbf{Ug}^+$, while $\int \frac{1}{4}$ is in $\mathbf{Ug}^+$ but not $\mathbf{Ug}_{[0,]}$. The structure of $\mathbf{Ug}_{[0,]}$ appears interesting but is beyond the scope of this paper. The games in Bargo are a proper subset of those in $\mathbf{Ug}_{[0,]}$.

The set $\mathbf{Ug}_{[0,2]}$ of games for which all integer endpoints are 0, 1, or 2 is not closed under normal addition, but is closed under *collapsed addition* $\oplus$, defined by $G \oplus H = c(G + H)$. With this operation, $\mathbf{Ug}_{[0,2]}$ and $\mathrm{Bargo}_{[0,2]}$ each form a partially ordered abelian semigroup with identity 0. Since the games in $\mathrm{Bargo}_{[0,2]}$ are a subset of $\mathbf{Ug}_{[0,2]}$, $\oplus$ seems to be the natural addition operation for $\mathrm{Bargo}_{[0,2]}$. The clipping of values to be ≥ 0 is symmetric with the clipping of values to be ≤ 2, so $\mathbf{Ug}_{[0,2]}$ has a reflective symmetry about 1, that is, the mapping $f(x) = 2 - x$ is a self-inverse isomorphism for $\mathbf{Ug}_{[0,2]}$. For each (collapsed) G, there is a unique H such that $H = f(G)$, $G = f(H)$, and $G + H = 2$. We call H the *complement* of G. Even though neither semigroup can have inverses (negatives) for elements other than 0, the complement acts like an inverse in some ways. In fact, by subtracting 1, we can map from complements in $\mathbf{Ug}_{[0,2]}$ to inverses in $\mathbf{Ug}_{[-1,1]}$, which is an abelian group (under its own, appropriately clipped, addition).

It is not clear whether $\mathrm{Bargo}_{[0,2]}$ has the same reflective symmetry as $\mathbf{Ug}_{[0,2]}$, since we have no proof that if G is a possible eyespace value in Go, then $2 - G$ is also. There may exist elements in $\mathrm{Bargo}_{[0,2]}$ that do not have complements in $\mathrm{Bargo}_{[0,2]}$. The value of "seki" (see next section) is possibly such an element.

3. Examples of Single-Group Values

This section gives examples of most of the known eyespace values for finite games in $\mathrm{Bargo}_{[0,2]}$, as well as of some values for simple loopy games. For each value, one has been worked out in detail, and the others are left as exercises.

3.1. $\frac{1}{2}\mathcal{E}$. In each of the examples in Figure 4, Black moving first can make one secure eye, and White moving first can eliminate Black's eye potential. Some of the examples are simple; in others it is more difficult to see how Black or White should play.

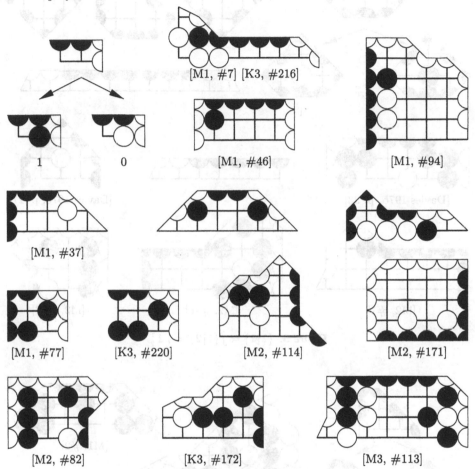

Figure 4. In each example, black moving first can go to a position of value 1, and white moving first can go to 0. We have $\{1|0\} = \int\{0|1\} = \int \frac{1}{2}$. The labeled examples are taken from the Go literature; thus [M1, #37] means Problem 37 in [M1] (also called [Maeda 1965a]).

3.2. $1\frac{1}{2}\mathcal{E}$. We saw a few examples of $1\frac{1}{2}\mathcal{E}$ earlier. Many of the classical unsettled eyeshapes have this value, including all of the three- to six-point nakade shapes [Davies 1975, pp. 13–14, 22–27; Berlekamp and Wolfe 1994, p. 156], and "seven on the second line" [Segoe 1960, p. 10; Davies 1975, pp. 18–19]. It occurs frequently in the small closed and open corridors whose values are given later, and can also be constructed as the sum of two smaller eyespaces. See Figure 5.

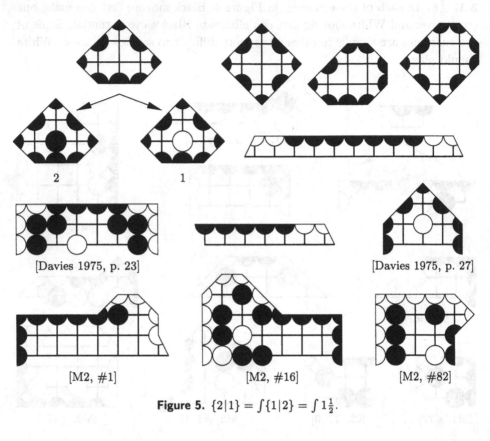

Figure 5. $\{2|1\} = \int\{1|2\} = \int 1\frac{1}{2}$.

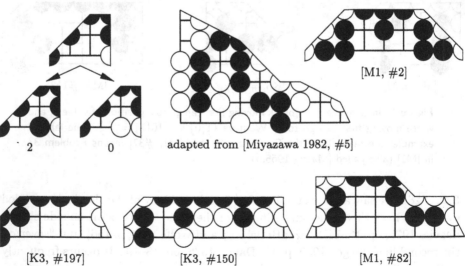

Figure 6. $\{2|0\} = \int\{1|1\} = \int 1*$.

3.3. $1*\mathcal{E}$. Both $\frac{1}{2}\mathcal{E}$ and $1\frac{1}{2}\mathcal{E}$ were simple gote games, what Yedwab [1985] calls switches. The largest possible switch is between two eyes and no eyes, and is written $1*\mathcal{E}$. See Figure 6.

3.4. $\frac{3}{4}\mathcal{E}$. When Black can make $1\frac{1}{2}\mathcal{E}$ in one move, it is better than only being able to make one eye in gote ($\frac{1}{2}\mathcal{E}$), and worse than being able to make two eyes in gote ($1*\mathcal{E}$). This game is $\frac{3}{4}\mathcal{E}$. In each of the examples, it is not enough for Black to guarantee one eye with his first move; he must also create an additional half eye. See Figure 7.

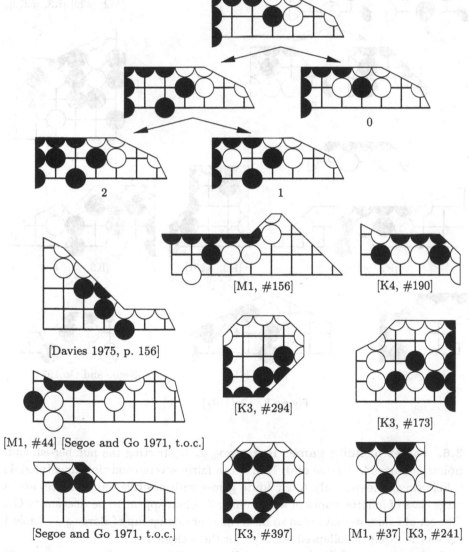

Figure 7. $\{\{2|1\}|0\} = \int \frac{3}{4}$. Here and in following figures t.o.c. stands for "table of contents".

3.5. $1\frac{1}{4}\mathcal{E}$. The complement of $\frac{3}{4}\mathcal{E}$ is $1\frac{1}{4}\mathcal{E}$. In this situation, Black can make two eyes in one move, but White can reduce him to only $\frac{1}{2}\mathcal{E}$. See Figure 8.

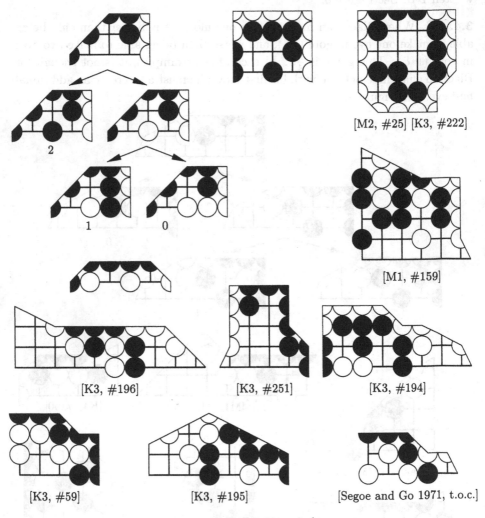

[M2, #25] [K3, #222]

[M1, #159]

[K3, #196]   [K3, #251]   [K3, #194]

[K3, #59]   [K3, #195]   [Segoe and Go 1971, t.o.c.]

Figure 8. $\{2\,|\,\{1\,|\,0\}\} = \int 1\frac{1}{4}$.

3.6. Values of finite games in Bargo$_{[0,2]}$. Restricting the number-ish endpoints of games to be one of 0, 1, or 2 is a fairly severe condition that (modulo infinitesimals) leaves only a few finite games with positive temperature, and a large class of infinite games of which only a fraction appear to be relevant to Go. The set of values we have seen so far form a sub-semigroup of Bargo$_{[0,2]}$. Table 1 shows the result of collapsed addition for these elements.

Early printings of [Berlekamp and Wolfe 1994] erroneously say on page 108 that $1 + \int \frac{3}{4} = \int 1\frac{1}{4}$; but $\int 1\frac{3}{4}$ actually collapses to $\int 1\frac{1}{2}$.

	0	$\int \tfrac{1}{2}$	$\int \tfrac{3}{4}$	$\int 1*$	1	$\int 1\tfrac{1}{4}$	$\int 1\tfrac{1}{2}$	2
0	0	$\int \tfrac{1}{2}$	$\int \tfrac{3}{4}$	$\int 1*$	1	$\int 1\tfrac{1}{4}$	$\int 1\tfrac{1}{2}$	2
$\int \tfrac{1}{2}$	$\int \tfrac{1}{2}$	1	$\int 1\tfrac{1}{4}$	$\int 1\tfrac{1}{4}$	$\int 1\tfrac{1}{2}$	$\int 1\tfrac{1}{2}$	2	2
$\int \tfrac{3}{4}$	$\int \tfrac{3}{4}$	$\int 1\tfrac{1}{4}$	$\int 1\tfrac{1}{2}$	$\int 1\tfrac{1}{2}$	$\int 1\tfrac{1}{2}$	2	2	2
$\int 1*$	$\int 1*$	$\int 1\tfrac{1}{4}$	$\int 1\tfrac{1}{2}$	2	$\int 1\tfrac{1}{2}$	2	2	2
1	1	$\int 1\tfrac{1}{2}$	$\int 1\tfrac{1}{2}$	$\int 1\tfrac{1}{2}$	2	2	2	2
$\int 1\tfrac{1}{4}$	$\int 1\tfrac{1}{4}$	$\int 1\tfrac{1}{2}$	2	2	2	2	2	2
$\int 1\tfrac{1}{2}$	$\int 1\tfrac{1}{2}$	2	2	2	2	2	2	2
2	2	2	2	2	2	2	2	2

Table 1. Collapsed addition of values in $\mathrm{Bargo}_{[0,2]}$.

Note that Table 1 is taking some (infinitesimal) liberties with game arithmetic. For example, $\int \tfrac{1}{2} + \int \tfrac{3}{4} = \{1|0\} + \{\{2|1\}|0\} = \{2|1, \int \tfrac{1}{2}\}$, whereas we call it $\int 1\tfrac{1}{4} = \{2| \int \tfrac{1}{2}\}$. In normal game arithmetic, the move to 1 is not dominated by the move to $\int \tfrac{1}{2}$, because in some circumstances 1 is infinitesimally better (for White) than $\int \tfrac{1}{2}$. However, giving Black a sure eye (value 1) is always at least as bad for White as giving Black a chance to make an eye in gote $(\int \tfrac{1}{2})$. Ignoring the infinitesimals left after the number of eyes is decided makes some moves effectively dominated that would not otherwise be so. We can justify this sort of reduction by considering the same game cooled by a tiny amount $d > 0$. Then it is clear that 1 is always worse for White than $\{1-d|0+d\}$, no matter how small we make d, and regardless of any infinitesimal modifications to either game. So our method of simplifying game G can be viewed as mapping $G \mapsto G_{0+}$, the limit of G_δ (G cooled by δ) as δ approaches zero. This is equivalent to taking the *thermal dissociation* of G [Berlekamp et al. 1982, p. 164] and discarding the term of temperature zero, which contains all the infinitesimals.

Wolfe [Berlekamp and Wolfe 1994, p. 107] states that the only (unchilled) numbers that can occur are 0, 1, and 2, and that therefore the only chilled values that can occur in this context are 0, $\tfrac{1}{2}$, $\tfrac{3}{4}$, $1*$, 1, $1\tfrac{1}{4}$, $1\tfrac{1}{2}$, and 2. The next section, however, shows that this need not be the case.

3.7. Seki as an eyespace value. Sometimes an eyespace can be partly filled with enemy stones in such a way that neither player wants to play next, in one form of the situation known as *seki*:

If Black tries to capture the White stones, he is left with only one eye; on the other hand, if White moves inside this eyeshape, Black can capture and gets 2

eyes. Formally this gives a simple seki-eye like this the value $S = \{1|2\} = 1\frac{1}{2}$. This is a number, not a hot game, and neither player wants to move first in it. Thus an eyeshape of value S is enough to prevent Black from being killed by White, even if Black has no other eyes or liberties.

Under some circumstances, living in seki may be almost as good as living with two eyes. We could construct a consistent theory in which seki is arbitrarily assigned the value 2, which would model this. That theory would then only have the finite eyespace values found in the previous section. However, there seems to be no reason (other than simplicity) to rob ourselves of the extra resolving power of treating living in seki as different from living with two eyes.

Once we allow S as a value, there are also games where one of the number-ish endpoints is S. They have values such as the following:

We have not determined the complete set of seki-related values, but it includes at least S, $2|S$, $S|1$, $S|\int\frac{1}{2}$, $S|0$, $2\|S\|\int\frac{1}{2}$, $2\|S|1$, $2\|S|\int\frac{1}{2}$, $2\|S|0$, and $\{2|\int\frac{3}{4}, \{S|0\}\}$.

3.8. Ko, (1 − Ko), (1 + Ko), (2 − Ko). The cyclical situation known as *ko* gives rise to a number of loopy games. If we denote as Ko the value in a simple ko fight over 1 eye, where Black can take the ko and White can win the ko, the position after Black takes has value $(1 - \text{Ko})$. These values can also be translated by adding one eye to give $(1 + \text{Ko})$ and $(2 - \text{Ko})$.

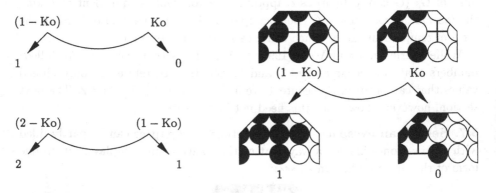

Since Ko + Ko + Ko = 1, we also have Ko + Ko = (1 − Ko) and $\mu(\text{Ko}) = \frac{1}{3}$. These games can also be viewed as the multiples of Ko: 1·Ko = Ko, 2·Ko = (1 − Ko), 3·Ko = 1, 4·Ko = (1 + Ko), 5·Ko = (2 − Ko), 6·Ko = 2.

3.9. Ko_2 and $(2 - Ko_2)$. It is also possible to have a ko fight over two eyes. The structure of the game is the same as for the one-point ko, but all the values are twice as large. Ko_2 has temperature and mean value of $\frac{2}{3}$.

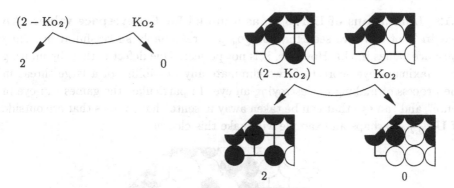

3.10. Other loopy values. There are other loopy values that derive from simple ko fights. They include:

- *two-stage ko*, where winning the first stage ko advances Black to a second stage ko;
- *approach move ko* or *multi-step ko*, where one side must ignore ko threats to fill outside liberties;
- *mannen-ko* (10,000 year ko).

Examples are given in [Davies 1975, pp. 9–10]. Some loopy games in Go have periods that are longer than two moves, including:

- *chosei* or eternal life [Haruyama 1979], with period four. Under current Japanese (Nihon Kiin) rules, if chosei occurs the game is declared "no result" and must be played over; this has happened once [Nakayama 1989]. Under the generalized ko rule, chosei plays much like a normal ko: after three moves in the cycle, the fourth move is forbidden, so one player makes a ko threat. If the threat is answered, three more moves can be played in the cycle, and then it is the other player's turn to make a ko threat.
- *rotating ko* [Haruyama 1979], a loop of period eight, where either player has the option at certain positions to convert the game to a seki.

Given the bewildering variety of loopy games in Go hinted at by these last two examples, we cannot hope to provide an exhaustive categorization of their values, even for the single-group case. Indeed, Robson has shown that the family of life-and-death problems involving multiple simple kos is Exptime-complete [Robson 1981; 1982; 1983; 1985].

3.11. Connecting out. Sometimes a group can live, not by creating eyes within its "own" eyespaces, but by connecting out to another group of the same color. In most tsume-go problems, it is assumed that such a connection makes the group completely alive; this implies that we should treat it as being worth two eyes. This makes a connection in gote worth $1*\mathcal{E}$. A connection may also be made with another group that has G (less than two) eyes; such a potential connection in gote is worth $\{G|0\}$ eyes.

3.12. Limitations of $Bargo_{[0,2]}$ as a model for Go eyespace values. An eye in sente.

We've seen that $Bargo_{[0,2]}$ is reasonably successful at modeling eyespace values in Go. However, it is not perfect. One defect is that, by limiting the maximum eye score to 2, we eliminate any possibility of a large threat in the process of making or destroying an eye. In particular, the games "an eye in sente" and "an eye that can be taken away in sente" have values that are outside of $Ug_{[0,2]}$. Perhaps an example will make this clearer:

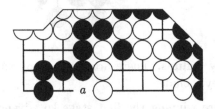

What is the value to the Black group in the corner of Black's eyespace around a? Within $Bargo_{[0,2]}$ the answer is $\frac{3}{4}\mathcal{E}$, since Black can make one eye while threatening to make another. But making the other eye kills the White group! In Bargo (without collapsing), killing the White group gets Black at least four more eyes, so the game is roughly $\{\{5|1\}|0\}$, which chills to $\{\{3|1\}|1\}$ or 1 plus miny-2, so the eyespace has a value something like $1 -_2 \mathcal{E}$. The precise value of this "eye in sente" depends on the size of the external threat; different threats give different (warmed) infinitesimals. However, all of these games collapse to $\frac{3}{4}\mathcal{E}$. The difference in external threats is lost.

We can also have an eye that can be taken away in sente, $\{1|\{0|-x\}\}$. In that case the pure game-theoretical value is a positive (warmed) infinitesimal that depends on x. But all of these collapse to $\frac{1}{2}\mathcal{E}$. Again the difference in threats is lost.

A serious problem arises when trying to add such values. Given the collapsed values above, one would expect the sum of an eye in sente and an eye that can be taken away in sente to be $\frac{1}{2}\mathcal{E} + \frac{3}{4}\mathcal{E} = 1\frac{1}{4}\mathcal{E}$. But $1\frac{1}{4}\mathcal{E}$ has Black stop two eyes and White stop one eye, whereas the actual result depends on which threat is greater. If White's threat is greater, the Black stop is only one eye, and the White stop zero eyes. Thus we see that some vital information can be lost through collapsing when large external threats are involved.

3.13. Values of open and closed corridors.

Figures 9–11 give the $Bargo_{[0,2]}$ values for all closed corridors of length one to five and all open (at one end) corridors of length one to six. All of the non-ko-related, non-seki-related values described previously $(0, \int\frac{1}{2}, \int\frac{3}{4}, 1, \int 1*, \int 1\frac{1}{4}, \int 1\frac{1}{2}, 2)$ occur in this context.

Every closed corridor is worth at least one eye to Black; this makes 0, $\int \frac{1}{2}$, $\int \frac{3}{4}$, $\int 1*$, and $\int 1\frac{1}{4}$ impossible, since they all have a 0-ish endpoint.

If the Black stones surrounding an open corridor are all considered immortal, then the corridor is what Moews [Moews a] calls a *hyper-Black room*. In that case, the eye value of the corridor is irrelevant since the group has two eyes elsewhere, and the best point-making move for either player is at the mouth of the corridor (blocking for Black, pushing in for White). However, when the number of eyes matters, this is no longer so; in some corridors the move at the mouth is not optimal for either player.

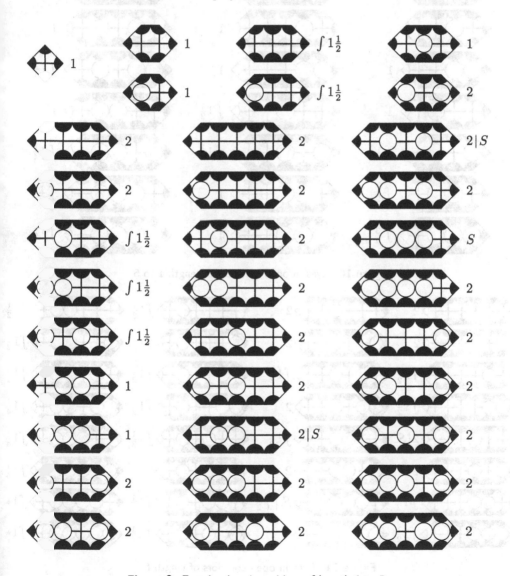

Figure 9. Eyes in closed corridors of length 1 to 5.

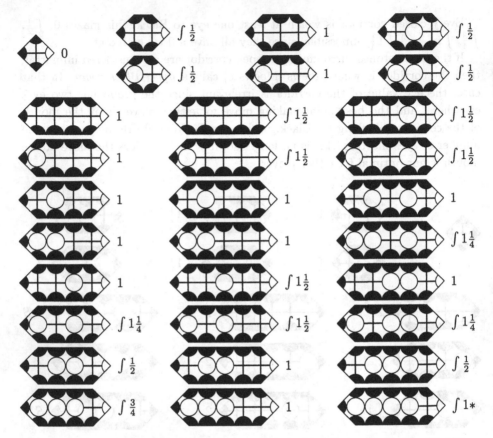

Figure 10. Eyes in open corridors of length 1 to 5.

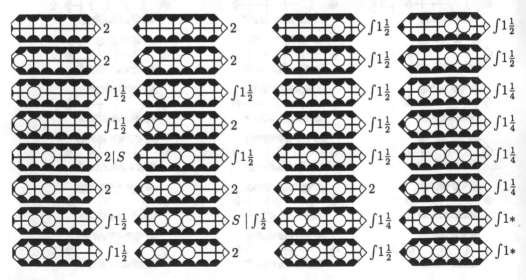

Figure 11. Eyes in open corridors of length 6.

4. A Mathematical Definition of Miai

One of the more subtle concepts of Go is *miai*, from *miru* (see, look at, observe) and *au* (meet, fit, be appropriate for). In non-Go Japanese, miai means "a marriage interview" or "an exchanging of glances". Nagahara devotes a chapter to miai [1972, pp. 3–4], and defines the Go sense of the word: "Miai means 'seeing together'. It refers to two points that are related in such a way that if one of them is occupied by a player, his opponent can handle the situation by taking the other." Davies [1975, p. 12] says that "Two points are miai if they represent two independent ways of accomplishing the same thing, so that if the enemy deprives you of one of them, you can always fall back on the other of them." As long as we view miai in terms of points or intersections, it is hard to apply game theory to it, for single intersections in general are not usually subgames, let alone independent ones. However, miai is also frequently applied to independent subgames, either in the context of scoring points, or in the context of making eyes. For example, Davies [1975, p. 71] describes a and b in the following position as miai:

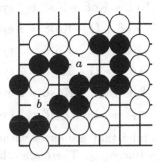

A little analysis shows that the eyespaces around a and b (call them A and B) are independent of each other, and that A is $\frac{1}{2}\mathcal{E}$ and B is $1\frac{1}{2}\mathcal{E}$. Their sum is exactly two eyes. If White plays a, Black can live by playing b; and if White plays b, Black can live at a. Since $A + B$ is a number (in terms of eyes), the Number Avoidance Theorem [Berlekamp et al. 1982, pp. 144 and 179] tells us that neither player needs to move in this sum while there are still nonnumber games to be played. This lack of urgency is characteristic of miai. Nagahara writes "An important point to notice about miai is that the two moves involved are often not urgent. That is, they are in a state of equilibrium. ... it is not necessary for [Black] to rush to play either 'a' or 'b' since either point will give him life." It thus seems possible to formalize the concept of miai as follows:

DEFINITION. A set of games is *miai* if none of them are numbers but their sum is a number.

The proviso that the sum not contain numbers prevents using miai to describe a single number-valued eyespace, which seems outside the spirit of the Go usage. In practice, it is possible to extend this definition slightly, and to call sums of unchilled games miai if they are infinitesimally close to a number. In that case, the chilled games will be miai by the above definition, since even a tiny amount of cooling makes the infinitesimals vanish.

Since any sum of miai is also miai, just knowing that a set of games is miai doesn't say much about the relationships among specific games in the set. We can address this by tightening the definition a bit:

DEFINITION. A set of games is *irreducible miai* if it is miai and no proper subset of it is miai.

The irreducible miai in $\text{Bargo}_{[0,2]}$ include (but are not limited to):

$$\tfrac{1}{2}\mathcal{E} + \tfrac{1}{2}\mathcal{E} = 1 \text{ eye}$$
$$\tfrac{1}{2}\mathcal{E} + 1\tfrac{1}{2}\mathcal{E} = 2 \text{ eyes}$$
$$\tfrac{1}{2}\mathcal{E} + \tfrac{3}{4}\mathcal{E} + \tfrac{3}{4}\mathcal{E} = 2 \text{ eyes}$$
$$\tfrac{3}{4}\mathcal{E} + 1\tfrac{1}{4}\mathcal{E} = 2 \text{ eyes}$$
$$1*\mathcal{E} + 1*\mathcal{E} = 2 \text{ eyes}$$
$$\text{Ko}\mathcal{E} + \text{Ko}\mathcal{E} + \text{Ko}\mathcal{E} = 1 \text{ eye}$$
$$\text{Ko}\mathcal{E} + (1 - \text{Ko})\mathcal{E} = 1 \text{ eye}$$
$$\text{Ko}\mathcal{E} + (2 - \text{Ko})\mathcal{E} = 2 \text{ eyes}$$
$$(1 + \text{Ko})\mathcal{E} + (1 - \text{Ko})\mathcal{E} = 2 \text{ eyes}$$
$$\text{Ko}_2\mathcal{E} + \text{Ko}_2\mathcal{E} + \text{Ko}_2\mathcal{E} = 2 \text{ eyes}$$
$$\text{Ko}_2\mathcal{E} + (2 - \text{Ko}_2)\mathcal{E} = 2 \text{ eyes}$$

The above only includes miai that are exact without collapsing, and hence are miai in Bargo as well as $\text{Bargo}_{[0,2]}$. There are others that do not meet this criterion, such as:

$$1\tfrac{1}{2}\mathcal{E} \oplus 1*\mathcal{E} = c(2\tfrac{1}{2}*\mathcal{E}) = 2 \text{ eyes}.$$

The sum of any (noninteger) game and its complement is necessarily a miai for two eyes.

One surprising consequence of the above definitions is that a set of three or more games may be miai even though no subset of them is! This possibility has not, to my knowledge, been considered by professional Go players or writers; this may be because ai is usually used to indicate a relationship between two objects.

Since the temperature of a number is less than the temperature of any non-number (except for infinitesimals, which we are ignoring anyway), miai involves a kind of mutual cancellation of temperature. The temperature of the sum is less than the temperature of any summand. Such cancellation may occur more weakly than in miai:

DEFINITION. A set of games is *partial miai* if it is not miai but the sum of the games has lower temperature than any game in the set.

We can also define *irreducible partial miai*; an example is $\tfrac{3}{4}\mathcal{E} + \tfrac{3}{4}\mathcal{E} = 1\tfrac{1}{2}\mathcal{E}$.

5. Semeai: Two-group Life and Death

5.1. Semeai. Sometimes life and death problems in Go involve more than a single group at risk. One common situation is called *semeai*, defined by Bauer [Segoe 1960, p. 6] as "a localized situation where only one of the two opposing groups of stones can live (unless the result is a seki), and therefore each must try to kill the other without any reference to or connection with other stones on the board".

To analyze semeai we must take account of both Black and White eyes, which will be the motivation for introducing the game Argo below. However, before proceeding, it is necessary to discuss the manner in which eyes and liberties affect the outcome of a semeai.

5.2. Semeai where no eyes are possible. Before considering the role of eyemaking moves in a semeai, it will help to understand the values of semeai where the eyes have already been decided. The simplest case is one where neither Black nor White can make any eyes, there are no kos, and all liberties are simple dame. In this case, life or death is entirely a function of liberty count.

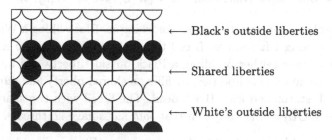

←— Black's outside liberties

←— Shared liberties

←— White's outside liberties

If the number of shared liberties is less than two, seki is not possible; either the Black group or the White group must die. This means that there are no zero games, and all values are either positive or negative or fuzzy. Whichever group has the most outside liberties wins, or if the outside liberties are equal then the player moving first wins. Letting BOL and WOL stand for the number of Black outside liberties and White outside liberties respectively, we can summarize the above by saying that the game's outcome class is the same as that of the game $(BOL - WOL + *)$. Whether the number of shared liberties is zero or one makes no difference. If the number of shared liberties is two or more, then seki is a conceivable result. For Black to capture White, Black must fill all White's outside liberties as well as all the shared liberties. Black must also have at least one outside liberty remaining when he fills the next-to-last shared liberty, else White will capture him.

Table 2 summarizes this in formulas and in a grid. We can see that the result space is partitioned into three large cool areas (W, S, and L), with thin hot boundaries (WS, WL, and SL) between them. Cool cells adjacent to the boundaries imply that there are ko threats for the side that is one move away

Black wins if (BOL − WOL) ≥ Shared.

Black wins moving first if (BOL − WOL) ≥ Shared − 1.

White wins if (WOL − BOL) ≥ Shared.

White wins moving first if (WOL − BOL) ≥ Shared − 1.

If neither Black nor White wins, the result is seki.

		BOL − WOL										
		5	4	3	2	1	0	−1	−2	−3	−4	−5
Shared Liberties	5	W	WS	S	S	S	S	S	S	S	SL	L
	4	W	W	WS	S	S	S	S	S	SL	L	L
	3	W	W	W	WS	S	S	S	SL	L	L	L
	2	W	W	W	W	WS	S	SL	L	L	L	L
	1	W	W	W	W	W	WL	L	L	L	L	L
	0	W	W	W	W	W	WL	L	L	L	L	L

Table 2. Semeai outcomes when no eyes are possible. BOL = Black outside liberties; WOL = White outside liberties; W = Black wins, White loses; S = Seki; L = Black loses, White wins; WL = {W|L}; WS = {W|S}; SL = {S|L}.

from being able to improve its status. For example, Black filling a White outside liberty and moving left from an L cell to an SL cell threatens to make a seki; White must answer (either by filling a Black outside liberty and moving right, or, if Black has no outside liberties, by filling a shared liberty and moving down) to return to L status and keep Black dead. The S cell with two shared liberties and (BOL − WOL) = 0 is unique in that *both* sides have ko threats.

5.3. Semeai where each side has one eye. When each side in a semeai has one single-point eye, the situation is similar to that for no eyes. The pair of opposing eyes behaves somewhat like a single shared liberty. The results are shown in Table 3.

Note that, for a seki to be possible between two one-eyed groups, they must also have at least one shared liberty.

		BOL − WOL										
		5	4	3	2	1	0	−1	−2	−3	−4	−5
Shared Liberties	5	WS	S	S	S	S	S	S	S	S	S	SL
	4	W	WS	S	S	S	S	S	S	S	SL	L
	3	W	W	WS	S	S	S	S	S	SL	L	L
	2	W	W	W	WS	S	S	S	SL	L	L	L
	1	W	W	W	W	WS	S	SL	L	L	L	L
	0	W	W	W	W	W	WL	L	L	L	L	L

Table 3. Semeai outcomes when each side has one eye.

	BOL – WOL										
	5	4	3	2	1	0	−1	−2	−3	−4	−5
Shared Liberties 5	W	W	W	W	W	W	W	W	W	W	W
4	W	W	W	W	W	W	W	W	W	W	WL
3	W	W	W	W	W	W	W	W	W	WL	L
2	W	W	W	W	W	W	W	W	WL	L	L
1	W	W	W	W	W	W	W	WL	L	L	L
0	W	W	W	W	W	W	WL	L	L	L	L

Table 4. Semeai outcomes when black has one eye and white has none.

5.4. Semeai with unbalanced eyes. We have already looked at cases with 0 versus 0 eyes and 1 versus 1 eyes. Semeai where one group has two eyes are not semeai at all; the group with two eyes is alive, and if the other group cannot make two eyes it is simply dead, regardless of liberty count.

The only remaining possibility with integer eyes is 1 eye versus 0 eyes. As in all two-group semeai with unequal number of eyes, seki is not possible; one group or the other must die. For the case where Black has one eye and White has none, the results are shown in Table 4. The case where White has one eye and Black has none is symmetric.

To a first approximation, the advantage of one eye converts the "seki region" in the previous tables into wins for Black. This is so beneficial that one might be tempted to infer that it is always better to make an eye than to worry about liberties. In fact there is a Go proverb that states "The semeai where only one player has an eye is a fight over nothing" [Segoe 1960, p. 76]. However, making an eye sometimes uses up more than one liberty (one converted to the eye, and one used up by the play), as well as taking a move (which might otherwise be used to fill an enemy liberty). In the borderline cases, there are counterexamples to this generalization.

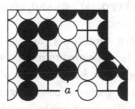

In this example, if Black makes an eye at a White will have enough outside liberties to kill him. The locally correct line of play is for Black to fill one of White's outside liberties, which forces White to play a, and then Black fills the other outside liberty to leave a seki. If Black had even one more liberty, either outside or shared, then the locally correct line of play would be to make the eye, which kills White. Segoe [1960, p. 79] observes: "Even in those cases where one player has one eye and the other has none the number of dame available to each player must be carefully analyzed or the semeai may be lost."

This interaction between eyes and liberties is awkward, and a full analysis of it is beyond the scope of the present paper. Such an analysis might also require, as a prerequisite, a better understanding of the subgames of Go where making or destroying liberties is the main objective, which seems to be a fairly rich topic in its own right. However, if the number of shared liberties is sufficiently greater than the absolute value of BOL − WOL, these complications do not occur. In order to simplify the analysis in what follows, we will usually assume that the number of shared liberties is sufficiently large that the outside liberty count can be ignored, so that whichever side makes the most eyes wins the battle or, if the eyes made are equal and less than two, both sides live in seki. Table 5 shows the possible results of a semeai as a function of the number of Black and White eyes, in the fully general case and after the simplifying assumption has been made. After this simplification, the result depends only on the difference between the number of Black eyes and White eyes.

| | White eyes | | |
	0	1	≥ 2
Black eyes ≥ 2	W	W	both live
Black eyes 1	W, L	W, S, L	L
Black eyes 0	W, S, L	W, L	L

| | White eyes | | |
	0	1	≥ 2
Black eyes ≥ 2	W	W	both live
Black eyes 1	W*	S*	L
Black eyes 0	S*	L*	L

Table 5. Possible results as a function of the number of eyes, under no assumptions (left) and under the assumption of enough shared liberties (right). Stars mark cases that depend on the assumption.

5.5. Modeling Semeai: Argo. When two groups are locked together in a semeai, the one to make the most eyes usually wins (as we have seen above). Under appropriately restrictive conditions (neither group can make more than two eyes, it is not possible for both groups to make two eyes, and there are enough shared liberties), these struggles can be solved simply by counting eyes, with the value of each eyespace being a game whose integer endpoints are between -2 and 2. Since both White and Black can make eyes in such a semeai, the natural score is the number of Black eyes minus the number of White eyes. If this is ≥ 1, Black wins; if it's ≤ -1, White wins; if it's 0, both groups are alive in seki; otherwise, the result depends on who moves first.

To model this we define a Go-like game that I call Argo (after Argus Panoptes, who had a hundred eyes; the ship Argo which the Argonauts sailed was named after a different Argus, son of Phrixus and quite human). The rules of Argo are the same as Go, with the exception of scoring: in Argo Black and White each get one point for every eye they have at the end of the game. That is, we count each Black eye as $+1$ and each White eye as -1.

In the case of Bargo (whose name, we can now reveal, stands for Black Argo), we wanted to restrict all values to have integer endpoints between 0 and 2 inclusive. For Argo, the natural limits are -2 and 2.

Using $\mathrm{Argo}_{[-2,2]}$ we can model situations where both Black and White eyes can be made. All of the values from $\mathrm{Bargo}_{[0,2]}$, their negatives, and many additional values such as $\frac{1}{4}\mathcal{E}$ and $*\mathcal{E}$ occur. Note that some of the games given above as examples for $\mathrm{Bargo}_{[0,2]}$ have different values in $\mathrm{Argo}_{[-2,2]}$, since the White eyes now count!

However, something seems not quite right. If we represent a result where Black has b eyes and White has w eyes by the ordered pair (b, w), the central problem is that there are real differences between $(0,0)$, $(1,1)$, and $(2,2)$ when it comes to adding them to other games; but, they are all represented in Argo as the game 0. And $(0,0) + G$ is always equal to G, but $(2,2) + G$ is equal to $(2,2)$ regardless of G.

The conclusion seems to be that adding negative White eyes to positive Black eyes is too simplistic; we really need to treat each separately (clipping against limits of 0 and 2). But that implies that the "integer" endpoints of each game need to be represented as an ordered pair of integers (x, y), or as a complex integer $x + iy$. Accepting that means that we must go outside of the established theoretical framework and construct a combinatorial game theory of complex- or (more generally) vector-valued games.

6. Directions for Future Research

6.1. Completely characterizing single-group seki-related values. The set of seki-related values in $\mathrm{Bargo}_{[0,2]}$ is probably finite, and not much larger than the set of values mentioned earlier. It should be possible to identify all elements of this set and provide examples of each.

The asymmetry introduced by S as a value could be eliminated if a Go position was found that had value $\{0|1\} = \frac{1}{2}$ eyes. I do not know of any examples of this value. However, there are several known Go positions other than seki that have the property that neither player wants to move first, such as "three points without capturing" [Berlekamp and Wolfe 1994, pp. 165–168], so its existence does not seem impossible.

6.2. Characterizing loopy game values. Even for the single-group case, there are clearly many loopy games that still need to be identified and their values characterized. The same holds true for multi-group problems. There does not appear to be much hope of a general solution for all loopy games in Go, since already, as mentioned earlier, the family of life-and-death problems

involving multiple simple kos is Exptime-complete [Robson 1981; 1982; 1983; 1985]. Robson's construction involves extremely complicated topologies that are unlikely to occur in real games, however. It may still be possible to understand the most common kos, which all have simple topologies.

6.3. Counting liberties. Since the number of liberties of a unit or group often has a critical effect on its survival, we can apply the theory to games in which making or eliminating liberties is the main goal. Liberties can be gained by connection, so the study of cut-or-connect problems will be relevant. Unlike eye-counting, liberty-counting has no built-in upper limit, so we should expect much hotter games to appear. In fact, making two eyes effectively supplies an infinite number (**on**) of liberties, so infinitely-hot games would not be unreasonable.

A firm theory of liberties, plus the vector-valued game theory mentioned earlier, might allow accurate analysis of situations where both eye-count and liberty-count are critical.

6.4. Ko threats. The existing theory, from [Conway 1976] to the present work, has the unfortunate effect of simplifying away ko threats. For example, if Black is topologically alive with two eyes, White has no threats against the Black group, but if Black is alive with exactly 1 eye $+ \frac{1}{2}\mathcal{E} + \frac{1}{2}\mathcal{E}$, White has at least one ko threat against Black (by moving one of the $\frac{1}{2}\mathcal{E}$ subgames to 0). Yet we treat both of these situations as identical, calling them two eyes.

Getting beyond this limitation would seem to require changes to the very foundations of the theory, altering the definition of equality and restricting the simplifications allowed.

Acknowledgements

The author would like to thank Elwyn Berlekamp and David Wolfe for sharing their early, unpublished research and providing many useful criticisms; David Fotland for asking me to analyze a tsume-go problem where (it turned out) Black had $1\frac{1}{4}\mathcal{E} + \frac{1}{2}\mathcal{E}$ and hence wasn't quite alive, thereby triggering the present investigations; Carol Seger and my wife Gelly for support and nudging; MSRI and the Londerville/Colella household for their hospitality; and many Go players, but especially Herb Doughty and Anton Dovydaitis, for their interest. This work is dedicated to Robert Gordon High (d. 1993), who taught me to play Go in 1967 and enthusiastically encouraged this line of research [High 1990], and to Robert Jordan Landman (b. 1993), whom I hope to teach in turn.

References

[Berlekamp et al. 1982] E. R. Berlekamp, J. H. Conway and R. K. Guy, *Winning Ways for Your Mathematical Plays*, vol. 1, Academic Press, London, 1982.

[Becker 1987] Ira Becker, "An algorithm for determining certain group properties in the game of Go", *Computer Go* **5** (Winter 1987–88), 14–19.

[Benson 1976] David B. Benson, "Life in the Game of Go", *Information Sciences* **10** (1976), 17–29.

[Benson 1980] David B. Benson, "A mathematical analysis of Go", pp. 55–64 in *Proc. 2nd Seminar on Scientific Go-Theory*, Mülheim an der Ruhr, 1979 (edited by K. Heine), Institut für Strahlenchemie, Mülheim an der Ruhr, 1980.

[Berlekamp 1988] E. R. Berlekamp, "Blockbusting and Domineering", *J. Combin. Theory* A**49** (1988), 67–116.

[Berlekamp and Wolfe 1994] E. R. Berlekamp and D. Wolfe, *Mathematical Go: Chilling Gets the Last Point*, A. K. Peters, Wellesley, MA, 1994. Also published in paperback, with accompanying software, as *Mathematical Go: Nightmares for the Professional Go Player*, by Ishi Press, San Jose, CA.

[Chen 1990] Ken Chen, "The move decision process of Go intellect", *Computer Go* **14** (Spring/Summer 1990), 9–17.

[Conway 1976] J. H. Conway, *On Numbers and Games*, Academic Press, London, 1976.

[Davies 1975] James Davies, *Life And Death*, Ishi Press, San Jose, 1975.

[Fotland 1986] David Fotland, "The program G2", *Computer Go* **1** (Winter 1986–87), 10–16.

[Good 1962] I. J. Good, "The mystery of Go", *New Scientist* **427** (21 Jan. 1962), pp. 172–174.

[Hanner 1959] O. Hanner, "Mean play of sums of positional games", *Pacific J. Math.* **9** (1959), 81–99.

[Haruyama 1979] Haruyama Isamu, "Strange and wonderful shapes", *Go World* **50** (Winter 1987–88), 16–22. Translation of article originally appearing in *Igo Club*, January 1979.

[Harker 1987] W. H. Harker, "An efficient chain search algorithm", *Computer Go* **3** (Summer 1987), 8–11.

[High 1990] Robert G. High, "Mathematical Go", *Computer Go* **15** (Fall/Winter 1990/1991), 14–24. Also reprinted as pp. 218–224 in *The Go Player's Almanac* (edited by Richard Bozulich), Ishi Press, San Jose, 1992.

[Hsu and Liu 1989] Hsu Shun-Chin and Liu Dong-Yeh, "The design and construction of the computer Go program Dragon II", *Computer Go* **10** (Spring 1989), 10–19.

[Ing 1991] Ing Chang-ki, *Ing's SST Laws of Wei-Ch'i* 1991 (translated by Richard Bozulich), Ing Chang-ki Wei-ch'i Educational Foundation, 1991.

[K3 = Kano 1987a] Kano Yoshinori, *Graded Go Problems for Beginners, 3: Intermediate Problems, 20-kyu to 15-kyu*, Nihon Kiin, 1987.

[K4 = Kano 1987b] Kano Yoshinori, *Graded Go Problems for Beginners, 4: Advanced Problems*, Nihon Kiin, 1987.

[Kierulf et al. 1989] Anders Kierulf, Ken Chen, and Jurg Nievergelt, "Smart game board and go explorer: a study in software and knowledge engineering", *Comm. ACM* **33**(2) (February 1990), 152–166.

[Kierulf 1990] Anders Kierulf, "Smart game board: A workbench for game-playing programs, with Go and Othello as case studies", Dissertation #9135, ETH Zürich; Verlag der Fachvereine an den schweizerischen Hochschulen und Techniken, Zürich, 1990.

[Kraszek 1988] Janusz Kraszek, "Heuristics in the life and death algorithm of a Go-playing program", *Computer Go* **9** (Winter 1988–89), 13–24.

[M1 = Maeda 1965a] Maeda Nobuaki, *Maeda Shokyuu Tsume-go (Maeda's Beginning-kyu Tsume-go)*, Tokyo Sogensha, 1965 (Showa 40).

[M2 = Maeda 1965b] Maeda Nobuaki, *Maeda Chuukyuu Tsume-go (Maeda's Mid-kyu Tsume-go)*, Tokyo Sogensha, 1965 (Showa 40).

[M3 = Maeda 1965c] Maeda Nobuaki, *Maeda Jookyuu Tsume-go (Maeda's Upper-kyu Tsume-go)*, Tokyo Sogensha, 1965 (Showa 40).

[Milnor 1953] J. Milnor, "Sums of positional games", pp. 291–301 in *Contributions to the Theory of Games*, vol. 2 (edited by H. W. Kuhn and A. W. Tucker), Ann. of Math. Stud. **28**, Princeton University Press, Princeton, 1953.

[Millen 1981] Jonathan K. Millen, "Programming the Game of Go", *Byte*, April 1981, pp. 102 ff.

[Miyazawa 1982] Miyazawa Goro, *Tsume-go Joi Bukkusu (Tsume-go Joy Books)* **10**, Nihon Kiin, Tokyo, 1982.

[Moews a] David Moews, "Coin-sliding and Go", to appear in *Theoret. Comp. Sci.* A.

[Müller 1995] M. Müller, "Computer Go as a Sum of Local Games: An Application of Combinatorial Game Theory", Ph.D. Dissertation, ETH Zürich, 1995.

[Nagahara 1972] Nagahara Yoshiaki, *Strategic Concepts of Go*, Ishi Press, 1972.

[Nakayama 1989] Nakayama Noriyuki, "Strange and wonderful shapes", *Go World* **64** (Summer 1991), 61–64. Translation of articles originally appearing in *Kido*, January 1989, and *Go Weekly*, February 1989.

[Remus 1963] H. Remus, "Simulation of a learning machine for playing Go", pp. 428–432 in *Information Processing* 1962: Proceedings of IFIP Congress (edited by Cicely M. Popplewell), North-Holland, Amsterdam, 1963.

[Robson 1981] J. M. Robson, "Another intractable game on Boolean expressions", Technical Report TR-CS-81-02, Australian National University, Canberra, 1981.

[Robson 1982] J. M. Robson, "Exponential time decision problems relating to ko-like transition rules", Technical Report TR-CS-82-02, Australian National University, Canberra, 1982.

[Robson 1983] J. M. Robson, "The complexity of Go", pp. 413–417 in *Information Processing* 1983: Proceedings of 9th IFIP Congress (edited by R. E. A. Mason), North-Holland, Amsterdam, 1983.

[Robson 1985] J. M. Robson, "Alternation with Restrictions on Looping", *Information and Control* **67**(1–3) (1985), pp. 2–11.

[Segoe 1960] Segoe Kensaku, "Go Proverbs Illustrated", Nihon Kiin, 1960.

[Segoe and Go 1971] Segoe Kensaku and Go Seigen, *Tesuji Jiten (Tesuji Dictionary)*, vol. 2, Seibundo Shinkosha, Tokyo, 1971.

[Thorp and Walden 1964] Edward Thorp and William E. Walden, "A partial analysis of Go", *Computer Journal* **7**(3) (1964), 203–207.

[Thorp and Walden 1972] Edward Thorp and William E. Walden, "A computer assisted study of Go on $M \times N$ boards", *Information Sciences* **4** (1972), 1–33.

[Yedwab 1985] L. J. Yedwab, "On playing well in a sum of games", M.Sc. Thesis, MIT, 1985. Issued as MIT/LCS/TR-348, MIT Laboratory for Computer Science, Cambridge, MA.

[Zobrist 1969] Alfred L. Zobrist, "A model of visual organization for the game of Go", pp. 103–111 in *Proc. AFIPS Spring Joint Computer Conference*, Boston, 1969, AFIPS Press, Montvale, NJ, 1969.

HOWARD A. LANDMAN
HAL COMPUTER SYSTEMS, INC.
1315 DELL AVENUE
CAMPBELL, CA 95008
landman@hal.com

Games of No Chance
MSRI Publications
Volume 29, 1996

Loopy Games and Go

DAVID MOEWS

ABSTRACT. Berlekamp, Conway and Guy have developed a theory of partizan loopy combinatorial games—that is, partizan combinatorial games that
allow infinite play—under disjunctive composition. We review this theory
of loopy games and show how it can be adapted to the two-person strategy
game of Go, which also has the feature that situations involving infinitely
long play often arise.

1. Introduction

In the two-player strategy game of Go, it can happen that an endgame position
splits up into several non-interacting subpositions. Since each player must then
move in just one of the subpositions on his turn, the whole position will then be
the so-called *disjunctive compound*, or *sum*, of the subpositions. As it turns out,
we can then apply to these Go endgames the theory of partizan combinatorial
games with finite play under disjunctive composition, as found in *Winning Ways*
[Berlekamp et al. 1982], Chapters 1–8, or *On Numbers and Games* [Conway
1976].

This paper assumes that the reader is already somewhat familiar with Go and
with the application of this theory to Go, as given in [Wolfe 1991; Berlekamp
and Wolfe 1994]. In Chapter 11 of *Winning Ways* there is a theory of partizan
combinatorial games with possibly infinite play under disjunctive composition.
These games are there called *loopy*, since what was a game tree in the finite play
case is now a game graph, perhaps with cycles. We review this theory of loopy
games and show how it can be applied to Go, which also has cycles.

2. Loopy Games

We review the definitions of loopy games and basic theorems about them,
as given in *Winning Ways*, pp. 314–357. A *loopy game* is played on a directed
pseudograph (that is, a directed graph that is allowed to have multiple edges
from one vertex to another, or edges from a vertex to itself). Each vertex of this

graph corresponds to a position of the game. We partition the edges into two sets, Left's and Right's, corresponding to our two players, Left and Right. Each player plays along his own edges only, changing the position from the vertex at the start of the edge to that at the end. Left and Right play alternately; if either is ever left without a move, he loses.

To specify a loopy game completely, we need to specify a start-vertex as well as the graph and its edge-partition. We also need to specify for each legal infinite sequence of moves—whether it has alternating play by Left and Right or not—whether it is won by Left, won by Right, or is drawn. We assume that two infinite plays that are the same except for a finite initial segment are similarly designated. In general, such games can be very complicated. For example, assuming the Axiom of Choice, there are games with no plays drawn for which there is a winning strategy for neither Left nor Right. However, our infinite play criteria will always be so simple that this sort of thing will never arise.

The distinctive features of this theory of loopy games are the partizan nature of the game—i.e., that the two players may have different moves—and the disjunctive composition that we will define now. Given two loopy games, G and H, we define their disjunctive compound, or sum, $G + H$, as in the case where play must end in a finite length of time: it is just G and H placed side by side, with each player moving on just one of the games—whichever he chooses—on his turn. If there is infinite play in a sum, to determine who wins, we need to look at the subplays in each component:

A player *wins* the sum just if he wins *all* the components in which there is *infinite* play. (*Winning Ways*, p. 315)

If we interchange the roles of Left and Right in a game G, we call the result $-G$. Moreover, G^+ will be G with draws redefined as wins for Left, and G^- will be G with draws redefined as wins for Right. We say that $G \geq H$ if Left wins or draws both $G^+ - H^+$ and $G^- - H^-$ moving second. This relation is reflexive and transitive, and $G \geq H$ implies that $G + K \geq H + K$ for all K (*Winning Ways*, pp. 328–330). As in the finite play case, we define 0 as the game where neither player can ever move, and $G = H$ just if $G \geq H$ and $G \leq H$. Under $+$, the loopy games form a monoid with additive identity 0, but unlike the finite play case, they do not form a group. For example, let **on** be the game where there is only one position and a move by Left from the position to itself, infinite play being drawn. For any G, Left can always move from **on** to **on** in **on** $+ G$, and at least draw, whoever starts. Hence **on** $+ G$ can never be 0.

3. Sidling

Sidling, defined in *Winning Ways*, pp. 318 ff., is a means of approximating loopy games by simpler games—hopefully, even *enders*, which are what partizan combinatorial games with finite play are called in this new context. The basic idea is that we cut play off after a finite number of moves.

Let's say we have a loopy game G and a set M of moves in G, and that infinite play is always won for Left if there are infinitely many moves in M. Let S be the set of positions of G. Now if we have a loopy game H whose set of positions includes S, let $G^*(H)$ be what we get when we put G next to H and redirect the moves in M from their destinations in G to the corresponding positions in H. Infinite play for $G^*(H)$ is handled by the criterion for H or for G, according to whether the play ever enters H or not. After doing this, we consider the set of positions of the altered copy of G within $G^*(H)$ to be S. Let K_v be the game K with the start vertex redefined to be the position v of K. Then, if $H_v \geq K_v$ for all $v \in S$, we have $G^*(H)_v \geq G^*(K)_v$ for all $v \in S$. (You can see this by means of a simple reflection strategy in the difference game.) In this sense, G^* is an order-preserving operator. Let N have set of positions S and a move by Left from each position to itself. We consider all infinite play in N to be won for Left. Then $N_v = \mathbf{on}^+$ for each $v \in S$, and $\mathbf{on}^+$ is a maximal game, that is, $\mathbf{on}^+ \geq H$ for all H. Hence $N_v \geq G^*(N)_v$ for all $v \in S$, and then since G^* is order-preserving it follows that

$$N_v \geq G^*(N)_v \geq G^*(G^*(N))_v \geq \cdots$$

for all $v \in S$. If we are in luck, the sequence N, $G^*(N)$, ... will reach a fixed point U—i.e., a U such that $U_v = G^*(U)_v$ for all $v \in S$. This is then a maximal fixed point, since we got it by iterating G^* on the maximal N. But $G^*(G)_v = G_v$ for all $v \in S$; another obvious reflection strategy proves this, once you recall that we required our infinite play criterion in G to be independent of any finite initial play in G. Hence $U_v \geq G_v$ for all $v \in S$. But we have the following result:

SIDLING THEOREM. *For all R such that $R_v \leq G^*(R)_v$ for all $v \in S$, we have $R_v \leq G_v$ for all $v \in S$.*

PROOF. See *Winning Ways*, pp. 351–353, for the case when M contains every move in G. The extension to the general case is easy [Moews 1993, Theorem 1]. □

Hence our U above must in fact have $U_v = G_v$ for all $v \in S$, and we have found a simplified form for G. Often, U will even be an ender. For example, let G be the game in Figure 1, where we take all infinite play to be drawn. G^+ will then

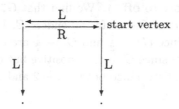

Figure 1. A loopy game.

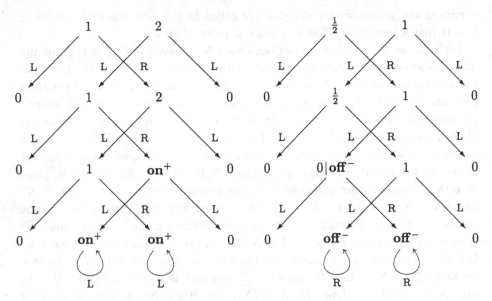

Figure 2. Left: Computing G^+. Right: Computing G^-.

have all infinite play won for Left. Taking M to be the set of the two horizontal moves in G, we can compute $U = G^*(G^*(G^*(N)))$ as shown in Figure 2, left. We have labelled the vertices of U with their values—that is, we have labelled each v with the name of a game equal to U_v. You can see from the figure that

$$G^*(G^*(N))_v = G^*(G^*(G^*(N)))_v$$

for all $v \in S$. It follows that $G^+ = 2$.

Similarly, suppose we start with the game N', which consists of moves for Right from positions to themselves, all infinite play being won for Right. We will then have $N'_v = \mathbf{off}^-$ for all v ($\mathbf{off} = -\mathbf{on}$ being a game with only one position, a move for Right from that position to itself, and all infinite plays drawn), and if we reach a fixed point by iterating G^* starting at N', it will be a minimal fixed point of G^*. If we negate everything and apply the Sidling Theorem, we can see that this fixed point contains a value for G^-. The result in our case can be seen in Figure 2, right (where $0 | \mathbf{off}^-$ is a game from which Left can move to 0 and Right can move to $\mathbf{off}^-$). We find that $G^- = 1$. From these values, we can find the outcome of $G + E$ for all enders E. For example, $G - \frac{1}{2}$ is won by Left moving first, since $G^+ - \frac{1}{2}$ and $G^- - \frac{1}{2}$ are both positive; $G - 1$ is a draw for Left moving first, since $G^+ - 1$ is positive but $G^- - 1$ is zero; and $G - 2$ is a loss for Left moving first, since both $G^+ - 2$ and $G^- - 2$ are less than or equal to zero.

4. Go and Kos

As we mentioned in the introduction, the board can split up into independently analyzable subpositions in Go endgames, and given such a subposition,
we can treat it as a partizan combinatorial game by identifying the integer scores
of the terminal positions arising from this subposition with the integer combinatorial games that occur in this theory [Wolfe 1991, Chapter 3; Berlekamp
and Wolfe 1994, Chapter 2; Moews 1993, Chapter 1]. (We identify Black and
White in Go with Left and Right in this theory, so that a negative final score is
favorable to White.) We will assume that these subpositions are separated by
stones, called *immortal* in [Berlekamp and Wolfe 1994, § 4.1] and [Moews 1993,
Chapter 1], which are uncapturable and will hence remain on the board until
the end of the game. In diagrams of these subpositions we show the separating
immortal stones on grid lines that extend off the edge of the subposition.

Consider the subposition shown at B in Figure 3. Right (White) can move on
the empty vertex to a zero position, but if Left (Black) moves, he will capture
a White stone and move to A. From A, Right can move back to a position
equivalent to B, and Left can move to a position with no moves and one White
stone captured, which is 1. This position is called *one-point ko* in Go, since the
total value at stake is $1 - 0 = 1$. The moves to 1 or 0 are said to *fill* the ko.

Naïvely, it looks as though play could continue forever in the ko, if both
players move back and forth without filling it. In Go, there are various *ko-ban*
rules used to prevent this. We will use the Japanese ko-ban rule, which states
that a player can't move back to the position that occurred immediately before
the last player's move (considering the whole board position, and not just our
subpositions.)

If G is a loopy game, we let $\phi(G)$ be the loopy game G with the constraint of
the Japanese ko-ban rule added. We take all infinite play to be drawn in $\phi(G)$.

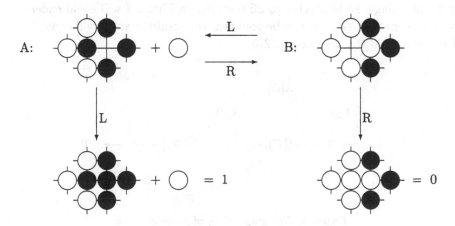

Figure 3. Ko.

A position of $\phi(G)$ will be either an ordered pair $\varepsilon[\{\lambda\}]$ of positions ε and λ of G, by which we mean the position ε of G with the position λ occurring before the last move, or $\varepsilon[\varnothing]$, by which we mean the position ε of G when there has been no previous move. It should be clear which moves are allowed in $\phi(G)$.

The difficulty with ϕ is that if H is an ender, we do not have the property $\phi(G + H) = \phi(G) + H$, even if $H = 0$. For example, let H be $\{100|0\|\}$, a game equal to 0. Right moving first wins on $\phi(A)$, since he moves from A to B and then Left has no move. Moving first on $\phi(A + H)$, Right's first move must still be from A to B. However Left can then move from H to $100|0$, and Right is then forced to move from $100|0$ to 0. Left can then move back from B to A without violating the ko-ban in the sum, and Right then has no move and loses. In Go terminology, H was a *ko-threat* for Left, since by using it Left was able to make an otherwise ko-banned move. We let $\phi_L(G)$ be $\phi(G)$ with ko-banned moves adjoined for Left, and similarly for $\phi_R(G)$ and Right. Since to make a ko-banned move in G Left must use up a ko-threat, and in a subposition of a real position there could only be finitely many ko-threats like H available, we make all infinite play in $\phi_L(G)$ with an infinite number of ko-banned moves a loss for Left, although all other infinite play remains a draw, as in $\phi(G)$. The rule for $\phi_R(G)$ is analogous.

Evidently, for all G,

$$\phi_L(G) \geq \phi(G) \geq \phi_R(G),$$

so ϕ_L and ϕ_R are bounds for ϕ, and for all G that can occur in Go, we have the desirable properties

$$\phi_L(G + H) = \phi_L(G) + H \qquad \text{for } H \text{ an ender,}$$
$$\phi_R(G + H) = \phi_R(G) + H \qquad \text{for } H \text{ an ender.}$$

In Figure 4, we see the graph of ϕ_L of the one-point ko of Figure 3. The graph of ϕ_L of a single ko is acyclic, so all positions in Figure 4 will equal enders. In fact, if we start at $A[\varnothing]$, it can be seen that the resultant value will equal 2, and if we start at $B[\varnothing]$, it will equal $2|0$.

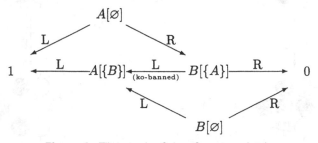

Figure 4. The graph of ϕ_L of a one-point ko.

5. An Example of Sidling in Go

When we have more than one ko in G, the game-graph of $\phi_L(G)$ or $\phi_R(G)$ may contain loops, so we will need to sidle to compute enders equal to them. For example, let T be the position on the right, consisting of the sum of a one-point and a seven-point ko. The decomposition of T as a sum of kos is shown in Figure 5. We wish to compute $\phi_L(T)$.

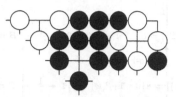

The game graph of $P = \phi_L(T)$ is the large connected component in the center-right of Figure 6. We have made certain simplifications, such as precomputing the values of subgraphs where only one ko is still alive. We have also omitted nodes of the form $\varepsilon[\varnothing]$, since their values can be shown to be equal to the value of $\varepsilon[\{\lambda\}]$, for any λ, and there are no moves back from the $\varepsilon[\{\lambda\}]$'s to the $\varepsilon[\varnothing]$'s. For reference, ϕ_L of the one-point ko is shown at the top and lower left. (Here, again, values of $\varepsilon[\varnothing]$ vertices are the same as those of $\varepsilon[\{\lambda\}]$ vertices, and the $\varepsilon[\varnothing]$ vertices are again omitted.) Also, ϕ_L of the seven-point ko is shown at the upper left and bottom. In the top half of P, a move on the one-point ko is shown by a horizontal line, and a move on the seven-point ko is shown by a vertical line. In the bottom half of P, this convention is reversed. The moves for Left violating the ko-ban are indicated as such and are always shown as vertical lines going from one half to the other. We have labelled the positions in the center A through H.

We try to approximate P's value by sidling. We will start by approximating P^+. We will let M be the set of Left's ko-banned moves. Remembering that all plays using infinitely many moves in M are lost for Left, and referring to Section 3, we see that we have to start sidling from $\mathbf{off}^-$. Let K_0 have moves for Right from every position to itself, with infinite play won for Right. The next step is to construct $(P^+)^*(K_0) = K_1$, which is itself loopy. In fact, if we let $\bar{P}$ (shown in Figure 7) be P with the moves in M removed, then any $(K_1)_v$ will be the same as $(\bar{P}^+)_v$, since Left will never want to take the moves in K_1 from the embedded P^+ to $\mathbf{off}^-$. So we need to sidle $\bar{P}$, and now we can let M be the set of all the moves in $\bar{P}$ that do not destroy a ko, that is, all the moves from one to another of the positions $A, \ldots, H$.

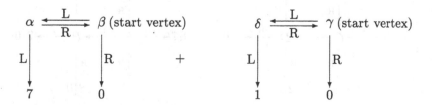

Figure 5. How the position shown above is the sum of two kos.

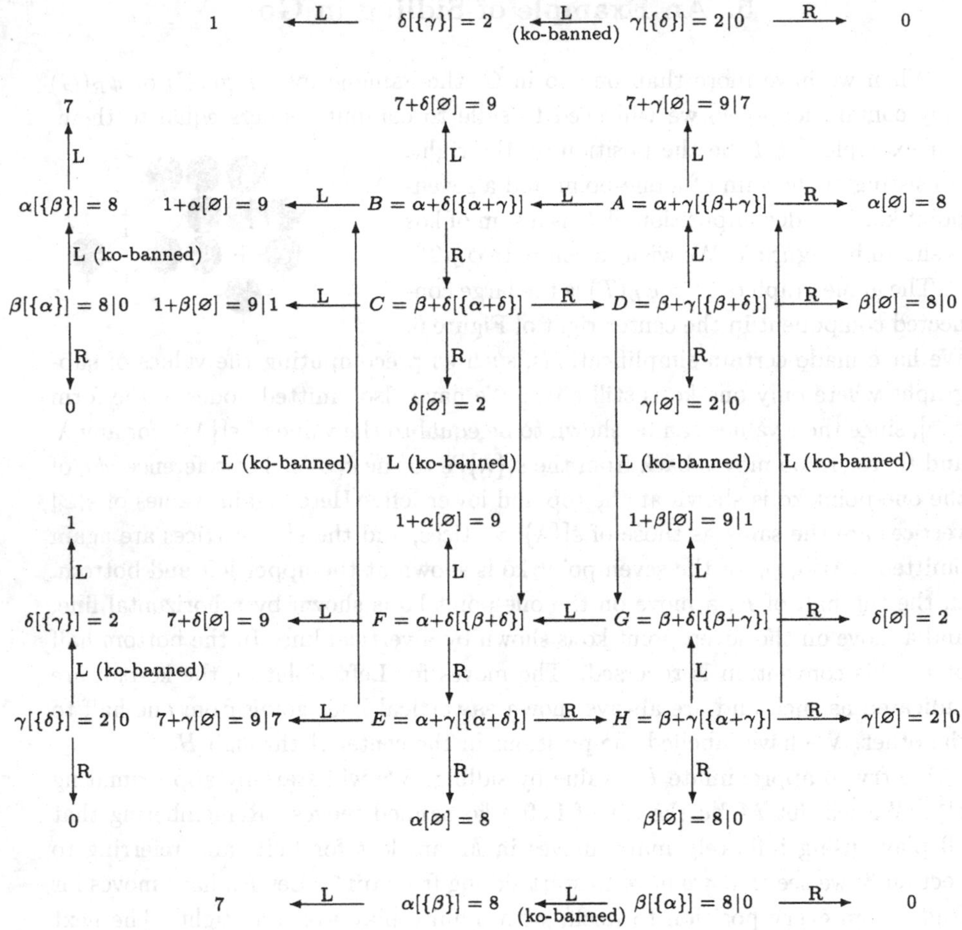

$$1 \xleftarrow{\text{L}} \delta[\{\gamma\}] = 2 \xleftarrow[\text{(ko-banned)}]{\text{L}} \gamma[\{\delta\}] = 2|0 \xrightarrow{\text{R}} 0$$

Figure 6. (graph)

7 $7+\delta[\varnothing] = 9$ $7+\gamma[\varnothing] = 9|7$

$\alpha[\{\beta\}] = 8$ $1+\alpha[\varnothing] = 9 \xleftarrow{\text{L}} B = \alpha+\delta[\{\alpha+\gamma\}] \xleftarrow{\text{L}} A = \alpha+\gamma[\{\beta+\gamma\}] \xrightarrow{\text{R}} \alpha[\varnothing] = 8$

L (ko-banned)

$\beta[\{\alpha\}] = 8|0$ $1+\beta[\varnothing] = 9|1 \xleftarrow{\text{L}} C = \beta+\delta[\{\alpha+\delta\}] \xrightarrow{\text{R}} D = \beta+\gamma[\{\beta+\delta\}] \xrightarrow{\text{R}} \beta[\varnothing] = 8|0$

0 $\delta[\varnothing] = 2$ $\gamma[\varnothing] = 2|0$

L (ko-banned)   L (ko-banned)   L (ko-banned)   L (ko-banned)

1 $1+\alpha[\varnothing] = 9$ $1+\beta[\varnothing] = 9|1$

$\delta[\{\gamma\}] = 2$ $7+\delta[\varnothing] = 9 \xleftarrow{\text{L}} F = \alpha+\delta[\{\beta+\delta\}] \xleftarrow{\text{L}} G = \beta+\delta[\{\beta+\gamma\}] \xrightarrow{\text{R}} \delta[\varnothing] = 2$

L (ko-banned)

$\gamma[\{\delta\}] = 2|0$ $7+\gamma[\varnothing] = 9|7 \xleftarrow{\text{L}} E = \alpha+\gamma[\{\alpha+\delta\}] \xrightarrow{\text{R}} H = \beta+\gamma[\{\alpha+\gamma\}] \xrightarrow{\text{R}} \gamma[\varnothing] = 2|0$

0 $\alpha[\varnothing] = 8$ $\beta[\varnothing] = 8|0$

$$7 \xleftarrow{\text{L}} \alpha[\{\beta\}] = 8 \xleftarrow[\text{(ko-banned)}]{\text{L}} \beta[\{\alpha\}] = 8|0 \xrightarrow{\text{R}} 0$$

Figure 6. An example of sidling. The large connected component in the center-right is P.

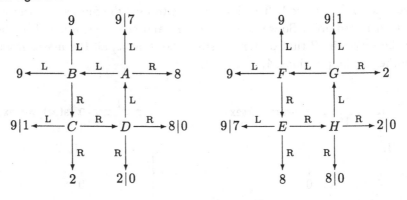

Figure 7. The graph $\bar{P}$.

Our first approximation to K_1, K_{10}, will have moves for Left from A, $\ldots$, H to themselves, with infinite play won for Left. (We can omit the other vertices since they are not moved to by moves in M.) We must then set $K_{11} = (\bar{P}^+)^*(K_{10})$. Given a sequence of approximate values for the vertices of $\bar{P}$, applying $(\bar{P}^+)^*$ refines it by taking just one more move in $\bar{P}$. For example, we have $(\bar{P}^+)^*(Q)_A = \{\{9|7\}, Q_B|8\}$, so $(K_{11})_A = \{\mathbf{on}^+|8\}$. The other $(K_{11})_v$'s are computed similarly. Their values are given in Table 1, and their graphs are shown in Figure 8, for $v \in \{A, B, C, D\}$. (The graphs for $v \in \{E, F, G, H\}$ are similar.)

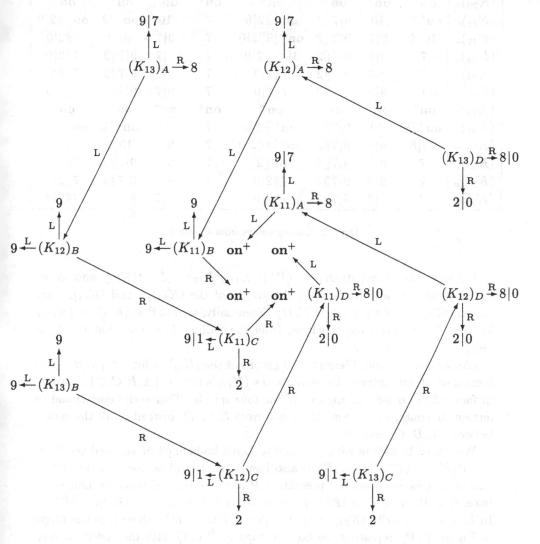

Figure 8. The graphs of $(K_{1i})_v$, for $i = 1, 2, 3$ and $v = A$, B, C, D.

v	A	B	C	D	E	F	G	H
$(K_0)_v$	off⁻	off⁻	off⁻	off⁻	off⁻	off⁻	off⁻	off⁻
$(K_{10})_v$	on⁺	on⁺	on⁺	on⁺	on⁺	on⁺	on⁺	on⁺
$(K_{11})_v$	on⁺|8	10	1	on⁺‖2|0	7	10	on⁺|2	on⁺‖2|0
$(K_{12})_v$	10|8	9|1	1	on⁺|8‖2|0	7	9|7	10|2	2
$(K_{13})_v$	7	9|1	1	10|8‖2|0	9|7‖2	9|7	9|7‖2	2
$(K_{14})_v$	7	9|1	1	7‖2|0	9|7‖2	9|7	9|7‖2	2
$(K_{15})_v$	7	9|1	1	7‖2|0	9|7‖2	9|7	9|7‖2	2
$(K_{20})_v$	on⁺	on⁺	on⁺	on⁺	on⁺	on⁺	on⁺	on⁺
$(K_{21})_v$	on⁺|8	10	9|7‖2	on⁺‖2|0	7	10	on⁺|2	on⁺‖2|0
$(K_{22})_v$	10|8	9|7	9|7‖2	on⁺|8‖2|0	7	9|7	10|2	7‖2|0
$(K_{23})_v$	7	9|7	9|7‖2	10|8‖2|0	7	9|7	9|7‖2	7‖2|0
$(K_{24})_v$	7	9|7	9|7‖2	7‖2|0	7	9|7	9|7‖2	7‖2|0
$(K_{25})_v$	7	9|7	9|7‖2	7‖2|0	7	9|7	9|7‖2	7‖2|0
$(K_{30})_v$	on⁺	on⁺	on⁺	on⁺	on⁺	on⁺	on⁺	on⁺
$(K_{31})_v$	on⁺|8	10	9|7‖2	on⁺‖2|0	7	10	on⁺|2	on⁺‖2|0
$(K_{32})_v$	10|8	9|7	9|7‖2	on⁺|8‖2|0	7	9|7	10|2	7‖2|0
$(K_{33})_v$	7	9|7	9|7‖2	10|8‖2|0	7	9|7	9|7‖2	7‖2|0
$(K_{34})_v$	7	9|7	9|7‖2	7‖2|0	7	9|7	9|7‖2	7‖2|0
$(K_{35})_v$	7	9|7	9|7‖2	7‖2|0	7	9|7	9|7‖2	7‖2|0

Table 1. Sidling values from Figure 6.

We must then compute $K_{12} = (\bar{P}^+)^*(K_{11})$, $K_{13} = (\bar{P}^+)^*(K_{12})$, and so on. These values are also in Table 1; the graphs of the $(K_{12})_v$'s and $(K_{13})_v$'s are shown in Figure 8 for $v \in \{A, B, C, D\}$. Eventually, we find that $(K_{14})_v = (K_{15})_v$ for all v. We may then conclude, from the Sidling Theorem, that $(\bar{P}^+)_v = (K_1)_v = (K_{15})_v$ for all v.

(As we can see from Figure 8, the graphs of the $(K_{1j})_v$'s for $v \in \{A, B, C, D\}$ form a set of four spirals. To compute the $(K_1)_v$'s for $v \in \{A, B, C, D\}$, it would in fact suffice to look at any one of the four spirals. This would correspond to letting M contain just, e.g., the move from C to D, instead of all the moves between A, B, C, and D.)

We return to sidling with M equal to Left's ko-banned moves, and set $K_2 = (P^+)^*(K_1) = (P^+)^*(\bar{P}^+)$. K_2 is also loopy. $(K_2)_v$ will be the same as $(\bar{P}^+)_v$, except that extra moves are present: if P has a ko-banned move for Left going from E to B (say), then $(K_2)_v$ has a move for Left from E to $(K_1)_B = (\bar{P}^+)_B$. In fact, we can write $(K_2)_v = Q_v$ for all $v \in \{A, \ldots, H\}$, where Q is the graph in Figure 9. By replacing the bottom copy of $\bar{P}$ in Q with the enders in K_{15}, we get the simplified graph Q_2 in Figure 10. To approximate K_2, we can then sidle with all those moves in $\bar{P}$ that do not destroy a ko, as we did with K_1. We

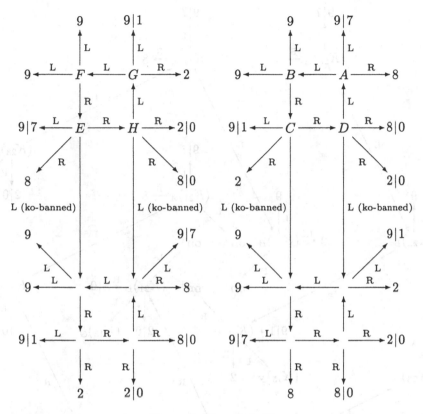

Figure 9. The graph Q.

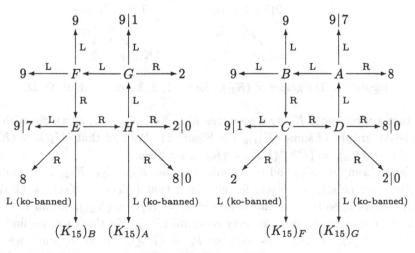

Figure 10. The graph Q_2.

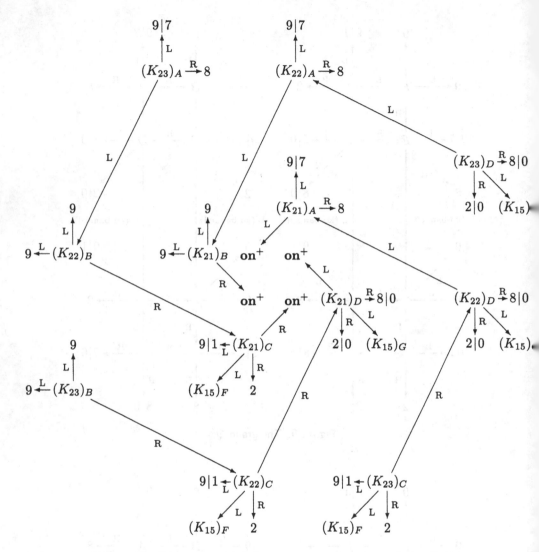

Figure 11. The graphs of $(K_{2i})_v$, for $i = 1, 2, 3$ and $v = A, B, C, D$.

can then approximate K_2 by a sequence $K_{20}, K_{21}, K_{22}, \ldots$, as we did with K_1; we display graphs of some $(K_{2i})_v$'s in Figure 11. We find that $(K_{24})_v = (K_{25})_v$ for all v, so $(K_2)_v = (P^+)^*(K_1)_v = (K_{25})_v$ for all v.

We can compute K_3, and its approximations, $K_{30}, K_{31}, K_{32}, \ldots$, similarly. We find that $(K_{34})_v = (K_{35})_v$ for all v and that $(K_3)_v = (K_{35})_v = (K_2)_v = (K_{25})_v$ for all v, so that we can write $(P^+)_v = (K_2)_v = (K_{25})_v$ for all v.

In an analogous manner, we may compute $(P^-)_v$. In this case, we find that $(P^-)_v = (K_{25})_v$ for all v as well, so $P_v = (K_{25})_v$ for all v, since we have $X^+ \geq X \geq X^-$ for all games X. As we remarked earlier, $\phi_L(T)$ will equal both P_H and P_D. Hence $\phi_L(T) = 7\|2\|0$.

6. Final Remarks

In the example T above, we found that $\phi_L(T)$ was an ender. The full power of the loopy theory was thus unnecessary, since our games effectively had finite play all along. However this will not always be so. Consider the *triple ko* position on the right, which we call A [Berlekamp and Wolfe 1994, Figure A.6]. As you can see, there are three kos, all involving the same pair of groups. Black's filling is suicidal, and if White fills, Black can immediately capture White's group. We will thus assume that no one ever fills a ko.

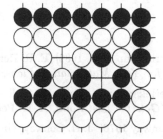

If White plays in the ko on the right, he will capture Black's group, plus one black stone. Assign the resultant position value W; then Right (White) has a move from A to W. If Left (Black) plays in a ko on the left, he reaches a position that we call E or F, depending on which he plays in; if Left plays again, he can capture White's group (plus two white stones); assign this position value V, so that Left has moves from E or F to V. We can apply similar reasoning to that concluding that no one fills in A to conclude that no one fills in E or F either. If we continue along these lines, we get the game graph in Figure 12, left.

In play, positions A, B, and C will all be equivalent, as will position D, E, and F, and players will thus be effectively moving on the game graph G of Figure 12, right, except that thereto E, Right can always move back to C, regardless of ko-ban. If we take G to have starting vertex ABC and infinite play drawn, sidling gives $G^+ = \{V\,|\,\} \,|\, W \neq G^- = \{\,|\,W\}$, so G is not an ender.

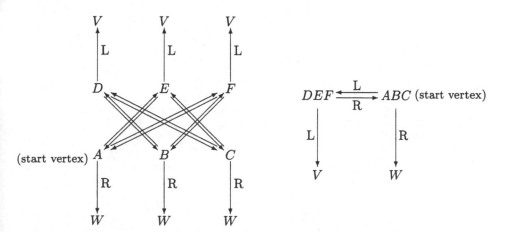

Figure 12. Left: A simplified game graph for the triple ko position. Arrows sloping upward are Left's moves; those sloping downward are Right's moves. Right: An equivalent graph G.

Acknowledgement

This material is substantially an abridgement of a part of my thesis [Moews 1993], which was supervised by E. Berlekamp.

References

[Berlekamp et al. 1982] E. R. Berlekamp, J. H. Conway and R. K. Guy, *Winning Ways for Your Mathematical Plays*, Academic Press, London, 1982.

[Berlekamp and Wolfe 1994] E. Berlekamp and D. Wolfe, *Mathematical Go: Chilling Gets the Last Point*, A. K. Peters, Wellesley, MA, 1994.

[Conway 1976] J. H. Conway, *On Numbers and Games*, Academic Press, London, 1976.

[Moews 1993] David Moews. "On some combinatorial games connected with Go", Ph.D. thesis, University of California, Berkeley, 1993.

[Wolfe 1991] David Wolfe, "Mathematics of Go: Chilling corridors", Ph.D. thesis, University of California, Berkeley, 1991.

DAVID MOEWS
34 CIRCLE DRIVE
MANSFIELD CENTER, CT 06250
    dmoews@xraysgi.ims.uconn.edu

Games of No Chance
MSRI Publications
Volume **29**, 1996

Experiments in Computer Go Endgames

MARTIN MÜLLER AND RALPH GASSER

ABSTRACT. Recently, the mathematical theory of games has been applied
to late-stage Go endgames [Berlekamp and Wolfe 1994; Wolfe 1991]. Based
upon this theory, we developed a tool to solve local Go endgames. We verified all exact game values in [Wolfe 1991] and analyzed some more complex
positions. We extended our method to calculate bounds for positions where
optimal play depends on Ko.

Our program Explorer uses this tool to play full board endgames. It
plays a nontrivial class of endgame positions perfectly. In a last section we
discuss heuristic play for a wider range of endgames.

1. Go Endgames

Towards the end of a Go game, the board position usually breaks down into
several local fights that can be analyzed individually (Figure 1). To find an optimal move globally, one needs to consider relations between these local subgames:
There is a conflict between maximizing the local score and gaining the initiative
to play in another part of the board. These relationships can be very complex,
even surpassing the abilities of professional Go players. Exhaustive analysis of

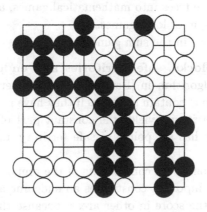

Figure 1. Late endgame position suitable for exact analysis.

full board endgames is only feasible for the most trivial problems, because of exponential explosion of the search.

The mathematical theory of games [Conway 1976; Berlekamp et al. 1982] can handle such complexity by divide and conquer: It defines an operation called a sum of games that builds up a game from its subgames (if no Ko fights exist, as explained below). This theory gives us algorithms to prune bad moves, compare the value of games and find an optimal move without exhaustive search on the full board.

To apply the theory to a Go endgame, we must assume that no Ko fights are left. (Extensions to handle some Ko situations will be discussed later.) By convention, positive scores are good for Black, negative scores are good for White. We use Japanese rules (though Chinese rules yield similar result). See [Berlekamp 1991; Berlekamp and Wolfe 1994] for a discussion of scoring rules.

The mathematical theory of games defines four basic types of games G:

$G > 0$: Black wins, no matter who plays first.

$G < 0$: White wins, no matter who plays first.

$G = 0$: Null game: the first player loses.

$G \parallel 0$: The first player wins.

Null games in Go are areas where no player can make a profitable move, like seki under Japanese rules. Games with an integer value—for example, safe territories—are also considered played out.

2. The Endgame Algorithm

We have implemented a six-step algorithm for endgame play:

- Board partition: find safe blocks, safe territories, and endgame areas.
- Generate local game trees in each endgame area.
- Evaluate local terminal positions (surrounded points plus prisoners).
- Transform local game trees into mathematical games, and simplify games.
- Calculate game as sum of local games.
- Find an optimal move in the sum game and play it.

Determining safe blocks, safe territories and endgame areas. We use an extension of the algorithm in [Benson 1980] to determine safe blocks and territories. Territory may contain prisoners if they have no chance of living. The difference in size of territories yields the constant part of the game value. See Figure 2, where Black has six points of safe territory and White has four, so $G_{\mathrm{const}} = 6 - 4 = 2$.

Next we compute connected components of the remaining unsafe blocks and empty points. Each component constitutes an endgame area. Play in one area cannot affect play or the score in other areas, because the areas are separated by safe blocks (Figure 3).

Figure 2. Safe blocks and territories.

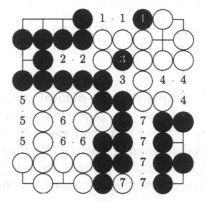

Figure 3. Endgame areas.

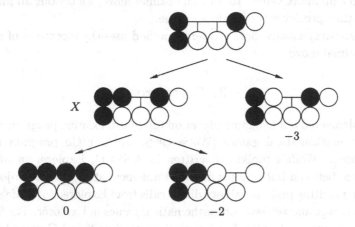

Figure 4. Local game tree for G_1.

Generating local game trees and evaluating terminal positions. In contrast to global game trees, local game trees must contain successive moves by the same color, because the opponent might tenuki (play in another local game). Move generation is locally exhaustive, with some rules for pruning dominated and reversible moves. Terminal positions are positions without a good move: all points are occupied or have become territory. We compute the local score for each terminal position (Figure 4).

Transforming game trees into mathematical games. Each local game tree is then transformed into a mathematical game, as follows. First sort all moves from a certain position by color to get the left and right options. For each move to another position, insert the value of that position. We have the numeric value of all terminal positions and compute the other games' values from them. In the example above we obtain $G_1 = \{X \mid -3\}$ in a first step. In the second step we evaluate X to get $X = \{0 \mid -2\}$, and therefore $G_1 = \{\{0 \mid -2\} \mid -3\}$. The complete values in our example are:

$$G_1 = \{0|\{-2|-3\}\}, \quad G_4 = \{\{0|-1\}|-2\}, \quad G_6 = \{\{0|-1\}|-2\},$$
$$G_2 = \{1|0\}, \qquad\qquad G_5 = \{\{0|-1\}|-2\}, \quad G_7 = \{5|\{4|\{3|\{2|0\}\}\}\}.$$
$$G_3 = \{0|-2\},$$

Calculating the sum of local games. The whole game is the sum of territory values plus all local games; in our running example, $G = G_{const} + G_1 + \cdots + G_7$. Even a sum of such short games can get very complicated: Printed out, G has several hundred characters. The operations can be simplified by chilling all endgames before computing the sum [Berlekamp and Wolfe 1994]. Chilling is a one-to-one mapping of endgames to simpler ones.

Finding an optimal move. Given a sum game and a player, it is easy to compute her minimax value. To find an optimal move, we try out all moves and select one that preserves the minimax value.

An alternative, usually more efficient, method uses the incentives of moves to find an optimal move.

3. Software

We implemented our endgame player on top of two existing programs: Wolfe's toolkit for mathematical games [Wolfe 1996] and our Go program Explorer [Müller 1995]. Wolfe's toolkit is written in ANSI C, Explorer in Modula-2. An interface between both programs was implemented as a term project [Fierz 1992]. The resulting program allows direct calls from Explorer to Wolfe's toolkit, as well as storage and retrieval of mathematical games in Explorer. For faster development we chose a low level of integration: Modula-2 and C parts keep their separate management of resources like memory blocks, lists and hash tables. The command line interface to the C toolkit was replaced by menu commands as needed.

4. Local Move Generation

Ignoring captures and illegal moves, the number of possible plays in an n-point area is approximately $2n$ (n for each player), and each play generates an $(n-1)$-area. A rough estimate for the size of the game tree is therefore $2^n n!$.

Due to combinatorial explosion, even fairly small endgames become prohibitively expensive to compute using this approach. We now turn to some techniques for reducing the size of the tree.

Hashing. There are at most 3^n different configurations on n points; this grows much more slowly than $2^n n!$. Thus, even a simple hash function will detect many transpositions in the game tree.

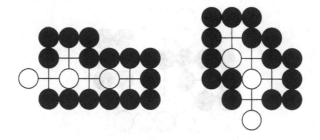

Figure 5. Isomorphic positions.

Figure 6. Here $a > b > c$.

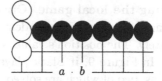

Figure 7. Only a and b appear useful.

An improved hash function might also recognize Go-specific graph isomorphisms, such as the one in Figure 5.

Pruning. The search tree can be further reduced by pruning. We can do this in either a rigorous or heuristic fashion. If we prune rigorously, we only eliminate moves that are provably worse than others (prune b if there is a such that $a \geq b$). For example, for corridors it can be shown that moves are better the closer they are to the entrance of the corridor. In Figure 6, we need therefore only consider move a for both players.

If we do not need the exact value of a position, heuristic pruning can be used to eliminate even more moves. For instance, in Figure 7, we may decide only to consider moves a or b for White.

Early Termination. If a position's value can be determined statically, the entire subtree can be pruned. For instance, any position that contains only dame points or safe territory can be easily evaluated in this fashion; Figure 8 shows an example. The value of n dame is 0 if n is even, and $\{0|0\} = *$ if n is odd.

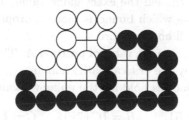

Figure 8. The value of this game is $2*$.

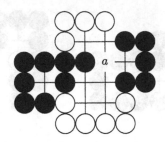

Figure 9. Black's play at a separates the area into three independent areas.

Splitting the local game. Often playing a move results in the local endgame splitting into several independent areas. In this case, we can analyze each area separately. The position's value is then the sum of the values of the independent areas. In Figure 9, if Black plays a the area separates into three independent positions, two of which are solved immediately, without further search (the black territory and the two dame points). Even if no areas could be pruned, this method still shrinks the game tree dramatically.

5. Bounds for Ko

Classical combinatorial game theory cannot handle loops in games. Unfortunately, loops often appear in Go—for example, in the case of Ko. We must therefore find a way of eliminating these loops from the game tree.

One way is to define the symbol K, which stands for the game at right. Whenever such a situation occurs in the game tree, we prune the loop and substitute the value K instead. This only masks the problem, since we still do not know how to play optimally. Also the simple Ko is not the only situation in Go that leads to loops. Assigning additional constants for all such situations does not seem feasible.

Another way of resolving this problem is to calculate bounds on the game value. This can be done by not considering all moves for a certain player. For instance, by suppressing any move by Black that would lead to a loop in the game tree, we calculate a lower bound. By suppressing White's options, we get an upper bound. If these two bounds are identical, optimal play does not depend on Ko and we have determined the exact game value. If the bounds differ, we may be able to determine which bound is more appropriate by considering the number of Ko threats still on the board.

In Figure 10, we determine the value of position A. The game tree has already been pruned, so that only the critical paths remain.

The following dependencies hold for the unknown game values:

$$A = \{2*|B\}, \qquad B = \{C+2|0\}, \qquad C = \{1|D-1\},$$
$$D = \{E|*, B\}, \qquad E = \{F+1|0\}, \qquad F = \{0|E-1\},$$

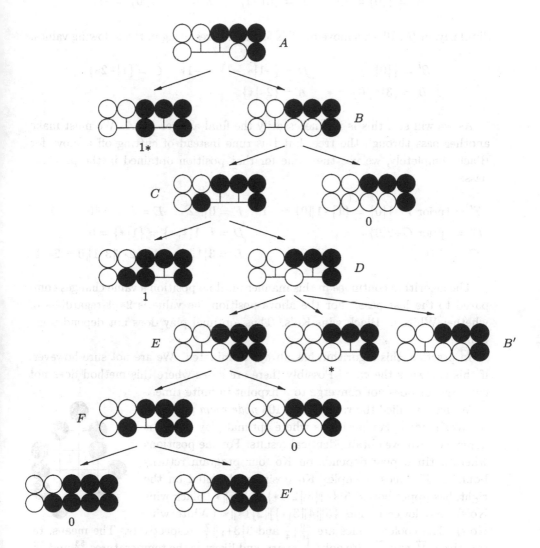

Figure 10. In this example, the calculation of bounds proves that Ko's do not affect the value of A.

We now determine a bound on the game value if White wins all Ko's. This is modeled by not allowing Black any moves that would lead into a repetition. So in E', Black is not allowed the move to F'; therefore

$$E' = \{|0\} = -1 \qquad F = \{0|-2\} \qquad E = \{1|-1\|0\} = -1.$$

Similarly, in B', Black's move to C' is forbidden, resulting in the following values:

$$B' = \{|0\} = -1 \qquad D = \{-1|*, -1\} = -1* \qquad C = \{1|-2*\}$$
$$B = \{3|*\|0\} = * \qquad A = \{2*|*\}$$

As we will see, this is not necessarily the final answer for A. We must make another pass through the tree, but this time instead of cutting off a move for Black completely, we use the value for that position obtained in the previous pass:

$$E' = \{\text{prior } F+1|0\} = \{1|-1\|0\} = -1 \quad F = 0|-2 \quad E = \{1|-1\|0\} = -1$$
$$B' = \{\text{prior } C+2|0\} = * \qquad\qquad D = \{-1|*, *\} = \{1|*\} = 0$$
$$C = 1|-1 \qquad\qquad\qquad\qquad B = 3|1\|0 \quad A = 2*\|\|3|1\|0 = 2*|1$$

The algorithm continues in this manner until no position's value changes compared to the last pass. For the above position the value is $2*|1$ regardless of whether White or Black wins Ko's. Thus, optimal play does not depend upon Ko.

In practice, this algorithm has always terminated. We are not sure however, if this is always the case. Possibly there are cases where this method does not converge, or does not converge to a fixpoint in finite time.

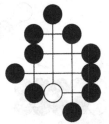

We have verified the values of all 103 *node room* positions in [Wolfe 1991]. For positions where optimal play does not depend on Ko, we obtain identical results. For the positions where optimal play depends on Ko, our program returns bounds. The most complex Ko position, depicted on the right, has upper bound $5|\{4\|\|3|2|1*\}\|\|2|1*\|*$ (Black wins Ko's), and lower bound $\{5\|\|\|4\|3|*\}\|\|2|1*\|*$ (White wins Ko's). The cooled values are $\frac{49}{16}|\frac{7}{4}$ and $3|3+_1\|\frac{7}{4}$, respectively. The means, respectively $\frac{77}{32}$ and $\frac{76}{32}$, are only $\frac{1}{32}$ apart, and likewise the temperatures $\frac{53}{32}$ and $\frac{52}{32}$.

6. Heuristic extensions to endgame algorithm

Many real endgame positions are not suitable for exact analysis: The endgame areas may be too big, or there may be Ko's on the board. In these cases we need to use heuristics.

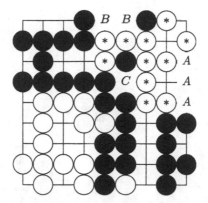

Figure 11. The stones marked * die if Black gets all moves in endgames A, B and C.

Relaxed definitions of safety. The rules for determining safe stones and territories are very restrictive. This leads to many unsafe blocks and big endgame areas.

Figure 11 shows one of Berlekamp's endgame problems. It is the same as our running example, except for the absence of a stone at the uppermost A point. The stones marked * are not completely safe: they will die if Black gets all moves in endgames A, B and C. Therefore there may be a weak relation between these endgames. In heuristic endgame play we can choose to ignore this relation, regard the marked stones as safe and independently analyze the areas A, B and C.

Heuristic move generation and local evaluation. In areas too big to analyze exhaustively, we need heuristic move generators which build a forward pruned subtree of an endgame. We also use a heuristic evaluation of nonterminal positions. By restricting the moves of only one player, we can still calculate correct upper and lower bounds for such positions, but they will be less tight than before.

Playing for a win. Playing for a win (a fixed number of points) is often much easier than optimal play. A program may not be able to solve an endgame position completely, but it may come up with bounds that are good enough to prove a win by, say, half a point. Human experts often do this kind of analysis in tournament games with limited time.

7. Mathematical Value, Mean and Temperature of Some Common Endgames

The diagrams on the next page show values or bounds for several corridors of width two and three. Expressions marked • and ○ indicate bounds obtained under the assumptions that Black and White, respectively, wins Ko's.

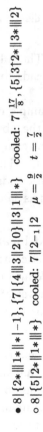

$3|*$ cooled: $2|1$ $\mu = \frac{3}{2}$ $t = \frac{3}{2}$

$5\|1*|0$ cooled: $4|\frac{3}{2}$ $\mu = \frac{11}{4}$ $t = \frac{9}{4}$

● $7|\{3*|0\}, \{7*|5*\|2*\||1*\}$ cooled: $6|\{3|2\}, 3_{-2*}$ $\mu = \frac{17}{4}$ $t = \frac{11}{4}$
○ $7|\{3*|0\}, \{\{8\||7\|2, \{5|1\}\|0\}|\{6*\|4|2\||1|-1\}, \{5, \{6,\{7|5\}|4\}|2*\||1*\}\|0\}$
    cooled: $6|\{3|2\}, \{6+_{4|4},\{4|2\}|\{5|4*\|2*\}, \{4,4\uparrow*|3\||3\}\|2\}$ $\mu = \frac{17}{4}$ $t = \frac{11}{4}$

● $8|\{2*|1*\||*|-1\}, \{7|\{4\||3\|2|0\}\|3|1\||*\}$ cooled: $7|\frac{17}{8}, \{5|3\uparrow2*\||3*\||2\}$ $\mu = \frac{73}{16}$ $t = \frac{55}{16}$
○ $8|\{5|2*\||1*\||*\}$ cooled: $7\|2_{-1}|2$ $\mu = \frac{9}{2}$ $t = \frac{7}{2}$

● $\{7\||5|\{4\||3\|2|0\}\|\{5|2*\||1*\||0|-2\}\}|0$ cooled: $\{5\||3\uparrow3\|2_{-1}|1*\}|1$ $\mu = \frac{9}{4}$ $t = \frac{9}{4}$
○ $5|\{4\||3\|2|0\}\|0$ cooled: $3\uparrow3|1$ $\mu = 2$ $t = 2$

● $\{4|3*\||2*\||1|*\}, \{\{6\||5\|1*|0\}|\{1*|0\}, \{4*\|3\||*|-1\}\}\|*$
    cooled: $\frac{3}{2}\uparrow, \{3+\frac{5}{2}|\frac{1}{2}, \{\frac{5}{2}|\frac{1}{2}\}\}|\frac{1}{2}$ $\mu = 1$ $t = \frac{3}{2}$
○ $\{3\||2\|1|*\}\|*|-1$ cooled: $\frac{9}{8}|\frac{1}{2}$ $\mu = \frac{13}{16}$ $t = \frac{21}{16}$

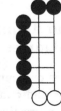
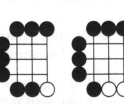
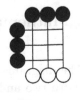

Figure 1.

• {7‖3*|2}, {7|{6‖1‖0|−3*}}
|{{6‖1‖0|−3*}‖1, {6, {7|1}|0}|−1‖−3*},
{{7‖4|2*|1}, {7‖4|{3‖1|−5‖−6}}
|{2*|1}, {3‖1*|0}, {3, {4|2*|1}
{{2*|1|{{1*|0‖*|−3‖−4}, {*, {3|2*|{3|1‖−2‖−3}‖|−3|−4}, {3‖1*|0‖|−5}‖−3}

cooled: {5|5/2}, 5+4₋₁|{5|1+1‖1, {5, {5|1}|1}|1‖0},
{{5‖3|5/2}, {5|3+6+4}|5/2, {3|5/2}, {3, {3|5/2}|{5/2}|1−1}, {3, {7/2‖1−2*}|1−1}, {3|5/2‖2|−1}‖−1}

μ = 3   t = 3

o 7‖3*|1*‖|{5|1‖0‖|*|−1‖−3}   cooled: 5|2*‖|1−2‖½|0   μ = 2   t = 19/8

● 7{5‖3*|2‖|{5|1‖0‖|{*|−2*}, {6‖0|−3}|−4, {−3|−5}}}
‖{*|−1}, {5|1‖0‖|{*|−2*}, {6‖0|−3}|−4, {−3|−5}},
{3*, {4|2*}|{2*|{1*‖0|−4‖−5}, {0, {5|3‖2‖|−3}‖|0|−4‖−5}},
{3*|1*‖|{1‖0|−4‖−6}}‖|{5*|{5|3|2‖|−3}|−3, {1‖0|−4‖−6}‖|−4}

cooled: 5{3|5/2‖|1−2}*, {5|1‖1|−1}|−1, −1*}
‖½, {1−2}*, {5|1‖1|−1}|−1, −1*},
{2, 2*|{2|2|−1−2}, {1, {3↓|−1}|−1−2}, {2*|1+2|−3}‖|{2‖1+2|−3}‖|{4‖3↓|−1}‖−1, {1+2|−3}‖−1}

μ = 3/2   t = 9/4

o 5‖3*|1*‖|{0|−1*‖−3}, {3*|{1|−1}, {2*|−4*}, {{2‖1‖0|−6}|−1*|−4}‖−1‖−3‖−4}
cooled: 3|2*‖|{−½|−1}, {2|1*, {2|−2}, {1|1+4‖0|−1}|−1}‖−1}   μ = 7/8   t = 21/8

● {3|{2, {3|1}|2|0|−7}}, {6‖5|*, {4*|2‖0|−1*‖−3}}‖|0|−1*‖−3
cooled: {1‖1, 1*‖1*|−5}, {3+₃|0, {2*|−½|−1}}‖|−½|−1}   μ = 1/8   t = 15/8

o 3‖1*|0‖|{{7‖0|−2|−2, {−1|−3}}|−3}, {*, {3‖1*|0‖|−2*|−4*}‖|*|−5‖−6}
cooled: 1½|{5|−1−*‖−1, −1‖−1}, {0, {1½|−2*}|−3−₃}   μ = −1/8   t = 15/8

References

[Benson 1980] D. B. Benson, "A mathematical analysis of Go", pp. 55–64 in *Proc. 2nd Seminar on Scientific Go-Theory* (edited by K. Heine), Institut für Strahlenchemie, Mühlheim a. d. Ruhr, 1979.

[Berlekamp 1991]   E. Berlekamp, "Introductory overview of mathematical Go endgames", pp. 73-100 in *Combinatorial Games* (edited by R. K. Guy), Proc. Symp. Appl. Math. **43**, Amer. Math. Soc, 1991.

[Berlekamp and Wolfe 1994] E. R. Berlekamp and D. Wolfe, *Mathematical Go: Chilling Gets the Last Point*, A K Peters, Wellesley (MA), 1994.

[Berlekamp et al. 1982] E. R. Berlekamp, J. H. Conway, and R. K. Guy, *Winning Ways For Your Mathematical Plays, I: Games In General*, Academic Press, London, 1982.

[Conway 1976] J. H. Conway, *On Numbers And Games*, Academic Press, London, 1976.

[Fierz 1992] W. Fierz, "Go Endgames", Semesterarbeit, ETH Zürich, 1992.

[Müller 1995] M. Müller, "Computer Go as a Sum of Local Games: An Application of Combinatorial Game Theory", Ph.D. Dissertation, ETH Zürich, 1995.

[Wolfe 1991] D. Wolfe, "Mathematics of Go: chilling corridors", Ph.D. Dissertation, Univ. of California, Berkeley, 1991.

[Wolfe 1996] D. Wolfe, "The Gamesman's Toolkit", pp. 93–99 in this volume.

MARTIN MÜLLER
RALPH GASSER
INSTITUT FÜR THEORETISCHE INFORMATIK
EIDGENÖSSISCHE TECHNISCHE HOCHSCHULE (ETH)
8092 ZÜRICH
SWITZERLAND

Taming the Menagerie

This section collects seven analyses of "simple" games in their dizzying variety: from the hitherto ignored *sowing games* (where counters are distributed from one pile to others, without ever being removed from the game) to *take-away games* (where there is only one pile, but the rule specifying how many counters can be removed may be fiendishly complicated). Not surprisingly, complete analyses are hard to come by, but several interesting general results are proved. There is also a description of a computer interface for Domineering, that archetype of partizan board games, amazing in its economy and elegance.

Games of No Chance
MSRI Publications
Volume **29**, 1996

Sowing Games

JEFF ERICKSON

ABSTRACT. At the Workshop, John Conway and Richard Guy proposed
the class of "sowing games", loosely based on the ancient African games
Mancala and Wari, as an object of study in combinatorial game theory.
This paper presents an initial investigation into two simple sowing games,
Sowing and Atomic Wari.

1. Introduction

Most well-studied combinatorial games can be classified into a few broad
classes.

TAKING AND BREAKING: Games played with piles of chips, in which the basic
move is to take some chips and/or split some piles. They include Nim, Kayles,
Dawson's Chess, other octal and hexadecimal games. Higher-dimensional
variants include Maundy Cake, Cutcake, Eatcake, and Chomp.

CUTTING AND COLORING: Games played on (colored) graphs, in which the basic move is to cut out a small piece of the graph of the appropriate color,
possibly changing other nearby pieces. They include Hackenbush, Col, Snort,
Domineering, and Dots and Boxes.

SLIDING AND JUMPING: Games played with tokens on a grid-like board, in which
the basic move is to move a token to a nearby spot, possibly by jumping over
opponent's pieces, which may then be (re)moved. Examples include Ski-
Jumps, Toads and Frogs, Checkers, and Konane.

At the Workshop, John Conway and Richard Guy suggested that studying entirely new classes of games, fundamentally different from all of these, might lead
to a more thorough understanding of combinatorial game theory, and proposed
a class of games loosely based on the African games Mancala and Wari, which
they called "sowing games". These games are played with a row of pots, each
containing some number of seeds. The basic move consists of taking all the
seeds from one pot and "sowing" them one at a time into succeeding pots. In
this paper, we offer a few introductory results on two simple sowing games.

Almost all the values given in this paper were derived with the help of David Wolfe's **games** package [Wolfe 1996].

2. Sowing

The first game in this family, simply called *Sowing*, was invented by John Conway. The basic setup is a row of pots, each containing some number of uncolored seeds. A legal move by Left consists of taking all the seeds out of any pot and putting them in successive pots to the right, subject to the restriction that *the last seed cannot go into an empty pot*. Right's moves are defined symmetrically.

We represent a Sowing position by a string of boldface digits, where each digit represents the number of seeds in a pot. For example, from the position **312**, Left's only legal move is to move the single seed in the second pot into the third pot, leaving the position **303**. Right has two legal moves, to **402** and **420**. Thus, this position can be evaluated as follows.

$$\mathbf{312} = \{\mathbf{303}\,|\,\mathbf{402}, \mathbf{420}\} = \{0\,|\,\{\,|\,\mathbf{510}\}, 0\} = \{0\,|\,\{\,|\,\{\,|\,\mathbf{600}\}\}, 0\} = \{0\,|\,-2\}$$

There is a natural impartial version of this game as well, in which either player can move in either direction. For example:

$$\mathbf{312} = \{\mathbf{303}, \mathbf{402}, \mathbf{420}\} = \{0, \{\mathbf{510}\}, 0\} = \{0, \{\{\mathbf{600}\}\}, 0\} = *$$

Simplifying Positions. There are some obvious ways of simplifying a Sowing position. First, if the first or last pot is empty, we can just ignore it, since it will always be empty.

We will call a pot *full* if the number of seeds is greater than the distance to either the first or last (nonempty) pot. Since neither player can ever move from a full pot, the exact number of seeds it contains is unimportant. Unlike empty pots, however, we can't simply ignore full pots at the ends, since it is still possible to end a turn by dropping a seed into a full pot. In this paper, we represent full pots with the symbol •. Thus, for example, **110451000** = **110451** = **110••1**.

Finally, sometimes Sowing positions can be split into independent components. For example, we can write **1200021** = **12**+**21** and **110••1** = **110•**+•**1**. Unfortunately, while in many cases, it is easy to detect such splits by hand, we do not know of a general method that always finds a split whenever one is possible. Clearly, any two positions separated by sufficiently many empty or full pots should be considered independent, but we don't know how many "sufficiently many" is!

The Towers of Hanoi Go to Africa: Sowing is Hard. Suppose we start with n pots, each with one seed, and we want to move all the seeds into the last pot. We can use the following algorithm to accomplish this task.

(i) Recursively move to the position $(n-1)\mathbf{0}^{n-2}\mathbf{1}$.

(ii) Sow the contents of the first pot, giving the position $\mathbf{1}^{n-2}\mathbf{2}$.

(iii) Recursively move to the position n, by pretending the last pot contains only one seed.

The reader should immediately recognize the recursive algorithm for solving the Towers of Hanoi! See p. 753 of *Winning Ways*. Since this algorithm requires $2^{n-1} - 1$ moves to complete, we conclude that the position 1^n has at least $2^{n-1} - 1$ distinct followers. This immediately implies the following theorem.

THEOREM 2.1. *Evaluating Sowing positions (by recursively evaluating all their followers) requires exponential time in the worst case.*

Despite this result, it is possible that a subexponential algorithm exists for determining the value of a Sowing position, by exploiting higher-level patterns. But this seems quite unlikely.

The algorithm we have just described is not the fastest way to get all the seeds into one pot. Consider the following alternate algorithm, which uses only polynomially many steps.

- If $n = 2m$:

    (i) Move recursively to $m0^{m-1}m$.

    (ii) Sow the first pot to $1^{m-1}(m+1)$.

    (iii) Move recursively to $(2m)$.

- If $n = 2m + 1$:

    (i) Move recursively to $m10^{m-1}m$.

    (ii) Move to $(m+1)0^m m$.

    (iii) Sow the first pot to $1^m(m+1)$.

    (iv) Move recursively to $(2m+1)$.

The number of moves $T(n)$ used by this algorithm obeys the following recurrence:

$$T(1) = 0,$$
$$T(2m) = 3T(m) + 1,$$
$$T(2m + 1) = 2T(m) + T(m + 1) + 2.$$

Asymptotically, $T(n) = O(n^{\log_2 3}) = O(n^{1.5850})$. Restricting our attention to powers of two, we get the exact expression $T(2^k) = \frac{1}{2}(3^k - 1)$.

Some Sowing Patterns. Very little can be said about the types of values that Sowing positions can take. The only general pattern we have found so far is that there are Sowing positions whose values are arbitrary integers and switches with arbitrarily high temperatures.

THEOREM 2.2. $(10)^m 03(01)^n = 0$ *for all m and n.*

PROOF. If Left goes first, she loses immediately, since she has no legal moves. If Right goes first, his only legal move is to the position $(10)^{m-1}211(01)^n$, from which Left can move to $(10)^{m-1}0220(01)^n$, which has value zero since there are no more legal moves. Thus, the second player always wins. □

THEOREM 2.3. $(01)^m2(01)^n = n+1$, *for all m and n except $m = n = 0$.*

PROOF. Right has no legal moves. If $n = 0$, Left has only one legal move, to $(10)^{m-1}03 = 0$ by the previous theorem. Otherwise, Left has exactly two legal moves, to $(01)^{m+1}2(01)^{n-1} = n$ by induction, and to $(10)^{m-1}3(01)^n = 0$, which is a terminal position. □

THEOREM 2.4. $11(01)^n = \{n+1 \mid 0\}$ *for all positive n.*

PROOF. Left has only one move, to $2(01)^n = n + 1$ by the previous theorem. Right has only one move, to the terminal position $20(01)^n$. □

THEOREM 2.5. $(10)^m2(01)^n = \{n \mid -m\}$ *for all positive m and n.*

PROOF. Left has only one move, to $(10)^m012(01)^{n-1}$, which, by a slight generalization of Theorem 3, has value n. Similarly, Right has only one move, to $(10)^{m-1}210(01)^n = m$. □

We have seen Sowing positions whose values are fractions, ups, tinies, higher-order switches, and even some larger Nim-heaps, but no other general patterns are known. Table 1 lists a few interesting values. Values for some "starting" positions, in which all pots have the same number of seeds, are listed in Table 2.

Even less is known about the impartial version. Table 3 lists the "smallest" known positions with values 0 through *9. Table 4 lists values for some impartial starting positions.

Open Questions. We close this section with a few open questions. For what values of n do Sowing positions exist with values 2^{-n}, and if they all exist, can we systematically construct them? What about $n \cdot \uparrow$? $+_n$? $*n$?

Is there a simple algorithm that splits Sowing positions into multiple independent components? Are there any other high-level simplification rules that would allow faster evaluation?

3. Atomic Wari

"Is it... *atomic?*"
"Yes! *Very* atomic!"
   —*The 5000 Fingers of Dr. T*

The second sowing game we consider, called *Atomic Wari*, is my invention. Atomic Wari is loosely based on a different family of African games, variously called wari or oware. The board is the same as in Sowing, but the moves are different. A legal move consists of taking all the seeds from one pot, and sowing them to the left or right, *starting with the original pot*. As in Sowing, Left moves seeds to the right; Right moves them to the left. To avoid trivial infinite play, it is illegal to start a move at a pot that contains only one seed. At the end

$211 = \frac{1}{2}$ $12202 = \frac{1}{4}$ $122011 = \frac{1}{8}$ $2121202 = \frac{1}{16}$	$2121 = 1\|*\|-1$ $41122 = 2\|0\|-2\|-4$
	$\bullet 2011 = +_1$ $\bullet 013 = +_2$ $\bullet 0114 = +_3$ $332011 = +_{1/2}$
$2202 = \uparrow$ $31011 = \uparrow*$	
$201321 = \Uparrow$ $22011 = \Uparrow*$ $122112 = \uparrow\uparrow\uparrow$	$11 = *$ $31\bullet 13 = *2$ $313005 = \uparrow*3 = *2\|0$

Table 1. Some interesting partisan Sowing values

of a move, if the last pot in which a seed was dropped contains either two or three seeds, those seeds are *captured*, that is, removed from the game. Multiple captures are possible: after any capture, if the previous pot has two or three seeds, they are also captured. The game ends when there are no more legal moves, or equivalently, when no pot contains more than one seed. As usual, the first player who is unable to move loses.

For example, consider the position **312**. Left can sow the contents of the first pot, then capture the contents of the other two pots, leaving the position **100 = 1**. Left can also sow the contents of the rightmost pot, leaving the position **3111**. Right can move to either **301** or **11112**. Thus, the Atomic Wari position **312** has the following value:

$$312 = \{1, 3111 \mid 301, 11112\}$$
$$= \{0, \{1001 \mid 111111\} \mid \{11 \mid 11101\}, \{111111 \mid 11101\}\}$$
$$= \{0, * \mid *, *\}$$
$$= \uparrow$$

Clearly, for every move by Left, there is a corresponding move by Right. Thus, Atomic Wari is an "all small" game, in the terminology of *Winning Ways*. All Atomic Wari positions have infinitesimal values, and in the presence of remote stars, correct play in a collection of Atomic Wari positions is completely determined by the position's atomic weights. (Hence, the name.)

There is also a naturally defined impartial version of the game. For example,

$$312 = \{1, 3111, 301, 11112\}$$
$$= \{0, \{1001, 111111\}, \{11, 11101\}, \{111111, 11101\}\}$$
$$= \{0, *, *, *\}$$
$$= *2$$

$1^2 = *$

$1^3 = 0$

$1^4 = \pm\frac{1}{2}$

$1^5 = 0$

$1^6 = 0$

$1^7 = *2$

$1^8 = *$

$1^9 = 0$

$\bullet 1^1 \bullet = *$

$\bullet 1^2 \bullet = 0$

$\bullet 1^3 \bullet = 0$

$\bullet 1^4 \bullet = \pm(0, \{1|0\})$

$\bullet 1^5 \bullet = 0$

$\bullet 1^6 \bullet = *$

$\bullet 1^7 \bullet = \pm(\{2|1\|+_1, +_3|_1\}, \{1|\{3|1\|0\|\|-1\}\})$

$\bullet 1^8 \bullet = 0$

$\leftarrow\bullet 1^1 \bullet\rightarrow = *$

$\leftarrow\bullet 1^2 \bullet\rightarrow = 0$

$\leftarrow\bullet 1^3 \bullet\rightarrow = 0$

$\leftarrow\bullet 1^4 \bullet\rightarrow = 0$

$\leftarrow\bullet 1^5 \bullet\rightarrow = \pm(4*|\frac{1}{2}, \{\frac{7}{2}|0\})$

$\leftarrow\bullet 1^6 \bullet\rightarrow = 0$

$\leftarrow\bullet 1^7 \bullet\rightarrow = \pm(\{4|3\|0\}, \{\frac{5}{2}*|\{2|-2\}, \{2|\frac{1}{2}\|\|0|-1\|-3\}\})$

$\leftarrow\bullet 1^8 \bullet\rightarrow = \pm(\frac{23}{4}|4, \{\{\{6*|\|\|6|5\|-4\}|1\}, 4+_{(2-5)}\|\| \pm(4+_{(2-5)}|\frac{1}{2}) \| \pm(4+_{(2-5)}|\frac{1}{2}),$
$$\{4|3, \{3\||*|-1\}\|*, \{1\||-2|-5\}\})$$

$2^3 = 2\bullet 2 = *$

$2^4 = *$

$2^5 = 0$

$2^6 = \pm(1\|\|\|1, \{1|\frac{7}{8}\}|\{1|-1*\}, \{*, \{1|0, *\}, \{1|\downarrow *\}|0, \{0|-1\}\}\|\|0|-1)$

$2^7 = *$

$\bullet 2^2 \bullet = *$

$\bullet 2^3 \bullet = *2$

$\bullet 2^4 \bullet = 0$

$\bullet 2^5 \bullet = *$

$\bullet 2^6 \bullet = 0$

$\leftarrow\bullet 2^2 \bullet\rightarrow = 0$

$\leftarrow\bullet 2^3 \bullet\rightarrow = \pm1$

$\leftarrow\bullet 2^4 \bullet\rightarrow = 0$

$\leftarrow\bullet 2^5 \bullet\rightarrow = \pm(\frac{3}{4}|\{\frac{1}{2}|0\}, \{0\|0, \{0|-1\}|-3\|\|0|-4\})$

$\leftarrow\bullet 2^6 \bullet\rightarrow = *$

$3^4 = 3\bullet\bullet 3 = *$

$3^5 = 33\bullet 33 = *$

$3^6 = *$

$3^7 = *$

$\bullet 3^3 \bullet = \bullet 3\bullet 3\bullet = *$

$\bullet 3^4 \bullet = \pm(1+_1)$

$\bullet 3^5 \bullet = \pm(\frac{1}{2}, \{1|*\})$

$\bullet 3^6 \bullet = *$

$\leftarrow\bullet 3^2 \bullet\rightarrow = 0$

$\leftarrow\bullet 3^3 \bullet\rightarrow = \pm1$

$\leftarrow\bullet 3^4 \bullet\rightarrow = *$

$\leftarrow\bullet 3^5 \bullet\rightarrow = \pm(0, \{\{1, \{1|0\}|0, \{1|0\}\},$
$$1-_2|0\})$$

Table 2. Values of "starting" positions in partisan Sowing. The symbols $\leftarrow\bullet$ and $\bullet\rightarrow$ denote full pots going forever to the left and right.

$$
\begin{aligned}
102 &= 0 = 102 \\
11 &= * = 11 \\
111 &= *2 = 111 \\
1112 &= *3 = 1112 \\
110111 &= *4 = 11131 \\
111121 &= *5 = 12113 \\
10111121 &= *6 = 111312 \\
11101112 &= *7 = 1111113 \\
11112111 &= *8 = 11112111 \\
111111122 &= *9 = 11132112
\end{aligned}
$$

Table 3. Simplest impartial Sowing positions with given Nim-values. The left column gives the position with the fewest seeds; the right column gives the position with the fewest pots.

	$\bullet 1^1 \bullet = *$	$\leftarrow 1^1 \bullet\rightarrow = *$
$1^2 = *$	$\bullet 1^2 \bullet = *2$	$\leftarrow 1^2 \bullet\rightarrow = 0$
$1^3 = *2$	$\bullet 1^3 \bullet = 0$	$\leftarrow 1^3 \bullet\rightarrow = *$
$1^4 = 0$	$\bullet 1^4 \bullet = 0$	$\leftarrow 1^4 \bullet\rightarrow = *$
$1^5 = 0$	$\bullet 1^5 \bullet = *$	$\leftarrow 1^5 \bullet\rightarrow = 0$
$1^6 = *$	$\bullet 1^6 \bullet = 0$	$\leftarrow 1^6 \bullet\rightarrow = 0$
$1^7 = *2$	$\bullet 1^7 \bullet = *4$	$\leftarrow 1^7 \bullet\rightarrow = 0$
$1^8 = 0$	$\bullet 1^8 \bullet = 0$	$\leftarrow 1^8 \bullet\rightarrow = *$
$1^9 = 0$	$\bullet 1^9 \bullet = *2$	$\leftarrow 1^9 \bullet\rightarrow = *$
$1^{10} = *$	$\bullet 1^{10} \bullet = 0$	$\leftarrow 1^{10} \bullet\rightarrow = 0$
$1^{11} = 0$		
$1^{12} = 0$		
$2\bullet 2 = *$	$\bullet 2^2 \bullet = *$	$\leftarrow 2^2 \bullet\rightarrow = 0$
	$\bullet 2^3 \bullet = *$	$\leftarrow 2^3 \bullet\rightarrow = *$
$2^4 = *2$	$\bullet 2^4 \bullet = *$	$\leftarrow 2^4 \bullet\rightarrow = *$
$2^5 = *3$	$\bullet 2^5 \bullet = 0$	$\leftarrow 2^5 \bullet\rightarrow = 0$
$2^6 = 0$	$\bullet 2^6 \bullet = 0$	$\leftarrow 2^6 \bullet\rightarrow = 0$
$2^7 = 0$	$\bullet 2^7 \bullet = *2$	$\leftarrow 2^7 \bullet\rightarrow = *$
$2^8 = *2$		
$3\bullet\bullet 3 = *$	$\bullet 3\bullet 3\bullet = *$	$\leftarrow 3^2 \bullet\rightarrow = 0$
		$\leftarrow 3^3 \bullet\rightarrow = *$
$33\bullet 33 = *2$	$\bullet 3^4 \bullet = *$	$\leftarrow 3^4 \bullet\rightarrow = 0$
	$\bullet 3^5 \bullet = *3$	$\leftarrow 3^5 \bullet\rightarrow = 0$
$3^6 = 0$	$\bullet 3^6 \bullet = 0$	$\leftarrow 3^6 \bullet\rightarrow = 0$
$3^7 = *3$	$\bullet 3^7 \bullet = *2$	
$3^8 = *3$		

Table 4. Values of "starting" positions in impartial Sowing. The symbols $\leftarrow\bullet$ and $\bullet\rightarrow$ denote full pots going forever to the left and right.

x	$y=1$	2	3	4	5	6	7	8	9	10
1	0	$*$	$\uparrow$	$\uparrow$	$\uparrow$	$\uparrow$	$\uparrow$	$\uparrow$	$\uparrow$	$\uparrow$
2	$*$	$*2$	$\uparrow(1\|0)$	$\uparrow^2*$	$\uparrow^2*$	$\uparrow^2*$	$\uparrow^2*$	$\uparrow^2*$	$\uparrow^2*$	$\uparrow^2*$
3	$\downarrow$	$\downarrow(1\|0)$	$*2$	$*2$	$\downarrow(1\|0)$	$\downarrow(1\|0)$	$\downarrow(1\|0)$	$\downarrow(1\|0)$	$\downarrow(1\|0)$	$\downarrow(1\|0)$
4	$\downarrow$	$\downarrow_2*$	$*2$	$*2$	$\downarrow(1\|0)$	$\uparrow^2*$	$\uparrow^2*$	$\uparrow^2*$	$\uparrow^2*$	$\uparrow^2*$
5	$\downarrow$	$\downarrow_2*$	$\uparrow(1\|0)$	$\uparrow(1\|0)$	$\uparrow(\pm1)$	$\uparrow(1\|0)$	$\uparrow^2*$	$\uparrow^2*$	$\uparrow^2*$	$\uparrow^2*$
6	$\downarrow$	$\downarrow_2*$	$\uparrow(1\|0)$	$\downarrow_2*$	$\downarrow(1\|0)$	$*2$	$*2$	$\uparrow^2*$	$\uparrow^2*$	$\uparrow^2*$
7	$\downarrow$	$\downarrow_2*$	$\uparrow(1\|0)$	$\downarrow_2*$	$\downarrow_2*$	$*2$	$*2$	$*2$	$\uparrow^2*$	$\uparrow^2*$
8	$\downarrow$	$\downarrow_2*$	$\uparrow(1\|0)$	$\downarrow_2*$	$\downarrow_2*$	$\downarrow_2\uparrow$	$*2$	$*2$	$*2$	$\uparrow^2*$
9	$\downarrow$	$\downarrow_2*$	$\uparrow(1\|0)$	$\downarrow_2*$	$\downarrow_2*$	$\downarrow_2*$	$\downarrow_2*$	$*2$	$*2$	$*2$
10	$\downarrow$	$\downarrow_2*$	$\uparrow(1\|0)$	$\downarrow_2*$	$\downarrow_2*$	$\downarrow_2*$	$\downarrow_2*$	$\downarrow_2*$	$*2$	$*2$

Table 5. Values of Atomic Wari positions of the form xy.

Except for deleting leading and trailing empty pots, there don't seem to be any clear-cut rules for simplifying Atomic Wari positions. The situation is similar to Sowing. Positions can often be split into sums independent components by hand, but no algorithm is known to find such splits in general. For example, **1231110101311 = 123111 + 1311**. Similarly, there are several cases where the first or last pot contains only one seed, where the position's value does not change when this pot is removed, but no algorithm is known for detecting such positions. For example, **1001321 = 1321**.

Simple Values. Table 5 lists the values for all Atomic Wari games with two adjacent nonempty pots, each containing ten or fewer seeds. We use the following notation from [Conway 1976]: For any game $G = \{G^L | G^R\}$, we recursively define $\uparrow G = \{*, \uparrow G^L | *, \uparrow G^R\}$ and $\downarrow G = \uparrow(-G) = -(\uparrow G)$. For all positive integers n, we define $\uparrow^n = \uparrow n - \uparrow(n-1)$, and $\downarrow_n = -(\uparrow^n)$.

We note that only ten different games appear in Table 5: 0, $*$, $\uparrow$, $\downarrow$, $*2$, and the "exotic" games

$$\uparrow(1|0) = \{\uparrow, *|0, *\} \qquad \uparrow(\pm1) = \{\uparrow, *|\downarrow, *\} \qquad \uparrow^2* = \{0, *|\downarrow\}$$
$$\downarrow(1|0) = \{0, *|\downarrow, *\} \qquad\qquad\qquad\qquad \downarrow_2* = \{\uparrow|0, *\}$$

The games $\uparrow$ and $\uparrow(1|0)$ have atomic weight 1; $\downarrow$ and $\downarrow(1|0)$ have atomic weight -1; all the other have atomic weight zero. These games are partially ordered as follows:

There are Atomic Wari positions with arbitrary integer atomic weights. For example, the position $(\mathbf{01300})^n$ has value $n \cdot \uparrow$ and atomic weight n. Even so, for positions arising in normal play, the atomic weight is almost always 0, 1, or

−1, occasionally 2 or −2, and in extremely rare cases, 3 or −3. We have yet to see even one "natural" position with any other atomic weight. It is an open question whether noninteger or nonnumeric atomic weights are possible.

Partisan Atomic Wari Is Partially Impartial. Even though Atomic Wari is a partisan game, there is a special case that can be analyzed as if it were impartial. We call an Atomic Wari position "sparse" if every pot has two or fewer seeds.

THEOREM 3.1. *Every sparse Atomic Wari position has the same value as the corresponding Impartial Atomic Wari position, and any such position can be split into independent components by removing all pots with fewer than two seeds.*

PROOF. We prove the claim by induction on the number of *deuces* (pots with two seeds). The base case, in which each pot contains either one seed or none, is trivial.

Consider a position X with n deuces, and let X' denote the sum of positions obtained by deleting pots with fewer than two seeds. Each of the options of X is a sparse position with either $n-1$ or $n-2$ deuces. For each move by Left, there is a corresponding move by Right in the same contiguous "string" of deuces that results in exactly the same position, once the inductive hypothesis is applied. For example, given the position **1222201**, the Left move to **12102201** $= 2 + 22$ is matched by the Right move to **12201201** $= 22 + 2$. Clearly, X and X' have the same options, once the inductive hypothesis is applied. The theorem follows immediately. $\square$

Sparse Atomic Wari is equivalent to the following take-away game. There are several piles of seeds. Each player can remove one seed from any pile, or remove two seeds from any pile and optionally split the remainder into two piles. In the octal notation of *Winning Ways*, this is the game **·37**. A computer search of the first 200,000 values of this game reveals no periodicity, and finds only thirteen $\mathcal{P}$-positions:

$$\{0, 3, 11, 19, 29, 45, 71, 97, 123, 149, 175, 313, 407\}.$$

It seems quite likely that these are in fact the only $\mathcal{P}$-positions. For a short list of Nim values, see page 102 of *Winning Ways*.

Theorem 3.1 implies that the values we see in Table 5 are the only values that a two-pot Atomic Wari position can have. By straightforward case analysis, we can classify all positions xy with $x \leq y$ as follows.

$x = 1$	$1 < x \leq y - 2$	$1 < x = y - 1$	$1 < x = y$
0 if $y = 1$	$\downarrow(1\|0)$ if $2^x = 0$	$\uparrow(1\|0)$ if $2^{x-2} = 0$	$*2$ if $x = 2$
$*$ if $y = 2$	$\uparrow^2 *$ otherwise	$\downarrow(1\|0)$ if $2^{y-2} = 0$	$\uparrow(\pm1)$ if $2^{x-2} = 0$
$\uparrow$ otherwise		$*2$ otherwise	$*2$ otherwise

Open Questions. Are noninteger or nonnumeric atomic weights possible in Atomic Wari? How can we systematically construct Impartial Atomic Wari positions with value $*n$ for any n? Is there a *simple* algorithm that splits Atomic Wari positions into multiple independent components? Are there Atomic Wari positions with exponentially many followers? (The answer is yes if we disallow capturing groups of three seeds.) Finally, are there any other high-level simplification rules like Theorem 3.1 that would allow faster evaluation?

4. Other Variants

An amazingly large number of other variations on these games are possible. Here is a short list of possible games, starting with common versions of the original African games, which might give interesting results. This list is by no means exhaustive!

Mancala is played on a two by six grid of pots, where each player owns one row of six. All moves go counterclockwise, but must begin at one of the moving player's pots. Each player has an extra pot called a store. Each player can drop seeds into his own store as if it were the seventh pot on his side, but not into his opponent's. Seeds never leave the stores. If a move ends by putting the last seed into the store, the same player moves again. If the last stone lands in an empty pot on the moving player's side, both that stone and the stones in the opponent's pot directly opposite are put into the moving player's store. If any player cannot move, the other player collects all the seeds on his side and puts them in his store, and the game ends. The winner is the player who has more seeds in his store at the end of the game. In the starting position, there are three (or sometimes six) seeds in each pot.

Wari (or *oware*) is also played on a two by six grid of pots, similarly to mancala, but with no stores. All moves go counterclockwise around the board, but each move must begin on the moving player's side. Otherwise, the rules are identical to Atomic Wari. In one version of the game, if one player cannot move, his opponent moves again; the game ends only when neither player can move. In other versions, the ending conditions are considerably more complicated. When the game ends, the player with more captured seeds wins. Typically, the game beings with four seeds in each pot.

Sowing can also be played in reverse. In *Reaping*, a legal move consists of picking up one seed from each of a successive string of pots and dropping them into the first available empty pot. One could also play Atomic Wari after reversing the movement rules, but one must be careful not to reverse the capturing rules as well!

Partisan sowing games could be played with either player moving in either direction, but with colored seeds. For example, a legal move by Left might consist of sowing all the blue seeds in any pot, in such a way that the last seed is put into a pot containing at least one other blue seed, and capturing all the red seeds in the last pot.

Finally, consider the following two-dimensional sowing game. The board consists of a two-dimensional grid of pots. Left can sow upwards or downwards; Right can sow to the left or right. Seeds are sown exactly as in Sowing; in particular, the last seed must be put into a nonempty pot. Whenever any seed lands in a nonempty pot, the contents of that pot, including the new seed, are captured. Thus, at least two seeds are captured on every turn. One special case of this game is already quite well-known!

References

[Berlekamp et al. 1982] Elwyn Berlekamp, John Conway, and Richard Guy, *Winning Ways for your Mathematical Plays*, Academic Press, New York, 1982.

[Conway 1976] J. H. Conway, *On Numbers and Games*, Academic Press, London, 1976.

[Wolfe 1996] David Wolfe. See pages 93–99 in this volume.

JEFF ERICKSON
COMPUTER SCIENCE DIVISION
UNIVERSITY OF CALIFORNIA
BERKELEY, CA 94720
jeffe@cs.berkeley.edu

Games of No Chance
MSRI Publications
Volume **29**, 1996

New Toads and Frogs Results

JEFF ERICKSON

ABSTRACT. We present a number of new results for the combinatorial game
Toads and Frogs. We begin by presenting a set of simplification rules, which
allow us to split positions into independent components or replace them
with easily computable numerical values. Using these simplication rules,
we prove that there are Toads and Frogs positions with arbitrary numerical
values and arbitrarily high temperatures, and that any position in which all
the pieces are contiguous has an integer value that can be computed quickly.
We also give a closed form for the value of any starting position with one
frog, and derive some partial results for two-frog positions. Finally, using
a computer implementation of the rules, we derive new values for a large
number of starting positions.

1. Introduction

Toads and Frogs is a two-player game, played on a one-dimensional board.
Left has a number of toads, and Right has a number of frogs, each on its own
square of the board. Each player has two types of legal moves: he may either
push one of his pieces forward into an adjacent empty square, or he may *jump*
one of his pieces over an adjacent opposing piece, into an empty square. Jumps
are never forced, and jumped-over pieces are not affected in any way. Toads
move to the right, frogs to the left. The first player without a legal move loses
the game.

Throughout the paper, we represent toads by T, frogs by F, and empty squares
by the symbol □.

Here is a typical Toads and Frogs game. Left moves first and wins.

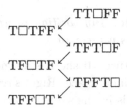

In this particular game, and in almost all previously analyzed positions, all the moves are forced, but this is certainly not typical.

Toads and Frogs was described very early in *Winning Ways* [Berlekamp et al. 1982], in order to introduce some simple concepts such as numbers, fractions, and infinitesimals. *Winning Ways* contains a complete analysis of a few simple types of Toads and Frogs positions.

We assume the reader is familiar with the terminology (value of a game, follower, temperature, atomic weight, ...) and notation ($\{x|y\}$, $*$, $\uparrow$, $+_x$, ...) presented in *Winning Ways*. Throughout the paper, we will use the notation X^n to denote n contiguous copies of the Toads and Frogs position X. For example, $(TF\square)^3F^4$ is shorthand for $TF\square TF\square TF\square FFFF$.

2. Simplifying Positions

We will employ a number of general rules to simplify the evaluation of board positions. These rules allow us to split positions into independent components, or replace positions with easily derived values.

The simplest rule is to remove dead pieces. A piece is *dead* if it cannot be moved, either in the current position or in any of its followers. For example, in the position $TFFT\square$, the first three pieces are dead, so we can simplify to the position $T\square$. Sometimes removing dead pieces causes the board to be split into multiple independent components. For example, in the position $T\square TTFF\square F$, the middle four pieces are dead. Furthermore, the pieces to the left of the dead group can never interfere with the pieces to the right of the dead group. Thus, this position is equivalent to the sum $T\square + \square F$. It's possible that after we split the board into components, some of the components have no pieces; obviously, they can be ignored. For example, $\square\square TTFF\square TF = \square\square + \square TF = \square TF$.

Fortunately, it's easy to tell what pieces are dead just by looking at the board.

Identifying Dead Pieces: *Any contiguous sequence of toads and frogs beginning with two toads (or the left edge of the board) and ending with two frogs (or the right edge of the board) is dead. Any piece that is not in such a sequence is alive.*

In *Winning Ways*, the "Death Leap Principle" was used to help analyze Toads and Frogs positions with only one space. We can generalize it to more complicated positions. We call a space *isolated* if none of its neighboring squares is empty.

The Death Leap Principle: *Any position in which the only legal moves are jumps into isolated spaces has value zero.*

PROOF. Suppose it's Left's turn. If she has no moves, Right wins. Otherwise, she must jump a toad into a single space. Right's response is to push the jumped-over frog forward. Now Left is in the same situation as before—her only moves are jumps into single spaces. (Right may have more moves at this point, but this only makes the situation better for him.) Eventually, Left has no moves, and Right wins. We can argue symmetrically if Right goes first.   □

We can also express this principle purely in syntactic terms, which makes it somewhat easier to implement on a computer.

The Death Leap Principle: *Any position that does not contain any of the four subpositions* TF□□, □□TF, T□, *or* □F *has value zero.*

If there are no frogs between a toad and the right edge of the board, we call the toad *terminal*. If there are no toads between a frog and the left edge of the board, we call the frog *finished*. Suppose Left has a terminal toad with three spaces in front of it. Intuitively, this toad is worth exactly three free moves for Left, since none of Right's pieces can ever interfere with it. If we change the position by moving this toad to the right end of the board (where it dies) and crediting Left with three free moves, we expect the value of the game to stay the same. After all, whenever she would have moved her terminal toad, she can just take one of her free moves instead.

This intuition suggests the following simplification principle.

The Terminal Toads Theorem: *Let* X *be any position. Then*

$$\text{XT}\square^n = \text{X}\square^n + \text{T}\square^n = \text{X}\square^n + n.$$

PROOF. The second player wins the difference $\text{XT}\square^n - (\text{X}\square^n + \text{T}\square^n)$ by the following mirror strategy. Initially, the last toad in the first component is marked T_*. Any move in either copy of X is answered by the corresponding move in the other copy. Any move in the third component is answered by moving T_*, and vice versa.

This is enough to show that Left loses if she moves first, but there are two special cases to consider when Right moves first. If Right moves a frog in the second component whose twin in the first component is blocked by T_*, then Left moves T_* instead, and then *moves the mark back* to the blocked toad. If Right moves in the third component, but there's a toad in front of T_*, then Left *moves the mark forward* and then pushes the new T_*. Any move by Right is answerable by a corresponding move by Left, so Right also loses going first.   □

Naturally, there is a symmetric version of this theorem for removing finished frogs.

The Finished Frogs Formula: *Let* X *be any position. Then*

$$\square^n\text{FX} = \square^n\text{X} + \square^n\text{F} = \square^n\text{X} - n.$$

Consider the position TFTFF□FT□□TTFF□TFF□□TT. At first glance it might look too complicated to evaluate the usual way. Fortunately, our simplification rules let us get a value quickly, without recursively evaluating any of its followers:

$$\mathrm{\overline{TFTFF}\square FT\square\square\overline{TTFF}\square TFF\square\square\overline{TT}} = \square\overline{\mathrm{FT}}\square\square + \square\mathrm{TFF}\square\square$$
$$= \boxminus\boxminus\boxminus + \square\mathrm{TFF}\square\square + 1$$
$$= \boxminus\overline{\mathrm{TFF}}\boxminus\boxminus + 1$$
$$= 1.$$

(i) We remove the three groups of dead pieces, splitting the board into two independent components.

(ii) The first component contains a terminal toad with two remaining moves, and a finished frog with one remaining move. We remove them, crediting Right with one free move and Left with two, for a net gain of one move for Left.

(iii) The first component is now completely empty, so we remove it.

(iv) We remove the remaining component, since it has value zero by the Death Leap Principle.

3. Arbitrary Numbers and Arbitrary Temperatures

We make two somewhat surprising observations. First, for any number, there is a Toads and Frogs position with that number as its value. Second, there are Toads and Frogs positions with arbitrarily high (half-integer) temperatures.

THEOREM 3.1. $(\mathrm{TF})^m \mathrm{T}\square(\mathrm{TF})^n = 2^{-n}$ for all m and n.

PROOF. We use induction on n. The base case $n = 0$ follows immediately from the Death Leap Principle. Suppose $n > 0$. Left has only one legal move, to the position $(\mathrm{TF})^m \square \mathrm{T}(\mathrm{TF})^n = (\mathrm{TF})^m\square = 0$ by the Death Leap Principle. Similarly, Right can only move to the position $(\mathrm{TF})^{m+1}\mathrm{T}\square(\mathrm{TF})^{n-1} = 2^{-n+1}$, by the induction hypothesis. $\square$

COROLLARY 3.2. For any dyadic rational number q, there is a Toads and Frogs position with value q.

PROOF. Write $q = (2k+1)/2^n$, and assume without loss of generality that $k \geq 0$. Then the position $(\mathrm{T}\square(\mathrm{TF})^n\mathrm{TTFF})^{2k+1}$ has value q. $\square$

THEOREM 3.3. $\mathrm{TF}\square\mathrm{F}\square^{n+3} = \{n|0\}$.

PROOF. Right has one legal move, to $\mathrm{TFF}\square^{n+4}$, which clearly has value 0. Left has one legal move, to $\square\mathrm{FTF}\square^{n+3}$, which, by the Finished Frogs Formula, is equivalent to $\square\mathrm{TF}\square^{n+3}-1 = \{n|n+4\}-1 = n$. $\square$

COROLLARY 3.4. For any integers $a \geq b$, there is a Toads and Frogs position with value $\{a|b\}$.

PROOF. One of $\mathrm{T}\square^b\mathrm{TTFFTF}\square\mathrm{F}\square^{a-b+3}$ or $\mathrm{TF}\square\mathrm{F}\square^{a-b+3}\mathrm{TTFF}\square^{-b}\mathrm{F}$ has value $\{a|b\}$, depending on whether b is positive or negative. $\square$

We do not know whether there are positions with arbitrary (dyadic rational) temperatures. The following theorem gives us positions whose temperatures are arbitrary *sufficiently large* multiples of $\frac{1}{4}$. We omit the proof, which follows from exhaustive case analysis. Every position eight moves away is an integer.

THEOREM 3.5. *For any $n \geq 6$, we have* $\text{T}\square\text{TF}\square\text{F}\square^n = \{a+\frac{1}{2}|1\}$.

4. Knots Have Integer Values

We now consider positions in which the toads and frogs form a single contiguous group. We call such positions *knots*, the collective term for toads [Lipton 1991]. Somewhat surprisingly, every knot has an integer value, which can be determined without evaluating any of the position's followers. We derive a series of rules that allow us to reduce every such position to one of a few simple cases. These rules are proved using relatively simple counting arguments, but disinterested readers are encouraged to skip the proofs.

Thanks to our simplification rules, we only need to consider positions that start with a toad and end with a frog, but not two toads and two frogs (or similar positions in which the knot is against one edge of the board). In the discussion that follows, unless otherwise stated, all superscripts are positive.

LEMMA 4.1. $\square^a\text{TF}\square^b = \{b-a-2|b-a+2\} = \begin{cases} b-a-1 & \text{if } b-a \geq 2, \\ 0 & \text{if } |b-a| \leq 1, \\ b-a+1 & \text{if } b-a \leq -2. \end{cases}$

PROOF. Immediate. $\square$

LEMMA 4.2. $\square^a\text{TFTF}\square^b = \begin{cases} b-1 & \text{if } 2a \leq b, \\ 2(b-a)-1 & \text{if } a < b \leq 2a, \\ 0 & \text{if } b=a, \\ 2(b-a)+1 & \text{if } b < a \leq 2b, \\ 1-a & \text{if } 2b \leq a. \end{cases}$

PROOF. This follows from Lemma 4.1 and the Terminal Toads Theorem by case analysis. See Figure 1. $\square$

LEMMA 4.3. $\square^a\text{TTF}\square^b = \begin{cases} 2b-a-2 & \text{if } a \leq b-1, \\ b-1 & \text{if } b-1 \leq a \leq b+1, \\ 2b-a & \text{if } b+1 \leq a \leq 2b, \\ 0 & \text{if } 2b \leq a. \end{cases}$

PROOF. This follows from Lemma 4.1 and the Terminal Toads Theorem by simple case analysis. Every position three moves away is an integer. $\square$

LEMMA 4.4. $\square^a\text{T}^b\text{F}\square^c = (b-2)(c-1) + \square^a\text{TTF}\square^c$.

PROOF. We use induction on b. The base case $b=2$ follows from Lemma 4.3. To complete the proof, we need to show that the second player wins the game $\square^a\text{T}^b\text{F}\square^c - \square^a\text{T}^{b-1}\text{F}\square^c - (c-1)$. This follows from a simple counting argument.

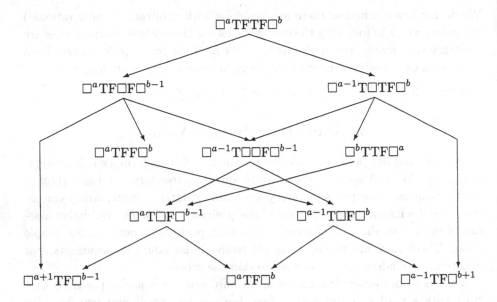

Figure 1. Followers of $\square^a\text{TFTF}\square^b$, with integer components (from the Terminal Toads Theorem) omitted.

Provided neither frog ever jumps, each player will move exactly $bc + a + b - 1$ times before the game ends. Moreover, as long as the second player's frog never jumps, the first player will run out of moves first. The second player can keep his frog from jumping by always pushing it immediately after the opponent's toad jumps over it. $\square$

LEMMA 4.5. *If neither player can move from the position* $\text{XF}\square$, *then*

$$\text{XT}^a\text{F}\square^b = a(b - 1).$$

PROOF. The base case $a = 0$ is trivial. Otherwise, the second player wins the game $\text{XT}^a\text{F}\square^b - a(b - 1)$ as follows.

Left moves first. Left is forced to jump her rightmost toad. Right responds by moving his frog forward, leaving the position $\text{XT}^{a-1}\text{F}\square^b - a(b-1) + (b-1)$ by the Terminal Toads Theorem. This position has value zero by induction. Thus, Right wins going second.

Right moves first. Left wins by a simple counting argument. Each of her a toads will move or jump at least b times before reaching the edge of the board, for a total of ab moves. Right's frog will move at most a times before hitting X, and he has $ab - a$ moves available in the integer component. If Right never jumps his frog, each player will have the same number of moves, but that's the best Right can do. So when Right uses up all his moves, Left will have at least one move left, to the winning position $\text{XF}\square^b = 0$. $\square$

The previous lemmas let us evaluate any knotted position from which only one player can legally move. Of course, we already knew that such positions had integer values, but we can now easily determine *what* integer. The only positions left to analyze are those in which both sides can legally move.

LEMMA 4.6. *If neither player can move from the position* $\square TXF\square$, *then*

$$\square^a TF^b XT^c F\square^d = \square^a TF^b + T^c F\square^d = c(d-1) - b(a-1).$$

PROOF. The second player wins $\square^a TF^b TXFT^c F\square^d + b(a-1) - c(d-1)$ by the same counting argument. Each player can move at most $ab + cd$ times, with equality if neither of the lone pieces ever jumps. Thus, as long as the second player's lone piece never jumps, which is easy to guarantee, the first player will run out of moves first. $\square$

LEMMA 4.7. $\square^a TF^b T^c F\square^d = \square^a TFTF\square^d + (c-1)(d-1) - (a-1)(b-1)$.

PROOF. The base case $b = c = 0$ follows immediately from Lemma 4.2. To complete the inductive proof, it suffices to prove that

$$\square^a TF^b T^c F\square^d = \square^a TF^b T^{c-1} F\square^d + (d-1).$$

The second player wins the difference game by the usual counting argument. If Right moves first, Left marks the first toad in the first component and the last frog in the second. If Left moves first, right marks the last frog in the first component and the first toad in the second. Either way, the second player moves so that the marked pieces never jump. This guarantees at least $ab + cd + a + b + c + d + 1$ moves to the second player, and at most that many to the first player, so the first player will always run out of moves first and lose.

This guarantees that Left will get at least $cd + b + d + 1$ moves in the first component and $ab + a + c$ in the second. Right gets at most $ab + a + c + 1$ moves in the first component, $cd - c + b + d + 1$ in the second, and $d - 1$ in the third. $\square$

THEOREM 4.8. *Every knot has an integer value, which can be determined directly, without evaluating its followers.* $\square$

5. One-Frog Starting Positions

Starting positions are those in which all toads are against the left edge of the board, and all frogs are against the right edge. In this section, we give a closed form for the value of a starting position in which there is only one frog.

LEMMA 5.1. $T\square^a F = a \cdot * = \begin{cases} * & \text{if } a \text{ is odd,} \\ 0 & \text{if } a \text{ is even.} \end{cases}$

PROOF. For any follower of the game $T\square^a F$, both players' moves are forced. Now suppose a is even. The second player's $(a/2)$-th move leaves the position $\square^{a/2} TF\square^{a/2}$, which equals 0 by Lemma 4.1, so the value of the original game is zero.

Suppose a is even. Then the second player wins the game $T\square^a F + *$ as follows. Suppose Left moves first. If Left moves in $*$ before the two pieces meet, then Right's $\lceil a/2 \rceil$-th moves leaves the winning position $\square^{\lfloor a/2 \rfloor} TF \square^{\lceil a/2 \rceil} = 0$ by Lemma 4.1. Otherwise, Right's $\lfloor a/2 \rfloor$-th move leaves the winning position $\square^{\lfloor a/2 \rfloor} T\square F \square^{\lfloor a/2 \rfloor} + * = 0$ by Lemma 4.1. $\square$

THEOREM 5.2. $T^a \square^b F = \begin{cases} \{\{a-2|1\}|0\} & \textit{if } b = 1, \\ (a-1)(b-1)* & \textit{if } b > 1 \textit{ is odd}, \\ (a-1)(b-1) & \textit{if } b \textit{ is even}. \end{cases}$

PROOF. The base case $a = 1$ follows immediately from the previous lemma, and the case $b = 1$ follows from the analysis in *Winning Ways* (p. 126). Otherwise, it suffices to prove that $T^a \square^b F = T^{a-1} \square^b F + (b - 1)$, since the theorem then follows by induction. The second player wins the difference by the usual counting argument. If the second player never jumps his frog, then he will get at least $ab+a-b+1$ moves, provided that all of his pieces eventually reach the appropriate end of the board. He can guarantee this by moving his frog forward on his first move. The first player gets at most $ab + a - b + 1$ moves, so the second player will win. $\square$

6. Partial Results for Two-Frog Starting Positions

We were unable to derive a closed form for two-frog starting positions, but we do have some interesting partial results.

LEMMA 6.1. $T^a F \square T^b FT^c \square = c$ *for all* $b \neq 0$.

PROOF. Follows immediately from the Terminal Toads Theorem and the Death Leap Principle. $\square$

LEMMA 6.2. $T^a F \square FT^c \square \leq c$.

PROOF. It suffices to show that Right can win the game $T^a F \square F \square$ if he moves second. Left's first move is forced, and Right responds by moving to $T^{a-1} F \square TF \square = 0$ by the Death Leap Principle. $\square$

LEMMA 6.3. $T^n \square F \square F = 0$ *for all* $n \geq 2$.

PROOF. The second player wins as follows.

Left moves first. Right moves his leftmost frog on his first move, and his rightmost frog on his second move. Left's moves are forced. Right's second move leaves the position $T^{n-1} F \square TF \square$, which has value zero by the Death Leap Principle. Thus, Right wins moving second.

Right moves first. Right has two options. If he moves the rightmost frog forward, Left will move her rightmost toad forward three times. This forces Right's next two moves. Left's third move leaves the winning position $T^{n-1} FF \square T \square = 1$.

If Right moves his leftmost Frog forward, Left must respond by jumping it, leaving $T^{n-1} \square FT \square F$. Again, Right has two options. If he moves his rightmost frog, Left responds by jumping it, leaving the positions $T^{n-1}\square F\square F$, which is zero by induction. (One can check the base case $n = 2$ directly, or see *Winning Ways*, p. 135.) Otherwise, Right moves to the position $T^{n-1}F\square T\square F$, and Left can force the following sequence of moves:

$$
\begin{array}{ccc}
 & & T^{n-1}F\square T\square F \\
 & T^{n-1}F\square\square TF \nearrow & \\
 & & \searrow T^{n-1}F\square FT\square \\
T^{n-2}\square FTFT\square \nearrow & & \\
 & & \searrow T^{n-2}F\square TFT\square \\
 & T^{n-2}F\square TF\square \nearrow & \\
\end{array}
$$

The final position is zero by the Death Leap Principle. Thus, Left wins moving second.   □

7. Known Values and Open Questions

We list here all known values for positions in which all toads are on the left edge of the board and all frogs are on the right edge of the board. These values, and the values of most of the positions used earlier in the paper, were derived with the help of David Wolfe's **games** package [Wolfe 1996]. I added code (about 600 lines of ANSI C) for evaluating Toads and Frogs positions, using all the simplification rules described in Section 2.

Here are all known values for symmetric games of the form $T^a\square^bF^a$, apart from the cases $a = 1$ and $b = 1$, for which all values were previously known (see *Winning Ways*).

			b			
a	1	2	3	4	5	6
1	$*$	0	$*$	0	$*$	0
2	$*$	$*$	$*$	$*$	0	0
3	$*$	$\pm\frac{1}{8}$	0	$*$	0	
4	$*$	$\cdots$		$*$	0	
5	$*$	$\cdots$		$*$		
6	$*$	$\cdots$				

$$T^4\square\square F^4 = \pm(\{1|*\}, \{\{\tfrac{3}{2}\,|||\,1\,||\,0\,|\,0, \downarrow *\}\,|\,0\})$$

$$T^5\square\square F^5 = \pm(\{2|*\}, \{\tfrac{5}{2}\,|\,\{2\,|||\,0\,||\,\{0\,|||\,\uparrow *\,|\,0\,||-1\}, \{0\,|\,\{0\,||-\tfrac{1}{32}\,|-2\,|||-\tfrac{1}{2}*\}\}\,|\,\{\uparrow *\,|\,0\,||-\tfrac{1}{2}\,|||-1*\}\}$$

$$\|0\})$$

$$T^6\square\square F^6 = \pm(\{3|*\}, \{\tfrac{7}{2}\,||\,3\,|\,\{0, G\,|||\,\{0, G\,|\,\{\tfrac{1}{2}\downarrow\,|\,m_1\,||\,0\,|||-2\}\},$$

$$\{0\,|\,\{0\,||\,m_2\,|-3\,|||-1\downarrow\}\}\,\|\,\{\tfrac{1}{2}\downarrow\,|\,m_1\,||\,0\,|||-\tfrac{1}{2}\}\,|-2*\}$$

$$\|||\,0\}),$$

where $m_1 = {}_{-2\|0\|+\frac{1}{64}}$, $m_2 = {}_{-1\|0\|+\Uparrow_*}$, and

$$G = \{\tfrac{1}{2}|0\}, \tfrac{1}{2}\{0\|\downarrow|-2\}|\{0\|m_1|-2\}, \{\tfrac{1}{2}\downarrow|m_1\|-\tfrac{3}{2}|-2\}.$$

Except for $\mathrm{TTT\square\square FFF} = \pm\tfrac{1}{8}$, all values are infinitesimals with atomic weight zero. The values for $b = 2$ are particularly interesting, since they seem to be totally patternless. The other values are certainly much tamer, but there are still no apparent patterns.

The table on the next page lists all known values for positions of the form $\mathrm{T}^a\square^b\mathrm{F}^c$ for $1 \le c < a \le 7$. The only really nasty value we get is for $\mathrm{T}^7\square^4\mathrm{F}^3$, which is $(7*|\tfrac{9}{2}\|4)$-ish.

These values naturally suggest a number of patterns, some of which we confirmed earlier in the paper. We list here several conjectures that we have been unable to prove, all of which are strongly supported by our experimental results.

CONJECTURE 7.1. $\mathrm{T}^a\square\square\mathrm{F}^b = \{\{a{-}3|a{-}b\}|\{*|3{-}b\}\}$ for all $a > b \ge 2$.

CONJECTURE 7.2. $\mathrm{T}^a\square\square\square\mathrm{FF} = (a-2)*$ for all $a \ge 2$.

CONJECTURE 7.3. $\mathrm{T}^a\square\square\square\mathrm{F}^3 = a - \tfrac{7}{2}$ for all $a \ge 5$.

CONJECTURE 7.4. $\mathrm{T}^a\square^a\mathrm{F}^{a-1} = 1$ or $1*$ for all $a \ge 1$.

CONJECTURE 7.5. $\mathrm{T}^a\square^b\mathrm{F}^a$ is an infinitesimal for all a, b except $(a, b) = (3, 2)$.

Many of our results seem to be special cases of a more general principle, which sometimes allows us to split board positions into multiple independent components. An important open problem is to characterize exactly what positions can be split this way.

We finish with a much more ambitious conjecture. Despite the existence of simple rules for special positions, Toads and Frogs positions are in general extremely hard to evaluate. It seems difficult, in general, to even determine the winner of a given Toads and Frogs position. We conjecture that Toads and Frogs is at least as hard as other seemingly richer games such as Red-Blue Hackenbush [Berlekamp et al. 1982].

CONJECTURE 7.6. *Toads and Frogs is NP-hard.*

To prove this conjecture, it would suffice to prove that for any integers $a > b > c$, there is a Toads and Frogs position whose value is $\{\{a|b\}|c\}$. The conjecture would then follow from a theorem of Yedwab and Moews. (See [Yedwab 1985, pp. 29–45] or [Berlekamp and Wolfe 1994, pp. 109–111].) The closest we can get is the following theorem, which can be verified by drawing out the complete game tree.

THEOREM 7.7. *For any* $a \ge 0$, $\mathrm{TF\square F\square F\square}^a = \{\{a{-}6|-1\}|-2\}$.

a	b 1	2	3	4	5
			$c=1$		
2	$\frac{1}{2}\|0$	1	$2*$	3	$4*$
3	$1*\|0$	2	$4*$	6	$8*$
4	$2\|1\|\|0$	3	$6*$	9	$12*$
5	$3\|1\|\|0$	4	$8*$	12	$16*$
6	$4\|1\|\|0$	5	$10*$	15	$20*$
7	$5\|1\|\|0$	6	$12*$	18	$24*$
			$c=2$		
3	$*$	$\frac{1}{2}\|0$	$1*$	$2*$	3
4	$*$	$\frac{3}{2}\|0$	$2*$	$4*$	$\frac{95}{16}$
5	$*$	$\frac{5}{2}\|0$	$3*$	$6\|\frac{11}{2}$	$\frac{17}{2}\|8$
6	$*$	$\frac{7}{2}\|0$	$4*$	$\frac{15}{2}\|7*$	$\frac{21}{2}$
7	$*$	$\frac{9}{2}\|0$	$5*$	9	13
			$c=3$		
4	$*$	$1*\|\downarrow$	$1\|\|\frac{1}{2}\|0$	$1*$	2
5	$*$	$2*\|\downarrow$	$\frac{3}{2}$	$3*\|\frac{5}{2}\|\|2$	4
6	$*$	$3*\|\downarrow$	$\frac{5}{2}$	$5*\|\frac{7}{2}\|\|3,3{\uparrow}*$	$\frac{41}{8}$
7	$*$	$4*\|\downarrow$	$\frac{7}{2}$	see below	
			$c=4$		
5	$*$	$2\|1\|\|*\|-1$	$1*$	$2*\|1*\|\|\frac{1}{2}\|0$	1
6	$*$	$3\|2\|\|*\|-1$	$2*$	$4*\|2*\|\|\frac{3}{2}\|1$	
7	$*$	$4\|3\|\|*\|-1$	$3*$		

$$T^7\square^4F^3 = 7*\|\tfrac{9}{2}\|\|4,4\{0\|\{\{\{2\|\tfrac{1}{2}*\|\{2\|\tfrac{3}{2}\|\|+\tfrac{1}{2}\|\|\|0,\{0,*\|\|*\|-\tfrac{1}{4}\}\}\},\{\tfrac{3}{2}*\|+\tfrac{1}{2}\}\|0,\{0,*\|\|*\|-\tfrac{1}{4}\}\},$$
$$\{\tfrac{3}{2}*\|+\tfrac{1}{2}\|\|0\}\|0\}\}$$

Known values for positions $T^a\square^bF^c$. See Theorem 5.2 for $c=1$.

Acknowledgments

The author thanks Dan Calistrate for verifying Theorems 3.5 and 7.7.

References

[Berlekamp et al. 1982] Elwyn Berlekamp, John Conway, and Richard Guy, *Winning Ways for your Mathematical Plays*, Academic Press, New York, 1982.

[Berlekamp and Wolfe 1994] Elwyn Berlekamp and David Wolfe, *Mathematical Go Endgames: Nightmares for the Professional Go Player*, Ishi Press, San Jose, 1994.

[Lipton 1991] James Lipton, *An Exaltation of Larks*, Viking, New York, 1991.

[Wolfe 1996] David Wolfe. See pages 93–98 in this volume.

[Yedwab 1985] Laura Yedwab, "On playing well in a sum of games", Master's Thesis, MIT, 1985. Available as Technical Report MIT/LCS/TR-348.

JEFF ERICKSON
COMPUTER SCIENCE DIVISION
UNIVERSITY OF CALIFORNIA
BERKELEY, CA 94720
jeffe@cs.berkeley.edu

Games of No Chance
MSRI Publications
Volume **29**, 1996

Xdom: A Graphical, X-Based Front-End for Domineering

DAN GARCIA

ABSTRACT. This article is an overview of Xdom, a mouse-oriented program
for playing Domineering. Xdom allows a user to input an initial position,
then play against the computer or another user. Optionally, it can determine the value and predicted winner of any position, show the available
moves of either or both players, and give hints.

Xdom is an X-based program for the player of Domineering, running under UNIX and written in Tcl/Tk. It uses David Wolfe's Gamesman's Toolkit
[Wolfe 1996], a powerful text-based game-analysis program that allows users to
explore combinatorial games, and in particular determine game values, but is
not designed to play out games interactively. The basic functionality of the
Gamesman's Toolkit consists of computing the value of a position specified by
the user:

```
unix% games
Type 'help' and 'help help'
> dom
Enter Domineering position followed by extra <cr>
* *** ***
* * * * *
*** ***

3/4
>
```

Xdom builds on top of this functionality, providing a nice interface for the game
of Domineering. Using Xdom, the user can play Domineering games against

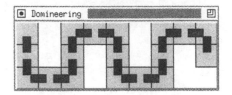

Figure 1. Starting domineering position loaded in from the file bigsnake.com. Each slot (square on the board) is 30 pixes wide for the example in the text.

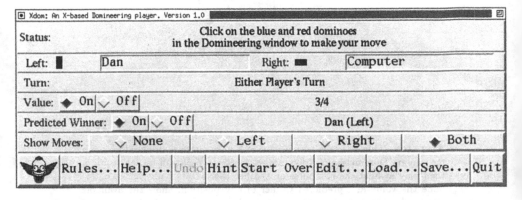

Figure 2. The main Xdom window.

the computer or another user, create new positions, read positions from and save positions into files, undo moves, etc. If the user wishes, Xdom will show the value and predicted winner of any position, show the available moves of either or both players, and propose good moves. It also documents the rules of Domineering and its own functions.

Xdom is called from a Unix shell, with two optional arguments: the name of a file containing a starting Domineering position, and the size, in pixels, that the slots on the board should have. For example, typing the the command

```
Xdom bigsnake.dom 30
```

to the Unix shell will result in the configuration shown in Figure 1, corresponding to the Toolkit example given above. (The contents of the initialization file are identical to the Toolkit input in the example above; the name of the file must end in .dom.) The defaults are a rectangle of 4×3 slots, with 60 pixels per slot.

Upon startup, Xdom creates two windows: the main window and the playing board. The main window (Figure 2) allows the user to interact with the program, show or hide values and predictions, show or hide the moves on the board, create new positions, load old positions from files, or save existing positions into files.

The playing board, shown in Figure 1, is made up of gray slots (squares) and possibly beige areas, indicating slots that are not, or are no longer, in play; they appear light gray in Figure 1.

Optionally, the playing board indicates Left's moves with miniature upright dominoes, and Right's moves with miniature horizontal dominoes. When the user clicks on such a domino, he is making a move: the two slots that together contain the domino are removed from play. On color monitors, Left and Right dominoes appear blue and red, respectively; in Figure 1 they appear black and dark gray.

When a move is made, Xdom calls the Gamesman's Toolkit to find the value of this new game, and who is expected to win. This value and the prediction are displayed in the main Xdom window, if that feature is turned on. Because the value calculation may not be doable in a reasonable amount of time for a board bigger than (say) 5 × 5, the feature should be turned off in such cases. Two human players can then play arbitrarily large games.

At any time, the user has the choice of viewing the moves available only to Left, those available only to Right, both or neither. By default, the program shows the moves of the player whose turn it is to play. The user can override this choice, and not only view the moves the player who has just played, but even play again for that player.

Hints (best possible moves) may be indicated at any time; one just clicks the Hint button. This feature can be used to simulate a computer opponent. The user simply makes a move, then clicks Hint to determine the move to make for the computer opponent. In this fashion, the user can get a feel for optimal strategies at various positions. (The Hint feature, and game evaluations in general, depends on Xdom obtaining a response from the Gamesman's Toolkit. These evaluations are not done in the background, so these features shouldn't be used for positions that take too long to evaluate with the Toolkit.)

Users can create their own Domineering positions or edit existing positions interactively, using a special window called up by clicking the Edit button. A user-created position, or a position reached in the course of a game, can be saved to a .dom file, using the Save button, and later reloaded, using the Load button. Since the internal format of dom files is very simple, such files can also be created manually, using a text editor.

Availability

Xdom is available for downloading at http://http.cs.berkeley.edu/~ddgarcia/software/xdom/. The current version is 1.0.

DAN GARCIA
COMPUTER SCIENCE DIVISION
535 SODA HALL
UNIVERSITY OF CALIFORNIA
BERKELEY, CA 94720
    ddgarcia@cs.berkeley.edu

Games of No Chance
MSRI Publications
Volume **29**, 1996

Infinitesimals and Coin-Sliding

DAVID MOEWS

ABSTRACT. We define and solve a two-player perfect information game, the *coin-sliding game*. One reason why this game is of interest is that its positions generate a large family of infinitesimals in the group of two-player partizan combinatorial games under disjunctive composition.

1. The Simplest Form of the Game

Consider the following game, played by two players, Left and Right: Coins of various (positive numeric) monetary values, colored red or blue, are placed on a semi-infinite strip. We call the red coins Right's and the blue coins Left's. The playing field and possible moves are indicated in Figure 1. Each player can, on his turn, either slide one of his coins down one square, or remove one of his opponent's coins from the strip. Each player gets to keep all the money he moves off the bottom of the strip. The winner is the player who ends up with the most money, or, in the event of a tie, the player who moved last.

Since each player moves just one coin at each turn, the overall game is a disjunctive composition of subgames corresponding to the individual positions. In fact, we can assign a partizan combinatorial game, in the sense of *Winning Ways* [Berlekamp et al. 1982], Chapters 1–8, or *On Numbers and Games* [Conway 1976], to each coin on a square, and the overall game will then be the sum of these individual games, which we call *terms*. Figure 1 has names for these terms [Berlekamp and Wolfe 1994, §§ 4.1, 4.11; Wolfe 1994, §§ 3.1, 3.10]. We have also numbered our squares, with square 0 at the bottom. When we refer to these squares by number, we will also call them *rows* (rows with more than one square will appear later.) In the figure and in the rest of the paper,

$$+_x \text{ means the game } 0 \,\|\, 0 \,|\, -x,$$
$$0^2 \,|\, +_x \text{ means the game } 0 \,\|\, 0 \,|\, +_x,$$
$$0^3 \,|\, +_x \text{ means the game } 0 \,\|\|\, 0 \,\|\, 0 \,|\, +_x,$$

and so on.

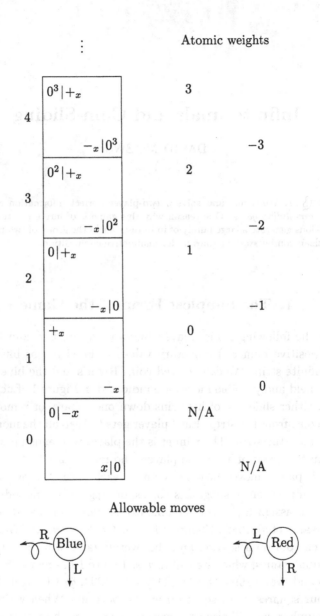

Figure 1. The simplest form of coin-sliding. The name for the term corresponding to a red coin of value x in a square is in the upper left corner of the square, and for a blue coin, the corresponding term name is in the lower right.

Replacing a coin on a square with a coin of the same monetary value and other color on the same square replaces a term by its negative. Hence

$$-_x \text{ is the negative of } +_x,$$
$$-_x|0^2 \text{ is the negative of } 0^2|+_x,$$
$$-_x|0^3 \text{ is the negative of } 0^3|+_x,$$

and so on.

2. Infinitesimals and Atomic Weight

A partizan combinatorial game (henceforth just a "game") G is *infinitesimal* if $-\varepsilon < G < \varepsilon$ for every positive rational number ε. The simplest nonzero infinitesimal is $* = 0|0$, but we will see many others in the sequel. All our coin-sliding terms are infinitesimal except for those coming from row 0—in fact, except for the row 0 terms, Right's terms (terms coming from Right's coins) are positive infinitesimals and Left's terms are negative infinitesimals. For the games we are considering, which always end in a bounded amount of time, infinitesimals are also called *small* (*On Numbers and Games*, Chapter 9), and a game is called *all small* if it and all its subpositions are small. All small games G that end in a bounded amount of time have an *atomic weight G''*, defined as follows (*Winning Ways*, Chapter 8; *On Numbers and Games*, Chapter 16):

$$G'' = \{G^{L''} - 2 | G^{R''} + 2\},$$

except when more than one integer is permitted by this definition. (By a number n being *permitted* we mean that it is $\rhd$ every $G^{L''} - 2$ and is $\lhd$ every $G^{R''} + 2$, where $G \rhd H$ when $G - H$ is positive or fuzzy, that is, is a first-player win for Left, and $G \lhd H$ when $G - H$ is negative or fuzzy, that is, is a first-player win for Right.) If more than one integer is permitted we pick the

smallest permitted, zero, or largest permitted

integer according to whether G is

less than, incomparable with, or greater than

any $*m$ not equal to a subposition of G—such a $*m$ is called a *remote star* (*Winning Ways*, Chapter 8; *On Numbers and Games*, Chapter 16).

Atomic weight is then an additive homomorphism—that is, $(G+H)'' = G'' + H''$, $(-G)'' = -G''$, and $0'' = 0$. The positive all small game $\uparrow = 0|*$ is the natural unit of atomic weight, with $\uparrow'' = 1$. We write $\downarrow = -\uparrow$, so $\downarrow'' = -1$. Positive atomic weights are favorable for Left. In fact, if $G'' \geq 2$, then $G > 0$, and if $G'' \rhd 0$, then $G \rhd 0$. Similarly, if $G'' \leq -2$, then $G < 0$, and if $G'' \lhd 0$, then $G \lhd 0$.

Given these facts, we can compute the atomic weight of $+_{n \cdot \uparrow}$, for n large:

$$\{0|-n \cdot \uparrow\}'' = \{0-2|-n+2\} = \{-2|2-n\}$$

and
$$+_{n\cdot\uparrow}'' = \{0{-}2\,|\,\{-2\,|\,2{-}n\}{+}2\} = \{-2\,\|\,0\,|\,4{-}n\},$$
and here 0 and -1 are permitted; since $+_{n\cdot\uparrow} \| *2$, we find that $+_{n\cdot\uparrow}'' = 0$. Now $+_x$ is a positive infinitesimal that decreases very rapidly as x increases. Hence it is reasonable to take $+_x'' = 0$ for the positive numbers x that we encounter in coin-sliding, and we do so. We can then compute the atomic weights for all our coin-sliding terms except those coming from row 0. The results are given in Figure 1, the upper left atomic weight being for Right's terms and the lower right being for Left's terms.

3. The Atomic Weight Strategy

Each player can observe that sliding down one of his own coins usually changes the atomic weight by 1 in his favor, and that removing one of his opponent's coins usually changes the atomic weight in his disfavor. In conjunction with the idea that moves that immediately give one money or prevent one's opponent from getting money are more important than other moves, this gives the *atomic weight strategy*:

(i) If the opponent has a coin in row 0, remove it.
(ii) Otherwise, if we have a coin in row 0, slide it down.
(iii) Otherwise, if we have a coin anywhere, slide it down.
(iv) Otherwise, remove any one of the opponent's coins.

Suppose that Left starts from an infinitesimal position (which we assume to have all coins in row 1 or above) and plays the atomic weight strategy, and that he has coins somewhere on the board. He will then usually increase the atomic weight by 1 on his move, and Right's return can decrease the atomic weight by no more than 1. The only exceptions to this are when coins are moved into row 0, in which case the position becomes non-infinitesimal. Right can repeatedly move his coins into row 0, which Left will repeatedly remove; after both players have moved, there is no change in the atomic weight, because row 1 coins had atomic weight 0 to begin with. Also, Left himself may slide a coin into row 0. Right can respond at first by sliding his own coins into row 0, which Left will immediately remove, but eventually he must remove Left's coin or lose the game. Again, there is no net change in atomic weight after the exchanges have ended.

If Left starts from an infinitesimal position with atomic weight at least 1, then, and plays the atomic weight strategy, he will eventually come to an infinitesimal position that also has atomic weight at least 1, but that has none of his own coins. There must then be an opponent's coin present, which he will remove, and Right's subsequent moves will slide his own coins down, which will always leave Left a coin to remove on his turn. Thus, Left will win.

If we start from an infinitesimal position, then, the atomic weight strategy will win for Left if the total atomic weight is at least 1 and Left moves first.

Since Right can decrease the atomic weight by no more than 1 on his turn, the atomic weight strategy will also win for Left if the total atomic weight is at least 2, whoever moves first. Similarly, it will win for Right if the total atomic weight is -2 or less, or if it is -1 or less and Right moves first.

The atomic weight strategy thus furnishes a proof of our assertions about atomic weight—namely, that $G'' \geq 2$ implies $G > 0$, $G'' \triangleright 0$ implies $G \triangleright 0$, $G'' \leq -2$ implies $G < 0$, and $G'' \triangleleft\!\mid 0$ implies $G \triangleleft\!\mid 0$—in the special case of simple coin-sliding. We need a more complicated strategy, the *desirability strategy*, to show who wins in the remaining cases: positions with atomic weight -1, Left starting, atomic weight 1, Right starting, and atomic weight 0, either player starting.

4. The Desirability Strategy

The atomic weight strategy takes no notice of the monetary value of the coins, but this will obviously influence our strategy. The desirability order in Figure 2 shows what terms we prefer to move on; it can be summarized as follows:

(i) A term coming from rows 0 or 1 is always more desirable than a term coming from row 2 or above.

(ii) Among terms coming from rows 0 or 1, terms with bigger monetary values are more desirable.

(iii) Among terms coming from row 2 or above, terms with smaller monetary values are more desirable.

(iv) If none of the above three rules apply, terms from a row closer to row 2 are more desirable.

(v) If none of the above four rules apply, the terms are equal or negatives of each other and are equally desirable.

$$\ldots;$$
$$+_3, \; -_3; \; \; 0|-3, \, 3|0; \; \ldots;$$
$$+_2, \; -_2; \; \; 0|-2, \, 2|0; \; \ldots;$$
$$+_1, \; -_1; \; \; 0|-1, \, 1|0; \; \ldots;$$
$$\ldots;$$
$$0|+_1, \, -_1|0; \; \; 0^2|+_1, \, -_1|0^2; \; \; 0^3|+_1, \, -_1|0^3; \; \ldots;$$
$$\ldots;$$
$$0|+_2, \, -_2|0; \; \; 0^2|+_2, \, -_2|0^2; \; \; 0^3|+_2, \, -_2|0^3; \; \ldots;$$
$$\ldots;$$
$$0|+_3, \, -_3|0; \; \; 0^2|+_3, \, -_3|0^2; \; \; 0^3|+_3, \, -_3|0^3; \; \ldots;$$
$$\ldots$$

Figure 2. Simple coin terms ordered from most desirable to least desirable (reading in the normal way, from left to right and from top down). A comma separates terms of equal desirability, and a semicolon those of unequal desirability.

Given this ordering, we can formulate the *desirability strategy*:

(i) If, on the board, there are coins that belong to us or are in row 0, move on the one with the most desirable term.

(ii) Otherwise, remove any one of the opponent's coins.

When applying this strategy, pairs of cancelling terms—that is, pairs of terms that come from a red and a blue coin of the same monetary value on the same square—must be ignored. (If the opponent makes a move in a pair of cancelling terms, the player should respond with the corresponding move in the other.) One can make a complete classification of who wins for all infinitesimal positions, with either player starting. Given this classification, it can be shown that the desirability strategy is optimal: if a player starts from an infinitesimal position won for him and plays it thereafter, it will win for him. (The classification is somewhat complicated, but we can say, for example, that an atomic weight 1 position is positive if the least desirable term in it (after cancelling pairs of terms) is Left's, or that in an atomic weight 0 position where Left has the least desirable term, Left can win provided that he moves first, provided that this least desirable term is not of the form $-_x$.)

5. A Generalization

We generalize the game: instead of playing on a semi-infinite strip, we now play on an infinite quarter-plane, as shown in Figure 3. Also, we allow coins to have zero monetary values as well as positive numeric monetary values. As before, the red coins are Right's and the blue are Left's. Each player can move each of his own coins from one row to the leftmost square on the next row down (or off the bottom of the board if the coin started in the bottom row.) Each player can move each of his opponent's coins one square to the left, or off the board if the coin starts in the leftmost column. Each player gets to keep the money he moves off the bottom (but not the left) of the board.

As before, each square on the board is marked with names [Berlekamp and Wolfe 1994, §§ 4.1, 4.11; Wolfe 1991, §§ 3.1, 3.10] of terms coming from coins on the square, a red coin of value x producing the term named in the upper left, and a blue coin producing the term named in the lower right. For a term $G = 0 \mid H$ coming from a red coin in the leftmost column, $G^{\rightarrow n}$ is the term for a red coin of the same monetary value $n - 1$ columns to the right; it is defined by taking $G^{\rightarrow 1} = 0 \mid H$ and $G^{\rightarrow n+1} = G^{\rightarrow n} \mid H$ for all $n \in \mathbb{Z}_{>0}$. For a leftmost-column blue coin term $G = H \mid 0$, $G^{\rightarrow n}$, the term for a blue coin of the same monetary value $n - 1$ columns to the right, is defined by $G^{\rightarrow n} = -\{0 \mid -H\}^{\rightarrow n}$. As before, except for the row 0 terms, Right's terms are positive infinitesimals and Left's terms are negative infinitesimals.

To make this version of the game easier to analyze, we alter the playing field and rules somewhat. Figure 4 shows our revision of the playing field. In the

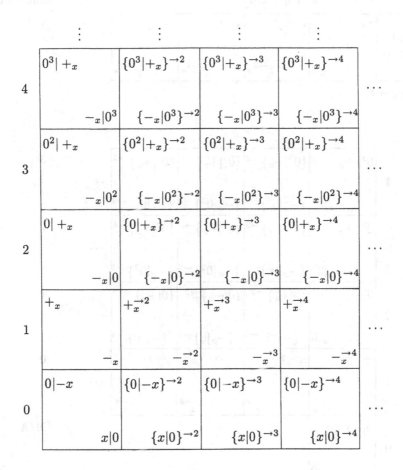

Players' possessions

Allowable moves

Figure 3. Another playing field for coin-sliding.

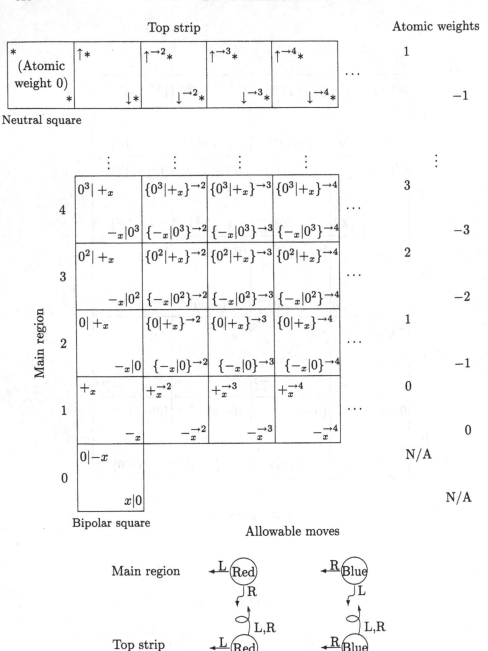

Figure 4. The revised playing field for coin-sliding.

main region, we have omitted the bottom row squares other than the leftmost. This is because a red (or blue) coin n squares to the right of the bipolar square behaves the same in play as a blue (or red) coin of the same monetary value n squares above the bipolar square together with a red (or blue) coin of the same monetary value that Right (or Left) already has; in other words, $\{0|-x\}^{\to n+1} = -x+\{x|0^{n+1}\}$, and $\{x|0\}^{\to n+1} = x+\{0^{n+1}|-x\}$.

We have also created a new row of squares at the top, the *top strip*. All zero monetary value coins are now placed in the top strip. The top strip behaves peculiarly in that both Left and Right are allowed to remove both red and blue coins from the strip at any time, as well as Left (or Right) being able to move red (or blue) coins left one square. The top strip terms $\uparrow^{\to n}*$ and $\downarrow^{\to n}*$ are defined by $\uparrow^{\to n}* = \uparrow^{\to n} + *$ and $\downarrow^{\to n}* = \downarrow^{\to n} + *$, where $\uparrow^{\to n}$ and $\downarrow^{\to n}$ are as above: $\uparrow^{\to 1} = \uparrow$, $\uparrow^{\to n+1} = \uparrow^{\to n}|*$ (for $n \geq 1$), and $\downarrow^{\to n} = -\uparrow^{\to n}$. It can be shown that a coin of zero value on the main board, m rows above and n columns to the right of the bipolar square, behaves the same as one coin on the neutral square, one coin $n+1$ squares to the right of the neutral square (if $m > 0$), and $m-1$ coins one square to the right of the neutral square (if $m > 1$); this fact is the reason for these definitions. (We assume here that the transformation in the last paragraph has already been made, so that $n = 0$ if $m = 0$.)

6. Ordinal Sums

The new terms we have introduced can be viewed as *ordinal sums* of our old terms with integers. The ordinal sum $G:H$ of games G and H, is defined (*On Numbers and Games*, Chapter 15) by

$$G:H = \{G^L, G:H^L | G^R, G:H^R\}.$$

We find that $G:0 = G$. In fact, $G:H$ is usually very close to G; Norton's Lemma (*On Numbers and Games*, Chapter 16) says that $G:H$ and G have the same order-relations with all games $K \neq G$ that do not contain a game equal to G as a subposition.

The significance here is that

$$G^{\to k} = \begin{cases} G:(k-1) & \text{if } G > 0, k \geq 1, \\ G:(-(k-1)) & \text{if } G < 0, k \geq 1, \end{cases}$$

so coin terms don't change much with a change in the column of the coin. We define more infinitesimals to tell us just how much they differ (see also [Berlekamp and Wolfe 1994, § 4.11; Wolfe 1991, § 3.10]):

$$G^1 = G,$$
$$G^k = G^{\to k} - G^{\to k-1} \quad \text{for } k \geq 2.$$

The order relation between $G:H$ and $G:K$ is always the same as that between H and K. Since $G^k = (G:(k-1)) - (G:(k-2))$ for $k \geq 2$ and $G > 0$, it follows that G^k is always positive for $G > 0$. After generalizing our desirability strategy to our new version of coin-sliding, we will be able to compare the G^k's with each other, where G can be $\uparrow$, $\{0^m|+_x\}$, or $+_x$.

7. The Desirability Strategy Generalized

The desirability ordering for our new terms is the same as for our original terms, except that

(i) all top strip terms are less desirable than all main region terms, and

(ii) among terms coming from coins in the same row and of the same monetary value, terms coming from coins on the right are more desirable than terms coming from coins on the left.

The resultant desirability ordering is shown in Figure 5. The desirability strategy that goes with this ordering is somewhat more complicated than in the simple case, and now runs as follows:

(i) If, on the board, there are coins not in the neutral square that belong to us or are in row 0, move on the one giving the most desirable term, unless it is "crucial", giving a term of $G^{\to k}$, say, and there is a $-G^{\to l}$ present, for some $l > k \geq 1$.

$$\ldots;$$
$$\ldots;\ +_3^{\to 3},\ -_3^{\to 3};\ +_3^{\to 2},\ -_3^{\to 2};\ +_3,\ -_3;\ 0|-3,\ 3|0;$$
$$\ldots;$$
$$\ldots;\ +_2^{\to 3},\ -_2^{\to 3};\ +_2^{\to 2},\ -_2^{\to 2};\ +_2,\ -_2;\ 0|-2,\ 2|0;$$
$$\ldots;$$
$$\ldots;\ +_1^{\to 3},\ -_1^{\to 3};\ +_1^{\to 2},\ -_1^{\to 2};\ +_1,\ -_1;\ 0|-1,\ 1|0;$$
$$\ldots;$$
$$\ldots;\ \{0|+_1\}^{\to 3},\ \{-_1|0\}^{\to 3};\ \{0|+_1\}^{\to 2},\ \{-_1|0\}^{\to 2};\ 0|+_1,\ -_1|0;$$
$$\ldots;\ \{0^2|+_1\}^{\to 3},\ \{-_1|0^2\}^{\to 3};\ \{0^2|+_1\}^{\to 2},\ \{-_1|0^2\}^{\to 2};\ 0^2|+_1,\ -_1|0^2;$$
$$\ldots;\ \{0^3|+_1\}^{\to 3},\ \{-_1|0^3\}^{\to 3};\ \{0^3|+_1\}^{\to 2},\ \{-_1|0^3\}^{\to 2};\ 0^3|+_1,\ -_1|0^3;$$
$$\ldots;$$
$$\ldots;\ \{0|+_2\}^{\to 3},\ \{-_2|0\}^{\to 3};\ \{0|+_2\}^{\to 2},\ \{-_2|0\}^{\to 2};\ 0|+_2,\ -_2|0;$$
$$\ldots;\ \{0^2|+_2\}^{\to 3},\ \{-_2|0^2\}^{\to 3};\ \{0^2|+_2\}^{\to 2},\ \{-_2|0^2\}^{\to 2};\ 0^2|+_2,\ -_2|0^2;$$
$$\ldots;\ \{0^3|+_2\}^{\to 3},\ \{-_2|0^3\}^{\to 3};\ \{0^3|+_2\}^{\to 2},\ \{-_2|0^3\}^{\to 2};\ 0^3|+_2,\ -_2|0^3;$$
$$\ldots;$$
$$\ldots;$$
$$\ldots;\ \uparrow^{\to 3}*,\ \downarrow^{\to 3}*;\ \uparrow^{\to 2}*,\ \downarrow^{\to 2}*;\ \uparrow*,\ \downarrow*;$$
$$*$$

Figure 5. Terms ordered from most desirable to least desirable (reading in the normal way, from left to right and from top down). A comma separates terms of equal desirability, and a semicolon those of unequal desirability.

(a) In this case, move on the coin as above giving the next most desirable term, if present.

(b) If such a coin is not present, move the $-G^{\rightarrow l}$ coin one square leftwards.

(ii) Otherwise, if there are exactly two coins in the top strip, move a coin in the top strip and not in the neutral square one square leftwards.

(iii) Otherwise, if there are coins in the top strip, remove any one.

(iv) Otherwise, move on any coin.

As before, pairs of cancelling coins must be ignored when applying this strategy. Since, on the neutral square, the color of a coin is irrelevant, we must also cancel a red coin with a red coin or a blue coin with a blue coin when more than one coin appears on the neutral square.

This desirability strategy can be proved to be optimal as before by first deriving a complete classification of who wins and loses for all infinitesimal positions, whoever starts. This classification, which provides the definition of "crucial", can be found in [Moews 1993, Chapter 5] or [Moews a]. Typically, no term is crucial. The classification is rather more complicated than in the simple case, but it is still the case that Left can win if he has an atomic weight 1 position and the least desirable term, or an atomic weight 0 position, the least desirable term, and the first move, provided that this least desirable term is not of the form $-\vec{}_x^{\rightarrow k}$. (Here, if there is a $*$ term present after cancellation, it should be counted as Left's, regardless of the color of coin it comes from.)

8. A Resultant Ordering of Some Infinitesimals

Our classification of positions lets us compare the infinitesimals $\uparrow^{k+1}$, $+_x^k$, and $\{0^m|+_x\}^{k+1}$ with each other. For positive x and positive integers k and m, these infinitesimals are all positive and have atomic weight 0. Also, it follows from our classification that $G^k \gg G^{k+1}$ for $k \geq 1$ and $G = \uparrow$, $+_x$ or $0^m|+_x$; here by $H \gg K$ we mean that $H > r \cdot K$ for all positive integers r. Define $H \, \|\|\| \, K$ to mean that $r \cdot H \, \| \, s \cdot K$ for all positive integers r and s. Then we can say as well that

$$
\begin{aligned}
\uparrow^k &\gg \{0^m|+_x\}^l && \text{for } l \geq 2, \\
\{0^m|+_x\}^k &\gg +_y^l && \text{for } m \geq 2, \\
\{0^m|+_x\}^k &\gg \{0^n|+_y\}^l && \text{for } x > y, \, m \geq 2, \, l \geq 2, \\
\{0^m|+_x\}^k &\gg \{0^n|+_x\}^l && \text{for } m > n, \, m \geq 2, \, l \geq 2, \\
+_x^k &\gg +_y^l && \text{for } x < y, \\
\{0^n|+_x\}^k &\, \|\|\| \, \{0|+_y\}^l && \text{for } x < y, \, k \geq 2, \, l \geq 2, \\
\{0|+_x\}^k &\, \|\|\| \, +_y^l && \text{for } x > y, \, k \geq 2, \\
\{0|+_x\}^k &\, \|\|\| \, +_x && \text{for } k \geq 2, \\
\{0|+_x\}^k &\gg +_x^l && \text{for } l \geq 2.
\end{aligned}
$$

Here are examples of some of these relations between infinitesimals:

$$\uparrow \gg \uparrow^2 \gg \uparrow^3 \gg \cdots$$
$$\gg \cdots$$
$$\gg \{0^4|+_3\}^2 \gg \{0^4|+_3\}^3 \gg \{0^4|+_3\}^4 \gg \cdots$$
$$\gg \{0^3|+_3\}^2 \gg \{0^3|+_3\}^3 \gg \{0^3|+_3\}^4 \gg \cdots$$
$$\gg \{0^2|+_3\}^2 \gg \{0^2|+_3\}^3 \gg \{0^2|+_3\}^4 \gg \cdots$$
$$\gg \cdots$$
$$\gg \{0^4|+_2\}^2 \gg \{0^4|+_2\}^3 \gg \{0^4|+_2\}^4 \gg \cdots$$
$$\gg \{0^3|+_2\}^2 \gg \{0^3|+_2\}^3 \gg \{0^3|+_2\}^4 \gg \cdots$$
$$\gg \{0^2|+_2\}^2 \gg \{0^2|+_2\}^3 \gg \{0^2|+_2\}^4 \gg \cdots$$
$$\gg \cdots$$
$$\gg \{0^4|+_1\}^2 \gg \{0^4|+_1\}^3 \gg \{0^4|+_1\}^4 \gg \cdots$$
$$\gg \{0^3|+_1\}^2 \gg \{0^3|+_1\}^3 \gg \{0^3|+_1\}^4 \gg \cdots$$
$$\gg \{0^2|+_1\}^2 \gg \{0^2|+_1\}^3 \gg \{0^2|+_1\}^4 \gg \cdots$$
$$\gg \cdots$$
$$\gg +_1 \gg +_1^2 \gg +_1^3 \gg \cdots$$
$$\gg +_2 \gg +_2^2 \gg +_2^3 \gg \cdots$$
$$\gg +_3 \gg +_3^2 \gg +_3^3 \gg \cdots,$$

$$\{0|+_3\}^k \| \| \{0^n|+_2\}^l, \{0^n|+_1\}^l, +_1^j, +_2^j, +_3 \quad \text{for } k \geq 2, l \geq 2,$$

$$\{0|+_2\}^k \| \| \{0^n|+_1\}^l, +_1^j, +_2 \quad \text{for } k \geq 2, l \geq 2,$$

$$\{0|+_1\}^k \| \| +_1 \quad \text{for } k \geq 2.$$

Acknowledgement

Much of this material was originally presented in Chapter 5 of my thesis [Moews 1993], which was supervised by E. Berlekamp.

References

[Berlekamp et al. 1982] E. R. Berlekamp, J. H. Conway and R. K. Guy, *Winning Ways for Your Mathematical Plays*, Academic Press, London, 1982.

[Berlekamp and Wolfe 1994] E. Berlekamp and D. Wolfe, *Mathematical Go: Chilling Gets the Last Point*, A. K. Peters, Wellesley, MA, 1994.

[Conway 1976] J. H. Conway, *On Numbers and Games*, Academic Press, London, 1976.

[Moews 1993] David Moews. "On some combinatorial games connected with Go", Ph.D. thesis, University of California, Berkeley, 1993.

[Moews a] David Moews, "Coin-sliding and Go", to appear in *Theoret. Comp. Sci.*

[Wolfe 1991] David Wolfe, "Mathematics of Go: Chilling corridors", Ph.D. thesis, University of California, Berkeley, 1991.

DAVID MOEWS
34 CIRCLE DRIVE
MANSFIELD CENTER, CT 06250
  dmoews@xraysgi.ims.uconn.edu

Games of No Chance
MSRI Publications
Volume **29**, 1996

Geography
Played on Products of Directed Cycles

RICHARD J. NOWAKOWSKI AND DAVID G. POOLE

ABSTRACT. We consider the game of Geography played on $G = C_n \times C_m$, the product of two directed cycles. The analysis is easy for $n = 2$ and in the case where both n and m are even. Most of the paper is devoted to solving the game on the graphs $C_n \times C_3$.

1. Introduction

The game called *Kotzig's Nim* in *Winning Ways* [Berlekamp et al. 1982] and *Modular Nim* in [Fraenkel et al. 1995] consists of a directed cycle of length n with the vertices labelled 0 through $n-1$, a coin placed initially on vertex 0, and a set of integers called the move set. There are two players, who alternate moves; a move consists of moving the coin from the vertex i on which it currently resides to vertex $i + m \bmod n$, where m is a member of the move set. However, the coin can only land on a vertex once. Thus, the game is finite. The last player to move wins. Most of the known results concern themselves with move sets of small cardinality and consisting of small numbers (see p. 481 of *Winning Ways*, and [Fraenkel et al. 1995]).

Obviously, this game can be extended to more general directed graphs, the move set being indicated by directed edges, for clarity. This has become known as *Geography* [Fraenkel et al. 1993; Fraenkel and Simonson 1993].

The *Grundy value* of a game G, denoted $\mathcal{G}(G)$, is either $\mathcal{P}$ for a previous-player win or $\mathcal{N}$ for a next-player win. Throughout, we call the first player Algois and the second Berol.

One general strategy for Geography can be identified quickly. A set A of independent edges in an undirected graph is called a *perfect matching* if every vertex is incident with an edge in A. A set of edges $\{(a_i, b_i) : i = 1 \ldots, n\}$ is a *directed perfect matching* of a directed graph G if the edges form a matching of the underlying undirected graph and if, for each $i = 1, \ldots, n$, the vertex b_i has

Nowakowski was partially supported by a grant from the NSERC.

no edges directed to b_j for any $j \neq i$. The following is a variant of results found in [Fraenkel et al. 1993] and [Bondy and Murty 1976, p. 71, problem 5.1.4].

THEOREM 1.1. *Let* $\{(a_i, b_i) : i = 1, \ldots, n\}$ *be a directed perfect matching of a graph* G. *If the coin is placed initially on* a_i *for some* i, *then* $\mathcal{G}(G) = \mathcal{N}$.

PROOF. Algois's strategy is always to move from the present position a along this matching edge. He has this option available on the first move since e is the start of a matching edge. This forces Berol to be the first to move to the start of a new matching edge. $\square$

In the sequel we concentrate on those graphs that are the products of two cycles.

2. Cartesian Products of Directed Cycles

We consider the cases where $G = C_n \times C_m$. Throughout, we will put $C_k = \{0, 1, 2, \ldots, k-1\}$. The graph $G = C_n \times C_m$ has vertex set

$$\{(i, j) : 0 \leq i \leq n-1,\ 0 \leq j \leq m-1\},$$

and (i, j) is adjacent to (k, l) (that is, there is an edge directed from (i, j) to (k, l)) if i is adjacent to k and $j = l$, or if $i = k$ and j is adjacent to l. Throughout this section (i, j) will be taken modulo n in the first coordinate and modulo m in the second.

When a vertex (a, i) is occupied, its Grundy value $\mathcal{G}(a, i)$ depends on the preceding moves. Since the context will indicate the history of the position, we do not add anything to our notation to take this into account. However, a useful notation is $[a, i]$, which will indicate that this is the first time that a vertex with a as the first coordinate has been visited. Let $\mathcal{G}[b, i]$ refer to the Grundy value of the position obtained by moving to $[b, i]$ from $(b-1, i)$. Again, the context will fill in the history.

THEOREM 2.1. *Let* $G = C_n \times C_m$. *If* $n = 2$, *or if both* n *and* m *are even, then* $\mathcal{G}(C_n \times C_m) = \mathcal{N}$.

PROOF. If $n = 2$, Algois follows a variant of the strategy given in Theorem 1.1. He plays moves that only change the first coordinate. He can do this on the first move. Berol has to move so as to change the second coordinate whereupon Algois can again change the first coordinate.

When both n and m are even, Algois always moves from a vertex where both coordinates are of the same parity to a vertex where the parities are different. Let $P = \{(ab, a(b+1)) : a \equiv b \bmod 2\}$. The set P is a directed perfect matching for G and the result follows from Theorem 1.1. $\square$

The interesting cases are therefore the ones where one or both of n and m is odd. The rest of the section is devoted to showing the following result:

THEOREM 2.2. *Let* $G = C_n \times C_3$, *where* $n \geq 3$. *Then the Grundy values for* G *are as follows, where* n *is taken modulo* 42:

n	$\mathcal{G}$	n	$\mathcal{G}$	n	$\mathcal{G}$	n	$\mathcal{G}$	n	$\mathcal{G}$	n	$\mathcal{G}$	n	$\mathcal{G}$
1	$\mathcal{P}$	7	$\mathcal{P}$	13	$\mathcal{N}$	19	$\mathcal{N}$	25	$\mathcal{N}$	31	$\mathcal{P}$	37	$\mathcal{P}$
2	$\mathcal{N}$	8	$\mathcal{P}$	14	$\mathcal{P}$	20	$\mathcal{P}$	26	$\mathcal{P}$	32	$\mathcal{N}$	38	$\mathcal{N}$
3	$\mathcal{P}$	9	$\mathcal{P}$	15	$\mathcal{N}$	21	$\mathcal{N}$	27	$\mathcal{N}$	33	$\mathcal{P}$	39	$\mathcal{P}$
4	$\mathcal{N}$	10	$\mathcal{N}$	16	$\mathcal{N}$	22	$\mathcal{N}$	28	$\mathcal{N}$	34	$\mathcal{N}$	40	$\mathcal{N}$
5	$\mathcal{P}$	11	$\mathcal{N}$	17	$\mathcal{N}$	23	$\mathcal{N}$	29	$\mathcal{P}$	35	$\mathcal{P}$	41	$\mathcal{P}$
6	$\mathcal{N}$	12	$\mathcal{P}$	18	$\mathcal{P}$	24	$\mathcal{P}$	30	$\mathcal{P}$	36	$\mathcal{N}$	42	$\mathcal{N}$

PROOF. The vertices of the directed graph will be denoted by (a, i), where a is taken modulo n and i modulo 3. During the game, the subgraph consisting of the unused vertices we call the *field*; this excludes the vertex occupied by the coin.

We define two move sequences:

$$S = (a, i), (a+1, i), (a+1, i+1), (a+1, i+2),$$
$$H = (a, i), (a, i+1), (a+1, i+1), (a+1, i+2).$$

Both are called *closing-off* sequences. Such a sequence changes the field graph from being strongly connected to just connected. This also means that the game can only go around (in the first coordinate) once more. Most of the proof revolves around whether and when one of the players should simplify the game by completing one of these closing-off sequences. The next lemma is technical but develops a useful tool for the later analysis.

LEMMA 2.3. *Suppose that player Y completes a closing-off sequence by moving to vertex (a, i). Then $\mathcal{G}(a-1, i) = \mathcal{P}$. Moreover, if (b, j) is in the field graph and player Y has previously played to $(b, j+1)$, then $\mathcal{G}(b, j) = \mathcal{P}$.*

PROOF. Suppose that a closing-off sequence has been completed by player Y moving to (a, i). Note that in either case, player X had moved to $(a-1, i+1)$. Then only $(a-1, i)$ has no outgoing edges and thus it is a $\mathcal{P}$ position.

The induced subgraph H of the field graph whose vertices have first coordinate b, for $0 \leq b \leq a$ (not taken modulo n), is bipartite. By working backwards and inducting on the distance from (b, j) to $(a-1, i)$, we see that all vertices in the same color class as $(a-1, i)$ are $\mathcal{P}$ positions, and all the other vertices are $\mathcal{N}$ positions. Adjoining to H the vertices used by the players but adding only the edges into these vertices results in another bipartite graph. All the vertices that X moved to are in one color class, and those moved to by Y are in the other. Since $(a-1, i)$ has an edge to a vertex once played to by X all the other such vertices are in the same color class and, thus, are $\mathcal{P}$ positions. □

The next result considers the other case in which the Grundy values can be easily found.

LEMMA 2.4. *If $\mathcal{G}[a, i] = \mathcal{P}$ and $\mathcal{G}[a, i+1] = \mathcal{N} = \mathcal{G}[a, i-1]$, it follows that $\mathcal{G}[a-1, i-1] = \mathcal{P}$ and $\mathcal{G}[a-1, i] = \mathcal{N} = \mathcal{G}[a-1, i+1]$.*

PROOF. Suppose that in a game it happens that $\mathcal{G}[a, i] = \mathcal{P}$ and $\mathcal{G}[a, i+1] = \mathcal{N} = \mathcal{G}[a, i-1]$. Then $\mathcal{G}(a-1, i) = \mathcal{N}$, since it has $[a, i]$ as a follower. The followers of $[a-1, i-1]$ are $[a, i-1]$ and $(a-1, i)$, both of which are $\mathcal{N}$ positions; thus $\mathcal{G}[a, i-1] = \mathcal{P}$.

The followers of $[a-1, i+1]$ are $[a, i+1]$, which is an $\mathcal{N}$ position, and $(a-1, i-1)$, if it exists. From this latter position the followers are both $\mathcal{N}$ positions; thus $\mathcal{G}[a, i+1] = \mathcal{N}$.   □

Algois has two possible first moves. The status of moving from $(0,0)$ to $(0,1)$ is the subject of the next lemma.

LEMMA 2.5. *Algois wins by moving from $(0,0)$ to $(0,1)$ if and only if $n \equiv 4$ mod 6.*

PROOF. Suppose that the first two moves are $(0,0) \to (0,1) \to (0,2)$. Then $\mathcal{G}[n-1, i] = \mathcal{P}$, since when the players reach that position there are exactly two moves left in the game.

Now, $\mathcal{G}[n-j, i]$ is a $\mathcal{P}$ position if j is odd and an $\mathcal{N}$ position otherwise. The proof is as follows. We assume it is X to move. If $\mathcal{G}[n-j, i] = \mathcal{P}$ with j odd, then $\mathcal{G}[n-j+1, i] = \mathcal{N}$, and X immediately moves to $(n-j, i)$. If $\mathcal{G}[n-j, i] = \mathcal{P}$ with j even, X has two options: (i) to move to $(n-j+1, i)$, which is not possible if $j = 1$ and otherwise, by induction, is an $\mathcal{N}$ position; or (ii) to move to $(n-j, i+1)$; but the good reply to the latter is for Y to move to $(n-j, i+2)$, leaving X with only a move to $(n-j+1, i+2)$, which again is an $\mathcal{N}$ position.

Thus Algois will move to $(0,1)$ only if n is even; otherwise he loses when Berol moves to $(0,2)$.

Assume n is even. Berol can also move to $(1,1)$ and Algois can close off by moving to $(1,2)$. At this point in the game, $\mathcal{G}(0,2) = \mathcal{P}$, and so $\mathcal{G}(n-1, 2) = \mathcal{N}$. If $(n-1, 1)$ is the first position in the $(n-1)$-st column to be occupied, its only follower is $(n-1, 2)$, and thus $\mathcal{G}[n-1, 1] = \mathcal{P}$. Similarly, it follows that $\mathcal{G}[n-1, 0] = \mathcal{N}$.

Thus Algois will win if $\mathcal{G}[2, 0] = \mathcal{P}$ and $\mathcal{G}[2, 1] = \mathcal{N} = \mathcal{G}[2, 2]$. From Lemma 2.4 the $\mathcal{P}$ positions are periodic with period 3 in the first coordinate. Since $\mathcal{G}[n-2, 0] = \mathcal{P}$, Algois wins when $n - 2 - 2 \equiv 0$ mod 3, that is, when $n \equiv 1$ mod 3. We have already determined that n is even; therefore Algois wins by moving from $(0,0)$ to $(0,1)$ just if $n \equiv 4$ mod 6.   □

The rest of the proof is devoted to the other case: that is, Algois moves from $(0,0)$ to $(1,0)$. The first part of the analysis deals with the question of when a player can move safely in the second coordinate.

LEMMA 2.6. *Suppose player X makes the non-closing-off move (a, i) to $(a, i+1)$. Player X will not lose to the closing-off move $(a, i+1)$ to $(a, i+2)$ if and only if either $n - a + i \equiv 0$ mod 3 and X is Algois, or $n - a + i \equiv 1$ mod 3 and X is Berol. If Y moves from $(a, i+1)$ to $(a+1, i+1)$ then in neither case can*

X win by completing the closing-off sequence by moving from $(a+1, i+1)$ *to* $(a+1, i+2)$.

PROOF. Suppose Berol has moved from $(a-1, i)$ to (a, i) and then Algois moves from (a, i) to $(a, i+1)$ and this does not complete a closing-off sequence. He will only do this if $(a, i+1)$ to $(a, i+2)$ is a losing move for Berol. If Berol does close off with this move, we have $\mathcal{G}(a-1, i+2) = \mathcal{P}$, since it has no followers.

Since Algois moved to $(a-1, i)$, Lemma 2.3 gives $\mathcal{G}(0, 1) = \mathcal{P}$. In turn, it follows that $\mathcal{G}(n-1, 0) = \mathcal{P}$ and that $\mathcal{G}[n-1, 1] = \mathcal{N} = \mathcal{G}[n-1, 2]$. By Lemma 2.4, $\mathcal{G}[a+1, i+1] = \mathcal{P}$ just if $(n-1) - a + i \equiv 2 \bmod 3$. Thus Algois will move from (a, i) to $(a, i+1)$ just if $(n-1) - a + i \equiv 2 \bmod 3$. Note that if indeed $(n-1) - a + i \equiv 2 \bmod 3$ then Berol will not close off but will move to $(a+1, i+1)$. The closing-off move to $(a+1, i+2)$ is now a bad move for Algois since from Lemma 2.3 and Lemma 2.4 we have that $\mathcal{G}[a+2, i+2] = \mathcal{P}$.

Suppose that Berol is player X. She will only move from (a, i) to $(a, i+1)$ (when this does not complete a closing-off sequence) if the move $(a, i+1)$ to $(a, i+2)$ is a losing move for Algois. As in the previous paragraph, Lemmas 2.3 and 2.4 show that this is the case when $(n-1) - a + i \equiv 0 \bmod 3$. Moreover, if Algois does not close off but moves from $(a, i+1)$ to $(a+1, i+1)$, then again from Lemma 2.3 and Lemma 2.4 Berol will lose by closing-off on her next move. □

This leads to a sequence of forcing moves that are actually three repetitions of a sequence going from (a, i) to $(a+7, i+1)$, one for each $i = 0, 1, 2$. See Figure 1.

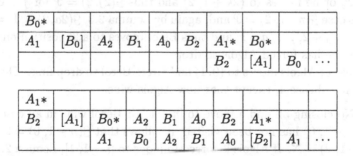

Figure 1. Forcing sequences for Algois (top) and Berol (bottom). The entry labels A and B indicate who moved to that vertex; the subscripts are $n - a + i$ mod 3. The brackets indicate the beginning and end of a sequence. In each case, the other player can determine part of the sequence, but an $*$ indicates the only time a choice is available.

Since neither can be forced into a disadvantageous closing-off move, it remains to analyze the situations when the players have to move to the last column before a closing-off move. In addition, we have to consider which player *drops* first, that is, makes a move in the second coordinate.

In what follows, the $\mathcal{N}$ and $\mathcal{P}$ positions are calculated using Lemma 2.3 as soon as the status of one position can be identified. All the information will be summarized in Table 1.

CASE 1. There is no closing-off, and there is a move to $[n-1, 0]$.

The following cases are independent of who dropped first:

$$
\begin{array}{cc|cc}
B\ A\ B & 0\ A \\
A \\
\mathcal{P}\ B & \mathcal{N}\ \mathcal{P}
\end{array}
\qquad
\begin{array}{cc|c}
B\ A & 0\ A \\
\mathcal{P}\ B & \mathcal{P} \\
A
\end{array}
\qquad
\begin{array}{cc|cc}
A\ B & 0\ A \\
\mathcal{P}\ A & \mathcal{N}\ \mathcal{P} \\
B\ \mathcal{N}
\end{array}
$$

subcase 1: B wins   subcase 2: A wins   subcase 3: A wins

The 0 indicates the initial vertex. In each case, the $\mathcal{P}$ position to the left of the vertical line is formed by the closing-off move. The other $\mathcal{P}$ and $\mathcal{N}$ positions are obtained from Lemma 2.3. No more need be said about these cases.

In the next two cases it matters who drops first.

$$
\begin{array}{cc|c}
A\ B\ A & 0\ A...B \\
B & A
\end{array}
\qquad\qquad
\begin{array}{cc|c}
A\ B\ A & 0\ A...A \\
B & B
\end{array}
$$

subcase 4: B wins   subcase 5: A wins

The situation presented in subcase 4 is a losing one for Algois. He has played to $[n-1, 0]$, so Berol is forced to move to $(n-1, 1)$. However, Berol does not have to move off the line $i = 1$: she can force Algois to have that doubtful privilege. Whenever he does, he either moves to $(n-1, 2)$ and Lemma 2.3 shows that $\mathcal{G}(0, 2) = \mathcal{P}$; or he moves to $(2k+1, 2)$ and thus $\mathcal{G}(2j, 2) = \mathcal{P}$ for $j = 0, 1, \ldots k$. But in this case $\mathcal{G}(n-1, 2) = \mathcal{P}$ and, again by Lemma 2.3, $\mathcal{G}(2a, 2) = \mathcal{P} = \mathcal{G}(2b, 2)$ for $b = j+1, j+2, \ldots, a-1$, and thus Berol moves to a $\mathcal{P}$ position on her next move after she has forced Algois to drop.

In subcase 5, Algois moves to $(0, 1)$ and forces Berol to drop first. The analysis follows as in subcase 4, except that now Algois wins.

CASE 2. No closing-off and a move to $[n-1, 1]$: Berol wins in all cases.

Since $(n-1, 0)$ has not been visited, it follows that $\mathcal{G}(n-1, 0) = \mathcal{P}$ and also that $\mathcal{G}(n-1, 2) = \mathcal{N}$. If Algois is forced to move to $[n-1, 1]$, Lemma 2.3 implies that $\mathcal{G}(0, 1) = \mathcal{P}$, and therefore Berol wins by moving to $(0, 1)$. If Berol moves to $[n-1, 1]$, Lemma 2.3 implies that $\mathcal{G}(0, 2) = \mathcal{P}$ and $\mathcal{G}(0, 1) = \mathcal{N}$. Thus Algois has no good move, and Berol wins again.

CASE 3. No closing-off, and a move to $[n-1, 2]$.

These cases are independent of who dropped first:

$$
\begin{array}{cc|c}
A & 0\ A \\
\mathcal{P}\ B & \mathcal{P} \\
B\ A\ B & \mathcal{P}
\end{array}
\qquad\qquad
\begin{array}{cc|c}
\mathcal{P}\ B & 0\ A \\
B\ A\ \mathcal{N} & \mathcal{P} \\
B\ A & \mathcal{P}
\end{array}
$$

subcase 6: A wins   subcase 7: B wins

Again, in each case, the $\mathcal{P}$ position to the left of the line is formed by the closing-off move. The other $\mathcal{P}$ and $\mathcal{N}$ positions are obtained from Lemma 2.3. No more need be said about these cases.

Here are the cases where it matters who drops first:

```
        | 0  A ... A  B                      | 0  A ... B  A
A B P   |    P  A              A B N   |  P        P  B
A B     | A  B ... B  P        B A     |  B        B  P

  subcase 8: A wins                   subcase 9: B wins
```

```
  B | 0  A          | 0  A ... A  B              | 0  A ... B  A
P A | N             |    P  A            N |  P        P  B
A B A               A B A | B  A    A N    A B A | B  A    B  N

subcase 10a: A wins    subcase 10b: A wins        subcase 10c: B wins
```

In subcase 8, Algois moves to $(0,2)$; then the players must remain on the bottom row, with Berol moving to $(2a+1, 2)$. Thus $\mathcal{G}(2a+1, 1) = \mathcal{P}$, and so $\mathcal{G}(n-1, 1) = \mathcal{P}$. From Lemma 2.3 we see that $\mathcal{G}(2a + 2, 2) = \mathcal{P}$, and thus Berol loses.

The analysis in subcase 9 is similar, but leads to the conclusion that Berol wins.

In subcase 10a, if Berol moves to $(n-1, 0)$ Algois is forced to move to $(n-1, 1)$. Now, $(n-2, 1)$ has no followers and so is a $\mathcal{P}$ position. By Lemma 2.3, $\mathcal{G}(0,2) = \mathcal{P}$ and so now Berol has no winning move. This is regardless of which player dropped first.

Suppose that Algois drops first, from $(2a,0)$ to $(2a,1)$. If Berol moves to $(0,2)$ instead of $(n-1, 0)$, the moves $(0,2) \mapsto (1,2) \mapsto (2,2) \ldots$ are forced until Algois moves to $(2a-1, 2)$. Thus $(2a-1, 1)$ is a $\mathcal{P}$ position, since it has no followers. Working back we see that $(n-1, 2)$ is a $\mathcal{P}$ position. Thus by Lemma 2.3, we conclude that $(2a, 2)$ is a $\mathcal{N}$ position and Berol loses.

In a similar way, we see that, if Berol drops first, from $(2a+1, 0)$ to $(2a+1, 1)$, and Berol moves to $(0,2)$, Berol wins by moving to $(0,2)$.

The preceding information is summarized in Table 1. We combine the information from these cases and the forcing sequence argument in the next tables. There are only three possible starting points for the sequences. Also, the dropping moves can only occur at those values of $n-a+i \bmod 3$ given in Lemma 2.6.

In Figure 2, we assume that Berol drops first. The arrows at the top (bottom) of the figure indicate in what positions of the top (bottom) forcing sequence Berol would win if that column were the column $n-1$. From this we see that the winning positions for Berol are 2, 3, 4, 5, 7, 9, 13, 15, 17, 18, 19 or 20 along the forcing sequences, thus repeating with period 21. The initial starting vertex could be any one of those indicated by $\Downarrow$. In any of those cases, Algois does not have a chance to drop before Berol.

		$[n-1, 0]$	$[n-1, 1]$	$[n-1, 2]$
A	ABA	B	B	A
drops	BAB	B	B	A
first	AB/AB	A	B	A
	BA/BA	A	B	B
B	ABA	A	B	B
drops	BAB	B	B	A
first	AB/AB	A	B	B
	BA/BA	A	B	B

Table 1. The results if there are no closing-off moves. The second column gives the sequence of moves, with / just before the move where the drop occurs.

The start adds 1, 3 or 5 to these values, except that there are extra congruence conditions to be satisfied: that is, since $a = i = 0$ then $n \equiv 2$, 1 or 0 modulo 3 as indicated by the number in the initial position. Thus Berol can win if $n = 21k + j$ for $j \in \{1, 3, 5, 7, 8, 9, 12, 14, 18, 20\}$ for $k \geq 0$.

In the other cases, Berol will not want to drop first. But Berol has little influence on Algois's forcing sequence. In Figure 3, we assume that Algois drops first. The arrows at the top (bottom) of the figure indicate in what positions of the top (bottom) forcing sequence Algois would win if that column were the column $n-1$. Algois will win when $n = 21k + j$ where $j \in \{11, 13, 14, 15, 16, 17, 18, 19, 21\}$.

Since the case of $n \equiv 4 \mod 6$ was settled first, the only cases left are $n \equiv 2, 6, 23, 25, 27 \mod 42$. In all of these cases neither player wishes to be the one who drops first, because they will then lose via the forcing sequences. Neither can drop until the congruence conditions are right, which happens six moves

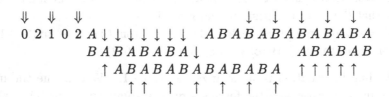

Figure 2. Berol's Winning Positions.

Figure 3. Algois's Winning Positions.

later. This allows an initial segement of up to length 12. This changes the cases
and they become equivalent to that of the case 6 less modulo 21. So modulo
42 they become respectively equivalent to the cases $n \equiv 36, 0, 17, 19, 21 \bmod 42$.
But all these are $\mathcal{N}$ positions.

To complete the proof, it is now enough to note that the cases $n = 2$ and
$n = 6$ were done by hand and were found to be $\mathcal{N}$ positions. Thus there are no
anomalous Grundy values for small values of n. $\qquad\qquad\qquad\qquad\qquad\square$

References

[Berlekamp et al. 1982] E. R. Berlekamp, J. H. Conway, and R. K. Guy, *Winning Ways For Your Mathematical Plays*, Academic Press, London, 1982.

[Bondy and Murty 1976]   A. Bondy and U. S. R. Murty, *Graph Theory with Applications*, North-Holland, Amsterdam, 1976.

[Fraenkel and Simonson 1993] A. S. Fraenkel and S. Simonson, "Geography", *Theoret. Comput. Sci. (Math Games)* **110** (1993), 197–214.

[Fraenkel et al. 1993] A. S. Fraenkel, E. R. Scheinerman and D. Ullman, "Undirected edge geography", *Theoret. Comput. Sci. (Math Games)* **112** (1993), 371–381.

[Fraenkel et al. 1995] A. S. Fraenkel, A. Jaffray, A. Kotzig and G. Sabidussi, "Modular Nim", *Theoret. Comput. Sci. (Math Games)* **143** (1995), 319–333.

RICHARD J. NOWAKOWSKI
DEPARTMENT OF MATHEMATICS
DALHOUSIE UNIVERSITY
HALIFAX, NS
CANADA B3H 3J5
    rjn@cs.dal.ca

DAVID G. POOLE
DEPARTMENT OF MATHEMATICS
TRENT UNIVERSITY
PETERBOROUGH, ON
CANADA K9J 7B8

Games of No Chance
MSRI Publications
Volume 29, 1996

Pentominoes: A First Player Win

HILARIE K. ORMAN

ABSTRACT. This article reports on a search-based computer proof that the game of pentominoes is a first-player win. Three search strategies were used in this proof, with dramatically different effects on the running time of the search. The two most effective strategies are compared and discussed.

The Short History of Pentominoes

Pentominoes is a two-player game involving twelve pieces—the regular 5-ominoes shown in Figure 1—and an 8×8 board. Players alternate placing pieces on the board, covering whole squares and without overlap. The player who cannot make a move loses. The game was first proposed by Solomon W. Golomb in the mid-fifties. Martin Gardner [1959] popularized it, also citing Golomb's work [1954] about polyominoes and checkerboards. See also [Golomb 1962; 1965].

In 1971, it was suggested that the game could be solved by computer search [Beeler et al. 1972], but no attempts to implement such a solution are known to me. In 1975 I wrote a computer program that played pentominoes, and used a PDP-11/45 computer to investigate the feasibility of a complete solution. Although the program was an excellent player against human opponents, at that time a complete solution was unattainable.

Today, high-speed workstations have changed the picture. A new search program exhaustively examined the game subtrees arising from certain two first moves. One of the moves was proved to be a win, and this assertion was verified by an independent program. The winning move was determined in about two weeks of execution time on a 64-bit 175 MHz DEC Alpha processor. The verification was done on a Sun IPC Sparcstation in about five days.

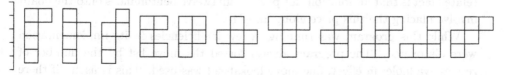

Figure 1. The twelve regular pentominoes.

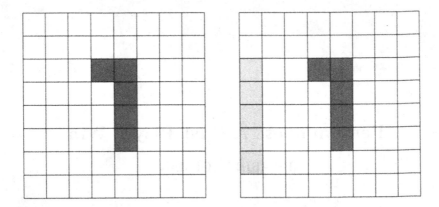

Figure 2. Left: One of the first-player moves that allows the minimum number of responses. Right: That move turns out to be losing, if the second player counters as shown.

The Search Begins

The computer program used to search the game tree is written in C and uses a simple backtracking search method. The moves and board state are represented as bit vectors. At the beginning of the game all possible moves are in the legal move list. At each ply, the legal move list is reduced, eliminating moves that are no longer possible. If all moves at depth n are losses, the value Win is returned to depth $n - 1$. If some move at depth n evaluates to Win, the value Loss is returned to depth $n - 1$.

A Losing Move. At the start of the game, there are 2308 possible moves, or 296 when symmetries are discounted. After the first move there are between 1181 and about 2000 replies. The search was originally conducted for one of the optimally restrictive moves, using the long "L" piece (Figure 2, left). There are 1181 replies to this move.

All replies to this first move were analyzed, the work being divided among several computers: an Intel Paragon, six Hewlett-Packard 720's, and four DEC RS3000 model 600 (Alpha) machines. The program ran for about two months. Of the 1181 replies, exactly two of them refute the opening move. In addition to the reply shown in Figure 2 (right), the straight piece can be moved down one square.

It is interesting to note the importance of the straight piece in this. A possibly related fact is that all solutions for packing all twelve pentominoes onto the board involve placing the this piece along an edge.

While this program was running, a few inefficiencies in its implementation were discovered. The program always sorted the move list by the number of replies available; in effect, one move lookahead was used. This is useful if there are a large number of moves available, but it is wasteful if there are few. Also,

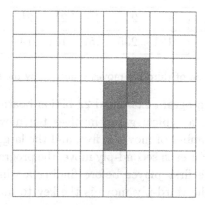

Figure 3. A winning move.

the program did not take advantage of the 64-bit wordsize of the DEC Alpha machines. With the 64-bit words, testing a possible move takes only one instruction. The program was rewritten to correct both deficiencies before proceeding. The sorting was restricted to the first few plies.

A Winning Move. Using the new program, the search was restarted with the opening move shown in Figure 3, one of the second most restrictive (1197 replies). Using two DEC Alphas, almost all replies were examined in two weeks of running time. The remaining replies were examined using other machines. This provided strong evidence that the move illustrated is a win for the first player.

To validate the move, Richard Schroeppel wrote an independent program to check the claimed win. The original program recorded the winning third-ply move that it found for each second-ply move, and the checking program took as input three plies and then solved the game with the usual backtracking search. The checking program ran on a Sun IPC Sparcstation for about five days.

Statistics on Strategies

The solving program and the checking program kept statistics about the amount of effort expended for each validation, which revealed interesting facts about the game tree. The most important contributor to the speed of the program is the ability to choose a winning move for the first player quickly at each first-player ply. Based on examining two opening moves, we might conjecture that about half of the possible opening moves are wins for the first player. But the fact that the first move is only very narrowly defeated by the second player indicates that it is unreasonable to expect the game tree to be well balanced.

The statistics recorded were the number of third-ply moves that were examined before finding a winning move for the first player, and the total number of board positions examined at the third and fourth plies.

HILARIE K. ORMAN

lower limit of bucket	1	2	2^2	2^3	2^4	2^5	2^6	2^7	2^8
frequency	121	235	252	181	198	134	80	17	1

Table 1. Distribution of third-ply moves examined by the solving program.

Overall, the solving program examined 22 billion board positions. Empirical evidence indicates that a typical game involves ten moves; it is known that the smallest possible number of moves is five, and the largest is twelve [Gardner 1959]. On the average, for each second-ply move, the program examined eighteen third-ply moves for the first player before finding a winning move. A good strategy for selecting the third-ply move is the key to minimizing the running time of the program.

Table 1 shows the distribution of third-ply moves examined by the solving program. Each column represents a bucket (range of trials); the bottom row gives the number of times that a winning move was found using a number of trials in the range. Although the program often found a win within seven tries (the first three buckets), it sometimes had to examine nearly half the available moves. The sum of the entries in the bottom row is slightly greater than the number of replies (1197) because there was some overlap in the ranges examined by the computers running the solving program.

Table 2 shows the distribution of number of board positions examined after each third-ply move. Each column represents a range of board positions, and the two bottom rows give the number of instances of counts in this range for the solving and checking programs. For example, the checking program needed to examine fewer than 1,048,576 positions in four cases; the solving program never examined this few. (Again, the counts reflect some overlap in the positions examined.)

It is interesting to note that the checking program did less work than the solving program as measured by the number of board positions examined. This means that it is more effective at finding a winning move for the first player. This was surprising, because the checking program does not have the moves sorted by number of replies; it tries the first piece at all possible board locations, then the next piece, etc., in a fixed pattern (pruned at each ply by eliminating illegal moves). This latter algorithm avoids some of the very long searches done by the first method. A third simple search strategy is to try each piece at board position $(0, 0)$, then each piece at position $(0, 1)$, etc. This turns out to be at

lower limit of bucket	2^{19}	2^{20}	2^{21}	2^{22}	2^{23}	2^{24}	2^{25}	2^{26}	2^{27}
solver frequency		5	41	256	481	282	127	24	3
checker frequency	4	9	58	289	527	279	49	3	

Table 2. Distribution of total evaluations for the solving and checking programs, which used different strategies.

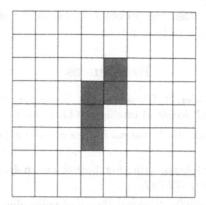

Figure 4. Another winning move, suggested by Golomb.

least ten times slower than the first two methods, and it is definitely not suitable for solving the game.

There are two plausible explanations for the checking program's search strategy being faster than that of the solving program. One reason might be that, in general, occupying a critical region of the board is more important than which piece is used for the occupation. The checking program does this by attempting to place pieces quickly without regard for which piece it is placing. A second explanation is based on noting that the program tries the thin pieces (having a linear block of four squares) first. The key to the game might be the placement of these pieces; the importance of the straight piece in refuting the most restrictive opening move is corroborating evidence.

Another Winning Move

Golomb (personal communication) suggested another winning move, one that comes closest to dividing the board symmetrically (Figure 4). This move indeed survives examination by the checking program. Schroeppel's validating program has not been used, because experience in double-checking the two moves described above has given us confidence in the first program.

The strategy of dividing the board symmetrically may be a sound one, and it would be interesting to see if it applies through the game tree.

Other Problems

It should be possible to determine all opening moves that are first-player wins, thus determining how much of an advantage the first player has. This task becomes much more difficult as less restrictive opening moves are examined. However, the checking program's strategy for move selection might offset this effect enough to make the overall running time reasonable.

A computationally challenging problem is to solve the similar game involving 35 hexominoes and a 15 × 15 board.

References

[Beeler et al. 1972] M. Beeler, R. Gosper, and R. Schroeppel, "HAKMEM", Technical Report Memo 239, MIT Artificial Intelligence Laboratory, 1972.

[Gardner 1959] M. Gardner, *Mathematical Puzzles and Diversions*, Simon and Schuster (sixth edition), 1959.

[Golomb 1954] S. W. Golomb, "Checker boards and polyominoes", *Amer. Math. Monthly* **61** (Dec. 1954), 672–682.

[Golomb 1962] S. W. Golomb, "General theory of polyominoes", *Recreational Math. Mag.* **4** (Aug. 1961), 3–12; **5** (Oct. 1961), 3–12 (see also Notes 12–14); **6** (Dec. 1961), 3–20; **8** (Apr. 1962), 7–16.

[Golomb 1965] S. W. Golomb, *Polyominoes*, Scribner, New York, 1965; second edition by Princeton Univ. Press, Princeton, 1994.

HILARIE K. ORMAN
1942 W. CAMINO BAJIO
TUCSON AZ, 85737
ho@cs.arizona.edu

Games of No Chance
MSRI Publications
Volume 29, 1996

New Values for Top Entails

JULIAN WEST

ABSTRACT. The game of Top Entails introduces the curious theory of entailing moves. In *Winning Ways*, simple positions are analysed and stacks of one and three coins are shown to be loony positions, since any move leaving such a stack is a losing move, regardless of other stacks that might also be present. We analyse all stacks of up to 600,000 coins and find three more loony stacks: those having 2403, 2505 and 33,243 coins.

The game of Top Entails is introduced in *Winning Ways* [Berlekamp et al. 1982, p. 376]. It is a nim-like, impartial game played with stacks of coins and the following two types of move. A player may either (1) split just one stack into two strictly smaller stacks; or (2) remove just one coin from just one stack; in this case the opponent is *entailed* to reply with a move (of either type) in the resulting stack.

An unentailed stack of zero coins can never really occur in this game, since neither type of move can create one. Nevertheless, we can posit one as a hypothetical object; indeed, it is clear that it has value $*0 = 0$, since it hardly matters whether there is such a stack on the table or not!

What about an entailed stack of zero coins? Such a position can certainly occur, by an entailing (type 2) move from any stack containing just one coin. Since the next player, N, is now forced to move in an empty stack, the game ends instantly with a win for the previous player, P. Indeed, this is the only way a game can actually end (short of the usual situation of players calculating the value of the position, agreeing on the winner, picking up their coins and going home). So what value z do we assign to the entailed zero-stack? Since the rest of the stacks on the table, G, constitute an impartial game, they must sum to some nimber $*n$; but $G + z$ can be seen to be 0 (a P-position) for any value of n. So it must be right to think of z as having whatever nim-value $*n$ seems convenient, or equivalently to ascribe it the set of all nim-values: $\{*0, *1, *2, \ldots\}$.

On the other hand, as soon as a player creates a stack of size 1, she has lost. Since the stack has just been created, her opponent can not possibly be entailed to move elsewhere, and may reply by leaving the entailed zero-stack z. Creating a stack of size 1 is thus a *loony* move. Since no nimber can be added to loony to get the $\mathcal{P}$-position 0, we can identify loony with $\{\ \}$, the set of no nim-values. Naturally, then, the set z of all nim-values must be called *sunny*.

In general, a stack will have a set of nim-values $\{a_0, a_1, \ldots\}$, of which only the smallest, a_0, is relevant if the stack is unentailed; if the stack is entailed, all are relevant. (Above we saw that an unentailed zero-stack had value 0, which is indeed the smallest nimber in the sunny set.) The complete set of nim-values for a position G consists of *all nimbers not among the relevant values* for options of G. This fairly simple rule enables us to compute new values. A little work by hand produces a table, the first 120 lines of which are reproduced in Table 1. The entries are of course sets of nimbers, but stars have been suppressed. An arrow indicates that all larger nimbers are included in the set.

For this paper, a program was written in C to compute game values. It computed 600,000 values in approximately 30 CPU hours on a Sun workstation. For the first 38,000 values, the complete set of nim-values was returned, exactly as in the above table; thereafter only the smallest member of the set, that is, the unentailed value of the stack.

The file of 38,000 values was searched for the first appearance of each integer. This is a rather arbitrary measure, because an integer n can appear in this table in two different ways, which were not distinguished: either n is in the set of nim values but some larger m is not; or else $n-1$ is the largest value to be absent (and so n begins the tail). The number 14 is the smallest to make its appearance in the first of these ways. The point of interest is that all values between $2^k + 1$ and 2^{k+1} tend to appear rather soon after 2^k appears as an unentailed value. (The value 16 first appears as the unentailed value for a 94-stack; 17 through 32 enter the table between lines 94 and 184; values larger than 32 do not appear until the unentailed 32 finally appears with the 534-stack.) This is not surprising, as this means that the k-th bit has now come into play in the nim-additions occasioned by stack-splitting moves. Indeed, a little work with the above table shows that if 2^k is the unentailed value for the T-stack, then $2^k + 1$ in general appears on line $T + 2$, $2^k + 2$ on line $T + 4$, $2^k + 3$ on line $T + 6$, $2^k + 4$ on line $T + 8$, and $2^k + 5$ on line $T + 12$.

Here is the list of the smallest stack to have each power of 2 as its unentailed value:

power of two	1	2	4	8	16	32	64	128	356	512	1024
stack size	4	6	12	32	94	534	2556	8062	35138	119094	293692

In *Winning Ways*, the only known loony values are for stacks of one and three coins. Among the first 600,000 values there are three more loony values, namely at 2403, 2505 and 33,243 coins.

The output enables us easily to determine the strategy for playing from a stack of 2403 coins, and we show it in Table 2. The value of this stack is loony because it is possible to move from it to any nim-value at all. For instance,

0: 0 →	40: 1 4 8 9 12 →	80: 3 4 6 15 →
1: ∅	41: 6 7 11	81: 12 13
2: 0 →	42: 2 5 8 10 12 →	82: 2 5 7 9 14 →
3: ∅	43: 9	83: 13
4: 1 →	44: 3 6 7 8 11 13 →	84: 4 11 12 16 →
5: 0	45: 0 2 4 10 12	85: 2 7 14
6: 2 →	46: 5 7 8 9 13 →	86: 13 15 →
7: 1	47: 3 6 10 11	87: 1 4 6 8
8: 3 →	48: 4 8 12 →	88: 11 14 →
9: 0 2	49: 2 7 9 10	89: 5 13
10: 1 4 →	50: 11 →	90: 8 15 →
11: 3	51: 3 4 6 8 9	91: 3 4 6 11
12: 4 →	52: 2 12 →	92: 12 14 16 →
13: 0 2	53: 5 7 10 11	93: 0 13 15
14: 3 5 →	54: 3 4 8 9 12 13 15 →	94: 16 →
15: 4	55: 6 14	95: 1 4 12 14
16: 2 6 →	56: 5 8 10 11 12 13 16 →	96: 2 9 17 →
17: 5	57: 0 9 14 15	97: 5 7 13 15 16
18: 1 3 4 7 →	58: 6 10 11 16 →	98: 1 4 8 12 18 →
19: 6	59: 1 2 4 9 12 13 14	99: 6 10 14 17
20: 2 5 8 →	60: 5 7 8 15 →	100: 2 5 11 12 13 16 19 →
21: 0 1 7	61: 10 11 13 14	101: 18
22: 3 4 6 8 →	62: 1 4 9 12 16 →	102: 6 10 14 16 17 20 →
23: 2 5	63: 14 15	103: 0 9 11 19
24: 7 →	64: 5 10 11 12 13 16 →	104: 7 15 16 18 20 →
25: 3	65: 4 6 8	105: 1 10 17
26: 1 4 6 8 →	66: 2 12 14 →	106: 4 8 9 12 16 19 21 →
27: 7	67: 0 10 11 13	107: 0 7 14 17 18 20
28: 3 5 8 →	68: 8 12 15 →	108: 2 5 8 10 12 15 16 21 →
29: 4 6	69: 1 3 6 10 11 14	109: 3 4 13 18 19
30: 2 7 →	70: 12 13 16 →	110: 11 15 16 17 20 →
31: 0 5	71: 0 7 8 9 14 15	111. 5 7 12 13 18
32: 8 →	72: 10 16 →	112: 15 16 19 →
33: 1 6	73: 1 4 6 12 13	113: 3 6 9 11 14 17 18
34: 5 9 →	74: 2 3 8 11 14 →	114: 5 12 16 19 →
35: 0 7 8	75: 7 13	115: 7 10 13 18
36: 1 6 10 →	76: 9 12 15 →	116: 4 6 16 17 20 21 23 →
37: 4 5 9	77: 3 4 6 11 14	117: 9 12 19 22
38: 7 8 11 →	78: 5 12 16 →	118: 2 7 16 17 18 20 21 24 →
39: 10	79: 9 10 13 14	119: 13 15 22 23

Table 1. Distribution of unentailed nim-values for stacks of up to 600,000 coins.

n	x	$v(x)$	$v(\bar{x})$	n	x	$v(x)$	$v(\bar{x})$	n	x	$v(x)$	$v(\bar{x})$
0	75	7	7	22	47	3	21	44	79	9	37
1	462	2	3	23	123	7	16	45	17	5	40
2	84	4	6	24	132	11	19	46	28	3	45
3	811	38	37	25	43	9	16	47	20	2	45
4	847	33	37	26	61	10	16	48	5	0	48
5	146	11	14	27	349	11	16	49	344	25	40
6	555	14	8	28	111	5	25	50	14†	3	49
7	26†	1	6	29	63	14	19	51	19	6	53
8	13	0	8	30	182	16	14	52	59	1	53
9	907	32	41	31	225	15	16	53	163†	1	52
10	2	0	10	32	68	8	4	54	8	6	48
11	1176	7	12	33	496	17	48	55	192†	3	52
12	913	44	32	34	37	4	38	56	365†	17	73
13	33	1	12	35	101	18	49	57	70	12	53
14	942	46	32	36	10	1	37	58	1097†	25	35
15	9	0	15	37	6	2	39	59	32	8	51
16	233	27	11	38	66	2	36	60	72	10	54
17	7	1	16	39	55	6	33	61	50†	11	54
18	533	7	21	40	49	2	42	62	90†	8	54
19	16	2	17	41	334	16	57	≥ 63	†		
20	34	5	17	42	88	11	33				
21	12	4	17	43	30	2	41				

Table 2. How to play when a position contains a stack of 2403 coins.

to move to $*2$, separate off a stack of $x = 84$ coins, leaving another stack of $\bar{x} = 2403 - x = 2319$ coins. The first of these stacks has the unentailed value $v(84) = *4$, the second $v(2319) = *6$. A † next to a number indicates that an entailing move is also possible. Thus to move to $*7$, either split into stacks of 26 and 2377, or make the entailing move by removing one coin. There may be other possibilities as well; 26 is the option with smallest x.

The histograms in Figure 1 show the distributions of unentailed nim-values among the first $100,000$ and the first $600,000$ stacks. It will be noted that values congruent to 0 or to 1 modulo a large power of 2 are particularly common, followed by a big drop and a general tapering off before the next large power of 2.

The program used to generate the above data is readily adaptable to any *octal* game (*Winning Ways*, chapter 4). For instance, it was used to generate 40 million values for the game called .611.... (This game is easy to analyse computationally because it has a particularly strong division into rare and common values). Plans are made to attack various outstanding octal games in the near future.

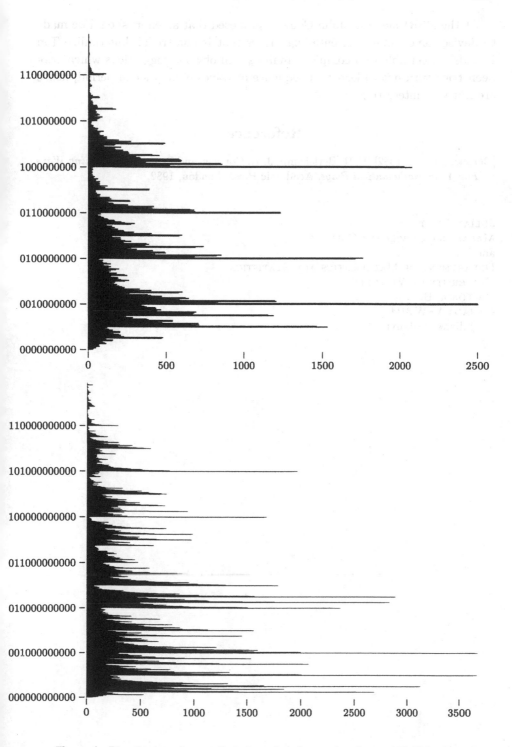

Figure 1. Distribution of unentailed nim-values for stacks of up to 100,000 coins.

At the MSRI meeting, John Conway proposed that an effort should be made to devise some game with entailing moves that is non-trivial, but (unlike Top Entails) susceptible to a complete analysis. All obvious suggestions which have been tried turn out to lead to a sequence of values of very small period, hence are not very interesting.

Reference

[Berlekamp et al. 1982] E. R. Berlekamp, J. H. Conway, and R. K. Guy, *Winning Ways For Your Mathematical Plays*, Academic Press, London, 1982.

JULIAN WEST
MALASPINA UNIVERSITY-COLLEGE
and
DEPARTMENT OF MATHEMATICS AND STATISTICS
UNIVERSITY OF VICTORIA
VICTORIA, B.C.
CANADA V8W 3P4
    julian@math.uvic.ca

Games of No Chance
MSRI Publications
Volume 29, 1996

Take-Away Games

MICHAEL ZIEVE

ABSTRACT. Several authors have considered take-away games where the players alternately remove a positive number of counters from a single pile, the player removing the last counter being the winner. On his initial move, the first player may remove any number of counters, so long as he leaves at least one. On each subsequent move, a player may remove at most $f(n)$ counters, where n is the number of counters removed by his opponent on the preceding move. We prove various results (improving all previously known results) about the sequence of losing positions when f is a linear function.

1. Introduction

Several works, including [Berlekamp et al. 1982; Epp and Ferguson 1980; Schwenk 1970], have studied take-away games where the players alternately remove a positive number of counters from a single pile, the player removing the last counter being the winner. On his initial move, the first player may remove any number of counters, so long as he leaves at least one. On each subsequent move, a player may remove at most $f(n)$ counters, where n is the number of counters removed by his opponent on the preceding move. Thus, any mapping f from the positive integers to themselves defines such a take-away game.

Epp and Ferguson [1980] considered the case where f is nondecreasing and $f(1) \geq 1$. For any such f, let $H_1 = 1$, H_2, ... be the sizes of the initial pile from which the first player has no winning strategy; we call these the *losing positions* (these are the $\mathcal{P}$-positions of [Berlekamp et al. 1982], the $\mathcal{P}$ standing for a Previous-player win). We will study this sequence of losing positions. We begin with a result from [Epp and Ferguson 1980]:

THEOREM 1.1. *If* $f(H_j) \geq H_j$, *then* $H_{j+1} = H_j + H_l$, *where*

$$H_l = \min_{i \leq j}\{H_i | f(H_i) \geq H_j\}.$$

If $f(H_j) < H_j$, *the sequence of losing positions is finite and* H_j *is the final term.*

PROOF. Assume $f(H_j) \geq H_j$; then $H_l = \min_{i \leq j}\{H_i : f(H_i) \geq H_j\}$ exists. For any losing position $H_i < H_l$, we have $f(H_i) < H_j$, so from an initial pile of size $H_j + H_i$ the first player can remove H_i counters and win (since this leaves the second player a pile of size H_j from which he cannot remove all counters, so he is in a losing position).

Now let $x < H_l$ be a winning position. Given a pile of size $H_j + x$, the first player can employ a winning strategy for a pile of size x whose final move removes y counters, where $f(y) < H_j$; this leaves the second player with a pile of size H_j from which he cannot remove all counters, and the first player wins. (He can always arrange for y to satisfy this property because, if the last move y of a winning strategy for x satisfies instead $f(y) \geq H_j$, then $y < H_l$ cannot be a losing position, and consideration of a winning strategy for y leads to a smaller final move.)

Finally, from a pile of size $H_j + H_l$, if the first player takes at least H_l counters the second player takes the rest and wins; if the first player takes fewer than H_l counters, we fall into the preceding paragraph's situation, with roles reversed. This proves the first statement of the theorem.

If $f(H_j) < H_j$, suppose we had $H_{j+1} = H_j + x$ for some x. As above, x cannot be any H_i, since the first player wins from $H_j + H_i$ by removing H_i counters (since $f(H_i) \leq f(H_j) < H_j$). Since $x < H_{j+1}$, x must be a winning position; thus, the first player can win from $H_j + x$ by employing a winning strategy for x whose final move is y, where $f(y) < H_j$. Thus H_{j+1} is not a losing position, which is a contradiction, so in fact there is no H_{j+1}. $\square$

A natural question is whether the sequence of losing positions eventually becomes a simple linear recursion $H_{i+1} = H_i + H_{i-k}$ for sufficiently large i. For the functions $f(n) = cn$, where $c \geq 1$, the answer is yes, as proved in [Schwenk 1970]; a positive answer for the functions $f(n) = cn - 1$, where $c \geq 2$, was claimed in [Berlekamp et al. 1982].

In this paper we modify the methods of [Schwenk 1970] to give a positive answer for functions $f(n) = cn - d$, where $c - 1 \geq d \geq 0$. Using entirely different methods we give a positive answer for functions $f(n) = cn + d$, where $c \geq 1$ and $d > 0$. We also derive results describing k, the degree of the recursion, in terms of c (our results do not depend on d); these are the first results about k that have been found. For certain small values of c we sharpen the general results to find k exactly. We also present algorithms for computing k and prove that they are valid. We have implemented these algorithms for several linear functions f, and we make several conjectures based on this data (in particular, about the dependence of k on d).

2. The Functions $f(n) = cn - d$

This section is devoted to the proof of the following result.

THEOREM 2.1. *In the game associated to the function* $f(n) = cn - d$, *where* $c - 1 \geq d \geq 0$, *there is a nonnegative integer* k *such that the sequence of losing positions* H_i *satisfies* $H_{i+1} = H_i + H_{i-k}$ *for all sufficiently large* i.

Note that $f(n) \geq n$ for all n, so Theorem 1.1 implies that the sequence of losing positions is infinite. The theorem follows from the following two lemmas.

LEMMA 2.2. *If* $H_i \leq f(H_j)$, *then* $H_{i+1} \leq f(H_{j+1})$.

PROOF. Decreasing j if necessary, we may assume $H_{i+1} = H_i + H_j$. Also, $H_{j+1} = H_j + H_l$ where $f(H_l) \geq H_j$. Thus,

$$f(H_{j+1}) = cH_{j+1} - d = cH_j + cH_l - d$$
$$\geq cH_j + H_j = f(H_j) + d + H_j \geq H_i + d + H_j = H_{i+1} + d,$$

and the lemma is proved. □

LEMMA 2.3. *There exists a positive integer* r *such that* $f(H_{n-r}) < H_n$ *for all* $n > r$.

PROOF. For any integer $i \geq 1$ we have $H_{i+1} = H_i + H_j$, where $cH_j - d = f(H_j) \geq H_i$; thus,

$$\frac{H_{i+1}}{H_i} = 1 + \frac{H_j}{H_i} \geq 1 + \frac{1}{c}.$$

Let r be any integer for which $(1 + 1/c)^r > c$; then, for any $n > r$,

$$\frac{H_n}{H_{n-r}} = \prod_{i=n-r}^{n-1} \frac{H_{i+1}}{H_i} \geq \left(1 + \frac{1}{c}\right)^r > c.$$

Thus, $f(H_{n-r}) \leq cH_{n-r} < H_n$. □

PROOF OF THEOREM 2.1. Each H_{i+1} equals $H_i + H_j$ for some $j \leq i$; Lemma 2.2 implies that the sequence of differences $i - j$ is nondecreasing, and Lemma 2.3 implies that this sequence is bounded above by $r - 1$, so together they show that the sequence must be constant for sufficiently large i. This limiting value is the k described in the theorem. □

3. The Functions $f(n) = cn + d$

The proof that the sequence of losing positions for the take-away game associated to $f(n) = cn - d$ eventually satisfies a simple recursion was quite simple: the sequence of differences $i - j$ (where $H_{i+1} = H_i + H_j$) was shown to be nondecreasing and bounded. The sequence of differences for $f(n) = cn + d$, however, is generally not monotonic, so one cannot hope for such a simple proof in this case.

THEOREM 3.1. *In the take-away game associated to the function* $f(n) = cn + d$, *where* $c \geq 1$ *and* $d > 0$, *there is a nonnegative integer* k *such that the sequence of losing positions* H_i *satisfies* $H_{i+1} = H_i + H_{i-k}$ *for all sufficiently large* i.

PROOF. First, any function f as in the theorem satisfies $f(n) \geq n$ for all n, so Theorem 1.1 implies that the sequence of losing positions is infinite. Now, each H_{i+1} equals $H_i + H_j$ for some $j \leq i$; we will show that the sequence of differences $i - j$ is nonincreasing for sufficiently large i, which implies that the sequence is eventually constant, finishing the proof.

Define $J_0 = \{1\}$, and inductively

$$J_{i+1} = \{\alpha : H_\alpha = H_{\alpha-1} + H_r \text{ for } r \in J_i\}.$$

Let r_i and s_i be the minimal and maximal elements of J_i, respectively. We claim that every J_i is a finite nonempty set of consecutive integers, $J_i = \{r_i, r_i + 1, \ldots, s_i\}$, and that $r_{i+1} = s_i + 1$; thus the sets $J_0, J_1, \ldots$ partition the positive integers into intervals. The proof of this claim is inductive: we have

$$J_1 = \{\alpha : H_\alpha = H_{\alpha-1} + 1\} = \{\alpha : H_{\alpha-1} \leq f(1)\} = \{2, 3, \ldots, s_2\}$$

and, for $i \geq 1$,

$$\begin{aligned} J_{i+1} &= \{\alpha : f(H_{r-1}) < H_{\alpha-1} \leq f(H_r), r \in J_i\} \\ &= \{\alpha : f(H_{r_i-1}) < H_{\alpha-1} \leq f(H_{s_i})\}, \end{aligned}$$

so J_{i+1} is a finite set of consecutive integers whose least element is $s_i + 1$; thus the claim is proved. Now, every positive integer j is in some J_i, and we define $\psi(j) = i$. Then ψ is a nondecreasing function.

Suppose that, for each $\alpha \in J_i$ (where $i > 0$, so that $\alpha > 1$), the unique r satisfying $H_\alpha = H_{\alpha-1} + H_r$ also satisfies $H_\alpha \geq f(H_r) - d\varepsilon$. Then, for any $\beta \in J_{i+1}$, we have $H_\beta = H_{\beta-1} + H_\alpha$ for some $\alpha \in J_i$, so $H_\alpha \geq f(H_r) - d\varepsilon$. Thus $f(H_{\alpha-1}) < H_{\beta-1} \leq f(H_\alpha)$. Now

$$\begin{aligned} H_\beta - f(H_\alpha) = H_{\beta-1} + H_\alpha - f(H_\alpha) &= H_{\beta-1} + H_\alpha - cH_\alpha - d \\ &> f(H_{\alpha-1}) + H_\alpha - cH_\alpha - d = cH_{\alpha-1} + H_\alpha - cH_\alpha = H_\alpha - cH_r \\ &= H_\alpha + d - f(H_r) \geq d - d\varepsilon, \end{aligned}$$

so $H_\beta > f(H_\alpha) - d(\varepsilon - 1)$.

Put $\varepsilon = (c + d - 2)/d$. Note that $J_1 = \{2, 3, \ldots, \lfloor c + d \rfloor + 1\}$, where $\lfloor c + d \rfloor$ denotes the greatest integer not exceeding $c + d$. For any $\alpha \in J_1$, we have $H_\alpha = H_{\alpha-1} + H_1$, and

$$f(H_1) - d\varepsilon = c + d - d\varepsilon = 2 \leq H_\alpha.$$

By the above paragraph, for each $\alpha \in J_i$, the unique r satisfying $H_\alpha = H_{\alpha-1} + H_r$ also satisfies $H_\alpha \geq f(H_r) - d(\varepsilon + 1 - i)$.

For any $j > 1$ such that $i = \psi(j) \geq \max\{2, 1 + \varepsilon\}$, we have $H_j = H_{j-1} + H_m$ where $m \in J_{i-1}$. Since $i \geq 1$, we have $H_m = H_{m-1} + H_t$, where

$$H_m \geq f(H_t) - d(\varepsilon + 2 - i) \geq f(H_t) - d.$$

Now, since $f(H_{m-1}) < H_{j-1} \le f(H_m)$, we get

$$
\begin{aligned}
H_j - f(H_m) &= H_{j-1} + H_m - cH_m - d \\
&> f(H_{m-1}) - d + H_m - cH_m \\
&= cH_{m-1} + H_m - cH_m = H_m - cH_t = H_m - f(H_t) + d \ge 0,
\end{aligned}
$$

so $H_j > f(H_m)$. Thus, if $H_{j+1} = H_j + H_l$, we have $f(H_l) \ge H_j$, so $l > m$, and thus the difference $j - l \le (j-1) - m$.

We have now shown that, for sufficiently large j, the sequence of differences $j - r$, where $H_{j+1} = H_j + H_r$, is nonincreasing; since each such difference is a nonnegative integer, this implies that the sequence is eventually constant, which completes the proof.   □

4. The Degree of the Recursion

The theorems of the previous two sections imply that, if $f(n) = cn + d$ where $c \ge 1$ and $d \ge 1 - c$, the sequence of losing positions H_i for the corresponding take-away game satisfies $H_{i+1} = H_i + H_{i-k}$ for all sufficiently large i. In this section we derive some general results about k. Our methods use only the ultimate behavior of the sequence, namely that $H_{i+1} = H_i + H_{i-k}$ for sufficiently large i, and disregard the early behavior.

THEOREM 4.1. *If $f(n) = cn + d$ where $c > 1$ and $d \ge 1 - c$, and the sequence of losing positions for the corresponding take-away game satisfies the recursion $H_{i+1} = H_i + H_{i-k}$ for all sufficiently large i, then*

$$
\frac{\log(c-1)}{\log c - \log(c-1)} \le k \le \frac{\log c}{\log(c+1) - \log c}.
$$

PROOF. We will only consider i that are "sufficiently large", so we may assume that $H_{i+1} = H_i + H_{i-k}$ for all i under consideration. The characteristic polynomial for this recursion relation is $g(x) = x^{k+1} - x^k - 1$. By Descartes' rule of signs, g has exactly one positive real root, which we will denote by r. Note that $g(t) \to +\infty$ as $t \to +\infty$, where t is a real variable; thus, for $t > r$ we must have $g(t) > 0$. Now, any complex root z of this polynomial satisfies

$$
1 = |z^{k+1} - z^k| = |z|^k \cdot |z - 1| \ge |z|^k \cdot (|z| - 1) = |z|^{k+1} - |z|^k,
$$

or equivalently $g(|z|) \le 0$. Thus, $|z| \le r$. If $|z| = r$ then $g(|z|) = 0$, so we have equality in the displayed equation above, implying that $|z - 1| = |z| - 1$; by the equality criteria for the triangle inequality, this last equality implies that z is a nonnegative real number, so $z = r$. Thus, any complex root of g other than r has absolute value less than r. Since r is necessarily a simple root of g, it follows by a standard result about linear recursion relations that

$$
\lim_{i \to \infty} \frac{H_{i+1}}{H_i} = r.
$$

Since $cH_{i-1} + d < H_{i+k} \le cH_i + d$, we have

$$\frac{H_{i-1}}{H_{i+k}} < \frac{1 - d/H_{i+k}}{c} \le \frac{H_i}{H_{i+k}};$$

taking the limit as $i \to \infty$, we see that

$$\frac{1}{r^{k+1}} \le \frac{1}{c} \le \frac{1}{r^k},$$

or, equivalently, $r^k \le c \le r^{k+1}$. Note that $r > 1$, since $f(1) < 0$; now, since $r^k = 1/(r-1)$, we have

$r^k \le c \le r^{k+1}$

$\quad \Leftrightarrow 1/(r-1) \le c \le r/(r-1)$

$\quad \Leftrightarrow 1/c \le r - 1 \quad \text{and} \quad (c-1)r \le c$

$\quad \Leftrightarrow (c+1)/c \le r \le c/(c-1)$

$\quad \Leftrightarrow g\left(\dfrac{c+1}{c}\right) \le 0 \le g\left(\dfrac{c}{c-1}\right)$

$\quad \Leftrightarrow \left(\dfrac{c+1}{c}\right)^k\left(\dfrac{1}{c}\right) \le 1 \le \left(\dfrac{c}{c-1}\right)^k\left(\dfrac{1}{c-1}\right)$

$\quad \Leftrightarrow k\Big(\log\left(c+1\right) - \log c\Big) - \log c \le 0 \le k\Big(\log c - \log\left(c-1\right)\Big) - \log\left(c-1\right)$

$\quad \Leftrightarrow \dfrac{\log\left(c-1\right)}{\log c - \log\left(c-1\right)} \le k \le \dfrac{\log c}{\log\left(c+1\right) - \log c},$

which completes the proof. $\square$

Define $\chi(c) = (\log c)/\big(\log\left(c+1\right) - \log c\big)$ for $c > 0$. Then the result of the above theorem is that $\chi(c-1) \le k \le \chi(c)$.

LEMMA 4.2. *The function $\chi(c)$ is increasing for $c > 1$.*

PROOF. Since

$$\chi(c) = \frac{1}{\log\left(c+1\right)/\log c - 1}$$

for $c > 1$, this function is increasing if and only if

$$\phi(c) = \frac{\log\left(c+1\right)}{\log c}$$

is decreasing. But its derivative satisfies

$$\phi'(c) = \frac{(\log c)/(c+1) - \big(\log\left(c+1\right)\big)/c}{(\log c)^2} = \frac{c\log c - (c+1)\log\left(c+1\right)}{c(c+1)(\log c)^2} < 0,$$

so indeed ϕ is decreasing, hence χ is increasing for $c > 1$. $\square$

COROLLARY 4.3. *For any integers c_1, c_2, d_1, d_2 such that $c_1 - 1 \geq c_2 \geq 2$ and $d_1 \geq 1 - c_1$, $d_2 \geq 1 - c_2$, we have $k_1 > k_2$, where the losing positions of the take-away game associated to the function $f(n) = c_j n + d_j$ satisfy $H_{i+1} = H_i + H_{i-k_j}$ for sufficiently large i.*

PROOF. We have $k_2 \leq \chi(c_2) \leq \chi(c_1 - 1) \leq k_1$; if $k_2 = k_1$ every equality would hold, so $\chi(c_2) = k_2$ would be an integer and $1 + 1/\chi(c_2) = \log(c_2 + 1)/\log c_2$ would be a rational number, implying that $c_2 + 1$ is a rational power of c_2, which cannot hold in light of the unique prime factorization theorem for the integers. $\square$

When $d_1 = d_2 = 0$, this corollary becomes:

COROLLARY 4.4. *For any integers $c_1 > c_2 \geq 2$, the degree of the recursion satisfied by the ultimate losing positions for $f(n) = c_1 n$ is greater than the corresponding degree for $f(n) = c_2 n$.*

References [Berlekamp et al. 1982; Schwenk 1970; Whinihan 1963] asked for results about these degrees; the above corollary and the preceding theorem are the first such results.

Theorem 4.1 gives an interval, in terms of c, in which the degrees corresponding to all $f(n) = cn + d$ lie; the length of this interval, $\chi(c) - \chi(c - 1)$, is asymptotic to $\log c$ as $c \to \infty$, since $\chi(c)$ is asymptotic to $c \log c$.

5. Special Values of c

In this section we derive sharp results about the degree corresponding to $f(n) = cn + d$ when c has certain special values.

PROPOSITION 5.1. *For any positive real number d, the sequence of losing positions for the take-away game corresponding to $f(n) = n + d$ satisfies $H_{i+1} = 2H_i$ for all sufficiently large i.*

Note that this could be stated: if $c = 1$ the degree is 0.

PROOF. Since the H_i form an increasing sequence of positive integers, there is some $H_i > 2d$. Then $H_{i+1} = H_i + H_r$, where $H_r + d \geq H_i$. Thus $H_i + d \leq H_r + 2d < H_r + H_i = H_{i+1}$, so $H_{i+2} = 2H_{i+1}$. Then $f(H_{i+1}) < H_{i+2}$, so $H_{i+3} = 2H_{i+2}$, and so on. $\square$

PROPOSITION 5.2. *For any positive real number d, the sequence of losing positions for the take-away game corresponding to $f(n) = 2n + d$ satisfies $H_{i+1} = H_i + H_{i-1}$ for all sufficiently large i.*

This could be stated: if $c = 2$ the degree is 1.

PROOF. By Theorem 3.1, we know that $H_{i+1} = H_i + H_{i-k}$ for all sufficiently large i. Theorem 4.1 implies that

$$k \leq \frac{\log 2}{\log 3 - \log 2} < 2,$$

so $k = 0$ or $k = 1$. For any i, we have $H_{i+1} = H_i + H_r$, so

$$f(H_i) > 2H_i \geq H_i + H_r = H_{i+1}.$$

Thus, we cannot have $k = 0$, so $k = 1$. $\qquad\qquad\qquad\qquad\qquad\qquad$ □

6. How to Compute the Degrees

In this section we derive theoretical results which provide algorithms for computing the degrees; in the next section we present data generated using these algorithms, and make several conjectures based on it.

PROPOSITION 6.1. *Suppose* $f(n) = cn - d$, *where* $c - 1 \geq d \geq 0$. *If* $H_{j+1} = H_j + H_{j-k}$ *for some* j, *and* $cH_{j+i-1-k} < H_{j+i}$ *for* $1 \leq i \leq k + 1$, *then* $H_{r+1} = H_r + H_{r-k}$ *for all* $r \geq j$.

PROOF. By Lemma 2.2, $f(H_{r-k-1}) < H_r \leq f(H_{r-k})$ implies $H_{r+1} \leq f(H_{r-k+1})$. Thus, since f is increasing, $H_r \leq f(H_{r-k})$ implies $H_{r+1} \leq f(H_{r-k+1})$. Since $H_j \leq f(H_{j-k})$, we have $H_r \leq f(H_{r-k})$ for all $r \geq j$. Now, since $cH_{j+i-1-k} < H_{j+i}$, we have

$$f(H_{j+i-1-k}) \leq cH_{j+i-1-k} < H_{j+i} \leq f(H_{j+i-k}),$$

so $H_{j+i+1} = H_{j+i} + H_{j+i-k}$ for $1 \leq i \leq k + 1$. In particular, $H_{j+2} = H_{j+1} + H_{j+1-k}$ and $H_{j+k+2} = H_{j+k+1} + H_{j+1}$. Thus

$$H_{j+k+2} = H_{j+k+1} + H_{j+1} > cH_j + cH_{j-k} = cH_{j+1}.$$

We have shown that $H_{j+2} = H_{j+1} + H_{j+1-k}$ and $cH_{j+1} < H_{j+k+2}$; thus, for $J = j + 1$ we have $H_{J+1} = H_J + H_{J-k}$ and, for $i = 1, 2, \ldots, k + 1$, we have $cH_{J+i-1-k} < H_{J+i}$. So, by induction, for all $r \geq j$ and all $1 \leq i \leq k+1$ we have $H_{r+1} = H_r + H_{r-k}$ and $cH_{r+i-1-k} < H_{r+i}$. $\qquad\qquad\qquad$ □

This suggests the following algorithm.

ALGORITHM 6.2. For $f(n) = cn - d$, where $c - 1 \geq d \geq 0$, compute successive terms H_i, by putting $H_1 = 1$ and

$$H_{i+1} = H_i + \min_{j \leq i}\{H_j : f(H_j) \geq H_i\}.$$

Stop when an integer j is found for which $H_{j+1} = H_j + H_{j-k}$ and $cH_{j+i-1-k} < H_{j+i}$ for each $1 \leq i \leq k + 1$, and output k.

We do not know, a priori, whether this algorithm will terminate. If it does, it will output the integer k for which $H_{r+1} = H_r + H_{r-k}$ for sufficiently large r. We implemented this algorithm and applied it for all pairs of integers (c, d) such that $90 \geq c > d > 0$; it terminated rather quickly in every case. Our data is considered in the next section.

Our algorithm for functions $f(n) = cn + d$ is a bit more complicated. We begin with a definition: For a given function $f(n) = cn + d$ with $c > 1, d > 0$, and any integer $j > 1$, let $\delta(j)$ be the unique positive integer such that $H_j = H_{j-1} + H_{\delta(j)}$.

LEMMA 6.3. *Suppose* $f(n) = cn + d$, *where* $c > 1$ *and* $d > 0$. *If* j *is an integer such that* $\delta(j) > 1$ *and that each* $i = \delta(j), \delta(j)+1, \ldots, j-1$ *satisfies* $H_i \geq cH_{\delta(i)}$, *the sequence of differences* $\big(i - \delta(i)\big)_{i \geq j}$ *is nonincreasing.*

PROOF. Put $m = \delta(j)$; then

$$H_j - cH_{\delta(j)} = H_{j-1} - (c-1)H_m$$
$$> f(H_{m-1}) - (c-1)H_m = cH_{m-1} + d - cH_m + H_m$$
$$= d + H_m - cH_{\delta(m)} \geq d,$$

so $H_j > f(H_{\delta(j)}) > cH_{\delta(j)}$. Now, $H_{j+1} = H_j + H_{\delta(j+1)}$ implies that $f(H_{\delta(j+1)}) \geq H_j$, so $\delta(j+1) > \delta(j)$. Since $\delta(j+1) \leq j$, we know that $H_i \geq cH_{\delta(i)}$ for each $i = \delta(j+1), \ldots, j$, so our hypotheses are satisfied if we replace j by $j+1$. By induction, our hypotheses are satisfied if we replace j by any larger integer r. As above, this implies that $\delta(r+1) > \delta(r)$, so $r - \delta(r) \geq r+1 - \delta(r+1)$ for any $r \geq j$, which is what we are trying to show. □

The proof of Theorem 3.1 shows that there do exist integers j satisfying the hypotheses of the above lemma.

PROPOSITION 6.4. *For* f *as in the preceding lemma, suppose that the sequence of differences* $\big(i - \delta(i)\big)_{i \geq j}$ *is nonincreasing. If* $H_j = H_{j-1} + H_{j-1-k}$ *and also* $H_{j+i} \leq cH_{j+i-k}$ *for each* $i = 0, 1, \ldots, k$, *then* $H_{r+1} = H_r + H_{r-k}$ *for all* $r \geq j-1$.

PROOF. First, $H_{j+i} \leq cH_{j+i-k}$ implies that $H_{j+i} < f(H_{j+i-k})$; but $H_{j+i+1} = H_{j+i} + H_{\delta(j+i+1)}$ implies that $H_{j+i} > f(H_{\delta(j+i+1)-1})$, so $\delta(j+i+1) \leq j+i-k$. Since the sequence of differences is nonincreasing,

$$k + 1 = j - \delta(j) \geq j+i+1 - \delta(j+i+1),$$

so $\delta(j+i+1) \geq j+i-k$. Thus, $\delta(j+i+1) = j+i-k$ for each $i = 0, 1, \ldots, k$, so $H_{j+i+1} = H_{j+i} + H_{j+i-k}$ for each such i. Now,

$$H_{j+k+1} - cH_{j+1} = H_{j+k} + H_j - cH_{j+1} \leq cH_j + cH_{j-k} - cH_{j+1} = 0,$$

so the hypotheses of this proposition are satisfied if we replace j by $j+1$. By induction, these hypotheses are satisfied if we replace j by any larger integer; thus, as above, $H_{r+1} = H_r + H_{r-k}$ for all $r \geq j-1$. □

ALGORITHM 6.5. For $f(n) = cn + d$, where $c > 1$ and $d > 0$, compute successive terms H_i by putting $H_1 = 1$ and

$$H_{i+1} = H_i + \min_{l \leq i}\{H_l : f(H_l) \geq H_i\}.$$

Find the least integer j_0 such that $\delta(j_0) > 1$ and that each i with $\delta(j_0) \leq i \leq j_0 - 1$ satisfies $H_i \geq cH_{\delta(i)}$. Stop when an integer $j \geq j_0$ is found for which $H_j = H_{j-1} + H_{j-1-k}$ and $H_{j+t} \leq cH_{j+t-k}$ for each $t = 0, 1, \ldots, k$. Output k.

By Lemma 6.3 and Proposition 6.4, if this algorithm terminates, we have $H_{r+1} = H_r + H_{r-k}$ for all $r \geq j - 1$. We implemented this algorithm and applied it to all pairs of positive integers (c, d) such that $c \leq 90$ and $d \leq 1000$; in every case the algorithm terminated rather quickly. The data is considered below.

7. Observations and Conjectures

Armed with the algorithms from the previous section, we computed the degrees for various functions $f(n) = cn + d$ for integers c, d. We computed the degrees for all pairs of integers $(c, -d)$ with $0 \leq -d < c \leq 90$, and also for all pairs of positive integers (c, d) with $c \leq 90$ and $d \leq 1000$. We now state some observations about these data, which lead us to make several conjectures.

Our first observation is that, for fixed c, positive integers d that are "close" to each other tend to produce the same degree, and similarly negative integers d that do not differ by much tend to produce the same degree. This is especially pronounced for large positive integers d: as it almost always happens that, for fixed c, all d between 50 and 1000 produce the same degree.

Next, for fixed c, the integers $d = 0, -1, \ldots, 1-c$ produce at most two different degrees, and if there are two they are consecutive integers. For fixed c and positive d, the degree tends to be smallest for $d < 20$, say, and the degree for $d = 1000$ exceeds the degree for $d = 1$ for all $c > 5$. Moreover, for $c > 5$, every $d > 30$ has larger degree than does $d = 1$. So, it seems that very small positive values of d produce a degree smaller than that produced by all larger positive d.

Finally, we observe that, for fixed c, every $d > 32$ produces a larger degree than any negative d. This suggests an interesting relationship between positive and negative values of d. Theorem 4.1 implies that, for fixed c, every d produces a degree in a certain interval whose length is something like $\log c$; our feeling is that the negative and only slightly positive values of d lead to degrees in the lower part of this interval, whereas larger positive values of d lead to degrees in the upper part of this interval. We shall not spoil the reader's fun by proving these conjectures.

References

[Berlekamp et al. 1982] E. R. Berlekamp, J. H. Conway, and R. K. Guy, *Winning Ways For Your Mathematical Plays*, Academic Press, London, 1982.

[Epp and Ferguson 1980] R. J. Epp and T. S. Ferguson, "A Note on Take-Away Games", *Fibonacci Quart.* **18** (1980), 300–303.

[Schwenk 1970] A. J. Schwenk, "Take-Away Games", *Fibonacci Quart.* **8** (1970), 225–234.

[Whinihan 1963] M. J. Whinihan, "Fibonacci Nim", *Fibonacci Quart.* **1** (1963), 9-12.

MICHAEL ZIEVE
DEPARTMENT OF MATHEMATICS
UNIVERSITY OF CALIFORNIA
BERKELEY, CA 94720
    zieve@math.berkeley.edu

New Theoretical Vistas

Many authors have examined games that do not satisfy one or more of the conditions on page 1, or have otherwise extended the framework of combinatorial games. No rule is sacred. In this section, Berlekamp introduces a methodology, based on the notion of auctions, to analyze a position in a possibly loopy game, in which play might continue forever. Blackwell relaxes the perfect-information condition. Loeb considers many-player games. Lazarus, Loeb, Propp, and Ullman replace the rule that players have alternate turns by an "auction-based" approach not unlike Berlekamp's. The remaining articles represent significant theoretical advances, while remaining within the traditional framework.

Games of No Chance
MSRI Publications
Volume **29**, 1996

The Economist's View of Combinatorial Games

ELWYN BERLEKAMP

ABSTRACT. We introduce two equivalent methodologies for defining and computing a position's mean (value of playing Black rather than White) and temperature (value of next move). Both methodologies apply in more generality than the classical one. The first, following the notion of a free market, relies on the transfer of a "tax" between players, determined by continuous competitive auctions. The second relies on a *generalized thermograph*, which reduces to the classical thermograph when the game is loop-free.

When a sum of games is played optimally according the economic rules described, the mean (which is additive) and the temperature determine the final score precisely.

This framework extends and refines several classical notions. Thus, finite games that are numbers in Conway's sense are now seen to have *negative* natural temperatures. All games can now be viewed as terminating naturally with integer scores when the temperature reaches −1.

Introduction

At every position of a game such as Go or Domineering, there are two very important questions:

Who is ahead, and by how much?

How big is the next move?

Following Conway [1976], classical abstract combinatorial game theorists answer these questions with a *value* and an *incentive*, as discussed in *Winning Ways* [Berlekamp et al. 1982]. These answers are precisely correct when the objective of the game is to get the last legal move. Values and incentives are themselves games, and can quickly become complicated.

Our ideal economist takes a different view. Following Hanner [1959] and Milnor [1953], he views the game as a contest to accumulate points, which can eventually be converted into cash. Our modern economist monetarizes the answers to our opening two questions into prices—real numbers that can be determined by competitive, free-market auctions. Specifically, after two gurus have completed their studies of a position, the economist might ask each to submit a sealed bid, representing the amount the guru is willing to pay in order to play Black. The bid may be any real number, including a negative one (if the position favors White). Assuming the bids differ, the referee computes their average μ. The player with the higher bid plays Left (= Black), in return for which he pays μ dollars to his opponent. The player with the lower bid plays Right (= White). Each player is necessarily happy with this assignment, since it is no worse for him than the bid he submitted.

For mnemonic purposes, the reader should think of μ as the *market value*. Later, we shall see that it can also be interpreted as the *mean value*, or, in thermography, as the *mast value*.

Our ideal economist also uses competitive auctions to determine the sizes of the moves. Throughout the game, the economist requires that each move made by either player must be accompanied by a fee or "tax", which the mover must pay to his opponent. At the beginning of the game, the tax rate is determined by a competitive auction. Each player submits a sealed bid. The bids are opened, and the higher one becomes the new tax rate. The player making this bid makes the first play at the new tax rate. Then, at each turn, a player may elect to either

- pay the current tax (to his opponent) and play a move on the board, or
- pass at the current tax rate.

After two consecutive passes at the current tax rate, a new auction is held to restart the game at a new and lower tax rate. Legal bids must be strictly less than the prior tax rate. There is also a minimum legal bid, t_{min}. A player unwilling to submit a bid $\geq t_{min}$ may pass the auction. The game ceases when both players pass the auction.

If the players submit equal bids, the referee may let an arbitrator decide who moves next. The arbitrator is disreputable: Each player suspects that the arbitrator may be working in collusion with his opponent. Alternatively, when there are equal bids for the tax rate, the referee may require that the next move be made by the player who did not make the prior move. In this way, the referee, at his sole discretion, can implement a "preference for alternating moves".

In some traditional games, including Japanese Go, the minimum tax rate t_{min} is defined as 0. When the game stops, the score is tallied according to the position of the board, and the corresponding final payment of "score" is made. This "score" payment is added to the tax payments that the players have paid each other, and to the payment(s) to buy the more desirable color(s) initially. Since all payments are zero-sum, one player's net gain is his opponent's net loss. The player who realizes an overall net gain is the winner. In Economic games, a tied outcome occurs whenever each player's total net cumulative payment to his opponent is zero.

For *Winning Ways*–style games, which include mathematical play on "numbers" beyond the conventional mathematical "stopping positions", t_{min} is defined to be -1. The minimum tax rate can also be defined as -1 for some non-Japanese versions of Go rules.

When such a game ceases, no score remains on the board: the outcome depends entirely on the net cumulative total of payments made between the players.

1. A Demonstration Game

We now hold a demonstration using the Economist's Rules. The players are Yonghoan Kim and David Wolfe.

The referee, EB, gives each player $5 initial capital, and they get change at the bank, in cash and chips valued as follows:

$$\$5, \$1 = \text{US notes}$$
$$Q = \text{US quarter}$$
$$G = \text{green chip, worth } 1/4 \text{ of a unit, now set at } \$1$$
$$Y = \text{yellow chip, worth } 1/16 \text{ of a unit}$$

The game they will play is the sum of a Go region and a Domineering region. The starting position of the Domineering region is the empty 4×9 board. The starting position of the Go region is shown on the right.

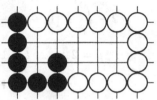

We begin by holding three concurrent auctions. Each player submits one bid for the Vertical dominoes, another for the Black Go stones, and a third for the first move on the sum of both games. All sealed bids are opened concurrently. Here are the results of our demonstration auction:

	YK	average	DW	result
Go Black	$-2\frac{1}{4}$	-2	$-1\frac{3}{4}$	DW plays Black (Left)
Domineering Vertical	$1\frac{1}{2}$	1	$\frac{1}{2}$	DW plays Horizontal (Right)
First Move	$1\frac{1}{4}$	$1\frac{1}{2}$	$1\frac{3}{4}$	DW moves first

For consistency, we decide to negate the Go board so that sides played by YK and DW in each game match our conventional uses of Left and Right, respectively. This is achieved by reversing all colors in the Go position. Here are the auction results restated in terms of $-$Go:

	YK	average	DW	result
$-$Go Black	$2\frac{1}{4}$	2	$1\frac{3}{4}$	YK plays Left
Domineering Vertical	$1\frac{1}{2}$	1	$\frac{1}{2}$	YK plays Left
First Move	$1\frac{1}{4}$	$1\frac{1}{2}$	$1\frac{3}{4}$	DW moves first

The following payments are then made:

YK pays DW $2 in cash to play Left (Black) on $-$Go
YK pays DW $1 in cash to play Left (Vertical) on Domineering
DW pays YK $1.50 in order to make the first move.[1]

---

[1]In order to pay the tax for the first move, DW first gets change from YK: $1 for 1Q + 2G + 4Y. He then puts $1.50 worth of chips (4G + 8Y) onto the table for YK.

1.1. Erdős' Book.

The mathematician Paul Erdős frequently mentions "The Book", a supernatural reference document that contains all mathematical truths in their most revealing forms. To expedite the remainder of this demonstration, we allow the participants and ourselves to peek into The Book. Figure 1 expresses

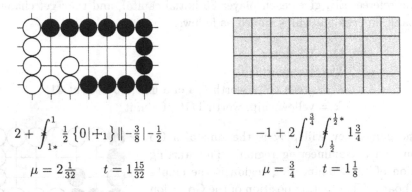

$$2 + \int_{1*}^{1} \tfrac{1}{2}\{0|+_1\}\| -\tfrac{3}{8}| -\tfrac{1}{2} \qquad\qquad -1 + 2\int_{\tfrac{1}{2}}^{\tfrac{3}{4}} \int^{\tfrac{1}{2}*} 1\tfrac{3}{4}$$

$$\mu = 2\tfrac{1}{32} \qquad t = 1\tfrac{15}{32} \qquad\qquad \mu = \tfrac{3}{4} \qquad t = 1\tfrac{1}{8}$$

Figure 1. Book starting values.

(albeit in perhaps unfamiliar symbols) the Book values for the two starting positions. The values also contain more information than we need to play optimally according to the Economist's Rules. In addition to a single good move for each player at the current tax rate, for each position we need only its fair market value, μ, and one of its fair tax rates, t (also called the "temperature"). These values are shown on the bottom line of Figure 1.

The remainder of our demonstration will illustrate how players able to peek into The Book can play optimally by following a few very simple common-sense rules, based on two important facts:

- The value of playing Left in any region is μ.
- Although the "size of a move" in any region is not necessarily unique, it may always safely be taken as the temperature of that region.

The *Book Strategy* is as follows:

0. Initially, bid μ for the privilege of playing Left.
1. At every stage of play, define the global temperature of the game as the maximum of the temperatures of its regions. If it is your turn, proceed as follows:

   (a) If opponent's prior move has created a region whose temperature exceeds the current tax rate, respond by playing in the same region as the opponent's prior move.

   (b) Otherwise, if the temperature of the game is at least as large as the current tax rate, play in a region of maximum temperature.

   (c) If the tax rate exceeds the maximum regional temperature, pass. If you have any canonical move(s), bid the maximum regional temperature.

The strategy is designed to answer directly this fundamental question:

Should I respond locally to my opponent's prior move?

Restated in Go jargon, that same question becomes

Does my opponent's prior move have *sente*?

Since the Book Strategy focuses on this question, it is also called *Sentestrat*. After the demonstration, we will discuss how Sentestrat compares with two other strategies, Hotstrat and Thermostrat.

Once Yonghoan Kim and David Wolfe obtain access to The Book, each of them elects to follow the Book Strategy.

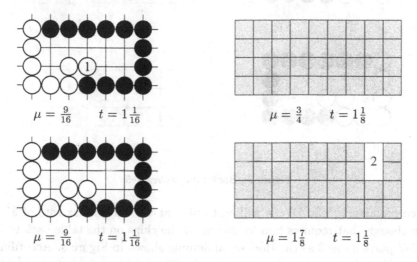

Figure 2. The demonstration game: moves 1 and 2 played according to The Book.

DW begins playing a White Go stone in the center of the bottom row. This is shown as move 1 in Figure 2. The cost of this move was $1.50, as determined by the initial auction. According to The Book, that bid was a slight error; the true Book value of the initial temperature was $1\frac{15}{32}$. However, White's move 1 has lowered the temperature of the Go region to $1\frac{1}{16}$. The Domineering region has temperature $1\frac{1}{8}$, which is larger. So, now that they are able to reference The Book, after move 1, both players decline to play at a tax rate of $1.50. Instead, they both bid $1.125.

Since the bids are tied, the play alternates, and YK is selected to play move 2 at a cost of $1.125. Since this is less than the $1.50 worth of chips he has on the table, YK pockets $0.375 of these chips, and moves the rest of the chips back to DW in payment of the tax on move 2. YK plays move 2 as the Vertical domino shown. According to The Book, this $1.125 move has increased μ for the Domineering region from $\frac{3}{4}$ to $1\frac{7}{8}$, but it has not changed the temperature,

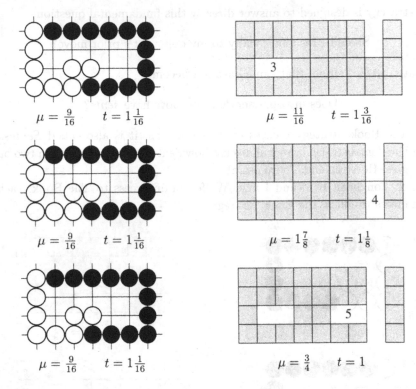

Figure 3. Book play, moves 3–5.

which remains $1\frac{1}{8}$. So DW is willing to play at the current tax rate of \$1.125, even though that requires him to give all of the chips on the table back to YK.

DW plays move 3 as the Horizontal domino shown in Figure 3. According to The Book, this move has changed μ for the Domineering region from $1\frac{7}{8}$ to $\frac{11}{16}$, a net decrease of $1\frac{3}{16} = \$1.1875$. As this exceeds the \$1.125 that Right paid for the move, Right may appear to have gained \$0.0625 at move 3. However, this apparent gain is ephemeral, because the move also increased the temperature to $1\frac{3}{16}$, which exceeds the current tax rate of $1\frac{1}{8}$. Therefore, following move 3, YK is eager to play move 4, because it is now possible for him to play so as to gain $1\frac{3}{16}$ at a cost of only $1\frac{1}{8}$. The tax on move 4 returns all of the chips on the table to DW, changes μ to $1\frac{7}{8}$, and lowers the temperature back to $1\frac{1}{8}$. DW then returns all of the chips on the table back to YK, and plays move 5. This lowers μ back to $\frac{3}{4}$.

The careful observer will notice that the four-move sequence 2–5 had no net effect on μ. And, although \$1.125 worth of chips moved back and forth across the table several times, after move 5 they all ended up back where they were after move 1.

After move 5, The Book reveals that the Domineering temperature is 1, which is now less than the Go temperature of $1\frac{1}{16}$. So both players then elect to pass

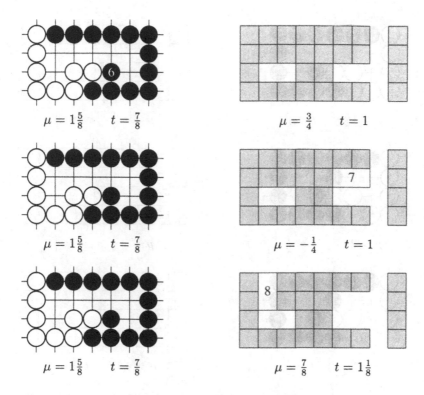

$$\mu = 1\tfrac{5}{8} \qquad t = \tfrac{7}{8}$$

$$\mu = \tfrac{3}{4} \qquad t = 1$$

$$\mu = 1\tfrac{5}{8} \qquad t = \tfrac{7}{8}$$

$$\mu = -\tfrac{1}{4} \qquad t = 1$$

$$\mu = 1\tfrac{5}{8} \qquad t = \tfrac{7}{8}$$

$$\mu = \tfrac{7}{8} \qquad t = 1\tfrac{1}{8}$$

Figure 4. Book play, moves 6–8.

at the tax rate of $1\tfrac{1}{8}$. The referee holds a new auction to determine who will move next, and both players bid $1\tfrac{1}{16}$. Because his opponent made the last move at the old tax rate, YK pockets chips worth $\tfrac{1}{16}$, leaving chips worth $1\tfrac{1}{16}$ on his side of the table. Since the auction was tied, the referee adopts the alternating move convention and assigns the next move to YK, who pays the new tax rate by moving all of the \$1.0625 worth of chips on the table to DW as he plays move 6 on the Go position, as shown in Figure 4. According to The Book, move 6 increases the local μ to $\tfrac{9}{16} + 1\tfrac{1}{16} = 1\tfrac{5}{8}$, and decreases the Go temperature to $\tfrac{7}{8}$. The Domineering region is now the hottest game, with temperature $t = 1$. As the old tax rate was $1\tfrac{1}{16}$, both players elect to pass and bid $t = 1$. DW pockets chips worth $\tfrac{1}{16}$, because his opponent made the last move at the old tax rate. The value of the chips on the table is then \$1, on DW's side. The referee assigns the next move to DW, who pays the chips back to YK and plays move 7, followed by YK's move 8 and DW's move 9 (in Figure 5). The net effect of the three-move sequence 7–9 was a payment of \$1 of chips from DW to YK, in return for a \$1 change in the value of the Domineering μ in DW's favor.

After move 9, both players pass and bid $\tfrac{7}{8}$. The tax rate decreases by $\tfrac{1}{8}$, which YK pockets, leaving \$0.875 worth of chips on the table. These remaining chips

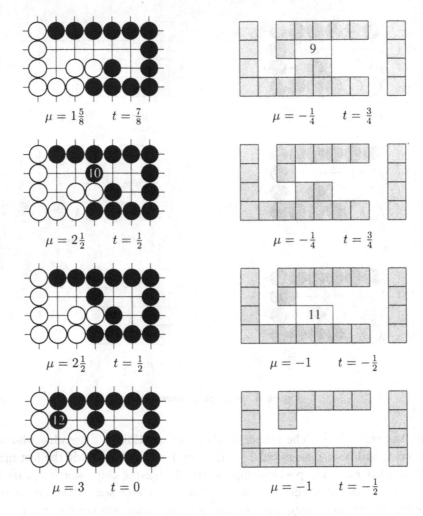

Figure 5. Book play, moves 9–12.

are transferred to DW as YK plays move 10. This move increases the Go value of μ by the same amount as the tax paid, $2\frac{1}{2} - 1\frac{5}{8} = \frac{7}{8}$.

After move 10, both players pass and bid $\frac{3}{4}$. The tax rate decreases by $\frac{1}{8}$, which DW pockets, leaving \$0.75 worth of chips on the table. These remaining chips are transferred to YK as DW plays move 11 on the Domineering board. This move decreases the Domineering value of μ by the same amount as the tax paid, $-1 - (-\frac{1}{4}) = -\frac{3}{4}$.

After move 11, both players pass and bid $\frac{1}{2}$. The tax rate decreases by $\frac{1}{4}$, which YK pockets, leaving \$0.50 worth of chips on the table. These remaining chips are transferred to DW as YK plays move 12 on the Go board. This move has the effect of increasing the Go value of μ by the same amount as the tax paid, $3 - 2\frac{1}{2} = \frac{1}{2}$.

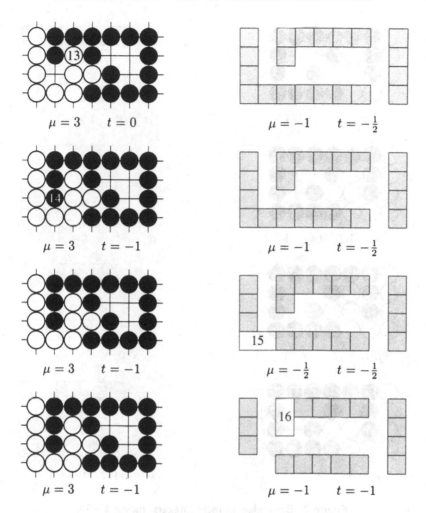

$\mu = 3 \qquad t = 0$

$\mu = -1 \qquad t = -\frac{1}{2}$

$\mu = 3 \qquad t = -1$

$\mu = -1 \qquad t = -\frac{1}{2}$

$\mu = 3 \qquad t = -1$

$\mu = -\frac{1}{2} \qquad t = -\frac{1}{2}$

$\mu = 3 \qquad t = -1$

$\mu = -1 \qquad t = -1$

Figure 6. Book play, moves 13–16.

According to The Book, after move 12, the temperature of the Go region is 0 and the temperature of the Domineering region is $-\frac{1}{2}$. So both players pass at the tax rate of $\frac{1}{2}$, and submit bids of 0 for the new tax rate. DW pockets all $0.50 worth of chips on the table. The new tax rate is now 0, which equals the value of the chips on the table.

DW plays move 13 in Figure 6, which is the first move at the tax rate of 0. YK responds with move 14, at the same tax rate of 0.

After move 14, The Book reveals that both the Go position and the Domineering position have negative temperatures. So both players would prefer not to play. In fact, neither will play unless his opponent pays him to move! So each player bids −$0.50. This means that neither player will move unless his opponent pays him $0.50. The new tax rate is negative! So DW *collects* $0.50

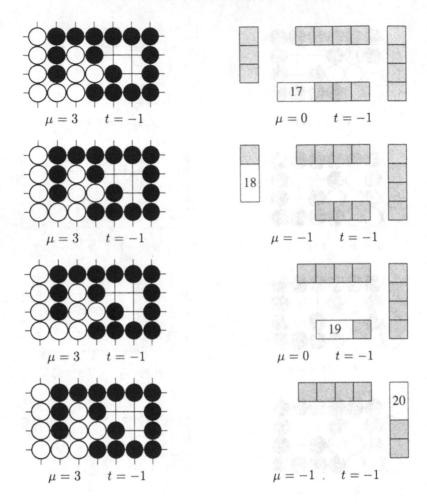

Figure 7. Book play on more integers, moves 17–20.

in cash from YK as he plays move 15. But YK responds with move 16, in return for which he gets his $0.50 back.

After move 16, The Book reveals that both positions have temperature equal to negative one. Both players pass at the tax rate of $-\frac{1}{2}$, and submit bids of -1 to set the new tax rate.

DW collects $1 cash from YK for move 17 in Figure 7, but YK collects it back for move 18. The players continue to "cash in" their positions, at the rate of $1/move, on moves 19 and 20 in Figure 7, then 21, 22, and 23 in Figure 8.

According to Japanese-style Go rules, the play ceases after move 23. The Go position is scored as $+3$ points for Black, and White is required to play Black $3 to settle this score.

According to the mathematized (Universalist) Go rules, the game continues after move 23. White submits no bid. Black submits the minimum bid that the

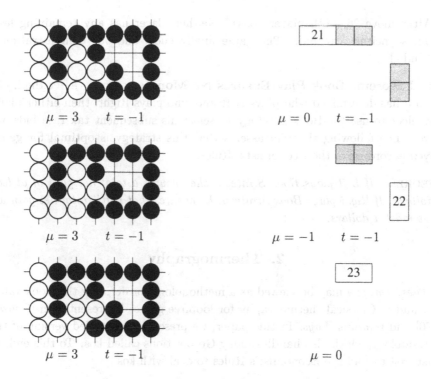

Figure 8. Book play on more integers, moves 21–23.

rules allow, which is −1. So Black wins the right to play at a tax rate of −1. In Figure 9, he plays move 24 as a Black Go stone in the middle of the three empty spaces, and collects \$1. (According to Universalist Go rules, the Black Go stones are now "immortal", and cannot be captured even if Black fills in both of the visible eyes.)

Again, White passes and submits no bid. So Black again submits the minimum allowable bid of −1, and plays move 25 as another Black Go stone, for which he again collects \$1 from White. Again White has no option but to pass and submit no bid. Black again submits the minimum allowable bid of −1, wins the auction, and plays move 26 as another Black Go stone, for which he collects his final \$1 from White.

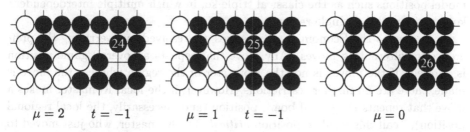

Figure 9. Book play on integers remaining after move 23.

After move 26, both players pass. Neither player has any remaining legal move, so neither can bid. The game finally ends, and our demonstration is concluded.

1.2. Theorem: Book Play Ensures No Monetary Loss. After the initial auction has determined who plays Left and who plays Right, then either player may elect to use the Book Strategy to select his subsequent tax rate bids and plays. The following theorem asserts that this strategy is optimal for games played according to the Economist's Rules:

THEOREM. *If Left plays Book Strategy, she ensures a total net gain of at least μ dollars. If Right plays Book Strategy, he ensures that Left's total net gain will be at most μ dollars.*

2. Thermography

Thermography may be viewed as a methodology for deriving the Book values of μ and t. Classical thermography for loopfree games is presented in [Conway 1976] and *Winning Ways*. In this paper, we present an extended version of this methodology, which also handles loopy Go positions called kos. To this end, we must first extend the Economist's Rules to deal with kos.

2.1. The Economist's Ko Rules. The game of Go includes some loopy positions. By far the most common such positions are 2-cycle loops. Locally, White captures a stone, Black captures it back, then White again captures Black's stone, etc. These positions are called kos. All popular dialects of Go rules prohibit such loopiness, by banning the immediate recapture of a single stone if it would return the board to the same position it reached after mover's prior turn.

The extension of thermography to cover kos is facilitated by the following extensions of the Economist's Rules:

For each region of the game, a specified player is designated as the local *Komaster*. In principle, the privilege of being Komaster might be determined by local auctions, held before the play begins. If the global game contains several regions with potential kos, then it is quite possible for Left to be Komaster of some regions, and Right to be Komaster of others. (These rules do not attempt to model positions such as the classical triple ko, in which multiple interdependent kos appear within a single region.)

According to the Economist's Rules, kos are resolved by overruling the preference for alternating moves. Any move that repeats an overall game position is allowed. If the mover is not the Komaster of the region wherein the move is made, there are no further restrictions. However, if the local Komaster makes a move that repeats the global board position (and, necessarily, the local regional position), I call this local ko position *critical*. The Komaster, who just moved to the critical ko position, is compelled to make another local move immediately.

This rule for resolving ko overrides the referee's normal preference for alternating moves. Instead, the Komaster makes two moves consecutively, to and from the critical ko position, and the Komaster pays two taxes (at the same tax rate), one tax for each move.

2.2. Bottom-Up Thermography.

As mentioned earlier, thermography may be viewed as a methodology for calculating the book values, μ and t. The version we now present handles kos. It also gives the classical results in the special case of loopfree games.

Trajectories. We call the basic data structure of thermography a *trajectory*. A trajectory is a piecewise linear real function of a real variable. Its input argument, t, is called the *temperature*, or the *tax rate*, and can range from 0 to $+\infty$ (or from -1 to $+\infty$, according to the extension made in Section 4.2.) The trajectory's output value is a real number, called v or μ. Within each linear piece of the trajectory, the value of the derivative dv/dt is constant and equal to an integer. In elementary calculus, it is conventional to plot the independent variable on the horizontal axis and the dependent variable on the vertical axis. However, in thermography it is conventional to rotate the conventional calculus plots counterclockwise by 90°, in conformity with a long-standing tradition introduced by Conway. The temperature, t, is plotted along the vertical axis, with increasing values of t corresponding to higher positions along the axis. The value, v, is plotted along the horizontal axis, with the convention that increasing positive values are plotted to the Left; large negative values, to the Right. This facilitates a more direct correspondence with the formal expressions for most hot games, in which Left moves toward more positive values and Right moves toward more negative values.

A thermograph is a pair of trajectories, called stops or scores or *walls*. In this exposition, we use the term "wall". The Left wall, LW, is always at least as great as the Right wall, RW.

Scaffolds, Hills, and Caves. The final stages of the construction of a thermograph begin with another pair of trajectories LS and RS, called *scaffolds* (Figure 10, left). In the most general case, the Left and Right scaffolds might cross several times. A temperature interval in which the Left scaffold exceeds the Right scaffold is a *hill*, or *solid region*. A temperature interval in which the Right scaffold exceeds the Left scaffold is a *cave*, or *gaseous region*. Within a hill, the Left wall equals the Left scaffold and the Right wall equals the Right scaffold. However, within a cave, the Left and Right walls coincide in a local trajectory called the *mast*.

The lowest region of a nondegenerate Thermograph is necessarily a hill. The interval along the v-axis, or "ground", between the walls of this hill is the *base* of the hill. The point at which the two scaffolds cross at the top of the hill is both the *summit* of the hill and the *drain* of the cave above it. If the scaffolds

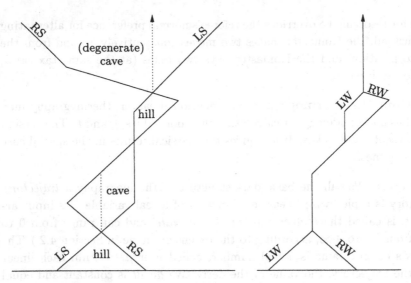

Figure 10. Scaffolds (left), and the thermograph constructed therefrom (right).

cross again at the top of the cave, that point is called the *chimney* of the cave, and the *fulcrum* of the hill above it.

Normal Masts. Within any cave, the normal mast can be easily constructed by following the path of an ideal balloon from the drain to the chimney. When free to do so, the balloon rises vertically. If it hits a scaffold, the balloon follows that roof scaffold upward until it is either again free to rise vertically or until it reaches the chimney, at which point the mast ends and the Left and Right walls again become different, starting at the fulcrum of the adjacent hill. This is shown on the right in Figure 10. Moving upward from $t = 0$, this thermograph has a hill, a cave, another hill, and then an infinite gaseous region (a degenerate cave) through which the mast rises vertically to infinity. All hot games have thermographs with a hill at $t = 0$ and a vertical mast as t approaches infinity.

The walls give the values of the economic game with a current tax rate of t.

If the thermograph is a hill region at temperature t, then each player desires to move first. By doing so, he can ensure a value given by the corresponding wall. If the thermograph is a cave region at temperature t, then at least one player prefers to pass until the tax rate becomes lower. If the mast is vertical, then both players prefer to pass, and neither will play until the tax rate is lowered. If the mast coincides with a roof of the cave which is a Left scaffold, then Left prefers to play and Right prefers to pass. If it is Right's turn, he will pass and propose a tax cut, but Left will then reject Right's proposal. Left will instead exercise his right to move at the old, higher tax rate. In a subsequent subsection we will see a specific example of this surprising phenomenon, where it is to a player's advantage to insist on playing at a higher tax rate rather than at a lower one.

DEFINITIONS. We define the *initial temperature of a game* G, as the temperature of the base of the highest mast of its thermograph. A game G is said to be *active at temperature t* unless its thermograph at t is a vertical mast. In that case, we say that G is *dormant at t*. If G is dormant at t, then the temperature of the base of this mast is called the *activation temperature of G at t*.

Pass-Banned Masts. The constructions of thermographs must reflect the Economic Rules that resolve kofights. As described in Section 2.1, these rules do not allow the Komaster to pass from a critical position. Under such circumstances, the masts within each cave degenerate into the maximum (Right) scaffold if Right is not allowed to pass, or the minimum (Left) scaffold if Left is not allowed to pass.

Recursion for Thermographs of Games with No Immediate Ko. Let G be a game represented in canonical form, $G = \{G^L \mid G^R\}$. where G^L and G^R are sets of games whose thermographs are already known. We first consider the case in which G itself is not part of any loop, even though some of the positions in G^L and G^R may contain kos.

Under these circumstances, the scaffolds of G are defined as

$$\text{Left scaffold of } G_t = -t + \max_{G^L}\left(\text{Right wall of } G_t^L\right),$$

$$\text{Right scaffold of } G_t = t + \min_{G^R}\left(\text{Left wall of } G_t^R\right)$$

Thermographs of Kos. Figure 11 shows a simple 33-point ko and the graph of its legal moves.

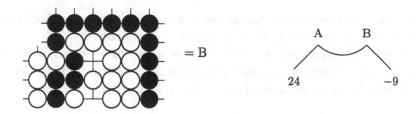

Figure 11. Simple 33-point Ko.

Let G and H be a pair of games that represent the two states of a ko. G is a Left follower of H, and H is a Right follower of G. In the general case, G and H may also have other followers; thus we may have

$$G = \{V \mid W, H\}, \qquad H = \{G, X \mid Y\}.$$

For such a ko, we construct two pairs of thermographs. $\hat{G}$ and $\hat{H}$ assume that Left is the regional Komaster; $\check{G}$ and $\check{H}$ assume that Right is the regional Komaster. From $\hat{H}$, Left may play to a close relative of G, which we call $\hat{G}$.

The relevant pair of thermographs are computed by working through the following three equations, from the bottom up:

$$\hat{G} = \{V \mid W, \hat{H}\},$$
$$\hat{H} = \{\hat{G}, X \mid Y\},$$
$$\hat{G} = \{V \mid W\}, \text{ with proviso that Left cannot pass immediately.}$$

The motivation behind these equations is as follows: Since Left, as komaster, can win the ko, she may effectively prevent Right's move from $\hat{G}$ back to $\hat{H}$. However, the price of this prevention is the constraint that if Right elects to pass because of the koban, Left is forbidden the privilege of passing immediately from $\hat{G}$. Such a pass would permit Right to recapture the ko at a lower tax than Left paid to get there, and to return to the same board position as $\hat{H}$. However, this new position would be accompanied by a ko state less favorable to Left than $\hat{H}$. So if Left is competent, we should never see the sequence of play in which Left moves from $\hat{H}$ to $\hat{G}$, and later passes when position $\hat{G}$ is still available. If Left intends to do that, she could do at least as well by passing immediately from $\hat{H}$.

Similarly, to find the thermographs of $\check{G}$ and $\check{H}$, we may use the next three equations, again from the bottom up:

$$\check{H} = \{\check{G}, X \mid Y\},$$
$$\check{G} = \{V \mid W, \check{H}\},$$
$$\check{H} = \{X \mid Y\}, \text{ with proviso that Right cannot pass immediately.}$$

A similar methodology enables us to compute thermographs for iterated kos. For example, suppose thermographs A and E are known, and that

$$B = A \mid C, \qquad C = B \mid D, \qquad D = C \mid E.$$

Then, if we are interested in the games in which Left wins these kos, we may compute thermographs of $\hat{B}$, $\hat{C}$, $\hat{D}$, $\hat{C}$, and $\hat{B}$, in that order.

Here are the thermographs derived by this methodology for the positions of Figure 11, when White is komaster. (Naturally, $\check{A} = A$, with Right as komaster; $\check{B} = B$, with Right as komaster; $\check{B} = \check{B}$ with Right not allowed to pass.)

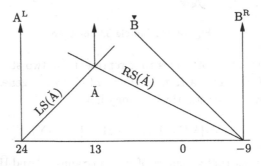

A^L is a vertical mast at $v = 24$. B^R is a vertical mast at $v = -9$. The pass-banned $\check{B}$ is a mast of slope $+1$ starting from $t = 0$, $v = -9$. The Left scaffold of Ǎ is obtained by subtracting t from A^L. It is a ray with slope -1, from $t = 0$, $v = 24$. The Right scaffold of Ǎ is obtained by adding t to $Ǎ^R = \check{B}$. It is a ray with slope $+2$ from $t = 0$, $v = -9$. The two scaffolds intersect at $t = 11$, $v = 13$. For $t \le 11$, $LW(Ǎ) = LS(Ǎ)$ and $RW(Ǎ) = RS(Ǎ)$. For $t \ge 11$, $LW(Ǎ) = RW(Ǎ) = 13$.

The thermograph for B̌ of Figure 11 is constructed as follows:

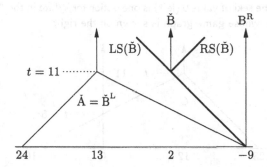

$RS(B̌)$ is a ray of slope -1 from $t = 0$, $v = -9$. $LS(B̌)$ is obtained by subtracting t from $RW(B̌^L) = RW(Ǎ)$. For $t \ge 11$, $LS(B̌) = RS(B̌)$, but $LS(B̌)$ changes direction at $t = 11$. Above $t \ge 11$, $LS(B̌)$ and $RS(B̌)$ form a cave, in which $LW(B̌) = RW(B̌) = 2$. For $0 \le t \le 11$, $LW(B̌) = RW(B̌) = -9 + t$.

The derivation of thermographs for Â and B̂ is directly analogous. Here are the results:

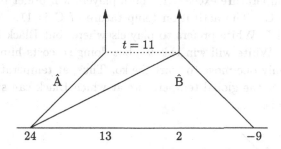

Figure 12, left, shows a Go position called a *seki*. This position is terminal because competent players will pass. Mathematically, its value is

$$24 \,|\, {-10} \,\|\, 26 \,|\, {-8} = 0.$$

Its thermograph is a vertical mast at $v = 0$.

Figure 12, middle, shows G, a famous position from traditional Go literature called the "thousand-year Ko". Its game graph is shown on the right in the same figure. From G, White can play either to the seki of value 0 just mentioned, or to B, the ko of Figure 11. Left has two similar options, whose thermographs are superimposed in Figure 13. From Ǧ, if $t < 5$, Left prefers to play to the seki; but if $t > 5$, Left prefers to start the ko. Figure 14, left, shows the result of

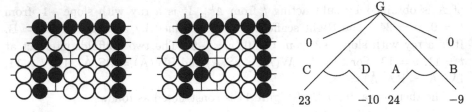

Figure 12. The seki of value 0 (left) is one option for White in the "thousand-year Ko" (middle), whose game graph is shown on the right.

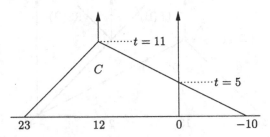

Figure 13. Thermographs of black's options from Ǧ.

translating the maximal Right wall of G^L into the Left scaffold of Ǧ. Figure 14, right, shows RS(Ǧ).

The thermograph of Ǧ, shown in Figure 15, left, illustrates some features of the thousand-year ko that most intermediate Go players find counterintuitive. If the global temperature exceeds 11, both players will prefer to play elsewhere rather than on Ǧ. The activation temperature of Ǧ is 11. At temperatures between 11 and 7, White prefers to play elsewhere, but Black is eager to start the ko from Ǧ. White will win this ko, but doing so costs him two moves and it costs Black only one move to start the ko. Thus, at temperatures between 11 and 7, the higher the global temperature at which Black can start the kofight, the happier she is.

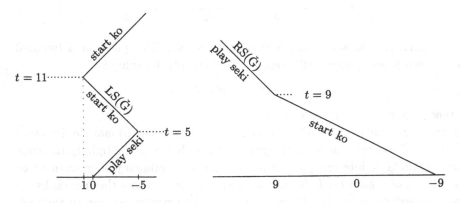

Figure 14. Left and Right scaffolds of Ǧ.

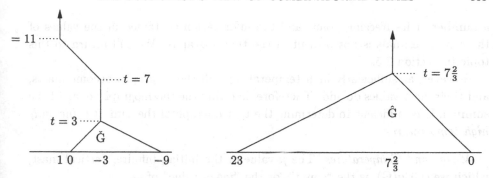

Figure 15. Left: Thermograph of Ǧ. Right: Thermograph of Ĝ.

At global temperatures between 3 and 7, Ǧ is again inactive. Both players prefer to play elsewhere.

At global temperatures below 3, both players are eager to play on Ǧ. White would like to start and win the ko, and Black is eager to avoid this loss by playing to the seki.

The thousand-year ko has two activation temperatures, and a region wherein one player prefers to pay a higher tax rate rather than a lower one. These features are unusual, and dependent on the assumption that White is Komaster. The thermograph of Ĝ, shown on the right in Figure 15, is much more common-looking. At high temperatures, the value of being Komaster on G is $7\frac{2}{3} - 1 = 6\frac{2}{3}$ points. This contrasts sharply with the common ko of Figure 11, where being komaster has zero value at high temperatures.

2.3. Overview of Thermography.

Combinations and Sums. Thermographs support combinations. The thermograph of A contains just enough information so that, when combined with the thermograph of B or C, one can obtain the thermographs of A|B and of C|A.

Thermographs do *not* support addition, as is evident from the following example. Let $A = 1|-1$, $B = \{1, 1 + A | -1\}$, $C = \{1, 1 + A | -1, -1 - A\}$.

Then A, B, −B, and C each has the thermograph shown on the left in Figure 16, which we call X. Yet A + B has thermograph Y; A − B has Z; A + C has X; and A + A has 0.

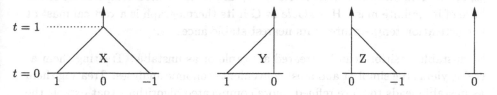

Figure 16. Thermographs don't behave well under addition.

For a general loopfree game, the shape of the thermograph at the base of its mast depends on the sign of the infinitesimal by which the game differs from a number at its freezing point, and the information contained in the values of these infinitesimals is not present in the thermographs. We will return to this topic in Section 2.5.

However, at sufficiently high temperatures, all thermographs become masts, and their mast values do add. Therefore, knowing the thermographs of all of the summands *is* sufficient to determine the thermograph of the sum *at sufficiently high temperatures.*

Means and Temperatures. The μ-value of the initial, infinite, vertical mast, which we call $\mu(G)$, is the "count", or the "mean value" of G.

The temperature at the base of this mast, which we call $t(G)$, is the "size of the move", or the "temperature" of G.

Placid Kos and Hyperactive Kos. A game G is said to be "hyperactive" if $\mu(\hat{G}) > \mu(\check{G})$. Such a game G necessarily contains a position K that is a ko. It is not the size of K itself that makes G hyperactive, but special circumstances which include the fact that K is a follower of some position of G that is cooler than K.

A game G is said to be "placid" if $\mu(\hat{G}) = \mu(\check{G})$. If G is placid, then when played according to Economic Rules, $\mu(G)$ does not depend on kothreats. Many kos, even many rather big kos, turn out to be placid.

Stable and Unstable Positions. Thermographs with more than a single hill and a single (degenerate) cave are relatively rare. In the common case, stable and unstable positions can be defined relatively simply, as follows:

DEFINITIONS. A position H of a game G is *stable* in G if all ancestors of H in G have temperatures greater than H's. H is *unstable* or *transient* in G if it has an ancestor of lower temperature. If H is not unstable, but has one or more ancestors of temperature equal to its own, it is called *semistable.*

In the more general case, in which some positions may have thermographs with multiple hills and caves, a single position may have multiple activation temperatures, and so a more refined definition is required:

REFINED DEFINITIONS. G is *stable in* G, and its activation temperature is the base of its infinite mast. H is *stable in* G if its thermograph is a vertical mast at the activation temperature of its nearest stable ancestor.

Semi-stable positions can be treated as stable or as unstable. Treating them as stable yields the simplest and most convenient economic analyses. Treating them as unstable leads to more refined, more complicated algorithms that exploit the calculus of infinitesimal games to get the last move at each t, whenever possible.

Sentestrat players will choose to play "in another region" "elsewhere" only when the region just played in has reached a stable position.

Sente and Gote. From the perspective of Sentestrat, an opponent's move that increases the temperature has sente; it requires our immediate response. However, this move and its response might begin a longer sequence of moves through several transient positions before the game reaches the next stable position. The question of whether the overall sequence had sente or gote depends on whether the length of this alternating sequence of moves was odd or even, and that question can usually be answered by examining the thermograph of the originating stable state. Generally speaking, if, just below the base of its relevant mast, both walls branch outward, then the initial move by either player is *gote*. If Left's wall drops vertically but Right's branches outward, then Left's initial move is *sente* but Right's is *reverse sente*. If Right's wall drops vertically but Left's branches outward, then Right's initial move is sente and Left's move is *reverse sente*.

These correspondences between the Go player's view of sente and gote and the thermograph are correct whenever the relevant positions have significantly different temperatures. However, when there are intermediate semi-stable positions, the distinction between sente and gote may become blurred.

Mainlines and Sidelines. Let n be a large integer, and suppose that two gurus play the sum of n copies of a game G. For each H that is a position of G, let $\#(H)$ be the number of copies of G that pass through the position H. If $\#(H)$ approaches infinity as n approaches infinity, then H is a "mainline" position of G. But if $\#(H)$ remains bounded as n approaches infinity, then H is a "sideline" position of G.

If G is a traditional gote position, then $\#(G^L) = \#(G^R) = n/2$, and both G^L and G^R are mainline positions. However, if G is a position in which Left's move is sente and Right's move is reverse sente, then, depending on who plays first from $n.G$, $\#(G^R) = 0$ or 1, and $\#(G^L) = n$ or $n-1$. In either case, G^R is a sideline position of G.

Thermographs of sideline positions have no effect on the means of any of their ancestors, although they often do affect the temperature(s) of some of their near ancestors.

We illustrate these notions with an example. Let A be the starting position on the right, before any of the numbered stones have been played. Then moves 1, 2, 3, and 4 show a main line of play, through positions we'll call B, C, D, and E. Position E is the terminal position after all four 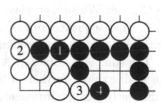 numbered stones have been played. F is the position if White rather than Black had played the first stone at 1. The means and temperatures of all dominant positions are shown in Table 1.

The only stable mainline positions are A, C, and E, each having mean 1. The mean of A is unaffected by any small change in the mean or temperature of any other position. Hence, to verify that the mean of A is 1, it is unnecessary to

Position	μ	t	Stable?	Mainline?
A	1	3	yes	yes
$B = A^L$	8	7	no	yes
B^L	15	-1	yes	no
$C = B^R$	1	1	yes	yes
C^L	2	-1	yes	no
$D = C^R$	$-\frac{1}{2}$	$1\frac{1}{2}$	no	yes
D^R	-2	0	yes	no
$E = D^L$	1	-1	yes	yes
$F = A^R$	-2	1	yes	no
F^L	-1	-1	yes	no
F^R	$-3\frac{1}{2}$	$1\frac{1}{2}$	no	no
F^{RR}	-5	0	yes	no
F^{RL}	-2	-1	yes	no

Table 1. Means and temperatures of dominant positions.

know the values of other means and temperatures precisely; appropriate bounds are sufficient.

In general, a stable mainline gote position will have both a mainline Black follower and a mainline White follower. On the other hand, stable mainline sente positions have only one mainline follower. In this example, the line of play shown is the unique main line. Notice that the mainline moves do not alternate colors; White plays both moves 2 and 3.

If we played a sum of n copies of A concurrently on n different Go boards, then the local line of play would follow the main line shown above on all but one or two boards. If White gets the first move overall, he plays one exceptional board from A to a sideline position, F. But Black then plays another board from A to B, and White responds by continuing to C. That happens on all $n - 1$ boards, successively. Black then might play a second exceptional board from C to C^L. But White then plays another C to D, and White responds by continuing to E. That sequence, from C to D to E, then happens on all boards, successively, until all positions are terminal.

The sum of n copies of A is only one example of a rich environment, which we shall explore further in Section 4.3. If a single copy of A is played in any sufficiently rich environment, the play within A will follow the main line shown above. Heuristically, the reason that Black rather than White will make the first move from A is that Black can move there at any tax rate between 7 and 3; White cannot afford to play A until the tax rate is 3. After A has been played to B and C, both players make several moves elsewhere if possible. Eventually, any time after the tax rate drops below 1.5, White can play at C. Since Black cannot afford to play there until the tax rate is 1, White will get in his move first, unless the background is so impoverished that there are no suitable moves "elsewhere".

Smaller Thermographic Databases of G. Given a game, G, and its purported thermograph, how much information about the positions of G is needed to verify that the purported thermograph of G is correct? Evidently, considerably less than the thermographs of all of the positions of G. Many positions of G are transient. Often the only information needed about many of the followers of transient positions is that they are "sufficiently hot", and this bound can easily be quantified. Precise thermographs of many sideline positions are also unnecessary; it is required only that portions of such thermographs fall between specified bounds.

How much information is needed to verify the purported mean of G? Evidently, even less than is needed to verify G's thermograph.

2.4. Top-Down Thermography.

"Top-down" thermography promises to be much more efficient than "bottom-up" thermography. The latter computes the details of many walls of positions that the former could ignore.

A proposed top-down approach to finding the thermograph of a position G starts by considering a reasonably large set of lines of play from G, such as $G^{LRL_1LRLRRLR}$. Most relevant lines do *not* follow alternating play. Indeed, empirical evidence indicates that Go positions, such as the thousand-year ko of Figure 12, which have different dominant Left options depending on the temperature, are fairly rare. So the main reason that there are typically many relevant lines of play from G is that there may be many relevant sequences of Left and Right plays, rather than choices between H^{L_1} and H^{L_2}.

The proposed approach to computing the "top-down" thermograph of G assumes that the given set of lines of play contains the "complete" set of G's stable positions. This assumption allows a program to infer that missing positions are sufficiently hot.

One is often interested in only the high-temperature portions of thermographs. For this reason, it would eventually be desirable to have programs that do *all* thermographic computations from a top-down perspective. Each trajectory might be stored as a list of slopes and breakpoints, listed in top-down order. A better programming strategy is to view each trajectory as a *function* whose outputs are a list of slopes and breakpoints, generated in the top-down order. This function is able to call other functions and itself, recursively, as needed, in order to pursue the top-down computation of thermographs. The programming goal is to get the next output as quickly as possible, by looking no further than necessary.

Thermography Justifies Some Go Maxims. Many of the standard Go maxims are equivalent to some of the common special cases that a top-down thermography program will encounter. For example, if we let $\tau = \frac{1}{2}(\mu(G^L) - \mu(G^R))$, and

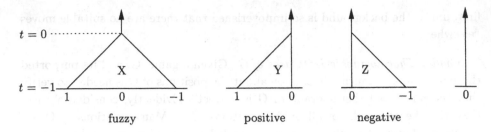

Figure 17. Thermographs of loop-free infinitesimals.

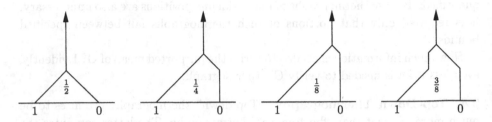

Figure 18. Thermographs of numbers.

if $t(G^L) \leq \tau$ and $t(G^R) \leq \tau$, then G might be said to be "gote", and we have $\mu(G) = \frac{1}{2}(\mu(G^L) + \mu(G^R))$, and $t(G) = \tau$. This condition is very common.

One type of situation in which thermography yields much more precision than traditional Go folklore is the "double sente" position, in which $t(G^L)$ and $t(G^R)$ are both substantially larger than $\text{LW}(G^L) - \text{RW}(G^R)$. Traditional Go maxims indicate only that a move from such a position G is very desirable, presumably relative to the naive approximation of movesize as $\text{LW}(G^L) - \text{RW}(G^R)$. This is true if the overall temperature is low and the position has just arisen.

2.5. Subzero Thermography. Although thermographs are conventionally plotted for temperatures ranging from 0 to infinity, they can also be plotted for negative temperatures. Temperatures infinitesimally less than zero appeared in *Winning Ways*, page 151, as "Foundations of Thermographs". Since the Economist's Rules allow temperatures as low as -1, thermographs can be extended into that same region. The four possible thermographs for an infinitesimal game are shown in Figure 17.

The astute reader will notice the correspondence between Figures 16 and 17.

At negative temperatures, the thermographs of noninteger numbers also become active, as illustrated in Figure 18.

The half-ish thermographs are shown in Figure 19.

If v is a number, at positive temperature, the thermograph of any v-ish game is a vertical line emating upward from the value v, which can be any rational number whose denominator is a power of 2. These values form a dense set on the line $t = 0$. However, the possible values of thermographs become sparse at negative temperatures. Figure 20 shows the superposition of the thermographs

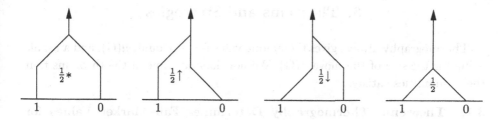

Figure 19. Thermographs of games infinitesimally close to $\frac{1}{2}$.

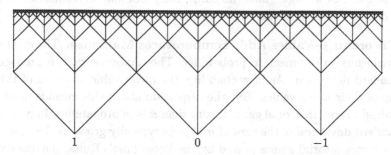

Figure 20. The foundations of number-ish thermographs.

of all number-ish loopfree games, in the temperature range between 0 and -1. Figure 21 illustrates where the thermographs of two particular numbers appear in Figure 20.

If one views an arbitrary conventional thermograph from a positive temperature perspective, it may appear to be resting on "solid ground", consisting of the v-axis at $t = 0$. However, if one probes deeper as in Figure 20, one finds that at $t = -1$ all thermographs rest on a discrete set of pilings located at the integer values. At $t = -\frac{1}{2}$, there are twice as many pilings located at half-integers, etc.

An earlier, less universal, version of Figure 20, based on *overheating* rather than on the Economist's Rules allowing negative temperature, appeared on page 172 of *Winning Ways*.

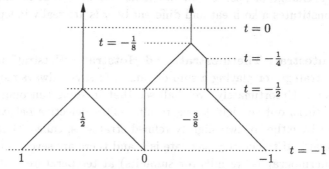

Figure 21. Underground thermographs of $\frac{1}{2}$ and $-\frac{3}{8}$.

3. Theorems and Strategies

Thermography always gives the unique value for the count, $\mu(G)$, and a usable value for the size of the move, $t(G)$. We now look at some of the theorems that these quantities satisfy.

3.1. Theorem: Thermography Determines Fair-Market Values for Games Played by Economic Rules. We assert that thermography determines fair economic values for μ and for the initial tax rate t. These values are "safe" to bid. For a single game (no sum), the proof follows recursively from the "Economist's Rules" and the way the thermograph was constructed.

3.2. Theorem: μ-Values Add; Temperatures Maximize. The mast-value, μ, is analogous to the mean in probability. The temperature, t, is analogous to the standard deviation. As in probability, the mean value of a sum of things is the sum of their mean values. But the dispersion about this mean is much more constrained in combinatorial game theory than it is in probability. In probability, the standard deviation of the sum of n objects typically grows as the square root of n. In combinatorial games played by the Economist's Rules, the dispersion is zero. In combinatorial games played according to the normal rule of alternating play, the dispersion of the sum about its mean is constrained by an upper bound independent of n. For sums containing no positions with hyperactive kos, this bound is proportional to t, and the constant of proportionality depends on the maximal iterative depth of any of the placid kos.

When multiple hyperactive kos are present, we need to define "independence" of regions. This can be done by constructing each local position on a different board, and then playing all boards concurrently. At each turn, a player selects any board and makes one move on that board. When the game ends, each player receives a single score, equal to the total score on all of the boards. Each ko that Left can win is placed on a board with a plentiful supply of Left kothreats; each ko that Right can win is placed on a board with a plentiful supply of Right kothreats. We further define the ko rule to prohibit any recapture of ko unless there is a change of position *on the same board*, so that no move on another board constitutes a kothreat and different boards are really independent of each other.

3.3. Sentestrat, Thermostrat, and Hotstrat. "Hotstrat" is an intuitively obvious strategy for playing a sum of games. It says "always move on a hottest summand". Unfortunately, we shall see that in some (uncommon) cases, this strategy turns out to yield losing results. This defective behavior of Hotstrat is evaded by either of two slightly refined strategies, called "Thermostrat" and "Sentestrat". If the current tax rate is t, and there are several current positions whose thermographs are hills (or summits) at temperature t, then these three strategies choose among those options in slightly different ways:

- *Hotstrat:* Move in a region whose temperature is maximal.

- *Thermostrat:* Move in a region whose hill is widest at temperature t.

- *Sentestrat:* If your opponent has just played in one of these hill regions, respond directly by playing in the same region in which he just moved.

All three strategies essentially agree under other circumstances. If all thermographs are caves at temperature t, then pick one of them whose mast touches your own scaffold at temperature t, if any such exist. When you are about to play a move of this type, you should *decline* any tax cut proposal that may be pending. If the only hill move at the current temperature is banned by ko, then *pass* but do not lower the bid tax rate. If there are no moves of any of the previously mentioned types, then you should *pass* and propose a lower tax rate. Sentestrat proposes the tax rate equal to the next lowest value at which you would be willing to move, namely, the drain of a cave or the point at which a vertical mast touches your (necessarily jagged) scaffold inside the cave. Thermostrat calculates other trajectories in hopes of getting a stronger result. Thermostrat computes the sum of all of your opponent's walls from Sentestrat's proposed tax rate on downwards, and to this sum it adds the width of the widest cave at each temperature. It then picks the temperature at which this trajectory is most favorable. Often, that is the same temperature as Sentestrat's, but on some examples (including one shown in *Winning Ways*), Thermostrat takes advantage of an opponent's recent mistake to outperform Sentestrat, even though both do equally well against a good opponent.

Hotstrat's Failure. Although it usually recommends the same moves as Sentestrat and Thermostrat, Hotstrat plays poorly in some situations. An example is the following sum $A + B$, as played by Left, going second:

$$A = 1\,\|\|\,10\,\|\,0\,|\,{-}20\,\|\|\,{-}21, \qquad B = 1\,\|\|\,0\,\|\,{-}18.$$

Both A and B are loopfree and have temperature 1. Right plays A to A^R. All three strategies play A^R to $A^{RL} = C = 10\,\|\,0\,|\,{-}20$. Right then plays B to $B^R = D = 0\,|\,{-}18$.

Left now faces the crucial decision, whether to play on C or on D. Sentestrat and Thermostrat both play D, after which Right plays C to C^R and Left to C^{RL}, giving a total score of 0. Hotstrat, on the other hand, computes the temperatures: $t(C) = 10$, and $t(D) = 9$. Ignoring the fact that both of these temperatures are far above the current tax rate (which has been 1 since the beginning of this example), Hotstrat plays C to 10, after which Right plays D to -18, giving a total score of -8. This failure of Hotstrat occurs under either the Economist's Rules, or under the Conventional Rules requiring alternating play without taxes.

4. An Economic Guru Kibitzes a Conventional Game

Sentestrat and Thermostrat both ensure perfect play according to economic rules. Either strategy can also be used in conventional games, in which there are no taxes and alternating play is required. In loopfree games, these strategies ensure that the first player will attain a stopping position at least as good as the mean.

When there are many summands and the temperature is t, a good player can often improve that score by about $\frac{1}{2}t$. To gain more understanding of this situation, it is helpful to view a conventional game through the eyes of a kibitzing economic guru.

4.1. Hard and Soft Currencies. It is convenient to refine the Economist's Rules by introducing dual currencies, cash and chips. The rules that specify which currency must be used for which types of debts are as follows:

Debt	Required Form of Payment
Privilege of playing Left	Cash
Privilege of playing First, at beginning of game	Chips, half of which must be bought from the opponent for cash
Tax on mover during game, while tax rate is positive	Chips
Payment to mover while tax rate is negative	Cash

Thanks to the footnote on page 368, all payments made during our demonstration followed these rules.

Except for kos, conventional play may be viewed as economic play with a preference for alternating moves, and with the proviso that chips are worthless.

In a typical economic game, the pile of chips on the table moves back and forth between the two players frequently. Only when the tax rate is lowered is a player able to move some of these chips off of the table and into his pocket. Economically, this is an apparent gain for the first player to move at the new and lower temperature. The player who made the last move at the old, higher temperature sees an immediate loss of chips. However, if she is is following Sentestrat, which ensures no net overall economic loss, then she will eventually receive a compensating payment later. If this occurs after the tax rate has become negative, then that payment will be in cash.

So the conventional player who gets the last move at some temperature would appear to receive a benefit equal to the change of temperature.

There is an alternative, heuristic view that leads to similar conclusions without assuming that the chips are worthless. In this view, the big pile of chips on the table at early stages of the economic game is very likely to be pocketed in many small decrements. These decrements are likely to be divided sufficiently evenly between the two players that, to a good approximation, the chips can be ignored.

4.2. The Economist's Forecast. In his role as a kibitzer of a conventional game, our economic guru computes its thermograph. He can also keep track of the temperature at each stage of the game. Before the game begins, the economist presumes the prehistoric (big bang?) temperature, $t = +\infty$. Then, at each stage of the game, the kibitzer determines whether the game is active or dormant at the temperature he has determined. If dormant, he lowers his temperature as little as needed to make the game active.

Our economic kibitzer also computes an economic *forecast* based on these observations and computations. If it is Left's turn to play from G and the current global temperature is t, then

$$\text{Forecast} = \text{LW}(G) + \tfrac{1}{2}t.$$

But if it is Right's turn to play from G and the current global temperature is t, then

$$\text{Forecast} = \text{RW}(G) - \tfrac{1}{2}t.$$

We note that if the players both follow Sentestrat, then the forecast does not change during any interval of ko-free play at which the temperature remains constant. But, when Left moves from G, of temperature t_{old}, to G^L, of temperature t_{new}, the forecast changes from

$$\text{LW}(G) + \tfrac{1}{2}t_{\text{old}} \qquad \text{to} \qquad \text{RW}(G^L) - \tfrac{1}{2}t_{\text{new}}.$$

If Left followed Sentestrat, then $\text{RW}(G^L) = \text{LW}(G) + t_{\text{old}}$, and the effect of Left's move was not only to lower the temperature from t_{old} to t_{new}, but also to increase the forecast by

$$\Delta\text{Forecast} = \tfrac{1}{2}|\Delta t|.$$

In general, whenever the temperature drops by Δt, the forecast may be revised by up to $\tfrac{1}{2}|\Delta t|$ in favor of the player who made the last move at the old temperature.

Since either player may choose not to follow Sentestrat, other moves may occur that cause the economist to revise his forecast. However, it is not hard to see that all such forecast revisions are adverse to the player who moves. Hence, moves that cause forecast revisions adverse to the mover are called *sacrificial*. From this viewpoint, all moves can be partitioned into three sets:

1. Temperature-lowering moves improve the forecast by up to $\tfrac{1}{2}|\Delta t|$.
2. Sacrificial moves degrade the forecast.
3. All other moves leave the forecast unchanged.

Sacrificial moves may be either voluntary or compulsory, depending on whether or not any alternative option was available. An examination of Sentestrat reveals the only condition under which a sacrifice can be compulsory:

4. A sacrifice is voluntary *unless* the only move at the current temperature is banned by the ko rule.

So far, we have assumed that our economic kibitzer is able to compute a single thermograph for the entire game. In practice, this assumption may entail too much computation to be practical. In particular, if the game is a sum of many components, then it may be feasible to compute the thermograph of each component even though it may not be feasible to compute the thermograph of the sum. Under such circumstances, our economic kibitzer may be forced to define temperature and make forecasts based only on the component thermographs. And the resulting temperature may be higher. In particular, if two components sum to zero, then their thermographs will have the same temperature, t, and this will be the temperature used by the kibitzer who sees only the thermographs of the individual component games. However, the better-informed kibitzer who also knows the thermograph of the combined game will see it to be a vertical mast of minimum temperature.

Nevertheless, we assert that our economic guru can still classify each move that changes the forecast as temperature-lowering or sacrificial, even though forecasters using different databases may take somewhat differing views of the same game.

And it is easy to see that as the game approaches its conclusion, the forecast converges to the final score.

Evidently, the forecast is a plausible prediction of the final score. Thermostrat is a greedy algorithm, which optimizes this forecast on a one-move time horizon.

In order to be a viable investment, our sacrificial move must prepare for some later payoff, and this payoff must eventually show up either as a temperature at which we get the last big move before a sizable drop in temperature, or as a ko that we can win while the opponent has only smaller (i.e., cooler) moves available.

If there are no foreseeable kos, then all changes in forecasts result from a sequence of plays in which the player who benefits from the forecast revision "gets the last big move". Since each such event lowers the temperature, it is clear that the forecast's absolute margin of error can be no worse than $t/2$. In very simple situations, the forecasting error is often precisely that. However, in a complicated game that is the sum of many components, it is reasonable to expect that the temperature will decline adiabatically in many small decrements, and that each side will be able to get about half of the total absolute value of these decrements. Under such conditions, the net of forecasting improvements and degradations will be about zero. We now concoct some models in which this result is precise and provable.

4.3. Enriched Environments. Let 2δ be the greatest common divisor of all temperatures of all positions of all summands of G. For example, if these temperatures are $\frac{1}{4}$, $\frac{1}{3}$, and $\frac{5}{8}$, then $2\delta = \frac{1}{24}$. Suppose the maximum initial temperature of all summands of G is T. Then we define the *minimally enriched environment* MEE for G as the sum of the elementary switches of temperatures δ, $2.\delta$, $3.\delta$, $4.\delta$, ..., $n.\delta$, where $n = T/\delta$. Let h be an upper bound on the total number of moves, in all positions of all summands of G, which might serve as a kothreat for any ko in G. (Any move that raises the temperature might serve as a kothreat). We then define SEE, the *standard enriched environment* for G, as $(2h + 1)$ copies of MEE. SEE is the sum of $(2h + 1).n$ distinct switches. An *over-enriched environment* is defined in the same way as the standard enriched environment, except that δ is taken to be some proper divisor of the greatest common divisor of all temperatures of all positions of all summands of G; the maximum temperature of OEE may be larger than the temperature of G, and the number of switches at each multiple of δ may be any odd number greater than $(2h - 1)$.

Theorem on Enriched Environments. Let G be a position that contains no hyperactive kos. Then:

THEOREM. *The economic forecasts for* $G + \text{MEE}$ *and* $G + \text{SEE}$ *are the same as the economic forecast for* G. *If* $G + \text{SEE}$ *is played by two gurus using Conventional Rules, the economic forecast of the final score will be precisely correct.*

If our economist provides us with all of the relevant means and temperatures, we can play $G + \text{SEE}$ against an opposing guru. Even in the conventional game, we can elect to play Sentestrat, subject to the proviso that whenever any move permitted by Sentestrat involves a ko, we will prefer the ko move, and if there are several permitted ko moves, we will play the same one (if any) that we previously played most recently.

It is not hard to see that this policy ensures that the overall result is at least as favorable as the forecast. If a placid ko has temperature t, then the standard enriched environment also contains a bountiful supply of switches at the same temperature. Our policy ensures that the ko will be resolved before all of these switches are played. So, whenever we are faced with a koban, we have another (switch) move at the same temperature that serves us just as well.

A Sequence of Universal Environments. Notice that it is possible to construct a sequence of over-enriched environments that are universal. We'll construct a specific sequence, U_1, U_2, U_3, ..., with the property that if G is *any* game that contains no hyperactive kos, then for all sufficiently large n,

- the Economist's forecast for $G + U_n$ is the same as the forecast for G, and
- the forecast of $G + U_n$ is precisely accurate.

To this end, it is sufficient to let

$$\delta_n = \frac{1}{\text{lcm}\{1, 2, 3, ..., n\}}.$$

Let U_n consist of $(2n+1)n/\delta_n$ switches, $(2n+1)$ copies each of temperatures δ, $2\delta, \ldots, n$.

It is easy to see that this construction has the specified property; it is a sequence of "Universal Enriched Environments". Any reasonable game G (one without hyperactive kos), when played in such an environment, behaves precisely according to the economist's forecast.

Universal environments, and other enriched environments, are artificial theoretical constructions intended to show why, when there are sufficiently many independent regions in the game, the economic forecast is a very plausible prediction of the outcome.

In fact, there are serious difficulties in realizing our theoretical enriched environments within the practical constraints of a game like Go, even if we allow it to be played as the sum of arbitrarily many boards, some of which may have arbitrarily large sizes. The first problem is that all final scores in Go are integers, and this Diophantine constraint prevents the direct construction of even simple switches, such as $\{\frac{1}{2} | -\frac{1}{2}\}$. This difficulty can be at least partially alleviated by working with "chilled Go", as described in [Berlekamp and Wolfe 1994]. But even in chilled Go, there is no switch of value such as $\{\frac{1}{3} | -\frac{1}{3}\}$; the denominators of all temperatures of switches must be powers of 2.

So, even though the enriched environments we have constructed are all sums of extremely simple (two-stop) abstract games, some of these games might not be realizable within the natural constraints of the class of games (e.g., Go) to which we want to apply our economic forecasts.

In more realistic environments, our economic forecasts are typically not precise. But they are still often quite close. Yonghoan [Kim 1995; Berlekamp and Kim 1996] has analyzed placid kos in several kinds of realistic environments, in which there are no other regions having exactly the same temperature as the ko. In such situations, kothreats have value, but the typical value of a kothreat is only a small multiple of δ, where (as in our constructions), δ is the typical difference between temperatures of other regions.

So, the total value of any fixed number of kothreats still approaches zero as the background becomes sufficiently rich. The rich environment must have many independent regions, whose temperatures are relatively dense in the vicinity of the temperature of the placid ko. Other details about the games in the background environment don't seem to matter much; they need not be switches. Even though the details of the background can affect the accuracy of the forecast by up to $t/2$, nearly all of these forecasting inaccuracies are due to inherent difficulties in predicting who can get the last move at each tax rate. Only a small part of the forecasting inaccuracies are due to placid kos, even though thermographs ignore kothreats entirely.

Hyperactive Kos. Even in a very rich environment, kothreats can have a big effect on the final score if the game contains a position that is a hyperactive ko.

Buyer	Item	Amt Bid	Book Value	Max $ Loss DW	YK	Max Chip Loss DW	YK
YK	Black in −Go	$2\frac{1}{4}$	$2\frac{1}{32}$		$\frac{7}{32}$		
DW	Black in −Go	$1\frac{3}{4}$	$2\frac{1}{32}$	$\frac{9}{32}$			
YK	Domin. Left	$1\frac{1}{2}$	$\frac{3}{4}$		$\frac{3}{4}$		
DW	Domin. Left	$\frac{1}{2}$	$\frac{3}{4}$	$\frac{1}{4}$			
DW	First move	$1\frac{3}{4}$	$1\frac{15}{32}$	$\frac{9}{32}$		$\frac{9}{64}$	
YK	First move	$1\frac{1}{4}$	$1\frac{15}{32}$		$\frac{7}{32}$		$\frac{7}{64}$
Total mistakes				$\frac{43}{64}$	$\frac{69}{64}$	$\frac{9}{64}$	$\frac{7}{64}$
Net gains due to bidding				$+\frac{13}{64}$	$-\frac{13}{64}$	$-\frac{1}{64}$	$+\frac{1}{64}$

Table 2. Accountant's error ledger for the demonstration game of Figures 2–6.

From the economist's perspective, the μ-value of such a ko depends on who is its master, and this, in turn, depends on a global tally of kothreats. And, en route to elegance and simplicity, combinatorial game theory suppresses much kothreat information from the canonical form. Then even more kothreat information is suppressed when the information in the canonical form is condensed into a thermograph. So, if there is a hyperactive ko, a precise analysis requires more data than the mathematics has retained. Of course, $\mu(\hat{G})$ and $\mu(\check{G})$ still provide firm bounds; but for some positions these bounds may be quite far apart.

4.4. Dollar Accounting. Economic Rules facilitate some very detailed accounting, which identifies the moves at which mistakes were made and the cost of each mistake. Table 2 shows an example.

In these auctions, every Book bid was intermediate between the bids submitted by the two players, and all maximum bidding losses were realized.

Altogether, in these auctions, the average of the two players made 1 point worth of bidding errors, $\frac{7}{8}$ in dollars, and $\frac{1}{8}$ in chips. DW was better than this average; YK was worse. The absolute difference of each from the average was $\frac{13}{64}$ dollars minus $\frac{1}{64}$ chips.

After all play was concluded, the final result (after converting chips back to cash) was that DW ended with \$5.1875; YK, with \$4.8125. Since both players played all moves according to perfect Book strategies, YK's net loss of \$$\frac{3}{16}$ resulted entirely from the fact that his bidding errors cost more than DW's.

The economist's accounting can be readily extended to itemize any errors that occur in play.

DW	YK		Total chips still on the table	Temper- ature
$-2G - 4Y$	$-2G - 4Y$	At start	$4G + 8Y$	$1\frac{1}{2}$ (bid)
$1G + 2Y$		(after move 1)	$3G + 6Y$	$1\frac{1}{8}$
$1Y$		(after move 5)	$3G + 5Y$	$1\frac{1}{16}$
	$1Y$	(after move 6)	$3G + 4Y$	1
$2Y$		(after move 9)	$3G + 2Y$	$\frac{7}{8}$
	$2Y$	(after move 10)	$3G$	$\frac{3}{4}$
$1G$		(after move 11)	$2G$	$\frac{1}{2}$
	$2G$	(after move 12)	0	0

$+ Y$	$- Y$		
$= + \frac{1}{16}$	$- \frac{1}{16}$	Net total chips gained from playing (both players used perfect Book play)	

Table 3. Pocketed chip ledger for the demonstration game of Figures 2–6.

4.5. Chip Accounting. The accuracy of economic forecasts is limited by a crucial approximation that states that, in the course of a sufficiently long and intricate well-played game, each player will eventually pocket about half of the chips that are placed on the table at the conclusion of the initial auctions.

It is instructive to see how the forecasts evolved during the (unenriched) demonstration game. At the start of the game, each player put chips of value $\frac{3}{4}$ onto the table ($Y = \frac{1}{4}G = \frac{1}{16}$). This total sum of $1\frac{1}{2}$ units ($4G + 8Y$) was then pocketed as shown in Table 3.

4.6. Forecasting Records. Economists studying history like the Big Picture, which, for the demonstration game, is shown in Figure 22. As usual, μ is plotted along the reversed horizontal axis; t, on the vertical. Each point plotted in this figure shows the global $[\mu, t]$ values of the full game after the numbered move.

Figure 22 records the values of the total μ and the maximum t after moves 0–12 of the demonstration game. The sequence terminates when the temperature reaches 0. The values of μ oscillate back and forth around the forecast, and each decrease in the tax rate is accompanied by a revision in the forecast, as illustrated in Table 4.

The magnitude of the revision is one half the decrease in the tax rate, and the sign of the revision favors the player who made the last move before the tax rate decrease.

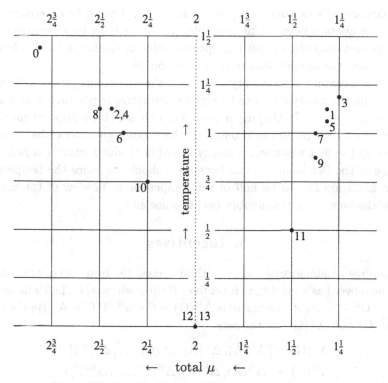

Figure 22. The Big Picture of the demonstration game.

Tax Rate	μ	Forecast	Move	Change
$1\frac{15}{32}$	$2\frac{25}{32}$	$2\frac{3}{64}$	1	$-\frac{11}{64}$
$1\frac{1}{8}$	$1\frac{5}{16}$	$1\frac{7}{8}$	2–5	$-\frac{1}{32}$
$1\frac{1}{16}$	$1\frac{5}{16}$	$1\frac{27}{32}$	6	$+\frac{1}{32}$
1	$2\frac{3}{8}$	$1\frac{7}{8}$	7–9	$-\frac{1}{16}$
$\frac{7}{8}$	$1\frac{3}{8}$	$1\frac{13}{16}$	10	$+\frac{1}{16}$
$\frac{3}{4}$	$2\frac{1}{4}$	$1\frac{7}{8}$	11	$-\frac{1}{8}$
$\frac{1}{2}$	$1\frac{1}{2}$	$1\frac{3}{4}$	12	$+\frac{1}{4}$
0	2	2	13 ...	

Table 4. Forecasting record.

4.7. The Komi in Go. Professional Go games begin with an empty board. By symmetry, $\mu = 0$. The player who makes the first move is required to give his opponent some number of points in return. This number of points is called the *komi*. To prevent ties, the komi is traditionally chosen as a half-integer. The komi is specified by the tournament directors. Values of the komi in modern tournaments have ranged from 5.5 to 8.5 points.

Our economist's viewpoint suggests that the komi could be determined by a competitive auction between the two players rather than by executive decree. Our economist conjectures that the optimal value of the komi, like his forecast, should be half the temperature of the empty board.

If the temperature of the empty board is t, then an expert player will find the first move on the empty board to be equally attractive to playing on an additive switch of value $+t|-t$. In Go jargon, such a switch is called a $2t$-point gote move. So, from the viewpoint of our economist, the n-point gote move that is just as tempting as the first move on an empty board is attained when n is *four* times the value of the fair komi. That's because n should be twice the temperature, and the fair komi should be half of the temperature. In view of the accepted range of the komi, n is presumably twenty-something.

5. Incentives

In loopfree combinatorial games without taxes, the best moves are those for which the mover has a maximum incentive. If the position is G, Left's incentive is $\Delta^L(G) = G^L - G$; Right's incentive is $\Delta^R(G) = G - G^R$. If $G = A+B+C+\cdots+F$, then $\Delta^L(G)$ and $\Delta^R(G)$ are the sets

$$\Delta^L(G) = \{\Delta^L(A), \Delta^L(B), \Delta^L(C), \ldots, \Delta^L(F)\}$$
$$\Delta^R(G) = \{\Delta^R(A), \Delta^R(B), \Delta^R(C), \ldots, \Delta^R(F)\}$$

It is helpful to consider the union of both sets:

$$\Delta = \{\Delta^L, \Delta^R\}.$$

When temperatures are negative, both incentives equal the temperature, and values are numbers that are equal to the means. However, when temperatures are nonnegative, both values and incentives are necessarily games that are more complicated than numbers. To manipulate these games, familiarity with classical combinatorial game theory is required [Berlekamp et al. 1982; Conway 1976].

The partial ordering of the incentives plays an important role in the mathematical analysis of various lines of play [Berlekamp and Kim 1996]. If $\Delta^L(A) > \Delta^L(B)$, Left will prefer to play on A rather than on B. If $\Delta^R(A) > \Delta^L(B)$, then Left's move on B is reversed through Right's response on A. Reversibility is a technical criterion well known in combinatorial game theory.

Every incentive is a game that has 0 as a right follower. Conversely, any game in which 0 is a right follower is one of its own right incentives. So no incentive can be 0 or positive.

Many pairs of incentives are incomparable. On the other hand, temperatures of incentives, being numbers, can be totally ordered (although ties will occur in cases of equal temperatures). Thus, a wise prelude to an investigation of the partial ordering of the incentives is to compute each incentive's temperature, and then sort them accordingly. If $\text{temp}(\Delta^L(A)) > \text{temp}(\Delta^L(B))$, this does not imply that $\Delta^L(A) > \Delta^L(B)$, but it certainly precludes the possibility that $\Delta^L(B) \geq \Delta^L(A)$.

In general, $\text{temp}(A - B) \leq \max\{\text{temp}(A), \text{temp}(B)\}$. However, in the special case of incentives, we can assert that for any G, there is at least one left incentive such that

$$\text{temp } \Delta^L(G) = \max\{\text{temp}(G), \text{temp}(G^L)\}.$$

This is true because if the temperature of the incentive were less, then we could pick a value of t between the temperature of the incentive and the temperature of G and G^L. Cooling the incentive-defining equation by t would yield

$$\Delta^L(G_t) = G_t^L - G_t,$$

where neither term on the right side is a number, but their difference is. But this contradicts another theorem of combinatorial game theory, which states that if $H = G_t$ is not a number, then it has at least one left incentive that exceeds all negative numbers.

The same arguments can be repeated for right incentives.

In earlier sections of this paper, we showed that for economic rules, when players bid to set a new tax rate, either player can safely select the temperature of the full board position as his bid, except in the case when he has no legal move. This leads directly to the important conclusion that initial tax rate bids may be submitted independently of the bids to play Left's side of the game. When the tax rate is fair and both players are competent, it makes no economic difference who moves first.

However, it is in many ways more natural to view the temperature as a function of the move rather than as a function of the initial position. In that sense, temperature is a natural function of the incentive. Sentestrat still works.

5.1. "Cooling" via Honorific Tie-Breaking and Prescribed Initial Temperatures.

The winner of an economic game is the player who realizes a net profit. His opponent necessarily incurs a net loss. If two gurus bid and play perfectly, then the net score should be zero and the economic outcome is a tie. We can break such ties by awarding an honorific victory to the player whose opponent made the first pass at the original temperature at which the game started.

We can also decree that the economic game begin with a prescribed finite tax rate, t, and with a prescribed player assigned to play first.

We temporarily assume that the sum is loopfree. If t is sufficiently large, this game is identical to the rules of our ideal economist, augmented only by the honorific tie-breaking convention. However, if t is 0, then this game is identical to a classical combinatorial game with the last mover winning. This is because at subzero temperatures, the economic rules and the conventional rules yield identical outcomes.

At intermediate values of the starting temperature, this game turns out to be identical to G_t, called "G cooled by t". Cooling is a well-known mapping of combinatorial games onto combinatorial games.

5.2. Implications for Computer Go and Similar Games

We focus on those endgames that are sums of numerous regional positions. Such problems best display the unique power of mathematics to cope with complexities that lie beyond the reach of alternative methods based only on heuristics and/or the judgment of human experts.

Combinatorial game theory provides the machinery to obtain precise analyses of sums of loopfree games played according to conventional rules. The recommended approach is to compute the incentives of all summands, then order the incentives as much as possible, and then search among all lines of play that are locally canonical. As exemplified by the $1000 Ko problem [Berlekamp and Kim 1996], this approach often yields a rigorous analysis relatively quickly, even for problems that are well beyond the scope of less sophisticated methods. However, it is known that, in general, even sums of simple games can be NP hard. The first such results were due to Lockwood Morris. Later, Laura Yedwab [1985] and David Moews [1993] showed that a precise determination of the outcome of even a sum of elementary three-stop games is NP-hard. So, for some problems in today's technology, the preferred methodology of combinatorial game theory (like any other method that finds only rigorously correct answers) must necessarily bog down in a combinatorial explosion of calculations that will not terminate until long after everyone has run out of patience.

On the other hand, thermography offers fast, polynomial-time algorithms that solve the same problems according to slightly different (economic) rules. The discrepancy between the ideal results of sums of games played according to economic rules and conventional rules is relatively small and easily bounded. Economic forecasting offers further small adjustments that usually yield further decreases in this discrepancy, driving it towards zero as the background environment included in the sum becomes sufficiently rich. However, in the most difficult problems, the background may still be "impoverished" despite the presence of many summands.

We have given an economic interpretation of the combinatorial game theory technique of *cooling*. This interpretation suggests a strategy by which computer Go buffs might achieve appropriate tradeoffs that would combine most of the precision of incentives with most of the speed of thermography. This approach begins by computing thermographs of positions and their incentives, top-down. This provides a solution of the frozen image of the original game. Then one refines these conclusions to obtain a solution to the original game cooled by a temperature that is not quite large enough to freeze it. This can then be further refined by further iterations, each of which yields a solution of a less-cooled version of the original.

5.3. Thermal Disassociation. Thermal disassociation [Berlekamp et al. 1982], first proposed by Simon Norton in the early 1970s, represents an arbitrary loopfree game as the sum of its mean and appropriate infinitesimals, each heated by a successively smaller temperature. Excellent approximations to the game, at various temperatures, can be obtained by ignoring the cooler terms in this expansion.

A ko can also be represented as the sum of its mean and a heated "infinitesimal ko". The study of infinitesimal kos is still in its infancy.

As a loopfree instructive example, we consider the incentives of the general 3-stop game, $G = x|y\|0$. G's stops are 0 and the numbers x and y, where $x \geq y \geq 0$. G's right incentive is $\Delta^{R}(G) = G - G^{R} = G$. Its left incentive is

$$
\begin{aligned}
\Delta^{L}(G) &= G^{L} - G \\
&= \{x|y\} + \{0\|-y|-x\} \\
&= \{x\|x{-}y|0, x|y\||0\} \\
&= \{x{-}y, x|y\|0\} \\
&= H.
\end{aligned}
$$

As shown in Figure 23, the thermographs of H can have any of three different forms, depending on how x compares with $2y$ and $3y$.

Corresponding to Figure 23, we have these expressions for the thermal dissociations of G and H, where $-_v = \{v|0\|0\}$ and $+_v = \{0\|0|-v\}$:

Cases 1 and 2: $\quad G = \dfrac{x+y}{4} + \displaystyle\int^{\frac{1}{4}(x+y)} * + \int^{\frac{1}{2}(x-y)} \varepsilon_1,$

where $\varepsilon_1 = \{*|\frac{1}{2}(x{-}3y)\} + \{\frac{1}{2}(3y{-}x)|0\}$ is fuzzy with 0

Case 3: $\quad G = y + \displaystyle\int^{y} -_{x-3y}$

Case 1: $\quad H = G$

Case 2: $\quad H = G + \displaystyle\int^{x-2y} \varepsilon_2,$

where $\varepsilon_2 = \{0, 3y{-}x|0\|x{-}3y\} - \{3y{-}x|0\|x{-}3y\} > 0$

Case 3: $\quad H = \dfrac{x-y}{2} + \displaystyle\int^{\frac{1}{2}(x-y)} * + \int^{y} +_{x-3y}$

(Note: We have $H > G$, since $v|0 > -_v -_v$.)

It is interesting to note that the Yedwab–Moews construction of NP-completeness requires no games of Case 2, and only their coolest game is of Case 3. All but one of their summands are of Case 1, in which left and right incentives are equal.

$$G = \quad \{x|y\|0\} = \Delta^R(G)$$
$$H = \{x{-}y, x|y\|0\} = \Delta^L(G)$$

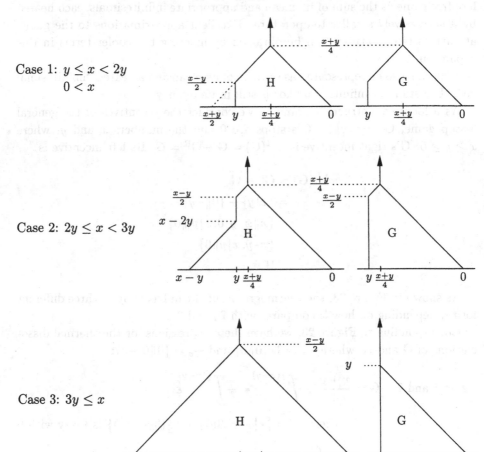

Case 1: $y \le x < 2y$
$\quad\quad\quad 0 < x$

Case 2: $2y \le x < 3y$

Case 3: $3y \le x$

Figure 23. Thermographs of incentives of three-stops.

References

[Berlekamp and Kim 1996] Elwyn R. Berlekamp and Yonghoan Kim, "Where is the thousand dollar ko?", pp. 203–226 in this volume.

[Berlekamp and Wolfe 1994] Elwyn R. Berlekamp and David Wolfe, *Mathematical Go: Chilling Gets the Last Point*, A. K. Peters, Wellesley, MA, 1994. Also published in paperback, with accompanying software, as *Mathematical Go: Nightmares for the Professional Go Player*, by Ishi Press International, San Jose, CA.

[Berlekamp et al. 1982] Elwyn R. Berlekamp, John H. Conway, and Richard K. Guy, *Winning Ways for Your Mathematical Plays*, Academic Press, London, 1982.

[Conway 1976] John H. Conway, *On Numbers and Games*, Academic Press, London, 1976.

[Hanner 1959] Olof Hanner, "Mean play of sums of positional games", *Pacific J. Math.* **9** (1959), 81–99.

[Kim 1995] Yonghoan Kim, "New values in Domineering and loopy games in Go", Ph.D. thesis, Mathematics Department, University of California at Berkeley, May 1995.

[Milnor 1953] John Milnor, "Sums of positional games", pp. 291–301 in *Contributions to the Theory of Games*, vol. 2 (edited by H. W. Kuhn and A. W. Tucker), Ann. of Math. Stud. **28**, Princeton University Press, Princeton, 1953.

[Moews 1993] David Moews, "On some combinatorial games connected with Go", Ph.D. thesis, Mathematics Department, University of California at Berkeley, December 1993.

[Yedwab 1985] Laura J. Yedwab, "On playing well in a sum of games", M.Sc. Thesis, MIT, 1985. Issued as MIT/LCS/TR-348, MIT Laboratory for Computer Science, Cambridge, MA.

ELWYN BERLEKAMP
DEPARTMENT OF MATHEMATICS
UNIVERSITY OF CALIFORNIA AT BERKELEY
BERKELEY, CA 94720
  berlek@math.berkeley.edu

Games of No Chance
MSRI Publications
Volume **29**, 1996

Games with Infinitely Many Moves and Slightly Imperfect Information

(Extended Abstract)

DAVID BLACKWELL

D. A. Martin in 1975 showed that all Borel Games with perfect information are determined. Question: are all Borel games with slightly imperfect information determined?

Let A, B be finite nonempty sets, let $C = A \times B$, and let W be the set of all infinite sequences $w = \{c_1, c_2, \ldots\}$ from C. Any subset S of W defines a game $G(S)$, whose n-th move, for $n = 1, 2, \ldots$, is played as follows: Player I chooses $a_n \in A$ and, simultaneously, Player II chooses $b_n \in B$. Each player is then told the other's choice, so that they both know $c_n = (a_n, b_n)$.

Player I wins $G(S)$ just if the play $w = \{c_1, c_2, \ldots\}$ is in S. We say that $G(S)$ is *determined* if there is a number v such that, for every $\varepsilon > 0$,

(a) Player I has a (random) strategy that wins for him with probability at least $v - \varepsilon$ against every strategy of Player II, and

(b) Player II has a (random) strategy that restricts his probability of loss to at most $v + \varepsilon$ against every strategy of Player I.

If S is *finitary*, i.e., depends on only finitely many coordinates of w, then $G(S)$ is a finite game, and the von Neumann minimax theorem says that $G(S)$ is determined.

If S is *open*, i.e., the union of countably many finitary sets, then it is well-known, and not hard to see, that $G(S)$ is determined (and that Player II has a good strategy).

If S is a G_δ-set, that is, the intersection of countably many open sets, again $G(S)$ is determined, but the calculation involves countable ordinal calculations. This complexity is probably necessary as, with a natural coordinatization of G_δ-sets, the value $v(S)$ is not a Borel function of S. Whether all $G_{\delta\sigma}$ games $G(S)$ are determined is not known.

Our restriction to finite A and B is essential. For A and B countable, the game "choosing the larger integer" is a special case that is *not* determined.

DAVID BLACKWELL
DEPARTMENT OF STATISTICS
BERKELEY, CA 94720-3860
davidbl@stat.berkeley.edu

Games of No Chance
MSRI Publications
Volume 29, 1996

The Reduced Canonical Form of a Game

DAN CALISTRATE

ABSTRACT. Cooling by $*$, followed by the elimination of the stars, is used
to define an operator $G \to \bar{G}$ on short games, having the following properties: $\bar{G}$ is the simplest game infinitesimally close to G; the operator is a
homomorphism; it can be used for recursive calculations, provided that the
games involved are not in a "strictly cold" form.

1. Introduction

We will use the classical definitions and facts about two-person, perfect information combinatorial games with the normal winning convention, as developed
in *Winning Ways* [Berlekamp et al. 1982] and *On Numbers and Games* [Conway
1976]. We recapitulate them briefly.

Formally, *games* are constructed recursively as ordered pairs $\{\Gamma^L \mid \Gamma^R\}$, where
Γ^L and Γ^R are sets of games, called, respectively, the *set of Left options* and
the *set of Right options* from G. We will restrict ourselves to *short games*, that
is, games where the sets of options Γ^L and Γ^R are required to be finite in this
recursive definition. The basis for this recursion is the game $\{\varnothing \mid \varnothing\}$, which is
called 0. We will often let G^L and G^R represent typical Left and Right options
of a game G, and write $G = \{G^L \mid G^R\}$.

The $\leq$ relation, defined inductively by

$$G \leq H \iff \text{there is no } H^R \text{ with } H^R \leq G \text{ and no } G^L \text{ with } H \leq G^L,$$

is a quasi-order (it is not antisymmetric). We identify G with H if $G \leq H$ and
$H \leq G$; we then say that G and H *have the same value*, and write $G = H$. The
relation $\leq$ becomes a partial order on the set of game values.

Two games G and H are *identical*, or *have the same form*, if they have identical
sets of left options and identical sets of right options. In this case we write $G \equiv H$.
Whenever the distinction between the value and the form of a game is essential,
we will specify it; otherwise, by G we will mean the form of G.

In the *normal winning convention*, the player who makes the last move wins.
The four possible classes of outcome for a game are determined by how the game's

value compares with 0, as shown in the following table (where $G \| 0$ means that G is not comparable to 0):

$$G \text{ is a win for Left whoever goes first} \quad \Longleftrightarrow \quad G > 0;$$
$$G \text{ is a win for Right whoever goes first} \quad \Longleftrightarrow \quad G < 0;$$
$$G \text{ is a win for whoever goes first} \quad \Longleftrightarrow \quad G \| 0;$$
$$G \text{ is a loss for whoever goes first} \quad \Longleftrightarrow \quad G = 0.$$

One can define an addition operation on the set of game values, making it into a group. Certain game values can be associated with real numbers; they are therefore called *numbers*. Since we only consider short games, all number values will be dyadic rationals. The *Left stop* $L(G)$ and *Right stop* $R(G)$ of a game G are the numbers recursively defined by: $L(G) = R(G) = x$ if $G = x$, where x is a number; otherwise $L(G) = \max_{G^L} R(G^L)$ and $R(G) = \min_{G^R} L(G^R)$. An *infinitesimal* is a game whose stops are both 0. Any infinitesimal lies strictly between all negative and all positive numbers.

Any game G admits a unique canonical form, that is, a form with no *dominated options* and no *reversible moves*. The canonical form has the *earliest birthday* among all the games that have the same value as G. (For details and proofs, see *Winning Ways*, pp. 62–65, or *On Numbers and Games*, pp. 110–112). Actually, the canonical form of a game G has the even stronger property of minimizing the size of the edge-set for the game-tree of G. For the purposes of this paper, a game G will be called *simpler than H* if the size of the edge-set of the game-tree of G is less than or equal to the size corresponding to H, so the canonical form is the simplest (in this sense) among all games with a given value.

The aim of this paper is to introduce a yet simpler form, called the *reduced canonical form*, by relaxing the condition that it should have the same value as the initial game to the condition that it should be infinitesimally close to the initial game. This new form should be the simplest possible subject to this condition, and the transformation $G \to \bar{G}$ should be linear. Algebraically, we will show that the reduced canonical forms form a subgroup **Rcf**, and the group of games (with the disjunctive compound operation) is the direct sum $\mathbf{I} \oplus \mathbf{Rcf}$, where $\mathbf{I}$ is the subgroup of infinitesimals. Often, the information provided by the **Rcf**-component of a game G is enough to decide the outcome class of G. For games of this type, it is important to know when knowledge of the reduced canonical forms of the options of G would imply knowledge of the reduced canonical form of G. This will be answered by Theorem 5.

2. Construction of the Reduced Canonical Form

The operation of cooling a game by a positive number is essential for the Theory of Temperature in the World of Games (see *Winning Ways*, Chapter 6, or *On Numbers and Games*, chapter 9). Similarly, one can define cooling by a non-number, and specifically by $* = \{0|0\}$:

DEFINITION. Given $G = \{G^L | G^R\}$, we recursively define a new game G_*, called G *cooled by* $*$, as follows:

$$G_* = \begin{cases} G & \text{if } G \text{ is a number,} \\ \{G^L_*+*|G^R_*+*\} & \text{otherwise,} \end{cases}$$

where, as usual, G^L and G^R are generic Left and Right options of G, and we write G^L_* for $(G^L)_*$.

It is easy to check that, if $G - H$ is a zero game, so is $G_* - H_*$. Thus the definition above is independent of the form of G, and G_* is well-defined for any game value G.

DEFINITION. If $H_0 \equiv \{H^L_0 | H^R_0\}$ is the canonical form of a game H, we recursively define $p(H)$, the $*$-*projection* of H, as follows:

$$p(H) = \begin{cases} x & \text{if } H = x \text{ or } x + *, \text{ where } x \text{ is a number,} \\ \{p(H^L_0)|p(H^R_0)\} & \text{otherwise.} \end{cases}$$

Because of the uniqueness of the canonical form, the definition of $p(H)$ is independent of the form of H.

DEFINITION. The *reduced canonical form* $\bar{G}$ of G is defined as $p(G_*)$.

Observe that $p(G_*)$ is a canonical form, because p is defined in terms of the canonical form of G_*, and it follows by induction that $p(G_*)$ is in canonical form as well.

EXAMPLE. Let $G = \{\{2|0\}, 1 \| 0\}$. This game is in canonical form. Then

$$G_* = \{\{2|0\}_*+*, 1+* \| *\} = \{\{2+*|*\}+*, 1+* \| *\}.$$

Now we use the translation principle for stars (*Winning Ways*, p. 123), which says that, for any numbers x, y, we have $\{x|y\}+* = \{x*|y*\}$ if $x \geq y$ and $\{x|y*\}+* = \{x*|y\}$ if $x > y$. We obtain $G_* = \{\{2|0\}, 1* \| *\}$. Since $G_* > 0$ (the game is a win for Left no matter who starts), the Left option $\{2|0\}$ is reversible, and can be replaced by all the Left options from 0. There are no Left options from 0, so $G_* = 1* | *$. It is easy to check that $1* | *$ has no reversible moves, so it is the canonical form of G_*. Hence

$$p(G_*) = \{p(1*)|p(*)\} = \{1|0\}.$$

We see that, in this example, the reduced canonical form of G is strictly simpler than the canonical form of G.

3. Properties of the Reduced Canonical Form

THEOREM 1. *The transformation $G \to \bar{G}$ is a homomorphism.*

PROOF. We will show that $G \to G_*$ and $H \to p(H)$ are homomorphisms, hence their composition is a homomorphism. It is a straightforward inductive check that none of the players can win going first in the game $(G-H)_*-G_*+H_*$; therefore $(G-H)_* = G_*-H_*$. If we consider first the more general case when G, H and $G-H$ are not numbers, we have:

$$(G-H_*)-G_*+H_* = \{(G^L-H)_*+*, (G-H^R)_*+*\|(G^R-H)_*+*, (G-H^L)_*+*\}$$
$$+\{(-G^R)_*+*\|(-G^L)_*+*\}+\{H_*^L+*\|H_*^R+*\}.$$

We can then see that, assuming the property true for pairs such as (G^L, H), (G^R, H), (G, H^L), and (G, H^R), every move in $(G - H)_* - G_* + H_*$ has an exact counter, that is, for any move of one player, there is a reply by the other player that brings the position to a value of 0.

If at least two of G, H, and $G - H$ are numbers, then all of them are numbers and the equality to be proved is trivial since we are in the first case of the definiton of cooling by $*$.

If precisely one of G, H is a number, say $H = x$, we need to show that $(G - x)_* = G_* - x$. We are in the second case of the definiton of cooling, so this is equivalent to

$$\{(G^L-x)_*+*|(G^R-x)_*+*\} = \{G_*^L+*|G_*^R+*\}-x.$$

Applying the translation principle (with x) one more time, we are done, because $(G^L-x)_* = G_*^L-x$ and $(G^R-x)_* = G_*^R-x$ by the induction hypothesis.

The proof that $H \to p(H)$ is a homomorphism is very similar: it is enough to show that $p(G)+p(H)+p(K) = 0$ if $G+H+K = 0$ and G, H, K are in canonical form.

Suppose that none of G, H, K is of the form x or $x*$ for some number x. If Left moves first in $p(G)+p(H)+p(K)$, he will leave for Right a position like $p(G^L)+p(H)+p(K)$. Since G, H and K were in canonical form and $G+H+K = 0$, there is a Right reply, H^R say, in a different component, so that $G^L+H^R+K \leq 0$ (otherwise G^{LR} would be reversible). Applying the induction hypothesis to this, we obtain $p(G^L)+p(H^R)+p(K) \leq 0$, which means that, going second in $p(G)+p(H)+p(K)$, Right wins. Thus $p(G)+p(H)+p(K) \leq 0$. Note that we have only assumed inductively (and proved) the inequality $p(G)+p(H)+p(K) \leq 0$. Because of symmetry, the opposite inequality can be obtained in the same way.

Finally, if exactly one of G, H, and K is of the form x or $x*$ for some number x, the implication $G+H+K = 0 \Rightarrow p(G)+p(H)+p(K) = 0$ is immediate, given the observation that if $\{G^L|G^R\}$ is a game in canonical form, so is $\{(G^L+x*)^c|(G^R+x*)^c\}$, where we are denoting by M^c the canonical form of a game M.   □

The following lemma shows one sense in which $\bar{G}$ approximates G.

LEMMA 2. *If G is any game, G and $\bar{G}$ have the same stops, that is, $L(\bar{G}) = L(G)$ and $R(\bar{G}) = R(G)$. In particular, they have the same stops as G^*.*

PROOF. We need to show separately that $L(G_*) = L(G)$ and $L(p(H)) = L(H)$. The second relation can be obtained inductively: If $H \neq x$ or $x*$, and H is in canonical form, $L(p(H)) = L(\{p(H^L)|p(H^R)\}) = \max R(p(H^L)) = \max R(H^L) = L(H)$, and similarly for the Right stops.

For the first relation, we observe that $R(G + *) = R(G)$ for any game G. Hence, if G is not a number,

$$L(G_*) = \max R(G_*^L + *) = \max R(G_*^L) = \max R(G^L) = L(G). \qquad \square$$

THEOREM 3. *A game G is an infinitesimal if and only if $\bar{G} = 0$.*

PROOF. The "if" direction follows from the lemma. Next, we will prove that, if G is an infinitesimal, then $G_* = 0$ (and hence $\bar{G} = 0$). We will do so by showing inductively that

$$L(G) \leq 0 \text{ and } R(G) \leq 0 \quad \text{imply} \quad G_* \leq 0,$$
$$L(G) \geq 0 \text{ and } R(G) \geq 0 \quad \text{imply} \quad G_* \geq 0.$$

Because of the symmetry of the definition of G_*, it is enough to prove the first of these implications. Thus we assume that $L(G) \leq 0$ and $R(G) \leq 0$. Suppose G_* is a number. This is an easy case because, from the lemma, G_* has the same stops as G, so $G_* \leq 0$ and we are done. Suppose G_* is not a number. Then, Left's move in G_* will lead to a position $G_*^L + *$. Now, if G_*^L is a number, we apply the lemma again to conclude that $G_*^L \leq 0$; hence Right's move from $G_*^L + *$ to G_*^L will force a loss for Left, so $G_* \leq 0$. Suppose now that G_*^L is not a number. Then

$$G_*^L + * = \{G_*^{LL} + * | G_*^{LR} + *\} + *.$$

Since $L(G) \leq 0$, there exists a Right-option G^{LR_0} in G^L such that $L(G^{LR_0}) \leq 0$. Therefore, Right can move from $G_*^L + *$ to $G_*^{LR_0} + * + * = G_*^{LR_0}$ and, applying the lemma one more time, we conclude that $L(G_*^{LR_0}) \leq 0$. Now, if $G_*^{LR_0}$ is a number, it cannot be strictly positive, so Left will lose going first in $G_*^{LR_0}$. Finally, if $G_*^{LR_0}$ is not a number, we have $R(G_*^{LR_0}) \leq 0$ (because we already know that $L(G_*^{LR_0}) \leq 0$), so $G_*^{LR_0}$ satisfies the conditions of the induction hypothesis, so $G_*^{LR_0} \leq 0$, so Left will lose going first in G_* in any case, so $G_* \leq 0$ and the proof is completed. $\qquad \square$

THEOREM 4. *The reduced canonical form $\bar{G}$ is infinitesimally close to G.*

PROOF. We will show first that G_* is infinitesimally close to G and then that $p(H)$ is infinitesimally close to H. We will establish inductively that $G_* - G - x \leq 0$ for every positive number x. This will be enough to ensure that $\bar{G}$ is infinitesimally close to G, because applying the induction assumption to $-G$ yields $(-G)_*+G-x \le 0$, so $-(-G)_*-G+x \ge 0$. Using the fact that $-(-G)_* = G_*$ (since G_* is a homomorphism), we get $G_*-G+x \ge 0$, so G_*-G will be greater than all negative numbers and smaller than all positive numbers.

We only have to consider the case when G is not a number. In this case,

$$G_*-G-x = \{G_*^L+*|G_*^R+*\}+\{-G^R-x|-G^L-x\}.$$

After Left makes his first move in this, Right can reply to one of the following:

$$G_*^L+*-G^L-x = (G_*^L-G^L-\tfrac{1}{2}x)+(*-\tfrac{1}{2}x),$$
$$-G^R-x+G_*^R+* = (G_*^R-G^R-\tfrac{1}{2}x)+(*-\tfrac{1}{2}x).$$

The induction hypothesis applies to $(G_*^L-G^L-\tfrac{1}{2}x)$ and $(G_*^R-G^R-\tfrac{1}{2}x)$, so they are both negative. Since $(*-\tfrac{1}{2}x)$ is also negative, Left loses, and we obtain $G_*-G-x \le 0$, as desired.

The proof that $p(G)-G-x \le 0$ follows precisely the same steps if we choose G to be in canonical form.   □

THEOREM 5. $\bar{G}$ is the simplest game infinitesimally close to G.

PROOF. Let H be infinitesimally close to G. We need to show that $\bar{G}$ is at least as simple as H. Since $G-H$ is an infinitesimal, we have $\overline{G-H} = 0$, and therefore $\bar{G} = \bar{H}$. Since both sides of this equation are in canonical form, we have $\bar{G} \equiv \bar{H}$, so all we need to show is that $\bar{H}$ is at least as simple as H for any game H. For this purpose, we can relax the definition of $\bar{G}$ in the sense that p is not applied to the canonical form of G_*, but directly to G_* (that is, to the form obtained after cooling G by $*$, without deleting any dominated options or bypassing any reversible moves). If we denote the result by $\widetilde{G}$, then $\widetilde{G}$ will be at least as simple as G; it can be seen inductively that the only thing achieved in the process of forming $\widetilde{G}$ is to replace $x*$ by x everywhere in (the form of) G, which is clearly a "simplification". Yet, $\bar{G}$ is at least as simple as $\widetilde{G}$, for the following reason. When p is applied to the canonical form K^c of a game K, the outcome will be at least as simple as when p is applied directly to K (consider the sequence $K, K_1, K_2, ..., K_n = K^c$, where each K_{i+1} is obtained from K_i by deleting a dominated option or by bypassing a reversible move; by induction, $p(K_{i+1})$ will be at least as simple as $p(K_i)$). We have thus proved that $\widetilde{G}$ is at least as simple as G, and $\bar{G}$ is at least as simple as $\widetilde{G}$, which implies that $\bar{G}$ is at least as simple as G.   □

NOTE. We needed this kind of argument because it can occur that G is simpler than G_*; for example, when $G = 1|*$, we have $G_* = 1*|*$. Here, the "simplification" is made by p to $p(G_*) = 1|0$.

DEFINITION. Let $G = \{G^L|G^R\}$. A number x is permitted by G if $G^L \not> x \not> G^R$ for every G^L and G^R.

THEOREM 6. *Let* $G = \{G^L | G^R\}$ *be such that* $\{\overline{G^L} | \overline{G^R}\}$ *permits at most one number. Then*

$$\bar{G} = \overline{\overline{G^L} | \overline{G^R}}.$$

PROOF. Suppose that at least one of G and $\{\overline{G^L} | \overline{G^R}\}$ is not a number. Then, for any positive number x, the translation principle can be applied to $H = \{G^L | G^R\} + \{-\overline{G^R} | -\overline{G^L}\} - x$, so that one of these equalities is satisfied:

$$H = \{G^L - x | G^R - x\} + \{-\overline{G^R} | -\overline{G^L}\},$$
$$H = \{G^L | G^R\} + \{-\overline{G^R} - x | -\overline{G^L} - x\}$$

Therefore, for any Left option in H, there is a Right response in the other component that leaves a negative game (applying Theorem 3). This means that Left, going first in H, loses, so $H \leq 0$. Similarly, $H' = \{G^L | G^R\} + \{-\overline{G^R} | -\overline{G^L}\} + x \geq 0$, so $G - \{\overline{G^L} | \overline{G^R}\}$ is an infinitesimal, so

$$0 = \overline{G - \{\overline{G^L} | \overline{G^R}\}} = \bar{G} - \overline{\{\overline{G^L} | \overline{G^R}\}}$$

and the result is proved in this case.

Suppose, now, that G and $\{\overline{G^L} | \overline{G^R}\}$ are both numbers. Denote by I^L and I^R the closures of the confusion intervals for $\overline{G^L}$ and $\overline{G^R}$, that is, $I^L = [L(\overline{G^L}), R(\overline{G^L})]$ and $I^R = [L(\overline{G^R}), R(\overline{G^R})]$. Since $\{\overline{G^L} | \overline{G^R}\}$ permits at most one number, and is a number itself, we must have $R(\overline{G^L}) = L(\overline{G^R}) = \{\overline{G^L} | \overline{G^R}\}$. Applying the lemma to $\overline{G^L}$ and $\overline{G^R}$ we find that G^L and G^R have the same closures of the confusion intervals as $\overline{G^L}$ and $\overline{G^R}$. Since G is a number as well, we obtain $G = \{\overline{G^L} | \overline{G^R}\}$, hence $\bar{G} = \overline{\{\overline{G^L} | \overline{G^R}\}}$.   □

NOTES. 1. The reduced canonical form operator can be used to exhibit, within small errors, the recursively obtained values for games such as $2 \times n$ Domineering [Berlekamp 1988], where the main complications are due to increasingly complex infinitesimals.

2. David Wolfe has implemented this approximation operator in his Gamesman's Toolkit [Wolfe 1996]: if the user types $G[e]$, the program will return the reduced canonical form of G.

References

[Berlekamp 1988] E. R. Berlekamp, "Blockbusting and Domineering", *J. Combin. Theory* (Ser. A) **49** (1988), 67–116.

[Berlekamp et al. 1982] E. R. Berlekamp, J. H. Conway, and R. K. Guy, *Winning Ways For Your Mathematical Plays*, Academic Press, London, 1982.

[Conway 1976] J. H. Conway, *On Numbers And Games*, Academic Press, London, 1976.

[Wolfe 1996] D. Wolfe, "The Gamesman's Toolkit", pp. 93–98 in this volume.

Dan Calistrate
The University of Calgary
2500 University Drive NW
Calgary, Alberta
Canada T2N 1N4
    calistra@ms2.math.ucalgary.ca

Games of No Chance
MSRI Publications
Volume **29**, 1996

Error-Correcting Codes Derived from Combinatorial Games

AVIEZRI S. FRAENKEL

ABSTRACT. The losing positions of certain combinatorial games constitute linear error-detecting and -correcting codes. We show that a large class of games, which can be cast in the form of *annihilation games*, provides a potentially polynomial method for computing codes (*anncodes*). We also give a short proof of the basic properties of the previously known *lexicodes*, which were defined by means of an exponential algorithm, and are related to game theory. The set of lexicodes is seen to constitute a subset of the set of anncodes. In the final section we indicate, by means of an example, how the method of producing lexicodes can be applied optimally to find anncodes. Some extensions are indicated.

1. Introduction

Connections between combinatorial games (simply *games* in the sequel) and linear error-correcting codes (*codes* in the sequel) have been established in [Conway and Sloane 1986; Conway 1990; Brualdi and Pless 1993], where lexicodes, and some of their connections to games, are explored. Our aim is to extend the connection between games and codes to a large class of games, and to formulate a potentially polynomial method for generating codes from games. We also establish the basic properties of lexicodes by a simple, transparent method.

Let Γ, any finite digraph, be the *groundgraph* on which we play the following general two-player game. Initially, distribute a positive finite number of tokens on the vertices of Γ. Multiple occupation is permitted. A move consists of selecting an occupied vertex and moving a single token from it to a neighboring vertex, occupied or not, along a directed edge. The player first unable to move loses and the opponent wins. If there is no last move, the play is declared a draw. It is easy to see (since Γ is finite) that a draw can arise only if Γ is *cyclic*, that is, Γ has cycles or loops. Games in this class—which includes Nim and Nim-like games for the case where Γ is acyclic—have polynomial strategies, in general [Fraenkel $\geq$ 1997]. It turns out that the P-positions (positions from which the player who just moved has a winning strategy) of any game in this class constitute a code.

It further turns out that, if Γ is cyclic, the structure of the P-positions is much richer if the above described game is replaced by an *annihilation game* (*anngame* for short). In such a game, when a token is moved onto a vertex u, the number of tokens on u is reduced modulo 2. Thus there is at most one token at any vertex, and when a token is moved to a vertex occupied by another, both are removed from the game.

If Γ is acyclic, it is easy to see by game-strategy considerations (or using the Sprague–Grundy function defined in Section 3) that the strategies of a non-annihilation game and the corresponding anngame are identical, so both have the same P-positions—only the length of play may be affected. Thus, for the prospect of constructing efficient codes and for the sake of a unified treatment, we may as well assume that all our games are anngames.

Summarizing, we can, without loss of generality, concentrate on the class of anngames. An anngame is defined by its groundgraph Γ, a finite digraph. There is an initial distribution of tokens, at most one per vertex. A move consists of selecting an occupied vertex and moving its token to a neighboring vertex u along a directed edge. If u was occupied prior to this move, the incoming and resident tokens on u are both annihilated (disappear from play). The player first unable to move loses and the opponent wins. If there is no last move, the outcome is a draw.

With an anngame A played on a groundgraph Γ, we associate its *annihilation graph* $G = (V, E)$, or *anngraph* for short, as follows. The vertex set V is the set of positions of A, and for $u, v \in V$ there is an edge $(u, v) \in E$ if and only if there is a move from u to v in A. We review the following basic facts, which can be found in [Fraenkel 1974; Fraenkel and Yesha 1976; 1979; 1982] (especially the latter), [Yesha 1978; Fraenkel, Tassa and Yesha 1978].

Like any finite digraph, G has a *generalized Sprague–Grundy function* γ. This function was first defined in [Smith 1966], and later expounded in [Fraenkel and Perl 1975]. See [Fraenkel 1996, p. 20] in this volume for its definition, and [Fraenkel and Yesha 1986] for full details. Let $V^f \subset V$ be the set of vertices on which γ is finite. If we make V into a vector space over GF(2) in the obvious way, then V^f is a linear subspace, and γ is a homomorphism from V^f onto $\mathrm{GF}(2)^t$, for some $t \in \mathbb{Z}^0 := \{k \in \mathbb{Z} : k \geq 0\}$, where we identify $\mathrm{GF}(2)^t$ with the set of integers $\{0, 1, \ldots, 2^t - 1\}$. The kernel $V_0 = \gamma^{-1}(0)$ is the set of P-positions of the annihilation game. This gives very precise information about the structure of G: its maximum finite γ-value is a power of 2 minus 1, and the sets $\gamma^{-1}(i)$ for $i \in \{0, \ldots, 2^t - 1\}$ all have the same size, being cosets of V_0. Moreover, V_0 constitutes an *anncode* (annihilation game code). Though G has 2^n vertices, it turns out that most of the relevant information can be extracted from an induced subgraph of size $O(n^4)$, by an $O(n^6)$ algorithm, which is often much more efficient.

If Γ is cyclic, γ is generally distinct from the (classical) Sprague–Grundy function g on Γ; in fact, g may not even exist on Γ. Also, A played on a cyclic Γ has a distinct character and strategy from the non-annihilation game played on Γ.

Annihilation games were suggested by John Conway. Ferguson [1984] considered misère annihilation play, in which the player first unable to move wins, and the opponent loses. A more transparent presentation of annihilation games is to appear in the forthcoming book [Fraenkel $\geq$ 1997].

Section 2 gives a number of examples, illustrating connections between games, anncodes and lexicodes, as well as exponential and polynomial digraphs and computations associated with them. Section 3 gives a short proof that the Sprague–Grundy function g is linear on the lexigraph associated with lexicodes, leading to the same kind of homomorphism that exists for anncodes. Some natural further questions are posed at the end of Section 3, including the definition of anncodes over $\mathrm{GF}(q)$, for $q \geq 2$. Section 4 indicates, by means of a larger example, how a greedy algorithm applied to an anncode can reduce a computation of a code by a factor of 2,000 compared to a similarly computed lexicode. The anncode method is potentially polynomial, whereas the lexicode method is exponential. But it is too early yet to say to what extent the potential of the anncode method can be realized for producing new efficient codes.

2. Examples

Given a finite digraph $G = (V, E)$, we define, for any $u \in V$, the set of *followers* $F(u)$ and *ancestors* $F^{-1}(u)$ by

$$F(u) = \{v \in V : (u, v) \in E\}, \qquad F^{-1}(u) = \{w \in V : (w, u) \in E\}.$$

If we regard the vertices of G as game positions and the edges as moves, we define, as usual, a *P-position* of the game as one from which the Previous player can win, no matter how the opponent plays, subject to the rules of the game; an *N-position* is one that is a Next-player win. Denote by $\mathcal{P}$ the sets of all P-positions of a game, and denote by $\mathcal{N}$ the set of all N-positions. The following basic relationships hold:

$$u \in \mathcal{P} \quad \text{if and only if} \quad F(u) \subseteq \mathcal{N},$$
$$u \in \mathcal{N} \quad \text{if and only if} \quad F(u) \cap \mathcal{P} \neq \varnothing.$$

If G has cycles or loops, the game may also contain dynamically drawn *D-positions*; the set $\mathcal{D}$ of such positions is characterized by

$$u \in \mathcal{D} \quad \text{if and only if} \quad F(u) \subseteq \mathcal{D} \cup \mathcal{N} \text{ and } F(u) \cap \mathcal{D} \neq \varnothing.$$

To understand the examples below we don't need γ or g; it suffices to know that $\mathcal{P}$ is the set of vertices on which γ or g is 0. Note that $\mathcal{P}$ can be recognized by purely game-theoretic considerations, as the set on which the Previous player

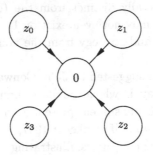

Figure 1. An acyclic groundgraph for annihilation (see Example 2.1).

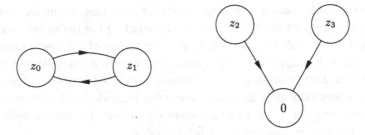

Figure 2. A cyclic groundgraph (see Example 2.2).

can win. In all these examples, we play an annihilation game A on the given groundgraphs Γ.

EXAMPLE 2.1. Let Γ be the digraph depicted in Figure 1. It is easy to see that with an odd number of tokens on the z_i the first player can win, and with an even number the second player can win in A played on Γ.

In this and the following examples, think of the z_i as unit vectors of a vector space V of dimension n, where $n - 1$ is the largest index of the z_i [Fraenkel and Yesha 1982]. In the present example, $z_0 = (0001), \ldots, z_3 = (1000)$. Encoded by the unit vectors, our anncode is

$$\mathcal{P} = \{(0000), (0011), (0101), (0110), (1001), (1010), (1100), (1111)\},$$

or, encoded in decimal, $\mathcal{P} = \{0, 3, 5, 6, 9, 10, 12, 15\}$. Note that $\mathcal{P}$ is a linear code with minimal Hamming distance $d = 2$.

(Recall that the *Hamming distance* between two vectors in $\mathrm{GF}(2)^n$ is the number of 1-bits of their difference. The number of 1-bits of a vector u is its *weight*, and is denoted by $w(u)$. Addition, or equivalently subtraction, over $\mathrm{GF}(2)$ is denoted by $\oplus$.)

EXAMPLE 2.2. Consider A played on the two-component graph Γ of Figure 2. If z_0 and z_1 host a token each, any move causes annihilation. Therefore the

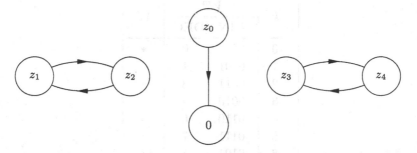

Figure 3. Another cyclic groundgraph (see Example 2.3).

position consisting of one token each on z_0, z_1, z_2 (or z_3 instead of z_2) is a P-position. Using decimal encoding, we then see that $\mathcal{P} = \{0, 7, 11, 12\}$, which is also a linear code with minimal distance 2.

EXAMPLE 2.3. Consider A played on a Nim-heap of size 5, i.e., Γ consists of the leaf 0 and the vertices $z_0, \ldots, z_4$, where $(z_j, z_i) \in E(\Gamma)$ if and only if $i < j$ and $(z_i, 0) \in E(\Gamma)$ for $i \in \{0, \ldots, 4\}$. It is not hard to see that then $\mathcal{P} = \{0, 7, 25, 30\}$, which is an anncode with minimal distance 3. Precisely the same code is given by the P-positions of the annihilation game A played on the ground graph Γ of Figure 3.

In order to continue with our examples, we now define lexicodes precisely. This is also needed for Section 3.

Let W be an $n \times n$ matrix over GF(2), of rank at least m, where $m \leq n$ is some integer. We will count the columns of W from the right and the rows from the bottom. Suppose the rightmost m columns of W constitute a basis of V^m, the m-dimensional vector subspace of V^n over GF(2). Then there are rows $1 \leq i_1 < \cdots < i_m \leq n$ of W such that the $m \times m$ submatrix W_m consisting of rows $i_1, \ldots, i_m$ and columns $1, \ldots, m$ of W has rank m. Construct the 2^m elements of V^m in lexicographic order:

$$V^m = \{0 = A_0, \ldots, A_{2^m - 1}\}.$$

Precisely, $A_k = WK$, where K is the column vector of the binary value of $k \in \{0, \ldots, 2^m - 1\}$, with the bits of K in positions $i_1, \ldots, i_m$, the least significant bit in i_1; and 0's in all the other $n - m$ positions. See Table 1 for an example with $m = n$.

For given $d \in \mathbb{Z}^+$, scan V^m from A_0 to $A_{2^m - 1}$ to generate a subset $V' \subseteq V^m$ using the following greedy algorithm. Put $V' \leftarrow 0$. If $A_{i_0} = 0$, $\ldots$, A_{i_j} have already been inserted into V', insert $A_{i_{j+1}}$ if $i_{j+1} > i_j$ is the smallest integer such that $H(A_{i_l}, A_{i_{j+1}}) \geq d$ for $l \in \{0, \ldots, j\}$, where H denotes Hamming distance. The resulting V' is the *lexicode* generated by W, with minimal distance d.

We remark that in [Brualdi and Pless 1993] the term "lexicode" is reserved for the code generated when W is the identity matrix, which is the case considered

k	V^m		V'
	BIN	DEC	
0	0000	0	*
1	0001	1	
2	0011	3	*
3	0010	2	
4	0110	6	*
5	0111	7	
6	0101	5	*
7	0100	4	
8	1100	12	*
9	1101	13	
10	1111	15	*
11	1110	14	
12	1010	10	*
13	1011	11	
14	1001	9	*
15	1000	8	

Table 1. Generating a lexicode (see Example 2.4).

in [Conway and Sloane 1986]; and "greedy codes" is used for the codes derived from any W whose columns constitute a basis. Actually, in both of these papers no matrices are used, but the ordering is done in an equivalent manner. It seems natural, in the current context, to use matrices (see the proofs in the next section) and "lexicode" for the entire class of codes.

EXAMPLE 2.4. Let

$$W = \begin{matrix} 4 & 3 & 2 & 1 \\ \begin{pmatrix} 1 & 0 & 0 & 0 \\ 1 & 1 & 0 & 0 \\ 0 & 1 & 1 & 0 \\ 0 & 0 & 1 & 1 \end{pmatrix} \end{matrix},$$

and $d = 2$, $m = n = 4$. We then get the ordered vector space depicted in Table 1. The vectors marked with an asterisk in column V' have been selected by our greedy algorithm, and constitute the lexicode. Note that this lexicode is precisely the same code as that found in Example 2.1 by using a small groundgraph with $O(n^2)$ operations rather than $O(2^n)$ for the lexicode.

EXAMPLE 2.5. Let

$$W = \begin{pmatrix} 1 & 0 & 0 & 0 \\ 1 & 0 & 1 & 0 \\ 0 & 1 & 1 & 0 \\ 0 & 1 & 1 & 1 \end{pmatrix},$$

and $d = 2$. The reader should verify that the lexicode generated by W is $(0, 7, 12, 11)$, in this order, which is identical to the code generated in Example 2.2.

EXAMPLE 2.6. Let

$$W = \begin{pmatrix} 1 & 0 & 0 & 0 & 0 \\ 1 & 1 & 0 & 0 & 0 \\ 0 & 1 & 1 & 0 & 0 \\ 0 & 0 & 1 & 1 & 0 \\ 0 & 0 & 0 & 1 & 1 \end{pmatrix},$$

and $d = 3$. The vector space now contains 32 entries, too large to list here. But the reader can verify that the lexicode generated by W is precisely the same as that generated by the two polynomial methods of Example 2.3.

3. The Truth About Lexicodes

We now study the Sprague–Grundy function of a certain game associated with a lexicode. Given a finite acyclic digraph $G = (V, E)$, the associated *Sprague–Grundy* function $g : V \to \mathbb{Z}^0$ is characterized by the property

$$g(u) = \operatorname{mex} g(F(u)), \tag{3.1}$$

where, for any finite subset $S \subseteq \mathbb{Z}^0$, we define $\operatorname{mex}(S) = \min(\mathbb{Z}^0 - S)$ and $g(S) = \{g(s) : s \in S\}$. This function exists uniquely on any finite acyclic digraph. See, for example, [Berge 1985, Ch. 14; 1989, Ch. 4; Conway 1976; Berlekamp, Conway and Guy 1982]. (When G has cycles or loops, g may not exist; a generalization of it, the γ-function mentioned in the introduction, can be used in this case.)

With a lexicode in V^m, with minimal distance d, associate a digraph $G = (V, E)$, called a *lexigraph*, as follows. The vertex set V is the set of all elements (vectors) of V^m, and $(A_k, A_j) \in E$ if and only if $j < k$ and

$$H(A_j, A_k) = w(A_j \oplus A_k) < d,$$

where, as before, H is the Hamming distance and w the weight. If $(A_k, A_j) \in E$, we have $A_j \in F(A_k)$ in the notation introduced at the beginning of Section 2. Note that G is finite and acyclic. (For other possibilities of orienting the lexigraph, see the homework problem towards the end of this section.)

Play a *lexigame* on G by placing a single token on any vertex. A move consists of sliding the token from a vertex to a neighboring vertex along a directed edge. The player first unable to play loses and the opponent wins. Note that any game with a single token on a digraph, and in particular the lexigame just introduced, can be considered an anngame. The P-positions of the lexigame constitute the lexicode; this is also the set of vertices of G on which $g = 0$. (Actually, the lexigame is not overly interesting, because the lexigraph is "analogous" to the game graph of a (more interesting) game played on a logarithmically smaller groundgraph with several tokens. The game graph of a game is not normally constructed, but used instead for reasoning about the game. In fact, we do this in the proof of Theorem 3.9 below.)

We point out that for lexicodes *per se* it suffices to consider the case $m = n$. It is only in Corollary 3.7 and in Section 4, where we apply a greedy algorithm on anncodes, that the case $m < n$ will be important. Incidentally, Brualdi and Pless [1993, § 2] define a function g and state, citing [Conway and Sloane 1986], that g is the Sprague–Grundy function of an associated heap game for the case where W is the unit matrix. It is easy to see that, in fact, g is the Sprague–Grundy function of the lexigraph defined above, for every matrix W.

For any positive integer s, let s^h denote the bit in the h-th binary position of the binary expansion of s, where s^0 denotes the least significant bit. Also, for any $a \in \mathbb{Z}^0$, write $\phi(a) = \{0, \ldots, a - 1\}$.

LEMMA 3.1. *Let* $a_1, a_2 \in \mathbb{Z}^0$, *and let* $b \in \phi(a_1 \oplus a_2)$. *Then there is* $i \in \{1, 2\}$ *and* $d \in \phi(a_i)$ *such that* $b = a_j \oplus d$ *for* $j \neq i$.

PROOF. Write $c = a_1 \oplus a_2$. Let $k = \max\{h : b^h \neq c^h\}$. Since $b < c$, we have $b^k = 0$ and $c^k = 1$. Hence there exists $i \in \{1, 2\}$ such that $a_i^k = 1$. Letting $d = a_i \oplus b \oplus c = a_j \oplus b$, we have $d \in \phi(a_i)$, since $b^h = c^h$ implies $d^h = a_i^h$ for $h > k$, and $d^k = 0$. $\qquad\square$

COROLLARY 3.2. *We have* $\phi(a_1 \oplus a_2) \subset a_1 \oplus \phi(a_2) \cup \phi(a_1) \oplus a_2$. $\qquad\square$

By the closure of V^m, for any j and k there exists l such that $A_j \oplus A_k = A_l$.

LEMMA 3.3. *We have* $A_j \oplus A_k = A_{j \oplus k}$.

PROOF. As noted above, $A_j \oplus A_k = A_l$ for some l. Then $A_j = WJ$, $A_k = WK$, $A_l = WL$. Thus

$$WL = A_l = A_j \oplus A_k = W(J \oplus K).$$

This matrix equation implies $W_m L_m = W_m(J_m \oplus K_m)$, where W_m was defined in Section 2, and any $m \times 1$ vector X_m is obtained from the $n \times 1$ vector X by retaining only the rows $i_1, \ldots, i_m$ of X and deleting the $n - m$ remaining rows, which contain only 0's for L, J and K. Since W_m is invertible, we thus get $L_m = J_m \oplus K_m$, so $l = j \oplus k$. $\qquad\square$

Here is the main lemma of this section.

LEMMA 3.4. *Let* $A_j, A_k \in V^m$. *Then, for the lexigraph on* V^m,

$$F(A_j \oplus A_k) \subseteq A_j \oplus F(A_k) \cup F(A_j) \oplus A_k \subseteq F(A_j \oplus A_k) \cup F^{-1}(A_j \oplus A_k).$$

PROOF. Let $A_l \in F(A_j \oplus A_k)$. By Lemma 3.3, $A_l \in F(A_{j \oplus k})$, so $w(A_l \oplus A_{j \oplus k}) = w(A_{j \oplus k \oplus l}) < d$ and $l < j \oplus k$. By Corollary 3.2, $l \in j \oplus \phi(k) \cup \phi(j) \oplus k$. Thus either there is $k' < k$ such that $l = j \oplus k'$, or there is $j' < j$ such that $l = j' \oplus k$. In the former case, $w(A_{j \oplus k \oplus l}) = w(A_{k \oplus k'}) < d$, so $A_l = A_j \oplus A_{k'} \in A_j \oplus F(A_k)$, and in the latter case we obtain, similarly, $A_l \in F(A_j) \oplus A_k$, establishing the left inclusion.

Now let $A_l \in A_j \oplus F(A_k) \cup F(A_j) \oplus A_k$. Then either $A_l = A_j \oplus A_{k'}$ for some $k' < k$ with $w(A_{k \oplus k'}) < d$, or $A_l = A_{j'} \oplus A_k$ for some $j' < j$ with $w(A_{j \oplus j'}) < d$. Without loss of generality, assume the former. Then $l = j \oplus k'$. Thus $w(A_{k \oplus k'}) = w(A_{j \oplus k \oplus l}) < d$. If $l < j \oplus k$, then $A_l \in F(A_j \oplus A_k)$, and if $l > j \oplus k$, then $A_j \oplus A_k \in F(A_l)$. $\qquad\square$

We now show that the g-function is linear on the lexigraph G.

THEOREM 3.5. *Let $G = (V, E)$ be a lexigraph. Then $g(u_1 \oplus u_2) = g(u_1) \oplus g(u_2)$ for all $u_1, u_2 \in V$.*

PROOF. Set
$$\mathcal{F}(u_1, u_2) = \{u_1\} \times F(u_2) \cup F(u_1) \times \{u_2\}, \qquad (3.2)$$
so that $(v_1, v_2) \in \mathcal{F}(u_1, u_2)$ if either $v_1 = u_1$ and $v_2 \in F(u_2)$, or $v_1 \in F(u_1)$ and $v_2 = u_2$: Thus $\mathcal{F}$ represents the set of followers in the sum game played on $G + G$. Let
$$K = \{(u_1, u_2) \in V \times V : g(u_1 \oplus u_2) \neq g(u_1) \oplus g(u_2)\},$$
$$k = \min_{(u_1, u_2) \in K} (g(u_1 \oplus u_2), g(u_1) \oplus g(u_2)).$$

If there is $(u_1, u_2) \in K$ such that $g(u_1 \oplus u_2) = k$, then $g(u_1) \oplus g(u_2) > k$. By Corollary 3.2 and the mex property (3.1) of g, there is $(v_1, v_2) \in \mathcal{F}(u_1, u_2)$ such that $g(v_1) \oplus g(v_2) = k$. Now (3.2) implies
$$v_1 \oplus v_2 \in u_1 \oplus F(u_2) \cup F(u_1) \oplus u_2 \subseteq F(u_1 \oplus u_2) \cup F^{-1}(u_1 \oplus u_2),$$
where the inclusion follows from Lemma 3.4. Since $g(u_1 \oplus u_2) = k$, it follows that $g(v_1 \oplus v_2) > k$, so $(v_1, v_2) \in K$. Let
$$L = \{(u_1, u_2) \in K : g(u_1) \oplus g(u_2) = k\}.$$

We have just shown that $K \neq \varnothing$ implies $L \neq \varnothing$.

Here we recall that g is the γ-function for the lexigame (see the first paragraph of this section). With a γ-function we can associate a monotonic counter function $c : V \to \mathbb{Z}^+$. We now pick $(u_1, u_2) \in L$ with $c(u_1) + c(u_2)$ minimal. For $(u_1, u_2) \in L$ we have $g(u_1 \oplus u_2) > k$. Then there is $v \in F(u_1 \oplus u_2)$ with $g(v) = k$. By the first inclusion of Lemma 3.4, there exists $(v_1, v_2) \in \mathcal{F}(u_1, u_2)$ such that $v = v_1 \oplus v_2$. So $g(v_1 \oplus v_2) = k$. Since $g(u_1) \oplus g(u_2) = k$, (3.2) implies $g(v_1) \oplus g(v_2) > k$, hence $(v_1, v_2) \in K$. As we saw earlier, this implies that there is $(w_1, w_2) \in \mathcal{F}(v_1, v_2)$ such that $(w_1, w_2) \in L$. Moreover, by property B in the definition of the γ-function (see [Fraenkel 1996, p. 20] in this volume), we can select (w_1, w_2) such that $c(w_1) + c(w_2) < c(u_1) + c(u_2)$, contradicting the minimality of $c(u_1) + c(u_2)$. Thus $L = K = \varnothing$. $\qquad\square$

Let $V_i = \{u \in V : g(u) = i\}$, for $i \geq 0$. We now state the main result of this section.

THEOREM 3.6. *Let $G = (V, E)$ be a lexigraph. Then $V_0 = V'$, where V' is a lexicode. Moreover, V_0 is a linear subspace of V. In fact, g is a homomorphism from V onto $GF(2)^t$ for some $t \in \mathbb{Z}^0$; its kernel is V_0, and the quotient space V/V_0 consists of the cosets V_i for $0 \le i < 2^t$}; in fact, $t = \dim V - \dim V_0$.*

PROOF. By definition, V is a vector space over $GF(2)$. Let t be the smallest nonnegative integer such that $g(u) \le 2^t - 1$ for all $u \in V$. Thus, if $t \ge 1$, there is some $v \in V$ such that $g(v) \ge 2^{t-1}$. Then the "1's complement" of $g(v)$, defined as $2^t - 1 - g(v)$, is less than $g(v)$. By the mex property of g, there exists $w \in F(v)$ such that $g(w) = 2^t - 1 - g(v)$. By Theorem 3.5, $g(v \oplus w) = g(v) \oplus g(w) = 2^t - 1$. Thus, again by the mex property of g, every value in $\{0, \ldots, 2^t - 1\}$ is the g-value of some $u \in V$. This last property holds trivially also for $t = 0$. Hence g is onto. It is a homomorphism $V \to GF(2)^t$ by Theorem 3.5, and since $g(1u) = g(u) = 1g(u)$ and $g(0u) = g(0 \ldots 0) = 0 = 0g(u)$.

By elementary linear algebra, $GF(2)^t \simeq V/V_0$, where V_0 is the kernel. Hence V_0 is a subspace of V. Clearly V_0 is also a graph-kernel of G. So is V', which, by its definition, is both independent and dominating. Since any finite acyclic digraph has a unique kernel, $V_0 = V'$. Let $m = \dim V_0$. Then $\dim V = m + t$. The elements of V/V_0 are the cosets $V_i = w \oplus V_0$ for any $w \in V_i$ and every $i \in \{0, \ldots, 2^t - 1\}$. $\qquad\square$

COROLLARY 3.7. *The greedy algorithm, applied to any lexicographic ordering of the subset $V_0 \subset V$, also produces a linear code.*

PROOF. Follows from Theorem 3.6, by considering the lexigraph $G = (V_0, E)$ instead of (V, E). $\qquad\square$

We remark that Algorithm B of [Fraenkel and Yesha 1982] yields a matrix Γ, whose bottom $n - m$ rows, padded with m bottom 0-rows, is the parity check matrix for the code (vectors where $\gamma = 0$). A much simplified version of this algorithm can be used to compute the parity check matrix for the present case (vectors where $g = 0$).

HOMEWORK 3.8. The lexigraph $G = (V, E)$ seems to exhibit a certain robustness, roughly speaking, with respect to E. That is, Theorem 3.6 seems to be invariant under certain edge deletions or reversions. In this direction, prove that Theorem 3.6 is still valid if E is defined as follows: $(A_k, A_j) \in E$ if and only if $A_j < A_k$ (rather than $j < k$) and $H(A_j, A_k) < d$.

THEOREM 3.9. *The set of lexicodes is a subset of the set of anncodes.*

PROOF. Let C be a lexicode with a given minimal distance. As we saw at the beginning of this section, C is the set of the P-positions of the lexigame played on the lexigraph G, or equivalently the set of vertices where the Sprague–Grundy function g is zero. The lexigame is played on G by sliding a single token, and as such it is an annihilation game; the anncode is the set of vertices where the generalized Sprague–Grundy function γ is zero. The two functions are the same, since the graph is acyclic. Thus C is an anncode. $\qquad\square$

The proof of Theorem 3.6 is actually a much simplified version of a similar result for annihilation games [Fraenkel and Yesha 1982], where also the linearity of γ (and hence of g) was proved for the first time, to the best of our knowledge. The simplification in the proof is no accident, since the lexigame played on the lexigraph (the groundgraph) can be considered as an anngame with a single token. It's an acyclic groundgraph, which makes the anngame theory much simpler than for cyclic digraphs.

It might be of interest to explore the subset of anncodes generated when several tokens, rather than only one, are distributed initially on a lexigraph.

Another question is: Under what conditions, and for what finite fields $\mathrm{GF}(p^a)$, where p is prime and $a \in \mathbb{Z}^+$, are there "anncodes"? The key seems to be to generalize annihilation games as follows. On a given finite digraph Γ, place nonzero "particles" (elements of $\mathrm{GF}(p^a)$), at most one particle per vertex. A move consists in selecting an occupied vertex and moving its particle to a neighboring vertex v along a directed edge. If v was occupied, then the "collision" generates a new particle, possibly 0 ("annihilation"), according to the addition table of $\mathrm{GF}(p^a)$. The special case $a = 1$, when the particles are $0, \ldots, p-1$, reduces to p-annihilation: the collision of particles i and j results in particle k, where $k \equiv i + j \bmod p$, for $k < p$; and this special case becomes anngames for $p = 2$. Such "Elementary Particle Physics" games, whose P-positions are collections of linear codes, thus constitute a generalization of anngames. These games and their applications to coding seems to be an as yet unexplored area.

4. Computing Anncodes

In this section we give one particular example illustrating the computation of large anncodes. One can easily produce many others. The present example also shows how anncodes and lexicodes can be made to join forces.

We begin with a family Γ_t of groundgraphs, which is a slightly simplified version of a family considered by Yesha [1978] for showing that the finite γ-values on an annihilation game played on a digraph without leaves can be arbitrarily large.

Let $t \in \mathbb{Z}^+$, and set $J = J(t) = 2^{t-1}$. The digraph Γ_t has vertex set $\{x_1, \ldots, x_J, y_1, \ldots, y_J\}$, and edges as follows:

$$F(x_i) = y_i \qquad \text{for } i = 1, \ldots, J,$$
$$F(y_k) = \{y_i : 1 \le i < k\} \cup \{x_j : 1 \le j \le J \text{ and } j \ne k\} \qquad \text{for } k = 1, \ldots, J.$$

Figure 4 shows Γ_3.

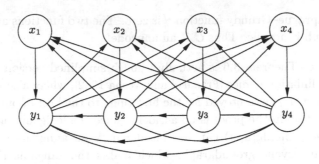

Figure 4. The cyclic groundgraph Γ_3.

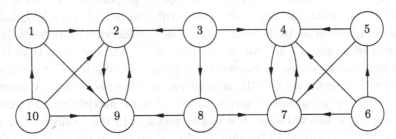

Figure 5. The cyclic groundgraph Γ'.

Since Γ_t has no leaf, $\gamma(x_i) = \gamma(y_i) = \infty$ for all $i \in \{1, \ldots, J\}$. The following facts about the anngraph $G_t = (V, E)$ of Γ_t are easy to establish, where $V^f = \{u \in V : \gamma(u) < \infty\}$.

(i) $\gamma(x_i \oplus x_j) = 0$ for all $i \neq j$.
(ii) $\gamma(x_i \oplus y_j) = j$ for all $i, j \in \{1, \ldots, J\}$.
(iii) $\gamma(y_i \oplus y_j) = i \oplus j$ for all $i \neq j$.
(iv) $\max_{\gamma(u) < \infty} \gamma(u) = \gamma(y_{J-1} \oplus y_J) = (J-1) \oplus J = 2^t - 1$.
(v) $V^f = \{u \in V : w(u) \equiv 0 \bmod 2\}$, $|V^f| = 2^{2J-1}$, $\dim V^f = 2J - 1$.

Thus, in the notation of Theorem 3.6, we have $m + t = 2J - 1$, hence

$$m = 2^t - t - 1.$$

For the family Γ_t of groundgraphs, the $O(n^6)$ algorithm for computing γ thus reduces to an $O(1)$ algorithm.

Now consider the groundgraph Γ' depicted in Figure 5. It is not hard to see that a basis for V_0 is given by the four vectors $(1, 2, 9, 10)$, $(4, 5, 6, 7)$, $(2, 3, 8, 9)$, $(3, 4, 7, 8)$. Each vector indicates the four vertices occupied by tokens.

We propose to play an annihilation game, say on $\Gamma = \Gamma_5 + \Gamma'$, which contains $32 + 10 = 42$ vertices. The vector space associated with the anngraph of Γ contains 2^{42} elements, and to find a lexicode on V^{42}, for any given d, involves 2^{42} operations. On the other hand, for Γ we have, since $t = 1$ for Γ',

$$\dim |V_0| = m = 2^5 - 5 - 1 + 4 + 1 = 31,$$

so the anncode defined by V_0, for which $d = 2$, has 2^{31} elements. By the results of Section 2, we can compute a *lexi-anncode* for any $d > 2$, by applying the greedy algorithm to a lexicographic ordering of V_0, which can be obtained by using any basis of V_0. This computation involves only 2^{31} operations.

HOMEWORK 4.1. Carry out this computation, and find lexi-anncodes for several $d > 2$ on $\Gamma = \Gamma_5 + \Gamma'$.

We note that the anncode derived from a directed complete graph, i.e., a Nim-heap, is identical to the code derived from certain coin-turning games as considered in [Berlekamp, Conway and Guy 1982, Ch. 14].

REMARK 4.2. The Hamming distance between any two consecutive P-positions in an annihilation game is obviously ≤ 4. Thus $d = 2$ for Γ_t and $d = 4$ for Γ'. For finding codes with $d > 4$, it is thus natural to apply the greedy algorithm to a lexicographic ordering of V_0. Another method to produce anncodes with $d > 4$ is to encode each vertex of the groundgraph, that is, each bit of the anngraph, by means of k bits for some fixed $k \in \mathbb{Z}^+$. For example, in a lexigraph, each vertex is encoded by n bits, and the distance between any two codewords is $\geq d$. In an Elementary Particle Physics game over $GF(p^a)$, it seems natural to encode each *particle* by a digits. A third method for producing anncodes with $d > 4$ directly seems to be to consider a generalization of anngames to the case where a move consists of sliding precisely k (or $\leq k$) tokens, where k is a fixed positive integer—somewhat analogously to Moore's Nim (see [Berlekamp, Conway and Guy 1982, Ch. 15], for example).

REMARK 4.3. Note that $\bigcup_{i=0}^{2^k-1} V_i$ is a linear subspace of V^f for every $k \in \{0, \ldots, t\}$. Any of these subspaces is thus also a linear code, in addition to V_0.

Acknowledgment

This paper is a direct result of Vera Pless's lecture at the Workshop on Combinatorial Games held at MSRI, Berkeley, CA, in July, 1994. I had intended to write it after I heard her lecture at the AMS Short Course on Combinatorial Games held in Columbus, OH, in the summer of 1990. But I put it off. This time, at H^sLordship's banquet in Berkeley, Herb Wilf challenged me with his insights and comments, rekindling my interest in this topic. I also had a shorter conversation about it with John Conway and Bill Thurston at that MSRI meeting. Actually, I had originally discussed the possibility of using codes derived from annihilation games with C. L. Liu and E. M. Reingold in 1978 or 1979, and again briefly with Richard Guy in the eighties and with Ya'acov Yesha in the early nineties.

References

1. [Berge 1985] C. Berge, *Graphs*, North-Holland, Amsterdam, 1985.

2. [Berge 1989] C. Berge, *Hypergraphs: Combinatorics of Finite Sets*, North-Holland and Elsevier, Amsterdam, 1989. Original French edition: *Hypergraphes: combinatoire des ensembles finis*, Gauthier-Villars, Paris, 1987.

3. [Berlekamp, Conway and Guy 1982] E. R. Berlekamp, J. H. Conway and R. K. Guy, *Winning Ways for Your Mathematical Plays*, Academic Press, London, 1982.

4. [Brualdi and Pless 1993] R. A. Brualdi and V. S. Pless, "Greedy codes", *J. Combin. Theory* A**64** (1993), 10–30.

5. [Conway 1976] J. H. Conway, *On Numbers and Games*, Academic Press, London, 1976.

6. [Conway 1990] J. H. Conway, "Integral lexicographic codes", *Discrete Math.* **83** (1990), 219–235.

7. [Conway and Sloane 1986] J. H. Conway and N. J. A. Sloane, "Lexicographic codes: error-correcting codes from game theory", *IEEE Trans. Inform. Theory* IT-**32** (1986), 337–348.

8. [Ferguson 1984] T. S. Ferguson, "Misère annihilation games", *J. Combin. Theory* A**37** (1984), 205–230.

9. [Fraenkel 1974] A. S. Fraenkel, "Combinatorial games with an annihiliation rule", pp. 87–91 in *The Influence of Computing on Mathematical Research and Education* (edited by J. P. LaSalle), Proc. Symp. Appl. Math. **20**, Amer. Math. Soc., Providence, RI, 1974.

10. [Fraenkel 1996] A. S. Fraenkel, "Scenic trails ascending from sea-level Nim to alpine chess", pp. 13–42 in this volume.

11. [Fraenkel ≥ 1997] A. S. Fraenkel, *Adventures in Games and Computational Complexity*, to appear in Graduate Studies in Mathematics, Amer. Math. Soc., Providence, RI.

12. [Fraenkel and Perl 1975] A. S. Fraenkel and Y. Perl (1975), "Constructions in combinatorial games with cycles", pp. 667–699 in *Proc. Intern. Colloq. on Infinite and Finite Sets* (edited by A. Hajnal et al.), vol. 2, Colloq. Math. Soc. János Bolyai **10** North-Holland, Amsterdam, 1975.

13. [Fraenkel, Tassa and Yesha 1978] A. S. Fraenkel, U. Tassa and Y. Yesha, "Three annihilation games", *Math. Mag.* **51** (1978), 13–17.

14. [Fraenkel and Yesha 1976] A. S. Fraenkel and Y. Yesha, "Theory of annihilation games", *Bull. Amer. Math. Soc.* **82** (1976), 775–777.

15. [Fraenkel and Yesha 1979] A. S. Fraenkel and Y. Yesha, "Complexity of problems in games, graphs and algebraic equations", *Discrete Appl. Math.* **1** (1979), 15–30.

16. [Fraenkel and Yesha 1982] A. S. Fraenkel and Y. Yesha, "Theory of annihilation games – I", *J. Combin. Theory* B**33** (1982), 60–86.

17. [Fraenkel and Yesha 1986] A. S. Fraenkel and Y. Yesha, "The generalized Sprague–Grundy function and its invariance under certain mappings", *J. Combin. Theory* A**43** (1986), 165–177.

18. [Smith 1966] C. A. B. Smith, "Graphs and composite games", *J. Combin. Theory* **1** (1966), 51-81. Reprinted in slightly modified form in *A Seminar on Graph Theory* (edited by F. Harary), Holt, Rinehart and Winston, New York, 1967.

19. [Yesha 1978] Y. Yesha, "Theory of annihilation games", Ph.D. Thesis, Weizmann Institute of Science, Rehovot (Israel), 1978.

AVIEZRI S. FRAENKEL
DEPARTMENT OF APPLIED MATHEMATICS AND COMPUTER SCIENCE
WEIZMANN INSTITUTE OF SCIENCE
REHOVOT 76100, ISRAEL
    fraenkel@wisdom.weizmann.ac.il
    http://www.wisdom.weizmann.ac.il/~fraenkel/fraenkel.html

Games of No Chance
MSRI Publications
Volume 29, 1996

Tutoring Strategies in Game-Tree Search

(Extended Abstract)

HIROYUKI IIDA, YOSHIYUKI KOTANI, AND
JOS W. H. M. UITERWIJK

Introduction. According to the analysis of grandmaster-like strategies in Shogi [Iida and Uiterwijk 1993], it is important for a teacher, at the beginning stages of teaching, to intentionally lose an occasional game against a novice opponent, or to play less than optimally in order to give the novice some prospects of winning, *without this being noticed by the opponent*. Such a strategy is called a *tutoring strategy* in game-tree search.

In this work we consider a *loss-oriented search strategy* (*LO-search* for short), by which a player attempts to lose a game, and we show an algorithm for LO-search based on the minimax strategy and OM-search (opponent-model search; see [Iida et al. 1993]). We further describe characteristics of LO-search, including the concept of *intentional error*. We next discuss the situation in which a player will notice such an intentional error. We then describe a *tutoring search* (*TU-search* for short) by which a player attempts to lose a game without this being noticed by the opponent.

LO-search is based on OM-search, since LO-search also takes the opponent model into account. OM-search is predicated on the assumption that one has complete knowledge of the opponent's strategy and that this strategy is completely contained in the opponent's evaluation function. In OM-search, two values are computed for all positions in a search tree. For clarity, the players are distinguished as a max player and a min player.

Loss-Oriented Search. In order to play games, the minimax strategy or variations thereof have been used by many game-playing programs. The minimax strategy will choose the best move according to the program's look-ahead, using some evaluation function. OM-search will also choose the best move while taking the opponent model into account. Note that these strategies back up the best

---

The full version of this article, including algorithms, example search trees and characteristics of the search strategies, will be published in the *ICCA Journal* vol. **18**, No. 4 (December 1995).

successor at any internal max node in a search tree. Thus, in order to lose a game against an opponent, it may be reasonable to choose a move at a max node that leads to a successor whose value is not the best. A definition of LO-search now is the following:

> a search strategy that chooses a move taking the opponent model into account at a min position and *maximizing* the value by OM-search at a max position, provided that the latter is less than the value by the minimax strategy from the max player's point of view. If there is no such successor at a max position, a move *minimizing* the value by OM-search is chosen.

LO-search thus plays an *intentional error*, provided that it exists, whose value by OM-search is maximal among intentional errors. When there are no intentional errors, that is, when every move leads to a better result than minimax due to the opponent's errors, LO-search plays a move whose value by OM-search is the minimum among the legal moves.

Note the difference between LO-search and misère play. In misère play, the goal for both players is to lose, whereas in LO-search the main goal for one player, called the *max player*, is to lose (or give the opponent some prospect of winning), while the *min player* is striving to win. Again we assume that the max player, who will adopt LO-search, has complete knowledge of the min player's evaluation function.

It is evident that, the worse the score of the move played, the higher the potential to lose. However, LO-search should not play moves that are too stupid, such as giving up a queen for free in chess. Therefore, it is important to examine at a given position whether LO-search is to be used or not.

Many variations on LO-search are possible. The above definition seems to be one of the most reasonable variations for playing an intentional error, since LO-search often plays only a little weaker than the program's real strength. But, in many tactical situations, the second-best move may be just as bad as the worst—for example, failing to recapture a piece or making the wrong response to a mate threat in chess. In such cases, an LO-search move will be very poor. A solution to this problem is to recognize when a non-optimal move should be played and when not.

An additional problem with LO-search is that the min player can be aware of an intentional error played by the max player.

Tutoring Search. To tackle the problems just mentioned, we define the concept of *loss* as the difference between the value of the move chosen and that of the optimal move in the OM-search sense, and the notion of *loss limit* ε, a bound for the loss beyond which the min player probably will be aware of the max player's intentional error. The value of ε should be small when the two players have comparable strength. When the min player's evaluation function is identical to the max player's, ε must be set to 0: the max player cannot play any move by TU-search other than the best move from his point of view. A definition of TU-search now is the following:

a search strategy that chooses a move taking the opponent model into account at a min position and *minimizing* the value by OM-search at a max position, provided that an intentional error will be unnoticed by the opponent.

Contrary to LO-search, tutoring search only will play an intentional error when this will be unnoticed by the opponent. It will not consider very stupid moves (from both player's point of view). On the other hand, since very dumb moves are avoided, it will be possible to choose the *lowest*-valued one among the candidate intentional errors.

References

[Iida and Uiterwijk 1993]   H. Iida and J. W. H. M. Uiterwijk, "Thoughts on grandmaster-like strategies", pp. 1–8 in *Proceedings of The Second European Shogi Workshop*, Shogi Deutschland and EMBL, Heidelberg, 1993.

[Iida et al. 1993] H. Iida, J. W. H. M. Uiterwijk, and H. J. van den Herik, "Opponent-model search", Technical Report CS 93-03, Department of Computer Science, University of Limburg, Maastricht, 1993.

HIROYUKI IIDA
YOSHIYUKI KOTANI
DEPARTMENT OF COMPUTER SCIENCE
TOKYO UNIVERSITY OF AGRICULTURE AND TECHNOLOGY
TOKYO, JAPAN
    iida@etl.go.jp

JOS W. H. M. UITERWIJK
DEPARTMENT OF COMPUTER SCIENCE
UNIVERSITY OF LIMBURG
MAASTRICHT, THE NETHERLANDS

Games of No Chance
MSRI Publications
Volume **29**, 1996

About David Richman

The following article is based on unpublished work of the late mathematician David Ross Richman (1956–1991).

David was a problem solver by temperament, with strong interests in number theory, algebra, invariant theory, and combinatorics. He did his undergraduate work at Harvard and received his Ph.D. in mathematics from the University of California at Berkeley, under the supervision of Elwyn Berlekamp. I met him at one of the annual convocations of the West Coast Number Theory Conference held at the Asilomar Conference Center in Monterey, California. His quick mind and unassuming manner made him a pleasant person to discuss mathematics with, and he was one of the people I most looked forward to seeing at subsequent conferences.

In one of our conversations in the mid-1980's, he mentioned his idea of playing combinatorial games under a protocol in which players bid for the right to make the next move. Over the course of the next few years, I urged him to write up this work, but he was too busy with other mathematical projects.

By the beginning of 1991, he had received tenure at the University of South Carolina, and was commencing his first sabbatical. He planned to spend the first half of 1991 in Taiwan and the second half at MSRI. He died on February 1, 1991, in a widely-reported accident at Los Angeles Airport in which many other people were killed. He left behind a wife, the mathematician Shu-Mei Richman; a daughter, Miriam; and his parents, Alex and Shifra. His wife was expecting their second child at the time of his death; the baby was born on Shu-Mei's birthday and named David Harry.

For those who knew him personally, David leaves behind fond recollections, and for those who knew him professionally, he leaves behind a tragic sense of promise incompletely fulfilled. This article is dedicated to his memory.

James G. Propp

Games of No Chance
MSRI Publications
Volume **29**, 1996

Richman Games

ANDREW J. LAZARUS, DANIEL E. LOEB,

JAMES G. PROPP, AND DANIEL ULLMAN

Dedicated to David Richman, 1956–1991

ABSTRACT. A Richman game is a combinatorial game in which, rather than
alternating moves, the two players bid for the privilege of making the next
move. We find optimal strategies for both the case where a player knows
how much money his or her opponent has and the case where the player
does not.

1. Introduction

There are two game theories. The first is now sometimes referred to as matrix
game theory and is the subject of the famous von Neumann and Morgenstern
treatise [1944]. In matrix games, two players make simultaneous moves and a
payment is made from one player to the other depending on the chosen moves.
Optimal strategies often involve randomness and concealment of information.

The other game theory is the combinatorial theory of *Winning Ways* [Berle-
kamp et al. 1982], with origins back in the work of Sprague [1936] and Grundy
[1939] and largely expanded upon by Conway [1976]. In combinatorial games,
two players move alternately. We may assume that each move consists of sliding
a token from one vertex to another along an arc in a directed graph. A player who
cannot move loses. There is no hidden information and there exist deterministic
optimal strategies.

In the late 1980's, David Richman suggested a class of games that share some
aspects of both sorts of game theory. Here is the set-up: The game is played
by two players (Mr. Blue and Ms. Red), each of whom has some money. There

1991 *Mathematics Subject Classification.* 90D05.

Key words and phrases. Combinatorial game theory, impartial games.

Loeb was partially supported by URA CNRS 1304, EC grant CHRX-CT93-0400, the PRC
Maths-Info, and NATO CRG 930554.

is an underlying combinatorial game in which a token rests on a vertex of some finite directed graph. There are two special vertices, denoted by b and r; Blue's goal is to bring the token to b and Red's goal is to bring the token to r. The two players repeatedly bid for the right to make the next move. One way to execute this bidding process is for each player to write secretly on a card a nonnegative real number no larger than the number of dollars he or she has; the two cards are then revealed simultaneously. Whoever bids higher pays the amount of the bid to the opponent and moves the token from the vertex it currently occupies along an arc of the directed graph to a successor vertex. Should the two bids be equal, the tie is broken by a toss of a coin. The game ends when one player moves the token to one of the distinguished vertices. The sole objective of each player is to make the token reach the appropriate vertex: at the game's end, money loses all value. The game is a draw if neither distinguished vertex is ever reached.

Note that with these rules (compare with [Berlekamp 1996]), there is never a reason for a negative bid: since all successor vertices are available to both players, it cannot be preferable to have the opponent move next. That is to say, there is no reason to part with money for the chance that your opponent will carry out through negligence a move that you yourself could perform through astuteness.

A winning strategy is a policy for bidding and moving that guarantees a player the victory, given fixed initial data. (These initial data include where the token is, how much money the player has, and possibly how much money the player's opponent has.) In section 2, we explain how to find winning strategies for Richman games. In particular, we prove the following facts, which might seem surprising:

- Given a starting vertex v, there exists a critical ratio $R(v)$ such that Blue has a winning strategy if Blue's share of the money, expressed as a fraction of the total money supply, is greater than $R(v)$, and Red has a winning strategy if Blue's share of the money is less than $R(v)$. (This is not so surprising in the case of acyclic games, but for games in general, one might have supposed it possible that, for a whole range of initial conditions, play might go on forever.)

- There exists a strategy such that if a player has more than $R(v)$ and applies the strategy, the player will win with probability 1, without needing to know how much money the opponent has.

In proving these assertions, it will emerge that a critical (and in many cases optimal) bid for Blue is $R(v) - R(u)$ times the total money supply, where v is the current vertex and u is a successor of v for which $R(u)$ is as small as possible. A player who cannot bid this amount has already lost, in the sense that there is no winning strategy for that player. On the other hand, a player who has a winning strategy of any kind and bids $R(v) - R(u)$ will still have a winning strategy one move later, regardless of who wins the bid, as long as the player is careful to move to u if he or she does win the bid.

It follows that we may think of $R(v) - R(u)$ as the "fair price" that Blue should be willing to pay for the privilege of trading the position v for the position u. Thus we may define $1 - R(v)$ as the *Richman value* of the position v, so that the fair price of a move exactly equals the difference in values of the two positions. However, it is more convenient to work with $R(v)$ than with $1 - R(v)$. We call $R(v)$ the *Richman cost* of the position v.

We will see that for all v other than the distinguished vertices b and r, $R(v)$ is the average of $R(u)$ and $R(w)$, where u and w are successors of v in the digraph that minimize and maximize $R(\cdot)$, respectively. In the case where the digraph underlying the game is acyclic, this averaging-property makes it easy to compute the Richman costs of all the positions, beginning with the positions b and r and working backwards. If the digraph contains cycles it is not so easy to work out precise Richman costs.

We defer most of our examples to another paper [Lazarus et al.], in which we also consider infinite digraphs and discuss the complexity of the computation of Richman costs.

2. The Richman Cost Function

Henceforth, D will denote a finite directed graph (V, E) with a distinguished blue vertex b and a distinguished red vertex r such that from every vertex there is a path to at least one of the distinguished vertices. For $v \in V$, let $S(v)$ denote the set of successors of v in D, that is, $S(v) = \{w \in V : (v, w) \in E\}$. Given any function $f : V \to [0, 1]$, we define

$$f^+(v) = \max_{w \in S(v)} f(w) \quad \text{and} \quad f^-(v) = \min_{u \in S(v)} f(u).$$

The key to playing the Richman game on D is to attribute costs to the vertices of D such that the cost of every vertex (except r and b) is the average of the lowest and highest costs of its successors. Thus, a function $R : V \to [0, 1]$ is called a *Richman cost function* if $R(b) = 0$, $R(r) = 1$, and for every other $v \in V$ we have $R(v) = \frac{1}{2}(R^+(v) + R^-(v))$. (Note that Richman costs are a curious sort of variant on harmonic functions on Markov chains [Woess 1994], where instead of averaging over all the successor-values, we average only over the two extreme values.) The relations $R^+(v) \geq R(v) \geq R^-(v)$ and $R^+(v) + R^-(v) = 2R(v)$ will be much used in what follows.

THEOREM 2.1. *The digraph D has a Richman cost function $R(v)$.*

PROOF. We introduce a function $R(v, t)$, whose game-theoretic significance will be made clearer in Theorem 2.2. Let $R(b, t) = 0$ and $R(r, t) = 1$ for all $t \in \mathbb{N}$.

For $v \notin \{b, r\}$, define $R(v, 0) = 1$ and

$$R(v, t) = \tfrac{1}{2}(R^+(v, t - 1) + R^-(v, t - 1))$$

for $t > 0$. It is easy to see that $R(v, 1) \leq R(v, 0)$ for all v, and a simple induction shows that $R(v, t+1) \leq R(v, t)$ for all v and all $t \geq 0$. Therefore $R(v, t)$ is weakly decreasing and bounded below by zero as $t \to \infty$, hence convergent. It is also evident that the function $v \mapsto \lim_{t \to \infty} R(v, t)$ satisfies the definition of a Richman cost function. $\square$

ALTERNATE PROOF. Identify functions $f : V(D) \to [0, 1]$ with points in the $|V(D)|$-dimensional cube $Q = [0, 1]^{|V(D)|}$. Given $f \in Q$, define $g \in Q$ by $g(b) = 0$, $g(r) = 1$, and, for every other $v \in V$, $g(v) = \tfrac{1}{2}(f^+(v) + f^-(v))$. The map $f \mapsto g$ is clearly a continuous map from Q into Q, and so by the Brouwer fixed point theorem it has a fixed point. This fixed point is a Richman cost function.
                                                                                                  $\square$

This Richman cost function does indeed govern the winning strategy, as we now prove.

THEOREM 2.2. *Suppose Blue and Red play the Richman game on the digraph D with the token initially located at vertex v. If Blue's share of the total money exceeds $R(v) = \lim_{t \to \infty} R(v, t)$, Blue has a winning strategy. Indeed, if his share of the money exceeds $R(v, t)$, his victory requires at most t moves.*

PROOF. Without loss of generality, money may be scaled so that the total supply is one dollar. Whenever Blue has over $R(v)$ dollars, he must have over $R(v, t)$ dollars for some t. We prove the claim by induction on t. At $t = 0$, Blue has over $R(v, 0)$ dollars only if $v = b$, in which case he has already won.

Now assume the claim is true for $t - 1$, and let Blue have more than $R(v, t)$ dollars. There exist neighbors u and w of v such that $R(u, t - 1) = R^-(v, t - 1)$ and $R(w, t-1) = R^+(v, t-1)$, so that $R(v, t) = \tfrac{1}{2}(R(w, t-1) + R(u, t-1))$. Blue can bid $\tfrac{1}{2}(R(w, t-1) - R(u, t-1))$ dollars. If Blue wins the bid at v, then he moves to u and forces a win in at most $t-1$ moves (by the induction hypothesis), since he has more than $\tfrac{1}{2}(R(w, t-1) + R(u, t-1)) - \tfrac{1}{2}(R(w, t-1) - R(u, t-1)) = R(u, t-1)$ dollars left. If Blue loses the bid, then Red will move to some z, but Blue now has over $\tfrac{1}{2}(R(w, t-1) + R(u, t-1)) + \tfrac{1}{2}(R(w, t-1) - R(u, t-1)) = R(w, t-1) \geq R(z, t-1)$ dollars, and again wins by the induction hypothesis. $\square$

One can define another function $R'(v, t)$ where $R'(b, t) = 0$ and $R'(r, t) = 1$ for all $t \in \mathbb{N}$, and $R'(v, 0) = 0$ and $R'(v, t) = \tfrac{1}{2}(R^+(v, t - 1) + R^-(v, t - 1))$ for $v \notin \{b, r\}$ for $t > 0$. By an argument similar to the proof of Theorem 2.2, this also converges to a Richman cost function $R'(v) \leq R(v)$ (with $R(v)$ defined as in the proof of Theorem 2.1). Thus, $R'(v, t)$ indicates how much money Blue needs to prevent Red from forcing a win from v in t or fewer moves, so $R'(v)$ indicates how much money Blue needs to prevent Red from forcing a win in any length of time.

For certain *infinite* digraphs, it can be shown [Lazarus et al.] that $R'(v)$ is strictly less than $R(v)$. When Blue's share of the money supply lies strictly between $R'(v)$ and $R(v)$, each player can prevent the other from winning. Thus, optimal play leads to a draw.

Nevertheless, in this paper, we assume that D is finite, and we can conclude that there is a unique Richman cost function $R'(v) = R(v)$.

THEOREM 2.3. *The Richman cost function of the digraph D is unique.*

The proof of Theorem 2.3 requires the following definition and technical lemma.

An edge (v, u) is said to be an *edge of steepest descent* if $R(u) = R^-(v)$. Let $\bar{v}$ be the transitive closure of v under the steepest-descent relation. That is, $w \in \bar{v}$ if there exists a path $v = v_0, v_1, v_2, \ldots, v_k = w$ such that (v_i, v_{i+1}) is an edge of steepest descent for $i = 0, 1, \ldots, k - 1$.

LEMMA 2.4. *Let R be any Richman cost function of the digraph D. If $R(z) < 1$, then $\bar{z}$ contains b.*

PROOF. Suppose $R(z) < 1$. Choose $v \in \bar{z}$ such that $R(v) = \min_{u \in \bar{z}} R(u)$. Such a v must exist because D (and hence $\bar{z}$) is finite. If $v = b$, we're done. Otherwise, assume $v \neq b$, and let u be any successor of v. The definition of v implies $R^-(v) = R(v)$, which forces $R^+(v) = R(v)$. Since $R(u)$ lies between $R^-(v)$ and $R^+(v)$, $R(u) = R(v) = R^-(v)$. Hence (v, u) is an edge of steepest descent, so $u \in \bar{z}$. Moreover, u satisfies the same defining property that v did (it minimized $R(\cdot)$ in the set $\bar{z}$), so the same proof shows that for any successor w of u, $R(w) = R(u)$ and $w \in \bar{z}$. Repeating this, we see that for any point w that may be reached from v, $R(w) = R(v)$ and $w \in \bar{z}$. On the other hand, $R(r)$ is *not* equal to $R(v)$ (since $R(v) \leq R(z) < 1 = R(r)$), so r cannot be reached from v. Therefore b can be reached from v, so we must have $b \in \bar{z}$.   □

PROOF OF THEOREM 2.3. Suppose that R_1 and R_2 are Richman cost functions of D. Choose v such that $R_1 - R_2$ is maximized at v; such a v exists since D is finite. Let $M = R_1(v) - R_2(v)$. Choose u_1, w_1, u_2, w_2 (all successors of v) such that $R_i^-(v) = R(u_i)$ and $R_i^+(v) = R(w_i)$. Since $R_1(u_1) \leq R_1(u_2)$, we have

$$R_1(u_1) - R_2(u_2) \leq R_1(u_2) - R_2(u_2) \leq M. \qquad (2.1)$$

(The latter inequality follows from the definition of M.) Similarly, $R_2(w_2) \geq R_2(w_1)$, so

$$R_1(w_1) - R_2(w_2) \leq R_1(w_1) - R_2(w_1) \leq M \qquad (2.2)$$

Adding (2.1) and (2.2), we have

$$(R_1(u_1) + R_1(w_1)) - (R_2(u_2) + R_2(w_2)) \leq 2M.$$

The left side is $2R_1(v) - 2R_2(v) = 2M$, so equality must hold in (2.1). In particular, $R_1(u_2) - R_2(u_2) = M$; i.e., u_2 satisfies the hypothesis on v. Since u_2 was any vertex with $R_2(u_2) = R_2^-(v)$, induction shows that $R_1(u) - R_2(u) = M$

for all $u \in \bar{v}$, where descent is measured with respect to R_2. Since $R_1(b) - R_2(b) = 0$ and $b \in \bar{v}$, we have $R_1(v) - R_2(v) \leq 0$ everywhere. That is, $R_1 \leq R_2$. The same argument for $R_2 - R_1$ shows the opposite inequality, so $R_1 = R_2$. □

The uniqueness of the Richman cost function implies in particular that the function R' defined after the proof of Theorem 2.2 coincides with the function R constructed in the first proof of Theorem 2.1. From this we deduce the following:

COROLLARY 2.5. *Suppose Blue and Red play the Richman game on the digraph D with the token initially located at vertex v. If Blue's share of the total money supply is less than* $R(v) = \lim_{t \to \infty} R(v, t)$, *then Red has a winning strategy.* □

It is also possible to reverse the order of proof, and to derive Theorem 2.3 from Corollary 2.5. For, if there were two Richman functions R_1 and R_2, with $R_1(v) < R_2(v)$, say, then by taking a situation in which Blue's share of the money was strictly between $R_1(v)$ and $R_2(v)$, we would find that both Blue and Red had winning strategies, which is clearly absurd.

Theorem 2.2 and Corollary 2.5 do not cover the critical case where Blue has exactly $R(v)$ dollars. In this case, with both players using optimal strategy, the outcome of the game depends on the outcomes of the coin-tosses used to resolve tied bids. Note, however, that in all other cases, the deterministic strategy outlined in the proof of Theorem 2.2 works even if the player with the winning strategy concedes all ties and reveals his intended bid and intended move before the bidding.

Summarizing Theorem 2.2 and Corollary 2.5, we may say that if Blue's share of the total money supply is less than $R(v)$, Red has a winning strategy, and if it is greater, Blue has a winning strategy (see [Lazarus et al.] for a fuller discussion).

3. Other Interpretations

Suppose the right to move the token is decided on each turn by the toss of a fair coin. Then induction on t shows that the probability that Red can win from the position v in at most t moves is equal to $R(v, t)$, as defined in the previous section. Taking t to infinity, we see that $R(v)$ is equal to the probability that Red can force a win against optimal play by Blue. That is to say, if both players play optimally, $R(v)$ is the chance that Red will win. The uniqueness of the Richman cost function tells us that $1 - R(v)$ must be the chance that Blue will win. The probability of a draw is therefore zero.

If we further stipulate that the moves themselves must be random, in the sense that the player whose turn it is to move must choose uniformly at random from the finitely many legal options, we do not really have a game-like situation anymore; rather, we are performing a random walk on a directed graph with two absorbing vertices, and we are trying to determine the probabilities of absorption at these two vertices. In this case, the relevant probability function is just the harmonic function on the digraph D (or, more properly speaking, the harmonic function for the associated Markov chain [Woess 1994]).

Another interpretation of the Richman cost, brought to our attention by Noam Elkies, comes from a problem about makings bets. Suppose you wish to bet (at even odds) that a certain baseball team will win the World Series, but that your bookie only lets you make even-odds bets on the outcomes of individual games. Here we assume that the winner of a World Series is the first of two teams to win four games. To analyze this problem, we create a directed graph whose vertices correspond to the different possible combinations of cumulative scores in a World Series, with two special terminal vertices (blue and red) corresponding to victory for the two respective teams. Assume that your initial amount of money is \$500, and that you want to end up with either \$0 or \$1000, according to whether the blue team or the red team wins the Series. Then it is easy to see that the Richman cost at a vertex tells exactly how much money you want to have left if the corresponding state of affairs transpires, and that the amount you should bet on any particular game is \$1000 times the common value of $R(v) - R(u)$ and $R(w) - R(v)$, where v is the current position, u is the successor position in which Blue wins the next game, and v is the successor position in which Red wins the next game.

4. Incomplete Knowledge

Surprisingly, it is often possible to implement a winning strategy without knowing how much money one's opponent has.

Define Blue's *safety ratio* at v as the fraction of the total money that he has in his possession, divided by $R(v)$ (the fraction that he needs in order to win). Note that Blue will not know the value of his safety ratio, since we are assuming that he has no idea how much money Red has.

THEOREM 4.1. *Suppose Blue has a safety ratio strictly greater than* 1. *Then Blue has a strategy that wins with probability* 1 *and does not require knowledge of Red's money supply. If, moreover, the digraph D is acyclic, his strategy wins regardless of tiebreaks; that is, "with probability* 1*" can be replaced by "definitely".*

PROOF. Here is Blue's strategy: When the token is at vertex v, and he has B dollars, he should act as if his safety ratio is 1; that is, he should play as if Red has R_{crit} dollars with $B/(B + R_{crit}) = R(v)$ and the total amount of money is $B + R_{crit} = B/R(v)$ dollars. He should accordingly bid

$$X = \frac{R(v) - R^-(v)}{R(v)} B$$

dollars. Suppose Blue wins (by outbidding or by tiebreak) and moves to u along an edge of steepest descent. Then Blue's safety ratio changes

$$\text{from} \quad \frac{\left(\dfrac{B}{B+R}\right)}{R(v)} \quad \text{to} \quad \frac{\left(\dfrac{B-X}{B+R}\right)}{R(u)},$$

where R is the actual amount of money that Red has. However, these two safety ratios are actually equal, since

$$\frac{B-X}{B} = 1 - \frac{X}{B} = 1 - \frac{R(v) - R(u)}{R(v)} = \frac{R(u)}{R(v)}.$$

Now suppose instead that Red wins the bid (by outbidding or by tiebreak) and moves to z. Then Blue's safety ratio changes

$$\text{from} \quad \frac{\left(\dfrac{B}{B+R}\right)}{R(v)} \quad \text{to} \quad \frac{\left(\dfrac{B+Y}{B+R}\right)}{R(z)},$$

with $Y \geq X$. Note that the new safety ratio is greater than or equal to

$$\frac{\left(\dfrac{B+X}{B+R}\right)}{R(w)},$$

where $R(w) = R^+(v)$. But this lower bound on the new safety ratio is equal to the old safety ratio, since

$$\frac{B+X}{B} = 1 + \frac{X}{B} = 1 + \frac{R(w) - R(v)}{R(v)} = \frac{R(w)}{R(v)}.$$

In either case, the safety ratio is nondecreasing, and in particular must stay greater than 1. On the other hand, if Blue were to eventually lose the game, his safety ratio at that moment would have to be at most 1, since his fraction of the total money supply cannot be greater than $R(r) = 1$. Consequently, our assumption that Blue's safety ratio started out being greater than 1 implies that Blue can never lose. In an acyclic digraph, infinite play is impossible, so the game must terminate at b with a victory for Blue.

In the case where cycles are possible, suppose first that at some stage Red outbids Blue by $\varepsilon B > 0$ and gets to make the next move, say from v to w. If Blue was in a favorable situation at v, the total amount of money that the two players have between them must be less than $B/R(v)$. On the other hand, after the payoff by Red, Blue has

$$B + X + \varepsilon B = \left(1 + \frac{R(v) - R(u)}{R(v)} + \varepsilon\right) B$$

$$= \left(\frac{2R(v) - R(u)}{R(v)} + \varepsilon\right) B$$

$$= \left(\frac{R(w)}{R(v)} + \varepsilon\right) B,$$

so that Blue's total share of the money must be more than $R(w) + \varepsilon R(v)$. Blue can do this calculation as well as we can; he then knows that if he had been in a winning position to begin with, his current share of the total money must exceed $R(w) + \varepsilon R(v)$. Now, $R(w) + \varepsilon R(v)$ is greater than $R(w, t)$ for some t, so Blue can win in t moves. Thus, Red loses to Blue's strategy if she ever bids more than he does. Hence, if she hopes to avoid losing, she must rely entirely on tiebreaking. Let N be the length of the longest path of steepest descent in the directed graph D. Then Blue will win the game when he wins N consecutive tiebreaks (if not earlier). $\qquad\square$

When D has cycles, Blue may need to rely on tiebreaks in order to win, as in the case of the Richman game played on the digraph pictured in Figure 4.

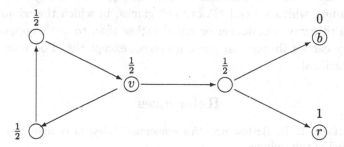

Figure 1. The digraph D and its Richman costs.

Suppose that the token is at vertex v, that Blue has $\boldsymbol{B}$ dollars, and that Red has $\boldsymbol{R}$ dollars. Clearly, Blue knows he can win the game if $\boldsymbol{B} > \boldsymbol{R}$. But without knowing $\boldsymbol{R}$, it would be imprudent for him to bid any positive amount $\varepsilon\boldsymbol{B}$ for fear that Red actually started with $(1 - \varepsilon)\boldsymbol{B}$ dollars; for if that were the case, and his bid were to prevail, the token would move to a vertex where the Richman cost is $\frac{1}{2}$ and Blue would have less money than Red. Such a situation will lead to a win for Red with probability 1 if she follows the strategy outlined in Theorem 2.2.

5. Rationality

For every vertex v of the digraph D (other than b and r), let v^+ and v^- denote successors of v for which $R(v^+) = R^+(v)$ and $R(v^-) = R^-(v)$. Then we have $R(b) = 0$, $R(r) = 1$, and $2R(v) = R(v^+) + R(v^-)$ for $v \neq b, r$. We can view this as a linear program. By Theorem 2.3, this system must have a unique solution. Since all coefficients are rational, we see that Richman costs are always rational numbers.

The linear programming approach also gives us a conceptually simple (though computationally dreadful) way to calculate Richman costs. If we augment our program by adding additional conditions of the form $R(v^-) \leq R(w)$ and $R(v^+) \geq R(w)$, where v ranges over the vertices of D other than b and r and where, for each v, w ranges over all the successors of v, then we are effectively adding in the constraint that the edges from v to v^- and v^+ are indeed edges of steepest descent and ascent, respectively. The uniqueness of Richman costs tells us that if we let the mappings $v \mapsto v^-$ and $v \mapsto v^+$ range over all possibilities (subject to the constraint that both v^- and v^+ must be successors of v), the resulting linear programs (which typically will have no solutions at all) will have only solutions that correspond to the genuine Richman cost functions. Hence, in theory one could try all the finitely many possibilities for $v \mapsto v^-$ and $v \mapsto v^+$ and solve the associated linear programs until one found one with a solution. However, the amount of time such an approach would take increases exponentially with the size of the directed graph. In [Lazarus et al.], we discuss other approaches.

We will also discuss in [Lazarus et al.], among other things, a variant of Richman games, which we call "Poorman" games, in which the winning bid is paid to a third party (auctioneer or bank) rather than to one's opponent. The whole theory carries through largely unchanged, except that Poorman costs are typically irrational.

References

[Berlekamp 1996] E. R. Berlekamp, "An economist's view of combinatorial games", pp. 365–405 in this volume.

[Berlekamp et al. 1982] E. R. Berlekamp, J. H. Conway, and R. K. Guy, *Winning Ways For Your Mathematical Plays*, Academic Press, London, 1982.

[Conway 1976] J. H. Conway, *On Numbers And Games*, Academic Press, London, 1976.

[Grundy 1939] P. M. Grundy, "Mathematics and Games", *Eureka* **2** (1939), 6–8. Reprinted in *Eureka* **27** (1964), 9–11.

[Lazarus et al.] A. J. Lazarus, D. E. Loeb, J. G. Propp, and D. Ullman, "Combinatorial games under auction play", submitted to *Games and Economic Behavior*.

[von Neumann and Morgenstern 1944] J. von Neumann and O. Morgenstern, *Theory of Games and Economic Behavior*, Wiley, New York, 1944.

[Sprague 1935–36] R. Sprague, "Über mathematische Kampfspiele", *Tôhoku Math. J.* **41** (1935–36), 438–444.

[Woess 1994] W. Woess, "Random walks on infinite graphs and groups: a survey on selected topics", *Bull. London Math. Soc.* **26** (1994), 1–60.

ANDREW J. LAZARUS
2745 ELMWOOD AVENUE
BERKELEY, CA 94705
  drlaz@aol.com

DANIEL E. LOEB
LaBRI
UNIVERSITÉ DE BORDEAUX I
33405 TALENCE CEDEX, FRANCE
  loeb@labri.u-bordeaux.fr
  http://www.labri.u-bordeaux.fr/~loeb

JAMES G. PROPP
DEPARTMENT OF MATHEMATICS
MASSACHUSETTS INSTITUTE OF TECHNOLOGY
CAMBRIDGE MA 02139-4307
  propp@math.mit.edu
  http://www-math.mit.edu/~propp

DANIEL ULLMAN
DEPARTMENT OF MATHEMATICS
THE GEORGE WASHINGTON UNIVERSITY
FUNGER HALL 428V
2201 G STREET, NW
WASHINGTON DC 20052-0001
  dullman@math.gwu.edu
  http://gwis2.circ.gwu.edu/~dullman

Games of No Chance
MSRI Publications
Volume **29**, 1996

Stable Winning Coalitions

DANIEL E. LOEB

ABSTRACT. We introduce the notion of a stable winning coalition in a multi-player game as a new system of classification of games. An axiomatic refinement of this classification for three-player games is also presented. These classifications are compared in light of a probabilistic model and the existing literature.

1. Introduction

Are multi-player combinatorial games essentially different from two-player combinatorial games? John Nash was recently awarded the Nobel prize in economics in part for his resolution of the analog question about classical or matricial games: matricial games have equilibria regardless of the number of players.

From a classical game-theoretic point of view, combinatorial games are a "trivial" sort of zero-sum game: one of the two players has a forced win in any finite two-player deterministic sequential-move perfect-knowledge winner-take-all game.

When a game has more than two players, it is no longer the case that one always has a forced win. In Section 2, we will study Propp's so-called "queer" three-player games [Propp a], in which no player can force a win, but rather one player chooses which of his two opponents will win.

For example, the game of nim is played with several piles of stones. Players take turns removing stones from a single pile of their choice. At least one stone and up to an entire pile may be taken. The player who makes the last move is the sole winner. With a pile of one stone and a pile of two stones, no player can force a win alone (see Figure 1).

The most obvious classification of two-player combinatorial games is according to who can force a win alone. One begins to understand a game upon discovering who can force a win. (After dividing games into types by "outcome", the next logical step is to refine these types so that the outcome of a sum of games is

Author partially supported by URA CNRS 1304, EC grant CHRX-CT93-0400, the PRC Maths-Info, and NATO CRG 930554.

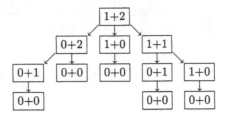

Figure 1. Three-player nim position $1 + 2$ game tree.

unchanged under replacement of one game with another in its class. For two-player impartial games, the two possible results, namely, next player wins and previous player wins, are further refined by Sprague–Grundy numbers [Grundy 1939; Sprague 1935]. All two-player impartial games are equivalent to one-pile nim. Similarly, in three-player nim, all nim positions are equivalent to one member of two countable families of nim positions and three sporadic positions [Propp a].)

However, the classification according to who can force a win is inappropriate to the vast majority of multi-player games, in which no player can force a win. Many authors have proposed various solutions to this problem. In Section 3, we classify multi-player combinatorial games according to which combinations of players can force a win. The set of all winning coalitions is itself a well-known combinatorial object: a maximal intersecting family of sets [Loeb and Meyerowitz]. In essence, an n-player game is reduced by the win classification into its 2^{n-1} two-player quotient games.

According to matricial game theory, the value of a combinatorial game to each player depends only on the win classification of the game [Shapley 1953]. For example, a three-player queer game has value $\frac{1}{3}$ to each player. Such fractional values arise because classical cooperative game theory allows binding side-agreements between players [von Neumann 1928].

However, in some contexts, such side-agreements may be unenforceable, and in other contexts, the players may not be allowed to communicate except through their moves and hence may not make any agreements whatsoever. Is there any hope for cooperative play without the use of side-payments, repeated play or binding agreements?

Just as a three-dimensional object is not completely described by its various two-dimensional projections, the win classification does not reveal all the essential information about a game. According to the probabilistic model introduced in Section 6.1, nearly all three-player games are classified together as queer games. The win classification does not distinguish between these games, and it does not distinguish between winning coalitions. Should players have preferences among queer games? Are all winning coalitions really equally effective?

The notion of a stable coalition is defined in Section 4 to answer these questions. In the example of Figure 1, two players form the only stable coalition.

They have a winning strategy by which either of them can win. The actual winner will be named by the third player. Despite the inability to share winnings, stable coalitions provide each member an incentive to cooperate. We will prove that stable winning coalitions exist in all multi-player combinatorial games.

This classification of multi-player games according to their stable coalitions contrasted with those of [Li 1978; Propp a; Straffin 1985]; however, we refer the reader to the bibliography for the full development of these promising systems for classifying multi-player games.

The fine classification introduced in Section 5 refines all the three-player game classifications above, with the exception of Li's. The fine type of a game is defined axiomatically via a number of tree rewriting rules similar to those of [Straffin 1985]. Thus, Section 5 makes heavy use of the language of graph grammars [Courcelle 1990].

As compared to the case of two-player games, we are confronted here with an embarrassing wealth of conflicting classification schemes. Before spending considerable effort classifying particular multi-player games, we must choose the most appropriate system of classification among the many alternatives. We hope that this paper is a step in that direction.

In order to focus on the general classification problem, we do not attempt an encyclopedic classification of particular games. However, we include examples for pedagogical reasons. To avoid unnecessary distractions, all examples are taken from one of the simplest combinatorial games of all: nim.

Although it is hardly mentioned in this paper, perhaps the most important concept in multi-player games is negotiation; it has close philosophical ties with many of the issues in this paper. Some relevant remarks on negotiation and how these classifications may be used to play a game can be found in [Loeb 1992, § 3].

2. Queer Games

We will consider finite deterministic sequential-move winner-take-all perfect-information games. That is, we make the following "combinatorial" assumptions about all games discussed here:

- They have a finite number of legal positions.
- There is no element of chance involved.
- There is no simultaneous movement.
- The players are at all times completely aware of the state of the game.
- The game necessarily terminates after a finite number of moves with a single unique winner.

We make no assumptions about the preferences of players among possible losses (compare [Li 1978]).

Every such game can be represented by a finite labeled game tree, like the one in Figure 2. All nodes are labeled with the name of a player. The label of

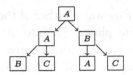

Figure 2. Example game.

an interior node indicates the player whose turn it is. This player selects one of
the node's children and play continues at that point. Play terminates at a leaf
node. The label of the leaf indicates who wins. Play begins at the root of the
tree.

In theory, the minimax algorithm can be used to play any two-player, deterministic, sequential-move game perfectly. (In practice, however, the size of the
game tree is too large to analyze directly via the minimax algorithm. The key
idea in combinatorial game theory [Berlekamp et al. 1982; Conway 1976] has
thus been to express games as a combination of simpler games, and relate them
to special families of games such as impartial games and cold games for which
rich mathematical theories have been discovered.)

However, what happens when a third player is added?

Let P be a node in a game tree corresponding to player A's turn. Suppose all
of P's children $\mathrm{Succ}(P)$ have been evaluated. How do we evaluate P. Obviously,
if P has a child whose value is equal to an A-win, then this child is a good
strategy for player A, and the value of P will be a win for A. Similarly, if *all* of
the children of P are wins for say player B, then P is itself a win for player B.
However, what happens if some of the children of P are wins for player B and
some are wins for player C? It is then impossible to evaluate P without some
psychological information regarding A's attitude to such situations. Given that
both choices are losing for A, we have no manner to predict how A will make
his choice: randomly or according to some unknown criterion. Nevertheless, A's
move is crucial to B and C.

DEFINITION 2.1 [Propp a]. A position in a three-player combinatorial game is
called *queer* if no player can force a win.

For example, consider the game of nim from the introduction, played with three
players: N next (first), F following (second), and P previous (third). The
position $0+0$ is a win for P. Thus, N wins $1+0$, $0+1$, $2+0$ and $0+2$ by taking
all the remaining stones from the remaining pile. Hence, $1+1$ is a forced win for
F. Therefore, the position $1+2$ is queer as it yields a choice by N betweens wins
for P and F (Figure 1).

Are queer positions very common? If not, we can hope to frequently apply the
traditional techniques of two-player game tree analysis. Unfortunately, we have
the following result (see Proposition 6.1 for similar probabilistic arguments):

PROPOSITION 2.2 [Propp a, Claim 11a,e,f]. *If G is an impartial three-player game, $2 + 2 + G$ is a queer game. In other words, any impartial game large enough to contain a $2 + 2$ component is queer.*

Recall that an n-player *impartial game* is an unlabeled tree, or rather a tree whose internal nodes have labels equal to their depth modulo n, and whose leaves have labels one less than their depth modulo n. Thus, players take turns moving. When playing the *sum* of two combinatorial games, players may play in either *component*. The game ends when a player has no available move in either component. The last player to move wins. See [Propp a] for a study of the win types of sums of three-player games.

PROOF. First we will show that $G + 2$ is not a win for P, and then that $G + 2 + 2$ is not a win for N or F.

TYPE P: If P can win $G + 2$, then F must win the options $G + 1$ and G. However, this is a contradiction since G is an option of $G + 1$.

TYPE N: Every move in $G + 2 + 2$ leaves a heap of size 2, so $G + 2 + 2$ has no P options. Thus, it has no winning moves for N.

TYPE F: Suppose F can win $G + 2 + 2$. Clearly, G is not the trivial game, since $2 + 2$ is queer. Thus, G has at least one option G'. Now, N must win $G' + 2 + 2$, which contradicts the preceding argument. $\square$

3. The Win Classification

Games with four or more players can also be classified according to which sets of players form winning coalitions. For four-player games, instead of four win types of games A, B, C, Q, twelve are needed (Table 1). In Types 1–4, one player can force a win. In Types 5–8, a coalition of three players is needed to stop a certain player from winning. In Types 9–12, one player plays no significant role, and the three remaining players participate in a "queer" game (in other words, any of the three players can be stopped by the other two).

To generalize this classification to any number of players, we first define the quotient of a game.

DEFINITION 3.1. Given a combinatorial game G with players V and a function $f : V \to W$ of its players, we define the *quotient game* G/f to be a new game with players W, and whose game tree is obtained by taking the game tree for the game G and replacing each label p with the label $f(p)$.

Thus, G/f represents a variant of the game G. Player $f(p)$ moves for the player p, and player $f(p)$ wins if p would have won the original game. If f is not surjective, some players in W will have no effect on the game, and no possibility of winning. If f is bijective, G and G/f are isomorphic as games.

DEFINITION 3.2. The set of players $A \subseteq V$ is said to be a *winning* (or *losing*) coalition in the game G if the quotient game G/χ_A is a win (or loss) for 1, where

win type	$\mathcal{F}(P)$	nim example	probability
1	A	1	
2	B	$1+1$	0
3	C	$1+1+1$	
4	D	$1+1+1+1$	
5	$AB\ AC\ BC$	$2+1$	
6	$AB\ AD\ BD$	$3+2+1+1$	$(4-\sqrt{10})/8 \sim 10.47\%$
7	$AC\ AD\ CD$	$2+1+1$	
8	$BC\ BD\ CD$	$2+2+2+2$	
9	$AB\ AC\ AD\ BCD$	$2+2+1$	
10	$AB\ BC\ BD\ ACD$	$2+2$	$(4+\sqrt{10})/8 \sim 14.53\%$
11	$AC\ BC\ CD\ ABD$	$3+3+3$	
12	$AD\ BD\ CD\ ABC$	$4+4+3+2+1$	

Table 1. Four-player win types, their probabilities in a random game, and examples from nim.

χ_A is the characteristic function of A (that is, $\chi_A(p) = 1$ if $p \in A$ and $\chi_A(p) = 0$ if $p \notin A$). The *win type* of G is given by its set of winning coalitions $\mathcal{F}(G)$.

The quotient defined above is compatible with the quotient in [Loeb a] in the sense that $\mathcal{F}(G/f) = \mathcal{F}(G)/f$.

Win types are in one-to-one correspondence with maximal intersecting families of sets [Shapley 1962] (Table 2).

DEFINITION 3.3 [Loeb and Meyerowitz]. Given a set V of players, a *maximal intersecting family of subsets of* V is a collection $\mathcal{F}$ of winning coalitions subject to the condition that if $V \supseteq B \supseteq A \in \mathcal{F}$ then $B \in \mathcal{F}$, and if $A = V - B$ then exactly one of A and B can be a member of $\mathcal{F}$.

In other words, a maximal intersecting family defines an order-preserving map f from the subsets of V to $\{0, 1\}$ such that $f(A) = 1 - f(V - A)$. The number of maximal intersecting families and thus the number of win types grows extremely rapidly as n increases [Korshunov 1991]. For example, there are 1,422,564 seven-player win types [Loeb 1992; Bioch and Ibarki 1994], and 229,809,982,112 eight-player win types [Conway and Loeb 1995]. See also Table 2.

$\mathcal{F}(G)$ is a maximal intersecting family, since adding players to a coalition can only strengthen it, and given a pair of opposing coalitions exactly one can win while the other must lose. It is from this connection that the terminology of strong/simple games is derived [Shapley 1953].

4. Stable Coalitions

All queer games have the same win type. Moreover, under the hypotheses of matricial game theory [von Neumann 1928], every three-player game is either strictly determined (forced win) with pay-offs $(1, 0, 0)$, or symmetric non-strictly determined (queer) with payoffs $(\frac{1}{3}, \frac{1}{3}, \frac{1}{3})$. However, when side-agreements are not allowed, are queer games always "symmetric"? Or are there queer games that favor certain winning coalitions over others?

For example, consider the game in Figure 2, where player A can choose between two queer positions: one where A would name B or C as winner, and one where B would name A or C as winner. Perhaps A would prefer the second choice even though both are queer games, since then he would retain some chance of winning [Straffin 1985, Axiom St2].

In this section, we propose a criteria by which certain "favored" coalitions of players may be deemed *stable*.

DEFINITION 4.1. Let P be a game tree. The set of stable coalitions $\mathrm{St}(P)$ or *stable type* is defined recursively as follows.

CASE 0: In a terminal position P, a player p has won the game. Player p himself forms the only stable coalition. $\mathrm{St}(P) = \{\{p\}\}$.

In a non-terminal position P resulting in a set $\mathrm{Succ}(P)$ of choices for the player p, one must consider two cases to determine if a set of players C is stable.

CASE 1 $(p \in C)$: $C \in \mathrm{St}(P)$ if and only if (1.1) there exists a choice that keeps C stable, and (1.2) there is no stable coalition strictly included in C:

$$(1.1) \quad \exists P' \in \mathrm{Succ}(P) : C \in \mathrm{St}(P'),$$
$$(1.2) \quad (D \subset C) \Rightarrow (D \notin \mathrm{St}(P)).$$

CASE 2 $(p \notin C)$: $C \in \mathrm{St}(P)$ if and only if (2.1) all choices keep some part of C stable, (2.2) for every player $q \in C$, there is some choice by which some subcoalition containing q is stable, and (2.3) there is no stable coalition strictly included in C:

$$(2.1) \quad P' \in \mathrm{Succ}(P) \Rightarrow \exists D \subseteq C : D \in \mathrm{St}(P').$$
$$(2.2) \quad q \in C \Rightarrow \exists P' \in \mathrm{Succ}(P), D \in \mathrm{St}(P) : q \in D \subseteq C.$$
$$(2.3) \quad D \subset C \Rightarrow D \notin \mathrm{St}(P).$$

In other words, a stable coalition must be a winning coalition; that is to say, it must have a strategy that guarantees a win for one of its members, regardless of the futile resistance of its opposition (Conditions 0, 1.1 and 1.2). However, it is the opposition who, via their choice of resistance, chooses which member of the coalition will actually win the game. In fact, every member of the coalition must be eligible to be elected winner by the opposition (Condition 2.2). Furthermore, we require that no member of the coalition can do better by joining a strictly included stable coalition (Conditions 1.2 and 2.3). (This last requirement seems the least plausible. However, it is not essential for certain of the results that follow; for example, in the proof of Theorem 4.2, nonminimal members of $E(P) \cup F(P)$ would also be considered stable.)

As a game proceeds, stable coalitions will sometimes disappear, as each player is confronted with a choice of conflicting strategies corresponding to different stable coalitions of which he is a member. Stable coalitions will never enlarge in size under ideal play, but they may decrease in size, as the opposition makes an arbitrary resistance. This reduction continues until the stable coalition consists of only a single player, at which point that player can force a win.

The following important result shows that our definition is not devoid of content. In fact, the proof gives a direct method by which $\text{St}(P)$ may be calculated.

THEOREM 4.2. *All games have at least one stable coalition.*

PROOF. Clearly this is true for terminal positions, so proceed by induction. Let P be a position with player p to move. Let $F(P)$ be the set of "friendly coalitions" $(p \in A)$ such that $A \in \text{St}(P')$ for some move $P' \in \text{Succ}(P)$. Let $E(P)$ be the set of "enemy coalitions" of the form

$$A = \bigcup_{P' \in \text{Succ}(P)} A_{P'},$$

where $p \notin A_{P'} \in \text{St}(P')$ for each choice $P' \in \text{Succ}(P)$. Note that $\text{St}(P)$ is the set of minimal elements of $E(P) \cup F(P)$.

It suffices to show that $E(P) \cup F(P) \neq \varnothing$. If $E(P)$ is empty, there must be some P' for which p belongs to all stable coalitions. By induction, P' must have at least one stable coalition A. Thus $A \in F(P)$. □

PROPOSITION 4.3. (i) *If the number of players is greater than one, the set of all players is not stable.*

(ii) *No stable coalition can contain another (in other words, $\text{St}(P)$ is an antichain).*

(iii) *Any pair of stable coalitions has a nonempty intersection.*

The set of all players is winning. Intuitively, however, it is not stable since there is no opposition to choose the winner.

PROOF. (i) By contradiction. Let P be a smallest counterexample. Thus, $V \in \text{St}(P)$. If P is of depth zero, Condition 0 is contradicted. Otherwise, by Condition 1.1, there must be a simpler position P' with $V \in \text{St}(P')$. This contradicts the minimality of P.

(ii) Conditions 1.2 and 2.3.

(iii) Stable coalitions are winning (Conditions 1.1 and 2.1). Subsets of their complements are losing. □

We have the following "converse" to Theorem 4.2 and Proposition 4.3.

PROPOSITION 4.4. *Let $\mathcal{F}$ be a nonempty family of intersecting subsets of X. Suppose that $X \notin \mathcal{F}$, and that $\mathcal{F}$ is an antichain. Then there is some game position P (with players X) such that $\text{St}(P) = \mathcal{F}$.*

PROOF. The game is played as follows. Players take turns eliminating all but one of the coalitions in $\mathcal{F}$ to which they belong (if any). After each player has

number of players	0	1	2	3	4	5	6	7
win types (Section 3)	0	1	2	4	12	81	2646	1,422,564
stable types (Section 4)	0	1	2	10	79	2644	1,422,562	229,809,982,110
Li types [Li 1978]	0	1	2	3	4	5	6	7
Straffin types [Straffin 1985]	0	1	2	6				
Fine types (Section 5)	0	1	2	$3\aleph$				
nim-sum types [Propp a]	0	1	2	$2\aleph + 3$				

Table 2. Systems of classification of multi-player games.

taken his turn there will be one set left. The first non-member of the set can then choose a member of the set to be the winner.   □

PROPOSITION 4.5. *The number of n-player stable types* St(P), *for* $n \geq 2$, *is equal to two less than the number of maximal intersecting families of subsets of an* $(n + 1)$-*element set.* (*See Table* 2.)

PROOF. The antichains $\mathcal{F}$ of pairwise intersecting subsets of $\{1, 2, \ldots, n\}$ are in bijection with maximal intersecting families $\mathcal{M}$ on $\{1, 2, \ldots, n + 1\}$:

$$\mathcal{F} = \min\{A : A \subseteq \{1, 2, \ldots, n\} \text{ and } A \in \mathcal{M}\}.$$

However, two maximal intersecting families must not be counted: the "dictatorship"

$$\mathcal{M} = \{A : n + 1 \in A\}$$

corresponds to $\mathcal{F} = \varnothing$, and the "constitutional monarchy"

$$\mathcal{M} = \{A : n + 1 \in A, |A| \geq 2\} \cup \{\{1, 2, \ldots, n\}\}.$$

corresponds to $\mathcal{F} = \{\{1, 2, \ldots, n\}\}$.   □

We study the stable classification, since a winning coalition itself does not win, but rather only one member of the coalition wins. Thus, in the end, many winning coalitions are "unstable." For example, in Table 2, if player A chooses the winner between players B and C, then the only stable coalition would be BC. A together with any other player wins, but would not be not stable. A would therefore prefer to let B choose between A and C (compare [Straffin 1985]).

Of the ten three-player stable types, three correspond to forced victories for one of the players. The other seven correspond to queer games. (See Table 3.) Nevertheless, the stable classification is not in general finer than the win classification (Table 2). For example, the four-player nim positions $5 + 1$ and $3 + 3$ both have stable type $ABD\ ACD$, whereas $5 + 1$ has win type $AB\ AC\ BC$ and $3 + 3$ has win type $AB\ BC\ BD\ ABD$. Note in particular that, according to the win classification, player D is a spectator in the game $5 + 1$; he has no chance to affect the game. However, according to the stable classification, player D is a member of the only two stable coalitions. Depending on the interpretation chosen, player D is either a spectator with no chance of winning, or an active

win type $\mathcal{F}(P)$ (§ 3)	stable type $St(P)$ (§ 4)	fine type (§ 5)	Straffin type [Straffin 1985]	nim example	prob.	$\mathcal{F}(P) =$ $St(P)$?
next N	N	n_0	N wins	1	0	yes
following F	F	f_0	F wins	$1+1$	0	yes
previous P	P	p_0	P wins	$1+1+1$	0	yes
queer	FP	n_1	N picks	$2+1$	3/23	no
NF NP FP	NP	f_1	F picks	$3+1$	3/23	no
	NF	f_1	P picks	$3+3$	3/23	no
	NF NP	p_1	N picks	$4+1$	2/23	no
	NF FP	n_2	F picks	$4+4$	2/23	no
	NP FP	f_2	P picks	$4+4+4$	2/23	no
	NF NP FP	p_2	? picks	$6+6$	8/23	yes

Table 3. Classification of three-player games along with probabilities in a random game, and examples from the game of nim.

participant with at least as good a chance of winning the game as any other player.

5. The Fine Classification

Consider three-player games. Although the classification by stable winning coalitions is an improvement over classification by winning coalitions, does the stable classification completely express the preferences of all players? For example, if three two-player coalitions are stable, is the game necessarily symmetrical? In this section we define a finer classification of three-player games. Two positions are in the same fine class if their game trees are identical after pruning dominated options from the game tree, and eliminating moves consisting of a single choice. Using a system of axioms similar to those used in [Straffin 1985], we reduce all game trees to a set of *basic trees*. Interestingly, none of the basic trees will be invariant under permutation of players.

Let n_0, f_0, and p_0 denote the one-node basic game trees $\boxed{N}$, $\boxed{F}$, and $\boxed{P}$. They represent immediate forced wins. Now,

$$n_1 = \begin{array}{c} \boxed{N} \\ \swarrow \quad \searrow \\ \boxed{F} \qquad \boxed{P} \end{array}$$

denotes the basic game tree giving a choice by the next player between f_0 and p_0. The basic trees f_1 and p_1 can be defined similarly. Player N has the following preferences:

$$n_0 > \underbrace{f_1, p_1 > f_0, p_0}_{n_1}.$$

Is he indifferent regarding f_0 and p_0? Can the position n_1 leading to such a choice be simplified by his opponents? Moreover, is N necessarily indifferent between f_1 and p_1? Unfortunately, we can not answer these questions until we clarify how a player makes a choice between symmetrical positions. We will propose a system of preference axioms below. In particular, PR3 states that the preferences of N are invariant under the exchange of F and P.

However, we first review several reasonable alternative hypotheses that have been proposed.

- One assumption is to make a random move in such a situation [V. Lefevre, personal communication]. In this model, the forced wins n_0, f_0 and p_0 are given Lefevre-values $(1, 0, 0)$, $(0, 1, 0)$, and $(0, 0, 1)$ in $\mathbb{R}^3$. Given a nontrivial game tree

$$\psi = \underset{\tau_1 \ \cdots \ \tau_n}{\swarrow \overset{\boxed{i}}{\cdots} \searrow}$$

in which the i-th player is to move, $v(\psi)$ is calculated by averaging the $v(\tau_j)$ having maximal i-th coordinate $v(\tau_j)_i$. Thus $v(\psi)$ represents the probabilities of victory for each of the three players according to this model.

  However, consider a game whose game tree has redundant subtrees. For example, consider chess but allow rooks to be moved by either hand and all other pieces to be moved only by the left hand. Is it then more likely that the player will move his rook? Although plausible, this is not a very pleasing model, since it is strongly affected by redundant branches from our game tree (compare [Straffin 1985, Axiom Bk1]).
- Another solution [von Neumann 1928; Shapley 1953] is to modify the game being analyzed, and allow *side-payments* by the other players in order to entice player N to agree to a *binding agreement*.
- Straffin [1985, axiom St3] proposed a rule of vengeance (attack the player who eliminated you).
- Li [1978] proposed that the players try to come in "second place". (According to Li's model, there is a forced win for some player assuming the players are ranked in function of position relative to the last player to move. To obtain this result, it is necessary to assume that the ranking of any one player determines the rankings of all players. The actual permutation used generalizes the two-player notation of misère variants of games. Sequential composition [Stromquist and Ullman 1993] of two Li games results in the composition of the misère permutation of the second game with the rankings of the first game.) See Table 2.

Instead, we propose rules by which game trees may be simplified. It will be shown that any position in a three-player game can be reduced to exactly one of the specific *basic games* n_i, p_i, or f_i, where i is an integer, n_{i+1} is recursively defined to be a choice by player N between the games p_i and f_i

$$n_{i+1} = \underset{f_i \qquad p_i}{\swarrow \overset{\boxed{N}}{\qquad} \searrow} ,$$

and f_{i+1} and p_{i+1} are defined similarly:

$$f_{i+1} = \underset{n_i \quad\ p_i}{\overset{\boxed{F}}{\diagdown}} \ , \qquad p_{i+1} = \underset{n_i \quad\ f_i}{\overset{\boxed{P}}{\diagdown}} \ .$$

For example, n_3 is the choice by player N of who will name the person who will choose the person who selects the winner.

The reduction rules we use are defined in terms of the preferences of players for game subtrees. Preferences are defined only for completely reduced game trees. Note that the preferences defined below are not total. Thus, for certain positions, we do not suppose that the player will make a certain preferred move; he may have a preference, he may move randomly, or he may move based on unknown criteria. We make no assumptions other than the axioms listed below.

Preference Rules

PR1: Let X and Y be distinct players. Player X prefers $x_0 = \boxed{X}$ to $y_0 = \boxed{Y}$.
(Players prefer a sure win to a sure loss.)

PR2: Let X and Y be distinct players. Let

$$\sigma = \underset{\tau_1 \qquad \tau_2}{\overset{\boxed{Y}}{\diagdown}}$$

be a completely reduced game tree. If player X prefers τ_1 to τ_2, player X also prefers τ_1 to σ and σ to τ_2. (Players prefer a good result to an uncertain result, and they prefer an uncertain result to a bad result.)

PR3: If player X prefers τ_1 to τ_2, player X also prefers τ_1 to τ_2' and τ_1' to τ_2, where τ_i' is the game subtree obtained by exchanging the two opponents of X wherever they occur in τ_i. (Players are impartial.)

PR4: Let $\psi = \underset{\tau_1 \ \cdots \ \tau_n}{\overset{\boxed{X}}{\diagup \cdots \diagdown}}$ be a completely reduced game tree.

    PR4a: Player X prefers one of $\tau_1, \ldots, \tau_n$ to σ if and only if he prefers ψ to σ.

    PR4b: Player X prefers σ to all of $\tau_1, \ldots, \tau_n$ if and only if he prefers σ to ψ.

    (A choice is as good as its best option.)

PR5: Let X and Y be distinct players, and let $\psi = \underset{\tau_1 \ \cdots \ \tau_n}{\overset{\boxed{Y}}{\diagup \cdots \diagdown}}$.

    PR5a: If player X prefers all of $\tau_1, \ldots, \tau_n$ to σ, player X prefers ψ to σ. (A tree is good if all its branches are good.)

    PR5b: If player X prefers σ to all of $\tau_1, \ldots, \tau_n$, player X prefers σ to ψ. (A tree is bad if all its branches are bad.)

PR6.: We assume no preferences other than those implied by PR1–PR5.

Reduction Rules

RR1: A game tree of the form $\boxed{X} \atop \underset{\tau}{\downarrow}$ can be reduced to τ. (We may collapse non-decisions [Straffin 1985, Axiom Bk2].)

RR2: A game tree of the form $\underset{\tau_1 \ \cdots \ \tau_1\tau_2 \ \cdots \ \tau_n}{\swarrow \cdots \downarrow\downarrow \cdots \searrow} \boxed{X}$ may be reduced to $\underset{\tau_1 \ \cdots \ \tau_n}{\swarrow \cdots \searrow}\boxed{X}$

(We may remove redundant game subtrees [Straffin 1985, Axiom Bk1].)

RR3: A game tree of the form

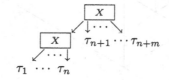

may be reduced to $\underset{\tau_1 \ \cdots \ \tau_{n+m}}{\swarrow \cdots \searrow}\boxed{X}$. (We may collapse sequential choices by the same player [Straffin 1985, Axiom Bk3].)

RR4: A game tree of the form $\underset{\tau_1 \ \cdots \ \tau_n}{\swarrow \cdots \searrow}\boxed{X}$ may be reduced to τ_1 if player X prefers τ_1 to all τ_i with $1 < i \leq n$. (We may assume that players will choose what they prefer [Straffin 1985, Axiom Bk3].)

Contrast PR2 with [Straffin 1985, Axioms St1 and St2]. Note that we do not adopt McCarthy's Revenge Rule [Straffin 1985, Axiom St3].

The reduction rules all strictly decrease the number of nodes in the game tree. Thus, no infinite reductions are possible, and the use of term "completely reduced" is justified. Since preferences are only defined on completely reduced game trees, application of RR4 must be carried out in a bottom-up fashion. Strategies in the reduced trees correspond in an obvious way to strategies in the original game tree.

Lefevre-values are not conserved by RR2 and RR3. However, the criteria of "preference" as defined above will be seen to be related to the simple comparison of Lefevre-values.

THEOREM 5.1. *By using the above reduction rules, all three-player games can be reduced to one and only one of n_i, f_i, or p_i.*

The *fine type* of a position P is defined to be the unique game tree n_i, f_i, or p_i to which it can be reduced.

We first establish the preferences (Lemma 5.2) and non-preferences (Lemma 5.3) for a given player, say N. We will then show that n_i, f_i, and p_i can not be further reduced (Lemma 5.4). It will then suffice to show that the set $\{n_i, f_i, p_i : i \geq 0\}$ is closed under tree construction.

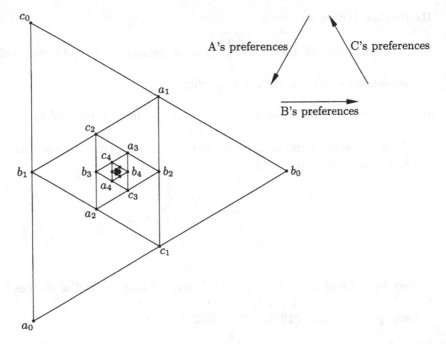

Figure 3. Preferences for basic-game trees.

		$\overbrace{n_2}$		$\overbrace{n_4}$			$\overbrace{n_5}$	$\overbrace{n_3}$	$\overbrace{n_1}$
preferences	$n_0 >$	$f_1, p_1 >$		$f_3, p_3 >$	$\cdots \quad \cdots >$		$f_4, p_4 >$	$f_2, p_2 >$	f_0, p_0
probability	1	$\frac{1}{2}$		$\frac{3}{8}$	$\frac{1}{3}$		$\frac{5}{16}$	$\frac{1}{4}$	0

Table 4. Player N's fine type preferences, along with the probability of player N winning given unbiased play by players F and P.

LEMMA 5.2. *The relations shown in the first row of Table 4 show preferences by player N for basic-game trees. (Preferences of other players can be determined mutatis mutandis; see Figure 3.)*

PROOF. Induction on k. The base case follows immediately from PR1. All of the required preferences for n_{2k+1} follow easily by hypothesis and the "only if" parts of PR4ab. We have $n_{2k} > f_{2k+1} > p_{2k}$ and $n_{2k} > p_{2k+1} > f_{2k}$ by hypothesis and PR2. Now, $f_{2k+1} > f_{2k}$ and $p_{2k+1} > p_{2k}$ by PR5. Thus, $f_{2k+1} > n_{2k+1}$ by PR4b. Furthermore, $f_{2k+1} < f_{2k-1}, p_{2k-1}$ by PR3 and the "if" part of PR4a. The remaining preferences for f_{2k+1} follow easily by hypothesis and PR5ab.

Preferences for $n_{2k+2}, p_{2k+2}, f_{2k+2}$ follows *mutatis mutandis* making use of the "if" part of PR4b to show that $f_{2k+2} > f_{2k}, p_{2k}$. □

LEMMA 5.3. *Let $\tau, \sigma \in \{n_i, f_i, p_i : i \geq 0\}$ be basic game trees. If $v(\tau)_1 \geq v(\sigma)_1$, the next player N does not prefer τ to σ.*

PROOF. It suffices by PR6 to suppose that a counterexample was derived from one of the axioms PR1–PR5 from earlier preferences, and deduce that at least one of the earlier preferences is another counterexample.

PR1: Only n_0, f_0, and p_0 apply. PR1 shows that $n_0 > f_0$ and $n_0 > p_0$, neither of which are counterexamples, since $v(n_0)_1 = 1 > 0 = v(f_0)_1 = v(p_0)_1$.

PR2: This rule shows for example that $f_{i+1} = \overset{\boxed{F}}{\underset{n_i \qquad p_i}{\diagdown\diagup}}$ lies between n_i and p_i in the preferences of player N. However, $v(f_{i+1})_1$ lies strictly between $v(p_i)_1$ and $v(n_i)_1$.

PR3: If $v(\tau) = (a, b, c)$ then $v(\tau') = (a, c, b)$.

PR4ab: In the "only if" direction, the only choice by player N that interest us here is n_{j+1}.

In the "if" direction, by RR4, the player N must have no preferences between the τ_i. The τ_i may not equal n_{j+1} by RR3, and they must be unequal. Thus we have $n = 2$, $\tau_1 = f_j$, and $\tau_2 = p_j$ for some $j \geq 0$, implying $\psi = n_{j+1}$. However, note that $v(f_j)_1 = v(p_j)_1 = v(n_{j+1})_1$.

PR5ab. Suppose, for example, that $v(\tau_i)_1 > v(\sigma)_1$. Now, $v(\psi)_1$ is the average of certain $v(\tau_i)_1$. Thus $v(\psi)_1 > v(\sigma)_1$. $\square$

LEMMA 5.4. *The basic game trees n_i, f_i, and p_i are irreducible.*

PROOF. In none of the game trees f_i, n_i, or p_i is there a node with a single child (by RR1), two identical subtrees depending on the same node (by RR2), or consecutive choices by the same player (by RR3). In all cases of a node with two children, we have shown that neither child is preferred by the player to move (by RR4). $\square$

PROOF OF THEOREM 5.1. Without loss of generality, we need only show how

$$\sigma = \overset{\boxed{N}}{\underset{\tau_1 \qquad \tau_2}{\diagdown\diagup}}$$

can be reduced for all basic game trees $\tau_1, \tau_2 \in \{n_i, f_i, p_i : i \geq 0\}$. If player N prefers one of τ_1 and τ_2 to the other, we are done by RR4. If $\tau_1 = \tau_2$, we are done by RR2 and RR1. The remaining possibilities for $\{\tau_1, \tau_2\}$ are as follows:

$\{f_i, p_i\}$: By definition, $\sigma = n_{i+1}$.

$\{n_i, p_{i-1}\}$: By RR3 and RR2, $\sigma \to n_i$.

$\{n_i, f_{i-1}\}$: Same as above. $\square$

It is clear how moves in the reduced game correspond to strategies in the original game. The analysis in [Propp a] does not assume that the other players will necessarily play well. However, unless some assumption about the intelligence of other players is made, there is no way to discriminate between queer games. Moreover, the mild degree of fairness imposed by PR3 is necessary in order to compare, say, f_0 with f_1.

(i,j)	0	1	2	3	4	5	6	7	8	9	10
0	p_0	n_0	n_0	n_0	n_0	n_0	n_0	n_0	n_0	n_0	n_0
1	n_0	f_0	n_1	f_1	n_2	n_2	n_2	n_2	n_2	n_2	n_2
2	n_0	n_1	f_1	n_2	n_2	n_2	n_2	n_2	n_2	n_2	n_2
3	n_0	f_1	n_2	p_1	p_1	p_1	p_1	p_1	p_1	p_1	p_1
4	n_0	n_2	n_2	p_1	f_2	n_3	f_3	n_4	n_4	n_4	n_4
5	n_0	n_2	n_2	p_1	n_3	f_3	n_4	n_4	n_4	n_4	n_4
6	n_0	n_2	n_2	p_1	f_3	n_4	p_3	p_3	p_3	p_3	p_3
7	n_0	n_2	n_2	p_1	n_4	n_4	p_3	f_4	n_5	f_5	n_6
8	n_0	n_2	n_2	p_1	n_4	n_4	p_3	n_5	f_5	n_6	n_6
9	n_0	n_2	n_2	p_1	n_4	n_4	p_3	f_5	n_6	p_5	p_5
10	n_0	n_2	n_2	p_1	n_4	n_4	p_3	n_6	n_6	p_5	f_6

Table 5. Fine classification of three-player two-pile nim position $i + j$. Regions associated with the four win types are outlined.

Without loss of generality we assume $i \leq j$.

$i = 0$, $j = 0$: Fine type p_0. The previous player has already won.

$j > 0$: Fine type n_0. The next player wins immediately.

$i = 1$, $j = 1$: Fine type f_0. Forced win for the following player.

$j > 1$: Fine type n_0. Stable type FP. Queer game.

$i = 2$, $j = 2$: Fine type f_1. Stable type NP. Queer game.

$j > 2$: Fine type n_2. Stable type $NF\,NP$. Queer game.

$i = 3$: Fine type p_1. Stable type NF. Queer game.

$i \geq 4$: Queer game. Fine type x_{n+2m}, where x_n is the fine type of the nim game $(i - 3m) + (j - 3m)$ and $m = \lfloor \frac{1}{3}(i-1) \rfloor$. Stable type $NF\,FP$ if $i = j = 4$, and $NF\,FP\,NP$ otherwise.

There is no two-pile nim position with stable type $NP\,FP$, or fine type p_{2k} for $k \geq 1$.

Table 6. Classification of the three-player two-pile nim position $i + j$.

The fine classification is so named since it is a refinement of the stable classification and Straffin classification [Straffin 1985], both of which refine the three-player win classification (Table 3). There are three countable families of fine types (Table 2). For $i > 0$, n_i corresponds to Straffin's "N decides". Note that the game n_2 favors player N according to our theory (N wins half of the time if F and P are not biased), but should be a loss for N according to Straffin's theory.

Calculations with fine types are rather complicated. For example, whereas two-pile two-player nim is completely trivial (the next player wins unless the two piles are of equal size), the fine type of the nim game $i + j$ can only be calculated by a complicated rule (see Tables 5 and 6).

No general pattern is known for three-player nim with an arbitrary number of piles. Moreover, any attempt to generalize this theory to four or more players, (even with a simplifying Revenge Rule) seems hopelessly complicated.

Rules PR1–PR6 and RR1–RR4 are sufficient to play a three-player game. However, to obtain numerical data concerning the probability of victory for a given player, say player N, we must make an additional "fairness" assumption (in addition to PR2). Namely, we must assume that in a position of fine type f_{i+1}, the option n_i will be chosen half of the time, and likewise in a position of fine type p_{i+1}. The probabilities that follow from these assumptions are given in Table 4. Note that this probability approaches $\frac{1}{3}$ but never attains this limit. Thus, we have the counter-intuitive result that *no* game is symmetric under the hypotheses above.

6. A Probabilistic Model

6.1. Queer Games. A simple heuristic argument shows that virtually all games are queer.

Any game can be represented by a tree, and without loss of generality a complete binary tree of some finite depth k. To generate a random game tree of depth k, we independently assign random labels to all nodes of a complete binary tree of depth k.

Let a_k, b_k, and c_k represent the probability that a random game of depth k is a forced win for player A, B, or C, respectively. Let q_k represent the probability that a random game of depth k is queer. For example, $a_0 = b_0 = c_0 = \frac{1}{3}$, and $q_0 = 0$.

Obviously, there are forced win games with arbitrarily large game trees. However, most large random games are queer (Table 3).

PROPOSITION 6.1. *As the depth of a random binary three-player game tree increases, the probability that the game is queer approaches a certainty:*

$$\lim_{k \to \infty} q_k = 1.$$

PROOF. Let $k > 0$. If player A moves first, he can force a win in the entire game tree if and only if he can force a win in one of the two principal subtrees; this occurs with probability $2a_{k-1} - a_{k-1}^2$. However, if player B or C moves first, in order to force a win A needs a forced win in both principal subtrees; this occurs with probability a_{k-1}^2. Thus, the a priori probability of A forcing a win is

$$a_k = \tfrac{1}{3}(2a_{k-1} - a_{k-1}^2) + \tfrac{2}{3}a_{k-1}^2 = a_{k-1}(\tfrac{2}{3} + \tfrac{1}{3}a_{k-1}).$$

However, by symmetry, $a_k = b_k = c_k$. Thus $a_k \leq \frac{1}{3}$. Hence $\frac{2}{3} + \frac{1}{3}a_{k-1} \leq \frac{7}{9}$, so

$$a_k \leq \tfrac{7}{9}a_{k-1} \leq \tfrac{1}{3}(\tfrac{7}{9})^k,$$

which tends exponentially to zero. Thus q_k rapidly converges to one. □

6.2. Win Types. In general, consider a large random n-player binary game tree G, and choose a subset $C \subseteq V$ of k players, with $k < \frac{1}{2}n$. What is the probability p that C is a winning coalition?

There are two cases: either a member of C must move (probability $x = k/n$), in which case C loses with probability $(1 - p_{k-1})^2$, or else a member of $V \setminus C$ must move, in which case C wins with probability p_{k-1}^2. Thus

$$p_k = x(2p_{k-1} - p_{k-1}^2)^2 + (1 - x)p_{k-1}^2 = p_{k-1}(2x + p_{k-1}(1 - 2x)).$$

Now, $p_{k-1} \leq \frac{1}{2}$ by symmetry, for $V \setminus C$ must have at least as great a chance of winning as C does. Thus $(2x + p_{k-1}(1 - 2x)) \leq 2x < 1$. Therefore,

$$p_k \leq 2xp_{k-1} \leq \tfrac{1}{2}(2x)^k.$$

Thus the probability of a win rapidly approaches zero for all minority coalitions. The probability approaches one for all majority coalitions. By symmetry, the probability is one-half for all $(\frac{1}{2}n)$-player coalitions.

We thus deduce the following theorem.

THEOREM 6.2. *Let $\mathcal{M}$ be a maximal intersecting family of subsets of $V = \{1, 2, \ldots, 2n + 1\}$. Let T_k be a random $(2n+1)$-player game tree of depth k. The limit win type probability, $\lim_{k=\infty} P(\mathcal{M} = \mathcal{F}(T_k))$, is one if*

$$\mathcal{M} = \{A \subseteq V : |A| > n\}$$

and zero otherwise. $\square$

Thus, the maximal intersecting family corresponding to a typical large combinatorial game with an odd number of players is a "democracy." Its winning coalitions are exactly the majority coalitions. Thus, of the 81 different five-player win types, and of the 1,422,564 seven-player win types [Loeb 1992] only the "democracy" is likely to occur as a large random game.

In the case of an even number of players $2n$, all winning coalitions are likely to contain at least n players. For example, of the twelve four-player win types listed on Table 1, the forced wins (win types 1–4) have probability zero, and of the 229,809,982,112 eight-player win-types, at most $2^{35} = 34,359,738,368$ may have non-zero probability. In order to compute the probabilities of the remaining

$$2^{\frac{1}{2}\binom{2n}{n}}$$

possible win types additional calculations are necessary.

For example, consider $n = 4$. Let p be the probability of win types 5–8, and q that for types 9–12. We consider the 64 possible pairs of win types that might arise for a player to choose from as left subtree and right subtree in a game, and determine the win type of the complete game tree. The resulting table leads to the system of equations

$$p = 10p^2 + 12pq + 6q^2,$$
$$q = 6p^2 + 20pq + 10q^2,$$

from which we deduce the win type probabilities

$$p = \tfrac{1}{2} - \tfrac{1}{8}\sqrt{10} \sim 10.47\%,$$
$$q = \tfrac{1}{2} + \tfrac{1}{8}\sqrt{10} \sim 14.53\%.$$

For six players, 1024 of the 2646 maximal intersecting families may have nonzero probabilities of occurring as the win type of a random game. Analysis of the 1,048,576 possible combinations of these games is still tractable. The result would be a system of thirteen quadratic equations in thirteen variables (once symmetries are taken into account, there are thirty classes of maximal intersecting families of subsets of a six-element set, seventeen of which involve at least one winning minority coalition, and thus have probability zero). This system of equations could be solved numerically, although probably not formally.

For $n = 2k \geq 8$, the calculation of win type probabilities is intractable, with over three million simultaneous quadratic equations to be solved.

The exact values of these probabilities should not be taken too seriously, since they are influenced by small changes in the model (such as using complete trinary tree instead of binary trees). In particular, this probabilistic model should not be used to study any particular game (such as nim).

6.3. Stable Types. The probability of a large random three-player game having a certain stable type was calculated as follows. The forced wins have probability zero and can be ignored. We consider the remaining $7 \times 7 = 49$ possible pairs of stable types (G_1, G_2) that might arise for a player to choose from a left subtree and right subtree in a game, and determine the stable type of the complete game. Identifying the probabilities of certain of the remaining seven stable types, the resulting table leads to the system

$$x = \tfrac{7}{3}x^2 + \tfrac{8}{3}xy + \tfrac{4}{3}xz,$$
$$y = \tfrac{2}{3}x^2 + \tfrac{10}{3}xy + \tfrac{7}{3}y^2 + \tfrac{2}{3}yz,$$
$$z = 2x^2 + 2xz + 4yz + z^2,$$

from which we deduce the probabilities $x = \tfrac{3}{23}$, $y = \tfrac{2}{23}$, and $z = \tfrac{8}{23}$. Thus, with probability $z = \tfrac{8}{23}$, the game will be symmetrical, and all three pairs of players will be able to form stable coalitions.

Four-player games can be divided into 79 stable types. Up to permutation of players, there are thirteen stable classes. Only five stable classes (25 stable types) have non-zero probability, which we can approximate numerically (Table 7).

Note that for twelve of the 79 stable types of four-player games, all winning coalitions are stable: $\mathcal{F}(P) = \mathrm{St}(P)$. However, unlike the case of three-player

representative class St(P)	types in class	type prob.	class prob.	$\mathcal{F}(P) =$ St(P)?	nim example
AB AC BC	4	0.0485%	0.194%	yes	2+2+2+2+2+1
AB AC BCD	12	1.1818%	14.182%	no	6+6+1+1
AB ACD BCD	6	7.1996%	43.197%	no	6+6+1
ABC AD BD CD	4	0.4326%	1.731%	yes	none known
ABC ABD ACD BCD	1	40.595%	40.595%	no	7+6

Table 7. The five "common" four-player stable classes, with examples from nim.

games, these stable types have relatively low probabilities (less than 2% in total) of occurring as the stable type of a large random game.

Numerical approximation of the 2644 five-player stable type probabilities is probably feasible. However, even that seems impractical for the 1,422,562 six-player stable type probabilities.

Acknowlegements

I would like to thank S. Levy, editor of the MSRI book series, and an anonymous referee for their helpful suggestions.

References

[Berlekamp et al. 1982] E. R. Berlekamp, J. H. Conway, and R. K. Guy, *Winning Ways For Your Mathematical Plays*, Academic Press, London, 1982.

[Bioch and Ibarki 1994] J. C. Bioch and T. Ibaki, "Generating and approximating non-dominated coteries", Rutcor Research Report 41–94, December 1994.

[Conway 1976] J. H. Conway, *On Numbers And Games*, Academic Press, London, 1976.

[Conway and Loeb 1995] A. R. Conway and D. E. Loeb, unpublished work.

[Courcelle 1990] B. Courcelle, "Graph rewriting: an algebraic and logic approach", pp. 195–242 in *Handbook of Theoretical Computer Science* (edited by J. van Leeuwen), Elsevier, New York, and MIT Press, Cambridge (MA), 1990.

[Grundy 1939] P. M. Grundy, "Mathematics and Games", *Eureka* **2** (1939), 6–8. Reprinted in *Eureka* **27** (1964), 9–11.

[Korshunov 1991] A. D. Korshunov, "Families of subsets of a finite set and closed class of boolean functions", pp. 375–396 in *Extremal problems for finite sets*, Visegrad, 1991 (edited by P. Frankl), János Bolyai Math. Soc., Budapest, 1994.

[Li 1978] S.-Y. R. Li, "N-person Nim and N-person Moore's games," *Internat. J. Game Theory* **7** (1978), 31–36.

[Loeb 1992] D. E. Loeb, "Challenges in Playing Multiplayer Games", presented at the Fourth Conference on Heuristic Programming in Artificial Intelligence, August 1992, London, organized by J. van den Herik and V. Allis. Also available as http://www.csn.net/~mhand/DipPouch/S1995M/Project.html.

[Loeb a] D. E. Loeb, "A new proof of Monjardet's median theorem", to appear in
  J. Comb. Th. A. Also available as http://www.labri.u-bordeaux.fr/~loeb/median/
  a.html.

[Loeb and Meyerowitz] D. E. Loeb and A. Meyerowitz, "The Graph of Maximal
  Intersecting Families of Sets", submitted to *J. Comb. Th. A.*

[von Neumann 1928] J. von Neumann, "Zur Theorie der Gesellschaftsspiele", *Math.
  Ann.* **100** (1928), 295–320. English translation by S. Bargmann: "On the Theory of
  Games of Strategy", pp. 13–42 in *Contributions to the Theory of Games* IV (edited
  by A. W. Tucker and R. D. Luce), Annals of Math. Studies **40**, Princeton University
  Press, Princeton, 1959.

[Propp a] J. G. Propp, "Three-Person Impartial Games", to appear in *J. Theoretical
  Comp. Sci.* (Math. Games). Also available as ftp://theory.lcs.mit.edu/pub/propp/
  three.ps.

[Shapley 1962] L. S. Shapley, "Simple Games: An Outline of the Descriptive Theory",
  Behavioral Science **7** (1962), 59–66.

[Shapley 1953] L. S. Shapley, "A value for *n*-person games", pp. 307–317 in
  Contributions to the Theory of Games II (edited by H. Kuhn and A. W. Tucker),
  Annals of Math. Studies **28**, Princeton University Press, Princeton, (1953.

[Sprague 1935] R. Sprague, "Über Mathematische Kampfspiele", *Tôkuku Math. J.* **41**
  (1935/36), 438–444. Reviewed in *Zentralblatt Math.* **13**, 290.

[Straffin 1985] P. D. Straffin, Jr., "Three-person winner-take-all games with McCarthy's
  revenge rule," *College J. Math.* **16** (1985), 386–394.

[Stromquist and Ullman 1993] W. Stromquist and D. Ullman, "Sequential Compounds
  of Combinatorial Games", *J. Theoretical Comp. Sci. (Math. Games)* **119** (1993),
  311–321.

Daniel E. Loeb
LaBRI
Université de Bordeaux I
33405 Talence Cedex, France
  loeb@labri.u-bordeaux.fr
  http://www.labri.u-bordeaux.fr/~loeb

Coda

Games of No Chance
MSRI Publications
Volume 29, 1996

Unsolved Problems in Combinatorial Games

RICHARD K. GUY

ABSTRACT. This periodically updated reference resource is intended to put
eager researchers on the path to fame and (perhaps) fortune.

As in our earlier articles, WW stands for *Winning Ways* [Berlekamp et al. 1982].
We say that the *nim-value* of a position is n when its value is the nimber $*n$.

1. Subtraction games are known to be periodic. Investigate the relationship
between the subtraction set and the length and structure of the period.

(For subtraction games, see WW, pp. 83–86, 487–498, and Section 4 of [Guy
1996] in this volume, especially the table on page 67. A move in the subtraction
game $S(s_1, s_2, s_3, \ldots)$ is to take a number of beans from a heap, provided that
number is a member of the *subtraction set* $\{s_1, s_2, s_3, \ldots\}$. Analysis of such a
game and of many other heap games is conveniently recorded by a *nim-sequence*,
$n_0 n_1 n_2 n_3 \ldots$, meaning that the nim-value of a heap of h beans in this particular
game is n_h: in symbols, $\mathcal{G}(h) = n_h$. Arbitrarily difficult subtraction games can
be devised by using infinite subtraction sets: the primes, for example.)

The same question can be asked about *partizan* subtraction games, in which
each player is assigned an individual subtraction set [Fraenkel and Kotzig 1987].

2. Are all finite **octal games** ultimately periodic? (See WW, pp. 81–115, [Guy
1996, Section 4], p. 67 in this volume. Such games are defined by a code
$d_0 \cdot d_1 d_2 d_3 \ldots$. If the binary expansion of d_k is $d_k = 2^{a_k} + 2^{b_k} + 2^{c_k} + \cdots$,
where $0 \leq a_k < b_k < c_k < \cdots$, then it is legal to remove k beans from a heap,
provided that the rest of the heap is left in exactly a_k or b_k or c_k or ... nonempty
heaps. Some specimen games are exhibited on page 69.)

Resolve any number of outstanding particular cases, e.g., ·6 (Officers), ·06,
·14, ·36, ·64, ·74, ·76, ·004, ·005, ·006, ·007 (One-dimensional tic-tac-toe,

---

An earlier version of this collection of problems appeared in *Combinatorial Games*, Proceedings of Symposia in Applied Mathematics, Vol. 43, 1991. Permission for use courtesy of
the American Mathematical Society. We have retained the numbering of problems present in
that list.

Treblecross), ·016, ·104, ·106, ·114, ·135, ·136, ·142, ·143, ·146, ·162, ·163, ·172, ·324, ·336, ·342, ·362, ·371, ·374, ·404, ·414, ·416, ·444, ·454, ·564, ·604, ·606, ·644, ·744, ·764, ·774, ·776 and Grundy's Game (split a heap into two unequal heaps). Find a game with a period longer than 149459. Explain the structure of the periods of games known to be periodic.

Gangolli and Plambeck [1989] established the ultimate periodicity of four octal games that were previously unknown: ·16 has period 149459 (a prime!), the last exceptional value being $\mathcal{G}(105350) = 16$. The game ·56 has period 144 and last exceptional value $\mathcal{G}(326639) = 26$. The games ·127 and ·376 each have period 4 (with cycles of values 4, 7, 2, 1 and 17, 33, 16, 32 respectively) and last exceptional values $\mathcal{G}(46577) = 11$ and $\mathcal{G}(2268247) = 42$.

In Problem 38 in *Discrete Math.* 44 (1983), 331–334, Fraenkel raises questions concerning the computational complexity of octal games. In Problem 39, he and Kotzig define *partizan octal games* in which distinct octals are assigned to the two players. In Problem 40, Fraenkel introduces *poset games*, played on a partially ordered set of heaps, each player in turn selecting a heap and removing a positive number of beans from all heaps that are greater or equal to the selected heap in the poset ordering.

3. Examine some **hexadecimal games** (games with code digits d_k in the interval from **0** to **15 = F**: see WW, pp. 115–116, and the last three examples in the table on page 67 of this book). Obtain conditions for arithmetic periodicity.

Most of the known arithmetically periodic hex games are fairly trivial: ·8 is a first cousin of Triplicate Nim, ·A, ·B, ·D and ·F are essentially She-Loves-Me-She-Loves-Me-Not, while ·1A and ·1B, after a couple of exceptional values, display period 2 and saltus 1. Kenyon's Game, ·3F in the table of page 69, whose saltus of 3 countered a conjecture of Guy and Smith that it should always be a power of two, is a little more interesting. There must be many others that are possibly accessible to present-day computers.

4. Extend the analysis of **Domineering** to larger boards. (Left and Right take turns to place dominoes on a checker-board. Left orients her dominoes North-South and Right orients his East-West. Each domino exactly covers two squares of the board and no two dominoes overlap. A player unable to play loses. See [Berlekamp 1988], WW, pp. 495–498, and the articles in this book starting on pages 85 and 311.)

An earlier version of this paper asked what are the values of 4×4 and 4×5 boards. David Wolfe has found them to be $\pm\{0, \{2|0, 2+_2 \| 2|0, -_2\} \|\| 0\}$ and 1, respectively. Berlekamp has shown that the value of a $4 \times n$ board, if n is odd, is twice that of the corresponding $2 \times n$ board plus a correction term of lower temperature. For the value g of the 4×6 board, Wolfe's program [1996] needed more than 200 characters to print it. g is incomparable with both $+_2$ and $-_2$, but he can show that $-4(+_2) < g < 3(+_2)$. Dan Calistrate observes that the value of a 6×6 board is ± 1 to within 'ish' (infinitesimally shifted).

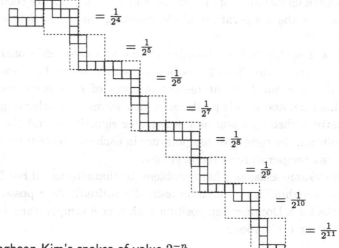

Figure 1. Yonghoan Kim's snakes of value 2^{-n}.

		$*$	$\ldots,\ \frac{1}{32},\ \frac{1}{16}$	$\frac18$	$\frac14$	$\frac12$	1e	1d
(3×3 grid)	$\int^{3/2}$	$1\frac12$						
$3\times\text{odd}\frac13$	$\int^{9/8}\int_{\frac14\ *}^{\frac14*}$	$1\frac38$		$1\frac{11}{32}$	$1\frac{5}{16}$	$1\frac14$		$1\frac18$
$2\times\text{odd}$	$\int^{3/4}\int_{\frac12\ *}^{\frac12*}$	$1\frac14$		$1\frac{3}{16}$	$1\frac18$	1		$\frac34$
Conway's ZigZags	$\int_1^1$ (ish)	1	$\ldots,\ \frac{31}{32},\ \frac{15}{16}$	$\frac78$	$\frac34$	$\frac12$		
Wolfe's Snakes [1993]	$\int_{\frac12*}^{\frac12}$	$\frac12$	$\ldots,\ \frac{31}{64},\ \frac{15}{32}$	$\frac{7}{16}$	$\frac38$	$\frac14$		
Kim's Snakes	$\int_{\frac14*}^{\frac14}$	$\frac14$	$\ldots,\ \frac{31}{128},\ \frac{15}{64}$	$\frac{7}{32}$	$\frac{3}{16}$	$\frac18$	0	
Kim's Numbers	Identity	0	$\ldots,\ -\frac{1}{32},\ -\frac{1}{16}$	$-\frac18$	$-\frac14$	$-\frac12$	-1	

Table 1. Temperatures in Domineering.

Yonghoan Kim [1996] announced that there are snakes of value 2^{-n} for all positive integers n (Figure 1). Various subsets of these snakes give a sequence of hot games (the penultimate row of the next table), another sequence of numbers, and several infinitesimals. Wolfe [1993] also has snakes with interesting values.

Table 1 shows temperatures of certain Domineering positions, as reported by Berlekamp. The rows are various overheating and warming operators associated with various (families of) positions, and the columns are their arguments. The entries are the temperatures of the resulting games (for references, see Problem 52 below).

Row 2 applies to $3 \times n$ boards with n odd and an additional square appended at one corner; row 3 to $2 \times n$ boards with n odd. The headings 1e, 1d refer to the argument 1 in the respective cases of *even* positions, which warm by overheating, and of *odd* positions, which warm by overheating and then adding *, distinguished by a star on the operator sign. If we omit the two entries in the 1d column, the rightmost temperature in each row (except the first) matches the leftmost temperature in the next row.

Berlekamp asks, as a hard problem, to characterize all hot Domineering positions to within "ish" (i.e., infinitesimally shifted). As a possibly easier problem he asks for a Domineering position with a new temperature, i.e., one not occurring in the table above.

5. Analyze positions in the game of **Go**. (Compare [Berlekamp and Wolfe 1994; Berlekamp and Kim 1996; Landman 1996].)

6. In an earlier edition of this paper I asked if **Go-Moku** (Five-in-a-row, Go-Bang, Pegotty) is a win for the first player. An affirmative answer for the unrestricted version and for the version where 6-in-a-row doesn't count as a win was given by Allis, van den Herik and Huntjens [1993]. Lustenberger [1967] showed that *Four-in-a-row* on a $4 \times n$ board is a first-player win for n sufficiently large. Selfridge states that $n = 28$ suffices. To find the least n should nowadays be a straightforward computer exercise.

7. Complete the analysis of impartial **Eatcakes**. (See WW, pp. 269, 271, 276–277. Eatcakes is an example of a *join* or *selective compound* of games. Each player plays in *all* the component games. It is played with a number of rectangles, $m_i \times n_i$; a move is to remove a strip $1 \times n_i$ or $m_i \times 1$ from each rectangle, either splitting it into two rectangles, or reducing the length or breadth by one. Winner removes the last strip.)

For fixed breadth the remoteness becomes constant when the length is sufficiently large. But 'sufficiently large' seems to be an increasing function of the breadth and doesn't, in the hand calculations already made, settle into any clear pattern. Perhaps computer calculations will reveal something.

8. Complete the analysis of **Hotcakes**. (See WW, pp. 279–282. Also played with integer-sided rectangles, but as a *union* or *selective compound* in which each player moves in *some* of the components. Left cuts as many rectangles vertically along an integer line as she wishes, and then rotates one from each pair of resulting rectangles through a right angle. Right cuts as many rectangles as he wishes, horizontally into pairs of integer-sided rectangles and rotates one rectangle from each pair through a right angle. The *tolls* for rectangles with one dimension small are understood, but much remains to be discovered.)

9. Develop a **misère theory for unions** of partizan games. (In a union of two or more games, you move in as many component games as you wish. In misère play, the last player *loses*.)

10. Extend the analysis of **Squares Off**. (See WW, pp. 299. Played with heaps of beans. A move is to take a perfect square (> 1) number of beans from any number of heaps. Heaps of 0, 1, 2 or 3 cannot be further reduced. A move leaving a heap of 0 is an overriding win for the player making it. A move leaving 1 is an overriding win for Right, and one leaving 2 is an overriding win for Left. A move leaving 3 doesn't end the game unless all other heaps are of size 3, in which case the last player wins.)

11. Extend the analysis of **Top Entails**. (WW, pp. 376–377. Played with stacks of coins. Either split a stack into two smaller ones, or remove the top coin from a stack. In the latter case your opponent's move must use the same stack. Last player wins. Don't leave a stack of 1 on the board, since your opponent must take it and win, since it's now your turn to move in an empty stack!)

Julian West [1996] wrote a program to check a student's work and calculated the first 38,000 values. He found loony positions at 2403 coins, 2505 coins, and 33,243 coins. The authors of WW did not know of a loony stack of more than 3 coins. These results are typical of the apparently quite unpredictable nature of combinatorial games, even when they have quite simple rules.

12. Extend the analysis of **All Square**. (WW, pp. 385. This game involves *complimenting moves* after which the same player has an extra *bonus move*. Note that this happens in Dots-and-Boxes when a box is completed. All Square is played with heaps of beans. A move splits a heap into two smaller ones. If both heap sizes are perfect squares, the player must move again: if he can't he loses!)

13. Analyze the **misère** version of the **octal games** of Problem 2. (See [Allemang 1984] and WW, pp. 411–421.)

William L. Sibert made a breakthrough by completing the analysis of misère Kayles; see the postscript to [Sibert and Conway 1992]. Plambeck [1992] has used their method to analyse a few other games, but there's a wide open field here.

14. Moebius, when played on 18 coins has a remarkable pattern. Is there any trace of pattern for larger numbers of coins? Can any estimate be made for the rate of growth of the nim-values?

(Played with a row of coins. A move turns 1, 2, 3, 4 or 5 coins, of which the rightmost must go from heads to tails, to make sure the game satisfies the ending condition. The winner is the player who makes all coins tails. See WW, pp. 432–435; Pless 1991; Curtis 1976; 1977; 1982).

15. Mogul has an even more striking pattern when played on 24 coins, which has some echoes when played on 40, 56 or 64 coins. Thereafter, is there complete chaos? (See Problem 14. A move turns 1, 2, ..., 7 coins.)

16. Find an analysis of **Antonim** with four or more coins. (WW, pp. 459–462. Played with coins on a strip of squares. A move moves a coin from one square to a smaller-numbered square. Only one coin to a square, except that square zero can have any number of coins. It is known that (a, b, c) is a $\mathcal{P}$-position in Antonim just if $(a + 1, b + 1, c + 1)$ is a $\mathcal{P}$-position in Nim, but for more than 3 coins much remains to be discovered.)

17. Extend the analysis of **Kotzig's Nim**. (WW, pp. 481–483. Players alternately place coins on a circular strip, at most one coin on a square. Each coin must be placed m squares clockwise from the previously placed coin, provided m is in the given *move set*, and provided the square is not already occupied. The complete analysis is known only for a few small move sets.)

In spite of recent advances [Fraenkel et al. 1995], a great deal remains to be discovered. Is the game eventually periodic in terms of the length of the circle for every finite move set? Analyze the misère version of Kotzig's Nim.

18. Obtain asymptotic estimates for the proportions of $\mathcal{N}$-, $\mathcal{O}$- and $\mathcal{P}$-positions in Epstein's **Put-or-Take-a-Square** game. (WW, pp. 484–486. Played with one heap of beans. At each turn there are just two options: to take away or add the largest perfect square number of beans that there is in the heap. Thus 5 is a $\mathcal{P}$-position, because 5 ± 4 are both squares; 2 and 3 are $\mathcal{O}$-positions, a win for neither player, since the best play is to go from one to the other, and not to 1 or 4, which are $\mathcal{N}$-positions.)

19. Simon Norton's game of **Tribulations** is similar to Epstein's game, but squares are replaced by triangular numbers. Norton conjectures that there are no $\mathcal{O}$-positions, and that the $\mathcal{N}$-positions outnumber the $\mathcal{P}$-positions in golden ratio. True up to 5000 beans.

20. Complete the analysis of **D.U.D.E.N.E.Y.** (Played with a single heap of beans. Either player may take any number of beans from 1 to Y, except that the immediately previous move mustn't be repeated. When you can't move you lose. Analysis easy for Y even, and known for 53/64 of the odd values of Y; see WW, pp. 487–489.)

Marc Wallace and Alan Jaffray have made a little progress here, but is the situation one in which there is always a small fraction of cases remaining, no matter how far the analysis is pursued?

21. Schuhstrings is the same as D.U.D.E.N.E.Y., except that a deduction of zero is also allowed, but cannot be immediately repeated (WW, pp. 489–490).

22. Analyze **nonrepeating Nim**, in which neither player is allowed to repeat a move. Assume that b beans have been taken from heap H, and pick your variant:

medium local: b beans may not be taken from heap H until some other move is made in heap H.

short local: b beans may not be taken from heap H on the next move.

long local: b beans may never again be taken from heap H.

short global: b beans may not be taken from any heap on the next move.

long global: b beans may never again be taken from any heap.

23. Burning-the-Candle-at-Both-Ends. John Conway and Aviezri Fraenkel ask us to analyze Nim played with a row of heaps. A move may only be made in the leftmost or in the rightmost heap. Of course, when a heap becomes empty, its neighbor becomes the end heap.

Conway has some recollection that the analysis was completed. For one or two heaps the game is the same as Nim. The $\mathcal{P}$-positions for three heaps are $\{n, m, n\}$ with $m \neq n$. If this game is indeed analyzed, then there is also **Hub-and-Spoke Nim**, proposed by Fraenkel. One heap is the hub and the others are arranged in rows forming spokes radiating from the hub. There are several versions:

(a) beans may be taken only from a heap at the end of a spoke;

(b) beans may also be taken from the hub;

(c) beans may be taken from the hub only when all the heaps in a spoke are exhausted;

(d) beans may be taken from the hub only when just one spoke remains;

(e) in (b), (c) and (d), when the hub is exhausted, beans may be taken from a heap at either end of any remaining spoke; i.e. the game becomes the sum of a number of games of Burning-the-Candle-at-Both-Ends.

24. Continue the analysis of **The Princess and the Roses.** (WW, pp. 490–494. Played with heaps of beans. Take one bean, or two beans, one from each of two different heaps. The rules seem trivially simple, but the analysis takes on remarkable ramifications.)

25. Analyze the Conway–Paterson game of **Sprouts** with seven or more spots, or the misère form with five or more spots. (WW, pp. 564–568. A move joins two spots, or a spot to itself by a curve that doesn't meet any other spot or previously drawn curve. When a curve is drawn, a new spot must be placed on it. The valence of any spot must not exceed three. Since this question was asked, Daniel Sleator of AT&T Bell Laboratories has pushed the normal analysis to about 10 spots and the misère analysis to about 8.)

26. Extend the analysis of **Sylver Coinage.** (WW, pp. 575–597. Players alternately name different positive integers, but may not name a number that is the sum of previously named ones, with repetitions allowed. Whoever names 1 loses. See [Nowakowski 1991, Section 3].)

27. Extend the analysis of **Chomp**. (WW, pp. 598–599. Players alternately name divisors of N, which may not be multiples of previously named numbers. Whoever names 1 loses.) David Gale offers a prize of US$100.00 for the first complete analysis of 3D-Chomp, i.e., where N has three distinct prime divisors, raised to arbitrarily high powers.

28. Extend Uléhla's or Berlekamp's analysis of **von Neumann's game** from diforests to directed acyclic graphs (WW, pp. 570–572; [Uléhla 1980]). Since Chomp and the superset game [Gale and Neyman 1982] can be described by directed acyclic graphs but not by diforests, the proposed extension could presumably throw some light on these two unsolved games.

29. Prove that Black doesn't have a forced win in **Chess**.

30. A **King and Rook versus King** problem. Played on a quarter-infinite board, with initial position WKa1, WRb2 and BKc3. Can White win? If so, in how few moves? It may be better to ask, "what is the smallest board (if any) that White can win on if Black is given a win if he walks off the North or East edges of the board?" Is the answer 9×11? In an earlier edition of this paper I attributed this problem to Simon Norton, but it was proposed as a kriegsspiel problem, with unspecified position of the WK, and with W to win with probability 1, by Lloyd Shapley around 1960.

31. David Gale's version of **Lion and Man**. L and M are confined to the non-negative quadrant of the plane. They move alternately a distance of at most one unit. For which initial positions can L catch M?

Conjecture: L wins if he is northeast of M. This condition is clearly necessary, but the answer is not known even if M is at the origin and L is on the diagonal.

Variation. Replace quadrant by wedge-shaped region.

32. Gale's **Vingt-et-un**. Cards numbered 1 through 10 are laid on the table. L chooses a card. Then R chooses cards until his total of chosen cards exceeds the card chosen by L. Then L chooses until her cumulative total exceeds that of R, etc. The first player to get 21 wins. Who is it?

(The rule can be interpreted to mean either "21 exactly" or "21 or more". Jeffery Magnoli, a student of Julian West, thought the second interpretation the more interesting, and found a first-player win in six-card Onze and in eight-card Dix-sept.)

33. Subset Take-away. Given a finite set, players alternately choose proper subsets subject to the rule that, once a subset has been chosen, none of *its* subsets may be chosen later by either player. Last player wins. David Gale conjectures that it's a second-player win; this is true for sets of less than six elements.

34. Eggleton and Fraenkel ask for a theory of **Cannibal Games** or an analysis of special families of positions. They are played on an arbitrary finite digraph. Place any numbers of "cannibals" on any vertices. A move is to select a cannibal and move it along a directed edge to a neighboring vertex. If this is occupied, the incoming cannibal eats the whole population (*Greedy Cannibals*) or just one cannibal (*Polite Cannibals*). A player unable to move loses. Draws are possible. A partizan version can be played with cannibals of two colors, each eating only the opposite color.

35. Welter's Game on an arbitrary digraph. Place a number of monochromatic tokens on distinct vertices of a directed acyclic graph. A token may be moved to any *unoccupied* immediate follower. Last player wins. Make a dictionary of $\mathcal{P}$-positions and formulate a winning strategy for other positions. See A. S. Fraenkel and Joseph Kahane, Problem 45, *Discrete Math.* **45** (1983), 328–329, where a paper is said to be in preparation.

36. Restricted Positrons and Electrons. Fraenkel places a number of Positrons (Pink tokens) and Electrons (Ebony tokens) on distinct vertices of a Welter strip. Any particle can be moved by either player leftward to any square u provided that u is either unoccupied or occupied by a particle of the opposite type. In the latter case, of course, both particles become annihilated (i.e., they are removed from the strip), as physicists tell us positrons and electrons do. Play ends when the excess particles of one type over the other are jammed in the lowest positions of the strip. Last player wins. Formulate a winning strategy for those positions where one exists. Note that if the particles are of one type only, this is Welter's Game; since a strategy is known for Misere Welter (WW, pp. 480–481), it may not be unreasonable to ask for a misère analysis as well. See Problem 47, *Discrete Math.* **46** (1983), 215–216.

37. General Positrons and Electrons. Like 36, but played on an arbitrary digraph. Last player wins.

38. Fulves's Merger. Karl Fulves, Box 433, Teaneck NJ 07666, sent the following game. Start with heaps of 1, 2, 3, 4, 5, 6 and 7 beans. Two players alternately transfer any number of beans from one heap to another, except that beans may not be transferred from a larger to a smaller heap. The player who makes all the heaps have an *even* number of beans is the winner.

The total number of beans remains constant, and is even (28 in this case, though one is interested in even numbers in general: a similar game can be played in which the total number is odd and the object is to make all the heaps odd in size). As the moves are nondisjunctive, the Sprague–Grundy theory is not likely to be of help. Fulves's particular game can be solved using Steinhaus's *remoteness function* [1925] as follows.

The number of odd heaps is even. If the number of odd heaps is two, there is an immediate win: such a position is an N-position of remoteness 1. In normal play, P-positions have even remoteness, and N-positions have odd remoteness.

To calculate the remoteness of a position:

- If there are no options, the remoteness is 0.
- If there is an option of even remoteness, add one to the *least* such remoteness.
- Otherwise, add one to the *greatest odd* remoteness.

WIN QUICKLY — LOSE SLOWLY.

If the number of odd heaps is zero, the position is terminal. All other P-positions must contain at least four odd heaps. Some examples, with their even remotenesses r, are:

P-position	r	P-position	r	P-position	r
1 1 1 25	2	1 1 1 1 1 23	4	1 1 1 1 2 4 18	10
1 3 3 21	4	1 1 1 2 2 21	6	1 1 1 1 3 5 16	12
1 5 5 17	6	1 1 1 3 3 19	8	1 1 1 1 6 8 10	18
1 7 7 13	8	1 1 1 4 4 17	10	1 1 1 2 3 8 12	14
1 9 9 9	10	1 1 1 5 5 15	12	1 1 1 2 5 9 9	20
3 5 9 11	10	1 1 1 6 6 13	14	1 1 1 3 6 7 9	22
		1 1 1 7 7 11	16	1 1 2 2 2 3 17	10
1 1 2 5 19	6	1 1 1 8 8 9	18	1 1 2 2 4 5 13	14
1 1 3 6 17	8	1 1 2 3 7 14	12	1 1 2 3 5 5 11	18
1 1 4 7 15	10	1 1 2 4 9 11	14	1 1 2 4 6 7 7	20
1 2 3 9 13	12	1 1 3 5 8 10	18	1 1 3 3 3 4 13	16
1 3 5 5 14	12	1 2 2 3 5 15	12	1 1 3 4 4 7 8	24
1 5 6 7 9	14	1 2 2 7 7 9	14	1 1 4 4 4 5 9	24
2 3 3 3 17	8	1 2 3 3 8 11	14	1 2 2 3 3 3 14	16
2 5 5 5 11	14	1 2 5 5 7 8	16	1 2 3 3 5 7 7	22
2 5 7 7 7	14	1 3 3 3 3 15	14		
3 3 4 5 13	14	1 3 3 4 7 10	16		
3 3 7 7 8	16	1 3 3 6 6 9	18		
		1 3 4 4 5 11	18		
		2 2 3 3 9 9	16		
		2 3 3 5 5 10	18		
		3 3 3 5 6 8	20		

Table 2 shows a winning strategy from Fulves's starting position.

Richard Nowakowski has investigated some variants of Fulves's game:

1. Take a row of consecutive coins. A move is to take one or more coins from a heap and put them on an adjacent heap of coins provided that the second heap is at least as large as the first. Assume that we start with a row of adjacent coins, that is, each heap consists of one coin. The nim-values are periodic with period seven and the values are 0 1 2 0 3 1 0 with no exceptional values. To prove

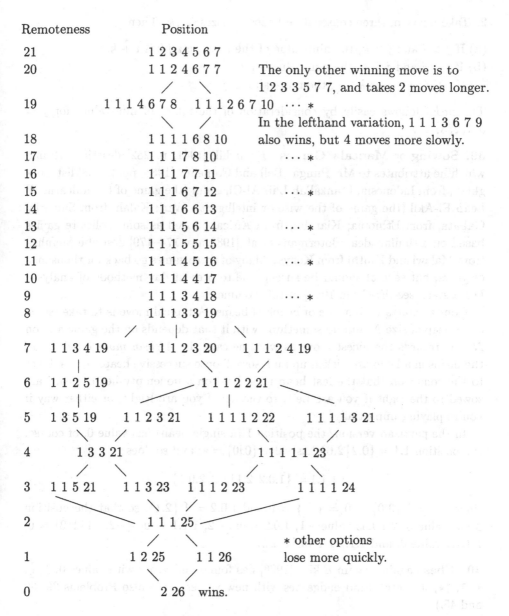

Remoteness	Position

21   1 2 3 4 5 6 7

20   1 1 2 4 6 7 7   The only other winning move is to
                                     1 2 3 3 5 7 7, and takes 2 moves longer.

19   1 1 1 4 6 7 8   1 1 1 2 6 7 10 · · · *
                                     In the lefthand variation, 1 1 1 3 6 7 9

18   1 1 1 1 6 8 10   also wins, but 4 moves more slowly.

17   1 1 1 7 8 10   · · · *

16   1 1 1 7 7 11

15   1 1 1 6 7 12   · · · *

14   1 1 1 6 6 13

13   1 1 1 5 6 14   · · · *

12   1 1 1 5 5 15

11   1 1 1 4 5 16   · · · *

10   1 1 1 4 4 17

9   1 1 1 3 4 18   · · · *

8   1 1 1 3 3 19

7   1 1 3 4 19   1 1 1 2 3 20   1 1 1 2 4 19

6   1 1 2 5 19   1 1 1 2 2 21

5   1 3 5 19   1 1 2 3 21   1 1 1 1 2 22   1 1 1 1 3 21

4   1 3 3 21   1 1 1 1 1 23

3   1 1 5 21   1 1 3 23   1 1 1 2 23   1 1 1 1 24

2   1 1 1 25

                                     * other options

1   1 2 25   1 1 26   lose more quickly.

0   2 26   wins.

Table 2. Winning strategy for Fulves's game.

this one needs to know the nim-values of the game with the first heap having two coins and also the game with both end heaps haing two coins. The latter is periodic with period seven: the nim-vales are 2 1 0 0 1 3 0 with no exceptions. The former also has period seven. The values are 5 5 4 2 5 6 4 but they do not start until the game 2.1^{15}.

The misère version of this variant might be interesting.

2. Take a row of three consecutive heaps, of sizes i, j, k. Then:

(a) If $j > i$ and $j > k$, the nim-value of the position $i.j.k$ is $i \overset{*}{+} k$.
(b) If $i = j$ and $j \geq k$, the nim-value is $i + k$.
(c) If either $i > j$ or $k > j$, the nim-value is j.

The proof follows easily by consideration of the tables of nim-values for $j = 0, 1, 2, 3, \ldots$.

39. Sowing or **Mancala Games.** Kraitchik [1941, p. 282] describes Ruma, which he attributes to Mr. Punga. Bell and Cornelius [1988, pp. 22–38] list Congklak, from Indonesia; Mankal'ah L'ib Al-Ghashim (the game of the unlearned); Leab El-Akil (the game of the wise or intelligent); Wari; Kalah, from Sumeria; Gabata, from Ethiopia; Kiarabu, from Africa; as well as some solitaire games based on a similar idea. Botermans et al. [1989, pp. 174–179] describe Mefuhva from Malawi and Kiuthi from Kenya. Many of these games go back for thousands of years, but several should be susceptible to present day methods of analysis. For a start, see [Erickson 1996] in this volume.

Conway starts with a *line* of heaps of beans. A typical move is to take (some of) a heap of size N and do something with it that depends on the game and on N. He regards the nicest move as what he calls the *African move* in which *all* the beans in a heap are picked up and 'sowed' onto successive heaps, *and* subject to the condition that the last bean must land on a nonempty heap. Beans are sowed to the right if you are Left, to the left if you are Right, or either way if you're playing impartially.

In the partizan version, the position 1 (a single bean) has value 0, of course; the position $1.1 = \{0.2 \mid 2.0\}$ has value $\{0 \mid 0\} = *$; and so does

$$1.1.1 = \{1.0.2, 2.1 \mid 1.2, 2.0.1\},$$

since $2.1 = \{\ \mid 3.0\}$, $3.0 = \{\ \mid\ \} = 0$ and $1.0.2 = \{\ \mid 2.1\}$, so that the position 3 has value 0, 2.1 has value -1, 1.0.2 value -2, 1.1.1 value $\{-2, -1 \mid 1, 2\} = 0$, 1.1.1.1 value 0, and 1.1.1.1.1 value $\pm \frac{1}{2}$.

40. Chess again. Noam Elkies [1996] has found endgames with values 0, 1, $\frac{1}{2}$, $*$, $\uparrow$, $\uparrow*$, $\Uparrow *$, etc. Find endgames with new values. (See also Problems 29, 30 and 45.)

41. Sequential compounds of games have been studied by Stromquist and Ullman [1993]. They mention a more general compound. Let $(P, <)$ be a finite poset and for each $x \in P$ let G_x be a game. Consider a game $G(P)$ played as follows. Moves are allowed in any single component G_x provided that no legal moves remain in any component G_y with $y > x$. A player unable to move loses. The sequential compound is the special case when $(P, <)$ is a chain (or linear order). The *sum* or disjunctive compound is the case where $(P, <)$ is an antichain. They have no coherent theory of games $G(P)$ for arbitrary posets.

They list some more specific problems that may be more tractable. Compare Problem 23 above.

42. Beanstalk and **Beans-Don't-Talk** are games invented by John Isbell and by John Conway [Guy 1986]. Beanstalk is played between Jack and the Giant. The Giant chooses a positive integer, n_0. Then J. and G. play alternately n_1, n_2, n_3, ... according to the rule $n_{i+1} = n_i/2$ if n_i is even, $= 3n_i \pm 1$ if n_i is odd; i.e. if n_i is even, there's only one option, while if n_i is odd there are just two. The winner is the person moving to 1. If the Giant chooses an odd number > 1, can Jack always win? Not by using the Greedy Strategy (always descend when it's safe to do so, e.g., play the moves that are <u>underlined</u> in the game below), as may be seen from $n_0 = 7$ (J's moves in parentheses).

7 (22) 11 (34) 17 (50) 25 (74) 37 (110) 55 (166) 83 (<u>248</u>) 124 (62) 31 (94) 47 (142 71 (214) 107 (322) 161 (482) 241 (722) 361 (1082) 541 (<u>1624</u>) 812 (406) 203 (<u>608</u>) 304 (152) 76 (38) 19 (<u>56</u>) 28 (14) 7 ...

In Beans-Don't-Talk, the move is from n to $(3n \pm 1)/2^*$ where 2^* is the highest power of two dividing the numerator; the winner is still the person moving to 1. Are there any drawn positions? There are certainly drawn *plays*, e.g., 7 (5) 7 (5) ..., but 5 is an $\mathcal{N}$-position because there is the immediate winning option $(5 \times 3 + 1)/2^4 = 1$, and 7 is a $\mathcal{P}$-position since the other option $(7 \times 3 + 1)/2 = 11$ is met by $(11 \times 3 - 1)/2^5 = 1$. What we want to know is: *are there any $\mathcal{O}$-positions* (positions of infinite remoteness)?

(For a definition of *remoteness* see Problem 38 above. There are several unanswered questions about the remotenesses of positions in these two games. Remoteness may also be the best tool we have for Problems 18 and 19 above.)

43. Inverting Hackenbush. John Conway turns Hackenbush (described on page 56 in this volume) into a hot game by amending the move to 'remove an edge of your color and everything thus disconnected from the ground, and then turn the remaining string upside-down and replant it'. The analysis replaces the 'number tree' (WW, p. 25) by a similar tree, but with the smaller binary fractions replaced by increasingly hot games. The game can be generalized to play on trees: a move that prunes the tree at a vertex V includes replanting the tree with V as its root.

44. Konane [Ernst and Berlekamp]. There is much to be discovered about this fascinating and eminently playable game, which exhibits the values 0, $*$, $*2$, $\uparrow$, 2^{-n}, and many other infinitesimals and also hot values of arbitrarily high temperature.

45. Elwyn Berlekamp asks: 'What is the **habitat of $*2$**?' This value, defined as $\{0, * \mid 0, *\}$, does not occur in Blockbusting, Hackenbush, Col or Go. It does occur in Konane and 6×6 Chess. What about Chilled Go, Domineering and 8×8 Chess?

46. There are various ways of playing **two-dimensional Nim**. One form is discussed on p. 313 of WW. Another is proposed by Berman, Fraenkel and Kahane in Problem 41, *Discrete Math.* *45* (1983), 137–138. Start with a rectangular array of heaps of beans. At each move a row or column is selected and a positive number of beans is taken from some of the heaps in that row or column [Fremlin 1973]. A variant is where beans may be taken only from contiguous heaps. Other variants are played on triangular or hexagonal boards; a special case of this last is Piet Hein's Nimbi, solved by Fraenkel and Herda [1980].

47. Many results are known concerning tiling rectangles with **polyominoes**. One can extend such problems to disconnected polyominoes. For instance, can a rectangle be tiled by □ □□? By □□ □□ □? If so, what are the smallest rectangles that can be so tiled?

48. Find all words that can be reduced to one peg in one-dimensional **Peg Solitaire**. (A move is for a peg to jump over an adjacent peg into an empty adjacent space, and remove the jumped-over peg: for instance, $1101 \rightarrow 0011 \rightarrow 0100$, where 1 represents a peg and 0 an empty space.) Examples of words that can be reduced to one peg are 1, 11, 1101, 110101, $1(10)^k1$. Georg Gunther, Bert Hartnell and Richard Nowakowski found that for an $n \times 1$ board with one empty space, n must be even and the space must be next but one to the end. If the board is cyclic, the condition is simply n even.

49. Elwyn Berlekamp asks if there is a game that has **simple**, playable rules, an intricate **explicit solution**, and is provably **NP** or harder.

50. John Selfridge asks: is **Four-File** a draw? Four-File is played on a chessboard with the chess pieces in their usual starting positions, but only on the a-, c-, e- and g-files (a rook, a bishop, a king, a knight and four pawns on each side). The moves are normal chess moves except that play takes place only on these four files; in particular, pawns cannot capture and there is no castling. The aim is to checkmate your opponent's king.

51. Elwyn Berlekamp asks for a complete theory of closed $1 \times n$ **Dots-and-Boxes**, those with starting position

A sample position is

(See WW, Chapter 16.) Are there more nimber decomposition theorems? Compile a datebase of nim-values.

52. Berlekamp notes that overheating operators provide a very concise way of expressing closed-form solutions to many games, and David Moews observes that monotonicity and linearity depend on the parameters and the domain. How does one play *sums* of games with varied overheating operators? Find a simple, elegant way of relating the operator parameters to the game. See WW, pp. 163–175, [Berlekamp 1988; Berlekamp and Wolfe 1994; Calistrate 1996].

References

[Allemang 1984] D. T. Allemang, "Machine Computation with Finite Games", M.Sc. thesis, Cambridge University, 1984.

[Allis et al. 1993] L. V. Allis, H. J. van den Herik, and M. P. H. Huntjens, "Go-Moku solved by new search techniques", pp. 1–9 in *Proc. AAAI Fall Symposium on Games: Planning and Learning*, AAAI Press Tech. Report FS93-02, Menlo Park, CA.

[Bell and Cornelius 1988] R. Bell and M. Cornelius, *Board Games Round the World*, Cambridge Univ. Press, Cambridge, 1988.

[Berlekamp 1988] E. R. Berlekamp, "Blockbusting and Domineering", *J. Combin. Theory* A**49** (1988), 67–116.

[Berlekamp and Kim 1996] E. Berlekamp and Y. Kim, "Where is the 'thousand-dollar Ko'?", pp. 203–226 in this volume.

[Berlekamp and Wolfe 1994] E. Berlekamp and D. Wolfe, *Mathematical Go: Chilling Gets the Last Point*, A. K. Peters, Wellesley, MA, 1994. Also published in paperback, with accompanying software, as *Mathematical Go: Nightmares for the Professional Go Player*, by Ishi Press International, San Jose, CA.

[Berlekamp et al. 1982] E. R. Berlekamp, J. H. Conway, and R. K. Guy, *Winning Ways For Your Mathematical Plays*, Academic Press, London, 1982.

[Botermans et al. 1989] ack Botermans, Tony Burrett, Pieter van Delft and Carla van Splunteren, *The World of Games*, Facts on File, New York and Oxford, 1989.

[Calistrate 1996] D. Calistrate, "The reduced canonical form of a game", pp. 409–416 in this volume.

[Curtis 1976] R. T. Curtis, "A new combinatorial approach to M_{24}", *Math. Proc. Cambridge Philos. Soc. 79* (1976), 25–42.

[Curtis 1977] R. T. Curtis, "The maximal subgroups of M_{24}", *Math. Proc. Cambridge Philos. Soc. 81* (1977), 185–192.

[Curtis 1984] R. T. Curtis, "The Steiner system $S(5, 6, 12)$, the Mathieu group M_{12} and the 'kitten'", pp. 353–358 in *Computational Group Theory*, Durham, 1982 (edited by M. D. Atkinson), Academic Press, London and New York, 1984).

[Elkies 1996] Noam D. Elkies, "On numbers and endgames", pp. 135–150 in this volume.

[Erickson 1996] J. Erickson, "Sowing Games", pp. 287–297 in this volume.

[Ernst and Berlekamp] Michael D. Ernst and Elwyn Berlekamp, "Playing (Konane) mathematically", to appear in volume of articles in tribute to Martin Gardner (edited by David Klarner), A K Peters, Wellesley, MA.

[Fraenkel and Kotzig 1987] A. S. Fraenkel and A. Kotzig, "Partizan octal games: partizan subtraction games", *Internat. J. Game Theory*, **16** (1987), 145–154.

[Fraenkel et al. 1995] A. S. Fraenkel, A. Jaffray, A. Kotzig and G. Sabidussi, "Modular Nim", *Theor. Comput. Sci.* (*Math Games*) **143** (1995), 319–333.

[Fremlin 1973] D. Fremlin, "Well-founded games", *Eureka* **36** (1973), 33–37.

[Gale and Neyman 1982] D. Gale and A. Neyman, "Nim-type games", *Internat. J. Game Theory* **11** (1982), 17–20.

[Gangolli and Plambeck 1989] A. Gangolli and T. Plambeck, "A note on periodicity in some octal games", *Internat. J. Game Theory* **18** (1989), 311–320.

[Guy 1986] R. K. Guy, "John Isbell's Game of Beanstalk and John Conway's Game of Beans-Don't-Talk", *Math. Mag.* **59** (1986), 259–269. (William Jockusch of Carleton College, MN, has corrected the tentative tables on p. 262.)

[Guy 1991] R. K. Guy (editor), *Combinatorial Games*, Proc. Symp. Appl. Math. **43**, Amer. Math. Soc., Providence, RI, 1991.

[Guy 1996] R. K. Guy, "Impartial Games", pp. 61–78 in this volume.

[Kim 1996] Y. Kim, "New values in domineering", *Theoret. Comput. Sci.* (*Math Games*) **156** (1996), 263–280.

[Kraitchik 1953] Maurice Kraitchik, *Mathematical Recreations*, 2nd ed., Dover, New York, 1953.

[Landman 1996] H. A. Landman, "Eyespace Values in Go", pp. 227–257 in this volume.

[Lustenberger 1967] Carlyle Lustenberger, M.S. thesis, Pennsylvania State Univ., 1967.

[Nowakowski 1991] R. J. Nowakowski, "... Welter's Game, Sylver Coinage, Dots-and-Boxes, ...", pp. 155–182 in [Guy 1991].

[Plambeck 1992] T. E. Plambeck, "Daisies, Kayles, and the Sibert–Conway decomposition in misère octal games", *Theoret. Comput. Sci.* (*Math Games*) **96** (1992), 361–388.

[Pless 1991] V. Pless, "Games and codes", pp. 101–110 in [Guy 1991].

[Sibert and Conway 1992] W. L. Sibert and J. H. Conway, "Mathematical Kayles" (with a postscript by R. K. Guy), *Internat. J. Game Theory* **20** (1992), 237–246.

[Steinhaus 1925] H. Steinhaus, "Definicje potrzebne do teorji gry i pościgu", *Myśl. Akad. Lwów*, **1**(1) (1925), 13–14. Reprinted and translated as "Definitions for a theory of games and pursuit", *Naval Res. Logist. Quart.* **7** (1960), 105–108.

[Stromquist and Ullman 1993] W. Stromquist and D. Ullman, "Sequential Compounds of Combinatorial Games", *J. Theoretical Comp. Sci.* (*Math. Games*) **119** (1993), 311–321.

[Uléhla 1980] J. Uléhla, "A complete analysis of von Neumann's Hackendot", *Internat. J. Game Theory* **9** (1980), 107–113.

[West 1996] J. West, "New Values for Top Entails", pp. 345–350 in this volume.

[Wolfe 1993] D. Wolfe, "Snakes in Domineering games", *Theoret. Comput. Sci.* **119** (1993), 323–329.

[Wolfe 1996] D. Wolfe, "The Gamesman's Toolkit", pp. 93–98 in this volume.

RICHARD K. GUY
MATHEMATICS AND STATISTICS DEPARTMENT
UNIVERSITY OF CALGARY
2500 UNIVERSITY AVENUE
ALBERTA, T2N 1N4
CANADA

Games of No Chance
MSRI Publications
Volume **29**, 1996

Combinatorial Games: Selected Bibliography with a Succinct Gourmet Introduction

AVIEZRI S. FRAENKEL

1. What are Combinatorial Games?
What are they Good For?

Roughly speaking, the family of *combinatorial games* consists of two-player games with perfect information (no hidden information as in some card games), no chance moves (no dice) and outcome restricted to (lose, win), (tie, tie) and (draw, draw) for the two players who move alternately. Tie is an end position such as in tic-tac-toe, where no player wins, whereas draw is a dynamic tie: any position from which a player has a nonlosing move, but cannot force a win. Both the easy game of Nim and the seemingly difficult chess are examples of combinatorial games. We use the shorter terms *game* and *games* below to designate combinatorial games.

Amusing oneself with games may sound like a frivolous occupation. But the fact is that the bulk of interesting and natural mathematical problems that are hardest in complexity classes beyond *NP*, such as *P*space, *E*xptime and *E*xpspace, are two-player games; occasionally even one-player games (puzzles) or even zero-player games (Conway's "Life"). Two of the reasons for the high complexity of two-player games are outlined below. Before that we note that in addition to a natural appeal of the subject, there are applications or connections to various areas, including complexity, logic, graph and matroid theory, networks, error-correcting codes, surreal numbers, on-line algorithms and biology.

But when the chips are down, it is this "natural appeal" that compels both amateurs and professionals to become addicted to the subject. What is the essence of this appeal? Perhaps it is rooted in our primal beastly instincts; the desire to corner, torture, or at least dominate our peers. An intellectually refined version of these dark desires, well hidden under the façade of scientific research, is the consuming strive "to beat them all", to be more clever than the most clever, in short — to create the tools to Math-master them all in hot combinatorial combat! Reaching this goal is particularly satisfying and sweet in the context of combinatorial games, in view of their inherent high complexity.

2. Why are Combinatorial Games Hard?

Decision problems such as graph hamiltonicity and Traveling Salesperson (Is there a round tour through specified cities of cost $\leq C$?) are *existential*: they involve a single existential quantifier ("Is there. . . ?"). In mathematical terms an existential problem boils down to finding a path, in a large "decision-tree" of all possibilities, that satisfies specified properties. The above two problems, as well as thousands of other interesting and important combinatorial-type problems are NP-*complete*. This means that they are *conditionally intractable*, i.e., the best way to solve them seems to require traversal of most if not all of the decision tree, whose size is exponential in the input size of the problem. No better method is known to date at any rate, and if an efficient solution will ever be found for any NP-complete problem, then all NP-complete problems will be solvable efficiently.

The decision problem whether White can win if White moves first in a chess game, on the other hand, has the form: Is there a move of White such that for every move of Black there is a move of White such that for every move of Black there is a move of White . . . such that White can win? Here we have a large number of alternating quantifiers rather than a single existential one. We are looking for an entire subtree rather than just a path in the decision tree. The problem for generalized chess on an $n \times n$ board is in fact Exptime-complete, which is a *provable intractability*. Most games are at least Pspace-hard.

Put in simple language, in analyzing an instance of Traveling Salesperson, the problem itself is passive: it does not resist your attempt to attack it, yet it is difficult. In a game, in contrast, there is your opponent, who, at every step, attempts to foil your effort to win. It's similar to the difference between an autopsy and surgery. Einstein, contemplating the nature of physics said, "Der Allmächtige ist nicht boshaft; Er ist raffiniert" (The Almighty is not mean; He is sophisticated). NP-complete existential problems are perhaps sophisticated. But your opponent in a game can be very mean!

Another reason for the high complexity of games is connected with the fundamental notion of *sum* (disjunctive compound) of games. A sum is a finite collection of disjoint games; often very basic, simple games. Each of the two players, at every turn, selects one of the games and makes a move in it. If the outcome is not a draw, the sum-game ends when there is no move left in any of the component games. If the outcome is not a tie either, then in *normal* play, the player first unable to move loses and the opponent wins. The outcome is reversed in *misère* play.

The *game-graph* of a game is a directed graph whose vertices are the positions of the game, and (u, v) is an edge if and only if there is a move from position u

to position v. It turns out that the game-graph of a sum has size exponential in the combined size of the input game-graphs! Since sums occur naturally and frequently, and since analyzing the sum entails reasoning about its game-graph, we are faced with a problem that is *a priori* exponential, quite unlike most present-day interesting existential problems.

3. Breaking the Rules

As the experts know, some of the most exciting games are obtained by breaking some of the rules for combinatorial games, such as permitting a player to pass a bounded or unbounded number of times, relaxing the requirement that players play alternately, or permitting a number of players other than two. But by far the most fruitful tampering with the rules seems to be to permit sums of games that are not quite fixed (which explains why misère play of sums of games is much harder than normal play) or not quite disjoint (Welter) or the game does not seem to decompose into a sum (Geography or Poset Games).

On the other hand, permitting a payoff function other than $(0, 1) = (\text{lose}, \text{win})$ or $(\frac{1}{2}, \frac{1}{2}) = (\text{draw}, \text{draw})$ usually, but not always, leads to games that are not considered to be combinatorial games, or to borderline cases.

4. Why Is the Bibliography Vast?

In the realm of existential problems, such as sorting or Traveling Salesperson, most present-day interesting decision problems can be classified into tractable, conditionally intractable, and provably intractable ones. There are exceptions, to be sure, e.g. graph isomorphism and primality testing, whose complexity is still unknown. But these are few. It appears that, in contrast, there is a very large set of games whose complexities are hard to determine. The set of these games is termed *Wonderland*, because we are wondering about the complexity classification of its members. Only few games have been classified into the complexity classes they belong to. Today, most games belong to Wonderland, and despite recent impressive progress, the tools for reducing Wonderland are still few and inadequate.

To give an example, many interesting games have a very succinct input size, so a polynomial strategy is often more difficult to come by (Octal games; Grundy's game). Succinctness and non-disjointness of games in a sum may be present simultaneously (poset games). In general, "breaking the rules" and the alternating quantifiers add to the volume of Wonderland. We suspect that the large size of Wonderland, a fact of independent interest, is the main contributing factor to the bulk of the bibliography on games.

5. Why Isn't it Larger?

The bibliography below is a partial list of books and articles on combinatorial games and related material. It is partial not only because I constantly learn of additional relevant material I did not know about previously, but also because of certain self-imposed restrictions. The most important of these is that only papers with some original and nontrivial mathematical content are considered. This excludes most historical reviews of games and most, but not all, of the work on heuristic or artificial intelligence approaches to games, especially the large literature concerning computer chess. I have, however, included the compendium Levy [1988], which, with its 50 articles and extensive bibliography, can serve as a first guide to this world. Also some papers on chess-endgames and clever exhaustive computer searches of some games have been included.

On the other hand, papers on games that break some of the rules of combinatorial games are included liberally, as long as they are interesting and retain a combinatorial flavor. These are vague and hard to define criteria, yet combinatorialists usually recognize a combinatorial game when they see it. Besides, it is interesting to break also this rule sometimes! Adding borderline cases is acknowledged in the "related material" postfixed to the title of this bibliography. We have included some references to one-player games, e.g., towers of Hanoi, n-queen problems and peg-solitaire, but hardly any on zero-player games (such as Life). We have also included papers on various applications of games, especially when the connection to games is substantial or the application is important.

In 1990, *Theoretical Computer Science* inaugurated a Mathematical Games Section whose main purpose is to publish papers on combinatorial games. The "Aims and Scope" and the names and addresses of the Mathematical Games Section editors are printed in the first issue of every volume of TCS. Prospective authors are cordially invited to submit their papers (in triplicate), to one of the editors whose interests seem closest to the field covered by the paper. This forum is beginning to become a focal point for high-class research results in the field of combinatorial games, thus increasing the bibliography at a moderate pace.

6. Cold and Hot Versions

The game bibliography below is very dynamic in nature. Previous versions have been circulated to colleagues for many years, intermittently, since the early 80's. Prior to every mailing updates were prepared, and usually also afterwards, as a result of the comments received from several correspondents. The listing can never be "complete". Thus also the present form of the bibliography is by no means complete.

Because of its dynamic nature, it is natural that the bibliography now became a "Dynamic Survey" in the Dynamic Surveys (DS) section of the *Electronic Journal of Combinatorics* (ElJC) and *The World Combinatorics Exchange* (WCE).

The ElJC and WCE are on the World Wide Web (WWW), and the DS can be accessed at http://ejc.math.gatech.edu:8080/Journal/Surveys/index.html. This document contains a copy of the *cold* version of the bibliography, together with the date of the latest modification.

Any document on the WWW may contain short text portions (underlined, or colored) that are *hypertext*, that is, that contain a hidden link to another relevant document. Clicking with your mouse on this *hot* hypertext brings up that document onto your screen, wherever in the world it may physically reside. Portions of that document may also be hot (clickable), and so the entire world is hyperlinked into a web, that is, virtualized into a complex mosaic, all at your fingertips. In fact, a good way to access the WWW is through mouse-activated browsers (Mosaic, Netscape, etc.)!

It is thus natural to have also a *hot* version of the bibliography. In it, the bibliographic items are hypertext, and so clicking on a hot item retrieves the document itself, displaying it on your screen for browsing, reading or downloading.

7. Hot and Cold Help

For the hot version to grow into a bibliography of practical value, we need the links to the bibliographic items. These links are mainly of two types.

- Links to papers published in refereed electronic journals, proceedings or books published by scientific societies or commercial enterprises directly from authors' TEX-files. At this time there are only few of these, but that is likely to change rapidly.
- Links to the documents in the authors' own home directories or ftp archives.

Authors and readers who have this information are requested to send it to me. Note that some copyright questions may be involved. Each author should clear those prior to submitting the hyper-links to me. Authors should send *updates* of the links to the Managing Editor of ElJC at calkin@math.gatech.edu. Updates should be sent whenever there is a change in the link, due, for example, to host and/or directory changes; even if a file is replaced by a compressed version of it, since its name changed!

Regarding the cold and hot versions alike, I wish to ask the readers to continue sending to me corrections and comments; and inform me of significant omissions, remembering, however, that it is a *selected* bibliography. I prefer to get reprints, preprints or URL's, rather than only titles, whenever possible.

8. Games on Web Sites

Material on games is mushrooming on the Web. Below we bring only some highlights. Following these URL's—which will become hot once this file does—

the reader will be lead to many others. Caution: links often get stale as people and files shift about. But the many interlinks included in the Web pages listed below will presumably enable you to overcome this problem. Please send me URL's of additional interesting game sites.

- "Topics in Mathematical Recreations" at:
  http://www.dcs.st-andrews.ac.uk/~ad/mathrecs/mathrectopics.html
  has very useful links to topics in recreational math, including a bibliography (mainly on puzzles) and some actual puzzles. It is maintained by Tony Davie.
- An excellent games page is maintained by David Eppstein of UC Irvine at:
  http://www.ics.uci.edu/~eppstein/cgt/
  Theory, bibliography, papers and many actual games can be accessed. Also much material on recreational math can be clicked on.
- A loaded page entitled "Mathematical Games, Toys, and Puzzles" is maintained by Jeff Erickson of UC Berkeley, at:
  http://http.cs.berkeley.edu/~jeffe/mathgames.html
  It is divided into "Theory", "Actual Games" (Connect, Othello,...) and "Fun Math".
- Daniel Loeb of University of Bordeaux, has attractive game theoretic material, including Multiplayer Combinatorial Games, Recreational Mathematics and Mathematical Education at:
  http://www.labri.u-bordeaux.fr/~loeb/game.html
- Andrew Plotkin of Carnegie Mellon University created "Zarf's List of Interactive Games on the Web" at:
  http://www.cs.cmu.edu/afs/andrew/org/kgb/www/zarf/games.html
  It contains large collections of games you can actually play on the Web, divided into 4 main categories: Interactive Games, Older Games, Interactive Toys and Older Toys.
- Dave Stanworth created the "Games Domain" at:
  http://www.gamesdomain.co.uk/
  It contains links to a huge *searchable* selection of games of all sorts, including video games. (The searcher should be modified: "Go" is not found by it, though it exists under "board games" and elsewhere.) In its "Games Information" alone there are hundreds of links.
- Mario Velucchi of University of Pisa at:
  http://www.cli.di.unipi.it/~velucchi/personal.html
  has extensive and very nice material on chess and its variations. Links to recreational math and mathematical games sites are also included.
- David Wolfe at UC Berkeley, a former Ph.D. student of Elwyn Berlekamp, created the Gamesman's Toolkit [Wolfe 1996] and has a link to a Postscript file of Unsolved Problems in Combinatorial Games of Richard Guy. See:
  http://http.cs.berkeley.edu/~wolfe/

9. Idiosyncrasies

Due to the changes announced in Section 7, hard copies of the bibliography will not be mailed out any more, with the possible exception of a few copies to individuals without access to the Internet.

Most of the bibliographic entries refer to items written in English, though there is a sprinkling of Danish, Dutch, French, German, Japanese, Slovakian and Russian, as well as some English translations from Russian. The predominance of English may be due to certain prejudices, but it also reflects the fact that nowadays the *lingua franca* of science is English. In any case, I'm soliciting also papers in languages other than English, especially if accompanied by an abstract in English.

On the administrative side, Technical Reports, submitted papers and unpublished theses have normally been excluded; but some exceptions have been made. Abbreviations of book series and journal names follow the *Math Reviews* conventions. Another convention is that de Bruijn appears under D, not B; von Neumann under V, not N, McIntyre under M not I, etc.

Earlier versions of this bibliography have appeared, under the title "Selected bibliography on combinatorial games and some related material", as the master bibliography for the book *Combinatorial Games* (*Proc. Symp. Appl. Math.* **43**, edited by R. K. Guy, AMS, 1991), with 400 items, and in the *Dynamic Surveys* section of the *Electronic J. of Combinatorics* in November 1994, with 542 items.

10. Suggestions and Questions

Two correspondents and myself have expressed the opinion that the value of the bibliography would be enhanced if it would be transformed into an annotated bibliography. Do you think this should be done?

One correspondent has suggested to include a list of combinatorial games people, with email addresses and URL's, where available.

One correspondent and myself think that the bibliography has, after the August 1995 update, become somewhat unwieldy:

1. Should henceforth new additions be kept in separate files? Disadvantage: machine searching will have to be done on several (.tex) files, and visual search on several .ps files. However, some of the older updates may be merged with the main part, after a while.

2. The bibliography could be broken up into sections of related papers. Thus, zero-person and one-person games could be in one section, games of imperfect information or those with chance moves in another, complexity papers in a third, partizan games in another, impartial games in still another, etc. The disadvantage here is that some papers will belong to several sections. For example, papers dealing with complexity aspects of impartial games, or of games with imperfect

information. This could be amended by using crossreferences to other sections, at the price of some further unwieldiness.

3. Any other suggestions?

Acknowledgments

Many people suggested additions to the bibliography, or contributed to it in other ways. Among them are: Akeo Adachi, Ingo Althöfer, Thomas Andreae, Adriano Barlotti, József Beck, Claude Berge, Gerald E. Bergum, H. S. Mac-Donald Coxeter, Thomas S. Ferguson, James A. Flanigan, Fred Galvin, Martin Gardner, Alan J. Goldman, Solomon W. Golomb, Richard K. Guy, Shigeki Iwata, David S. Johnson, Victor Klee, Donald E. Knuth, Anton Kotzig, Jeff C. Lagarias, Michel Las Vergnas, Hendrik W. Lenstra, Hermann Loimer, F. Lockwood Morris, Richard J. Nowakowski, Judea Pearl, J. Michael Robson, David Singmaster, Cedric A. B. Smith, Rastislaw Telgársky, Yōhei Yamasaki and many others. Thanks to all and keep up the game! Special thanks are due to Ms. Sarah Fliegelmann and Mrs. Carol Weintraub, who have been maintaining and updating the bibliography-file, expertly and skilfully, over several different TEX generations; and to Silvio Levy, who has, more recently, edited and transformed it into LATEX2e.

The Bibliography

1. B. Abramson and M. Yung [1989], Divide and conquer under global constraints: a solution to the n-queens problems, *J. Parallel Distrib. Comput.* **6**, 649–662.

2. A. Adachi, S. Iwata and T. Kasai [1981], Low level complexity for combinatorial games, Proc. 13th Ann. ACM Symp. Theory of Computing (Milwaukee, WI, 1981), Assoc. Comput. Mach., New York, NY, pp. 228–237.

3. A. Adachi, S. Iwata and T. Kasai [1984], Some combinatorial game problems require $\Omega(n^k)$ time, *J. Assoc. Comput. Mach.* **31**, 361–376.

4. H. Adachi, H. Kamekawa and S. Iwata [1987], Shogi on $n \times n$ board is complete in exponential time, *Trans. IEICE* **J70-D**, 1843–1852 (in Japanese).

5. W. Ahrens [1910], *Mathematische Unterhaltungen und Spiele*, Zweite vermehrte und verbesserte Auflage, Vol. I, Teubner, Leipzig.

6. M. Aigner and M. Fromme [1984], A game of cops and robbers, *Discrete Appl. Math.* **8**, 1–12.

7. M. Ajtai, L. Csirmaz and Zs. Nagy [1979], On a generalization of the game Go-Moku I, *Studia Scientiarum Math. Hungar.* **14**, 209–226.

8. E. Akin and M. Davis [1985], Bulgarian solitaire, *Amer. Math. Monthly* **92**, 237–250.

9. R. E. Allardice and A. Y. Fraser [1884], La tour d'Hanoï, *Proc. Edinburgh Math. Soc.* **2**, 50–53.

10. D. T. Allemang [1984], Machine computation with finite games, M. Sc. thesis, Cambridge University.

11. J. D. Allen [1989], A note on the computer solution of Connect-Four, *Heuristic Programming in Artificial Intelligence*, 1: *The First Computer Olympiad* (D. N. L. Levy and D. F. Beal, eds.), pp. 134–135, Ellis Horwood, Chichester, England.

12. L. V. Allis and P. N. A. Schoo [1992], Qubic solved again, *Heuristic Programming in Artificial Intelligence*, 3: *The Third Computer Olympiad* (H. J. van den Herik and L. V. Allis, eds.), pp. 192–204, Ellis Horwood, Chichester, England.

13. L. V. Allis, H. J. van den Herik and M. P. H. Huntjens [1993], Go-Moku solved by new search techniques, Proc. 1993 AAAI Fall Symp. on Games: Planning and Learning, AAAI Press Tech. Report FS93-02, pp. 1–9, Menlo Park, CA.

14. J.-P. Allouche, D. Astoorian, J. Randall and J. Shallit [1994], Morphisms, squarefree strings, and the tower of Hanoi puzzle, *Amer. Math. Monthly* **101**, 651-658.

15. S. Alpern and A. Beck [1991], Hex games and twist maps on the annulus, *Amer. Math. Monthly* **98**, 803–811.

16. I. Althöfer [1988], Nim games with arbitrary periodic moving orders, *Internat. J. Game Theory* **17**, 165–175.

17. I. Althöfer [1988], On the complexity of searching game trees and other recursion trees. *J. Algorithms* **9**, 538–567.

18. I. Althöfer [1989], Generalized minimax algorithms are no better error correctors than minimax is itself, in: *Advances in Computer Chess* **5** (D. F. Beal, ed.), Elsevier, Amsterdam, pp. 265–282.

19. I. Althöfer and J. Bültermann [1995], Superlinear period lengths in some subtraction games, *Theor. Comput. Sci. (Math Games)* **148**, 111–119.

20. M. Anderson and F. Harary [1987], Achievement and avoidance games for generating abelian groups, *Internat. J. Game Theory* **16**, 321–325.

21. R. Anderson, L. Lovász, P. Shor, J. Spencer, É. Tardós and S. Winograd [1989], Disks, balls and walls: analysis of a combinatorial game, *Amer. Math. Monthly* **96**, 481–493.

22. T. Andreae [1984], Note on a pursuit game played on graphs, *Discrete Appl. Math.* **9**, 111–115.

23. T. Andreae [1986], On a pursuit game played on graphs for which a minor is excluded, *J. Combin. Theory* (Ser. B) **41**, 37–47.

24. R. P. Anstee and M. Farber [1988], On bridged graphs and cop-win graphs, *J. Combin. Theory* (Ser. B) **44**, 22–28.

25. M. Ascher [1987], Mu Torere: An analysis of a Maori game, *Math. Mag.* **60**, 90–100.

26. I. M. Asel'derova [1974], On a certain discrete pursuit game on graphs, *Cybernetics* **10**, 859–864 (trans. of *Kibernetika* **10** (1974) 102–105).

27. J. A. Aslam and A. Dhagat [1993], On-line algorithms for 2-coloring hypergraphs via chip games, *Theoret. Comput. Sci. (Math Games)* **112**, 355–369.

28. M. D. Atkinson [1981], The cyclic towers of Hanoi, *Inform. Process. Lett.* **13**, 118–119.

29. J. M. Auger [1991], An infiltration game on k arcs, *Naval Research Logistics* **38**, 511–529.

30. R. Austin [1976], Impartial and partisan games, M. Sc. Thesis, Univ. of Calgary.

31. J. O. A. Ayeni and H. O. D. Longe [1985], Game people play: Ayo, *Internat. J. Game Theory* **14**, 207–218.

32. L. Babai and S. Moran [1988], Arthur–Merlin games: a randomized proof system, and a hierarchy of complexity classes, *J. Comput. System Sci.* **36**, 254–276.

33. C. K. Bailey and M. E. Kidwell [1985], A king's tour of the chessboard, *Math. Mag.* **58**, 285–286.

34. W. W. R. Ball and H. S. M. Coxeter [1987], *Mathematical Recreations and Essays*, 13th ed., Dover, New York, NY.

35. R. B. Banerji [1971], Similarities in games and their use in strategy construction, Proc. Symp. Computers and Automata (J. Fox, ed.), Polytechnic Press, Brooklyn, NY, pp. 337–357.

36. R. B. Banerji [1980], *Artificial Intelligence, A Theoretical Approach*, Elsevier, North-Holland, New York, NY.

37. R. B. Banerji and C. A. Dunning [1992], On misere games, *Cybernetics and Systems* **23**, 221–228.

38. R. B. Banerji and G. W. Ernst [1972], Strategy construction using homomorphisms between games, *Artificial Intelligence* **3**, 223–249.

39. R. Bar Yehuda, T. Etzion and S. Moran [1993], Rotating-table games and derivatives of words, *Theoret. Comput. Sci. (Math Games)* **108**, 311–329.

40. I. Bárány [1979], On a class of balancing games, *J. Combin. Theory* (Ser. A) **26**, 115–126.

41. J. G. Baron [1974], The game of nim — a heuristic approach, *Math. Mag.* **74**, 23–28.

42. R. Barua and S. Ramakrishnan [1996], σ-game, σ^+-game and two-dimensional additive cellular automata, *Theoret. Comput. Sci. (Math Games)* **154**, 349–366.

43. V. J. D. Baston and F. A. Bostock [1985], A game locating a needle in a cirular haystack, *J. Optimization Theory and Applications* **47**, 383–391.

44. V. J. D. Baston and F. A. Bostock [1986], A game locating a needle in a square haystack, *J. Optimization Theory and Applications* **51**, 405–419.

45. V. J. D. Baston and F. A. Bostock [1987], Discrete hamstrung squad car games, *Internat. J. Game Theory* **16**, 253–261.

46. V. J. D. Baston and F. A. Bostock [1989], A one-dimensional helicopter-submarine game, *Naval Research Logistics* **36**, 479–490.

47. J. Baumgartner, F. Galvin, R. Laver and R. McKenzie [1975], Game theoretic versions of partition relations, Colloquia Mathematica Societatis János Bolyai **10**, Proc. Internat. Colloq. on Infinite and Finite Sets, Vol. 1 (A. Hajnal, R. Rado and V. T. Sós, eds.), Keszthely, Hungary, 1973, North-Holland, pp. 131–135.

48. J. D. Beasley [1985], *The Ins & Outs of Peg Solitaire*, Oxford University Press, Oxford.

49. J. D. Beasley [1990], *The Mathematics of Games*, Oxford University Press, Oxford.

50. A. Beck [1969], Games, Chapter 5 in: A. Beck, M. N. Bleicher and D. W. Crowe, *Excursions into Mathematics*, Worth Publ., pp. 317–387.

51. J. Beck [1981], On positional games, *J. Combin. Theory* (Ser. A) **30**, 117–133.

52. J. Beck [1981], Van der Waerden and Ramsey type games, *Combinatorica* **1**, 103–116.

53. J. Beck [1982], On a generalization of Kaplansky's game, *Discrete Math.* **42**, 27–35.

54. J. Beck [1982], Remarks on positional games, I, *Acta Math. Acad. Sci. Hungar.* **40**(1–2), 65–71.

55. J. Beck [1985], Random graphs and positional games on the complete graph, *Ann. Discrete Math.* **28**, 7–13.

56. J. Beck and L. Csirmaz [1982], Variations on a game, *J. Combin. Theory* (Ser. A) **33**, 297–315.

57. R. C. Bell [1960, 1969], *Board and Table Games* (two volumes), Oxford University Press.

58. S. J. Benkoski, M. G. Monticino and J. R. Weisinger [1991], A survey of the search theory literature, *Naval Res. Logistics* **38**, 469–494.

59. G. Bennett [1994], Double dipping: the case of the missing binomial coefficient identities, *Theoret. Comput. Sci.* (*Math Games*) **123**, 351–375.

60. D. Berengut [1981], A random hopscotch problem or how to make Johnny read more, in: *The Mathematical Gardner* (D. A. Klarner, ed.), Wadsworth Internat., Belmont, CA, pp. 51–59.

61. B. Berezovskiy and A. Gnedin [1984], The best choice problem, *Akad. Nauk, USSR*, Moscow (in Russian).

62. C. Berge [1976], Sur lex jeux positionnels, *Cahiers du Centre Études Rech. Opér.* **18**, 91–107.

63. C. Berge [1977], Vers une théorie générale des jeux positionnels, Mathematical Economics and Game Theory, *Essays in honor of Oskar Morgentern*, (R. Henn and O. Moeschlin, eds.), *Lecture Notes in Economics* **141**, 13–24, Springer Verlag, Berlin.

64. C. Berge [1981], Some remarks about a Hex problem, in: *The Mathematical Gardner* (D. A. Klarner, ed.), Wadsworth Internat., Belmont, CA, pp. 25–27.

65. C. Berge [1985], *Graphs* (Chapter 14), North-Holland, Amsterdam.

66. C. Berge [1989], *Hypergraphs* (Chapter 4), Elsevier (French: Gauthier Villars 1988).

67. C. Berge and P. Duchet [1988], Perfect graphs and kernels, *Bull. Inst. Math. Acad. Sinica* **16**, 263–274.

68. C. Berge and P. Duchet [1990], Recent problems and results about kernels in directed graphs, *Discrete Math.* **86**, 27–31.

69. C. Berge and M. Las Vergnas [1976], Un nouveau jeu positionnel, le "Match-It", ou une construction dialectique des couplages parfaits, *Cahiers du Centre Études Rech. Opér.* **18**, 83–89.

70. E. R. Berlekamp [1972], Some recent results on the combinatorial game called Welter's Nim, Proc. 6th Ann. Princeton Conf. Information Science and Systems, pp. 203–204.

71. E. R. Berlekamp [1974], The Hackenbush number system for compresssion of numerical data, *Inform. and Control* **26**, 134–140.

72. E. R. Berlekamp [1988], Blockbusting and domineering, *J. Combin. Theory* (Ser. A) **49**, 67–116. An earlier version, entitled Introduction to blockbusting and domineering, appeared in: The Lighter Side of Mathematics, Proc. E. Strens Memorial Conf. on Recr. Math. and its History (R. K. Guy and R. E. Woodrow, eds.), Calgary, 1986, Spectrum Series, Math. Assoc. of America, Washington, DC, 1994, pp. 137–148.

73. E. R. Berlekamp [1991], Introductory overview of mathematical Go endgames, in: Combinatorial Games, *Proc. Symp. Appl. Math.* **43** (R. K. Guy, ed.), Amer. Math. Soc., Providence, RI, pp. 73–100.

74. E. R. Berlekamp [1996], The economist's view of combinatorial games, in: *Games of No Chance*, Proc. MSRI Workshop on Combinatorial Games, July, 1994, Berkeley, CA (R. J. Nowakowski, ed.), MSRI Publ. Vol. 29, Cambridge University Press, Cambridge, pp. 365–405.

75. E. R. Berlekamp, J. H. Conway and R. K. Guy [1982], *Winning Ways* (two volumes), Academic Press, London.

76. E. R. Berlekamp and Y. Kim [1996], Where is the "Thousand-Dollar Ko?", in: *Games of No Chance*, Proc. MSRI Workshop on Combinatorial Games,

July, 1994, Berkeley, CA (R. J. Nowakowski, ed.), MSRI Publ. Vol. 29, Cambridge University Press, Cambridge, pp. 203–226.

77. E. Berlekamp and D. Wolfe [1994], *Mathematical Go — Chilling Gets the Last Point*, A. K. Peters, Wellesley, MA 02181.

78. P. Berloquin [1976], *100 Jeux de Table*, Flammarion, Paris.

79. J. Bitar and E. Goles [1992], Parallel chip firing games on graphs, *Theoret. Comput. Sci.* **92**, 291–300.

80. A. Björner and L. Lovász [1992], Chip-firing games on directed graphs, *J. Algebraic Combin.* **1**, 305–328.

81. A. Björner, L. Lovász and P. Chor [1991], Chip-firing games on graphs, *European J. Combin.* **12**, 283–291.

82. D. Blackwell and M. A. Girshick [1954], *Theory of Games and Statistical Decisions*, Wiley, New York.

83. U. Blass and A. S. Fraenkel [1990], The Sprague–Grundy function of Wythoff's game, *Theoret. Comput. Sci. (Math Games)* **75**, 311–333.

84. M. Blidia [1986], A parity digraph has a kernel, *Combinatorica* **6**, 23–27.

85. M. Blidia, P. Duchet and F. Maffray [1993], On kernels in perfect graphs, *Combinatorica* **13**, 231–233.

86. H. L. Bodlaender [1991], On the complexity of some coloring games, *Intern. J. Foundations Computer Science* **2**, 133–147.

87. H. L. Bodlaender [1993], Complexity of path forming games, *Theoret. Comput. Sci. (Math Games)* **110**, 215–245.

88. H. L. Bodlaender and D. Kratsch [1992], The complexity of coloring games on perfect graphs, *Theoret. Comput. Sci. (Math Games)* **106**, 309–326.

89. K. D. Boklan [1984], The n-number game, *Fibonacci Quart.* **22**, 152–155.

90. E. Borel [1921], La théorie du jeu et les équations integrales à noyau symmetrique gauche, *C. R. Acad. Sci. Paris* **173**, 1304–1308.

91. C. L. Bouton [1902], Nim, a game with a complete mathematical theory, *Ann. of Math.* **3**(2), 35–39.

92. J. Boyce [1981], A Kriegspiel endgame, in: *The Mathematical Gardner* (D. A. Klarner, ed.), Wadsworth Internat., Belmont, CA, pp. 28–36.

93. A. Brousseau [1976], Tower of Hanoi with more pegs, *J. Recr. Math.* **8**, 169–178.

94. R. A. Brualdi and V. S. Pless [1993], Greedy codes, *J. Combin. Theory (Ser. A)* **64**, 10–30.

95. A. A. Bruen and R. Dixon [1975], The n-queen problem, *Discrete Math.* **12**, 393–395.

96. P. Buneman and L. Levy [1980], The towers of Hanoi problem, *Inform. Process. Lett.* **10**, 243–244.

97. D. W. Bushaw [1967], On the name and history of Nim, *Washington Math.* **11**, Oct. 1966. Reprinted in: *N. Y. State Math. Teachers J.* **17**, 52–55.

98. J-y. Cai, A. Condon and R. J. Lipton [1992], On games of incomplete information, *Theoret. Comput. Sci.* **103**, 25–38.

99. D. Calistrate [1996], The reduced canonical form of a game, in: *Games of No Chance*, Proc. MSRI Workshop on Combinatorial Games, July, 1994, Berkeley, CA (R. J. Nowakowski, ed.), MSRI Publ. Vol. 29, Cambridge University Press, Cambridge, pp. 409–416.

100. C. Cannings and J. Haigh [1992], Montreal solitaire, *J. Combin. Theory* (Ser. A) **60**, 50–66.

101. A. K. Chandra, D. C. Kozen and L. J. Stockmeyer [1981], Alternation, *J. Assoc. Comput. Mach.* **28**, 114–133.

102. A. K. Chandra and L. J. Stockmeyer [1976], Alternation, Proc. 17th Ann. Symp. Foundations of Computer Science (Houston, TX, Oct. 1976), IEEE Computer Soc., Long Beach, CA, pp. 98–108.

103. S. M. Chase [1972], An implemented graph algorithm for winning Shannon switching games, *Commun. Assoc. Comput. Mach.* **15**, 253–256.

104. B. S. Chlebus [1986], Domino-tiling games, *J. Comput. System Sci.* **32**, 374–392.

105. F. R. K. Chung [1989], Pebbling in hypercubes, *SIAM J. Disc. Math.* **2**, 467–472.

106. F. R. K. Chung, J. E. Cohen and R. L. Graham [1988], Pursuit-evasion games on graphs, *J. Graph Theory* **12**, 159–167.

107. F. Chung, R. Graham, J. Morrison and A. Odlyzko [1995], Pebbling a chessboard, *Amer. Math. Monthly* **102**, 113–123.

108. V. Chvátal [1973], On the computational complexity of finding a kernel, Report No. CRM-300, Centre de Recherches Mathématiques, Université de Montréal.

109. V. Chvátal [1981], Cheap, middling or dear, in: *The Mathematical Gardner* (D. A. Klarner, ed.), Wadsworth Internat., Belmont, CA, pp. 44–50.

110. V. Chvátal [1983], Mastermind, *Combinatorica* **3**, 325–329.

111. V. Chvátal and P. Erdős [1978], Biased positional games, *Ann. Discrete Math.* **2**: Algorithmic Aspects of Combinatorics, (B. Alspach, P. Hell and D. J. Miller, eds.), Qualicum Beach, B. C., Canada, 1976, North-Holland, pp. 221–229.

112. D. S. Clark [1986], Fibonacci numbers as expected values in a game of chance, *Fibonacci Quart.* **24**, 263–267.

113. A. J. Cole and A. J. T. Davie [1969], A game based on the Euclidean algorithm and a winning strategy for it, *Math. Gaz.* **53**, 354–357.

114. D. B. Coleman [1978], Stretch: a geoboard game, *Math. Mag.* **51**, 49–54.

115. A. Condon [1989], *Computational Models of Games*, ACM Distinguished Dissertation, MIT Press, Cambridge, MA.

116. A. Condon [1992], The complexity of Stochastic games, *Information and Computation* **96**, 203–224.

117. A. Condon, J. Feigenbaum, C. Lund and P. Shor [1993], Probabilistically checkable debate systems and approximation algorithms for PSPACE-hard functions, Proc. 25th Ann. ACM Symp. Theory of Computing, Assoc. Comput. Mach., New York, NY, pp. 305–314.

118. A. Condon and R. E. Ladner [1988], Probabilistic game automata, *J. Comput. System Sci.* **36**, 452–489.

119. I. G. Connell [1959], A generalization of Wythoff's game, *Canad. Math. Bull.* **2**, 181–190.

120. J. H. Conway [1972], All numbers great and small, Univ. of Calgary Math. Dept. Res. Paper No. 149.

121. J. H. Conway [1976], *On Numbers and Games*, Academic Press, London.

122. J. H. Conway [1977], All games bright and beautiful, *Amer. Math. Monthly* **84**, 417–434.

123. J. H. Conway [1978], A gamut of game theories, *Math. Mag.* **51**, 5–12.

124. J. H. Conway [1978], Loopy Games, *Ann. Discrete Math.* **3**: Proc. Symp. Advances in Graph Theory, Cambridge Combinatorial Conf. (B. Bollobás, ed.), Cambridge, May 1977, pp. 55–74.

125. J. H. Conway [1990], Integral lexicographic codes, *Discrete Math.*, **83**, 219–235.

126. J. H. Conway [1991], Numbers and games, in: Combinatorial Games, *Proc. Symp. Appl. Math.* **43** (R. K. Guy, ed.), Amer. Math. Soc., Providence, RI, pp. 23–34.

127. J. H. Conway [1991], More ways of combining games, in: Combinatorial Games, *Proc. Symp. Appl. Math.* **43** (R. K. Guy, ed.), Amer. Math. Soc., Providence, RI, pp. 57–71.

128. J. H. Conway [1996], The angel problem, in: *Games of No Chance*, Proc. MSRI Workshop on Combinatorial Games, July, 1994, Berkeley, CA (R. J. Nowakowski, ed.), MSRI Publ. Vol. 29, Cambridge University Press, Cambridge, pp. 3–12.

129. J. H. Conway and H. S. M. Coxeter [1973], Triangulated polygons and frieze patterns, *Math. Gaz.* **57**, 87–94; 175–183.

130. J. H. Conway and N. J. A. Sloane [1986], Lexicographic codes: error-correcting codes from game theory, *IEEE Trans. Inform. Theory* **IT-32**, 337–348.

131. M. Copper [1993], Graph theory and the game of sprouts, *Amer. Math. Monthly* **100**, 478–482.

132. H. S. M. Coxeter [1953], The golden section, phyllotaxis and Wythoff's game, *Scripta Math.* **19**, 135–143.

133. J. W. Creely [1987], The length of a two-number game, *Fibonacci Quart.* **25**, 174–179.

134. J. W. Creely [1988], The length of a three-number game, *Fibonacci Quart.* **26**, 141–143.

135. H. T. Croft [1964], 'Lion and man': a postscript, *J. London Math. Soc.* **39**, 385–390.

136. D. W. Crowe [1956], The n-dimensional cube and the tower of Hanoi, *Amer. Math. Monthly* **63**, 29–30.

137. L. Csirmaz [1980], On a combinatorial game with an application to Go-Moku, *Discrete Math.* **29**, 19–23.

138. L. Csirmaz and Zs. Nagy [1979], On a generalization of the game Go-Moku, II, *Studia Scientiarum Math. Hung.* **14**, 461–469.

139. P. Cull and E. F. Ecklund, Jr. [1982], On the towers of Hanoi and generalized towers of Hanoi problems, *Congr. Numer.* **35**, 229–238.

140. P. Cull and E. F. Ecklund, Jr. [1985], Towers of Hanoi and analysis of algorithms, *Amer. Math. Monthly* **92**, 407–420.

141. P. Cull and C. Gerety [1985], Is towers of Hanoi really hard?, *Congr. Numer.* **47**, 237–242.

142. J. Czyzowicz, D. Mundici and A. Pelc [1988], Solution of Ulam's problem on binary search with two lies, *J. Combin. Theory* (Ser. A) **49**, 384–388.

143. J. Czyzowicz, D. Mundici and A. Pelc [1989], Ulam's searching game with lies, *J. Combin. Theory* (Ser. A) **52**, 62–76.

144. G. Danaraj and V. Klee [1977], The connectedness game and the c-complexity of certain graphs, *SIAM J. Appl. Math.* **32**, 431–442.

145. M. Davis [1963], Infinite games of perfect information, *Ann. of Math. Stud.* **52**, 85–101, Princeton.

146. T. R. Dawson [1934], *Fairy Chess Review* (problem 1603, p. 94, Dec.).

147. T. R. Dawson [1935], Caissa's Wild Roses, Reprinted in: *Five Classics of Fairy Chess*, Dover, 1973.

148. N. G. de Bruijn [1972], A solitaire game and its relation to a finite field, *J. Recr. Math.* **5**, 133–137.

149. N. G. de Bruijn [1981], Pretzel Solitaire as a pastime for the lonely mathematician, in: *The Mathematical Gardner* (D. A. Klarner, ed.), Wadsworth Internat., Belmont, CA, pp. 16–24.

150. F. de Carteblanche [1970], The princess and the roses, *J. Recr. Math.* **3**, 238–239.

151. F. deCarte Blanche [1974], The roses and the princes, *J. Recr. Math.* **7**, 295–298.

152. A. P. DeLoach [1971], Some investigations into the game of SIM, *J. Recr. Math.* **4**, 36–41.

153. H. de Parville [1884], La tour d'Hanoï et la question du Tonkin, *La Nature* **12**, 285–286.

154. B. Descartes [1953], Why are series musical? *Eureka* **16**, 18–20. Reprinted *ibid.* **27** (1964) 29–31.

155. A. K. Dewdney, Computer Recreations, a column in Scientific American (since May, 1984).

156. C. G. Diderich [1995], Bibliography on minimax game theory, sequential and purallel algorithms, http://diwww.epfl.ch/~diderich/bibliographies.html

157. R. Diestel and I. Leader [1994], Domination games on infinite graphs, *Theoret. Comput. Sci.* (*Math Games*) **132**, 337–345.

158. A. P. Domoryad [1964], *Mathematical Games and Pastimes* (translated by H. Moss), Pergamon Press, Oxford.

159. P. Duchet [1980], Graphes noyau-parfaits, in: *Ann. Discrete Math.* **9**, 93–101.

160. P. Duchet [1987], A sufficient condition for a digraph to be kernel-perfect, *J. Graph Theory* **11**, 81–85.

161. P. Duchet [1987], Parity graphs are kernel-M-solvable, *J. Combin. Theory* (Ser. B) **43**, 121–126.

162. P. Duchet and H. Meyniel [1981], A note on kernel-critical graphs, *Discrete Math.* **33**, 103–105.

163. P. Duchet and H. Meyniel [1983], Une généralisation du théorème de Richardson sur l'existence de noyaux dans le graphes orientés, *Discrete Math.* **43**, 21–27.

164. P. Duchet and H. Meyniel [1993], Kernels in directed graphs: a poison game, *Discrete Math.* **115**, 273–276.

165. H. E. Dudeney [1958], *The Canterbury Puzzles*, 4th ed., Mineola, NY.

166. H. E. Dudeney [1989], *Amusements in Mathematics*, reprinted by Dover, Mineola, NY.

167. N. Duvdevani and A. S. Fraenkel [1989], Properties of *k*-Welter's game, *Discrete Math.* **76**, 197–221.

168. J. Edmonds [1965], Lehman's switching game and a theorem of Tutte and Nash–Williams, *J. Res. Nat. Bur. Standards* **69B**, 73–77.

169. A. Ehrenfeucht and J. Mycielski [1979], Positional strategies for mean payoff games, *Internat. J. Game Theory* **8**, 109–113.

170. N. D. Elkies [1996], On numbers and endgames: combinatorial game theory in chess endgames, in: *Games of No Chance*, Proc. MSRI Workshop on Combinatorial Games, July, 1994, Berkeley, CA (R. J. Nowakowski, ed.), MSRI Publ. Vol. 29, Cambridge University Press, Cambridge, pp. 135–150.

171. R. J. Epp and T. S. Ferguson [1980], A note on take-away games, *Fibonacci Quart.* **18**, 300–303.

172. R. A. Epstein [1977], *Theory of Gambling and Statistical Logic*, Academic Press, New York, NY.

173. M. C. Er [1982], A representation approach to the tower of Hanoi problem, *Comput. J.* **25**, 442–447.

174. M. C. Er [1983], An analysis of the generalized towers of Hanoi problem, *BIT* **23**, 429–435.

175. M. C. Er [1983], An iterative solution to the generalized towers of Hanoi problem, *BIT* **23**, 295–302.

176. M. C. Er [1984], A generalization of the cyclic towers of Hanoi, *Intern. J. Comput. Math.* **15**, 129–140.

177. M. C. Er [1984], The colour towers of Hanoi: a generalization, *Comput. J.* **27**, 80–82.

178. M. C. Er [1984], The cyclic towers of Hanoi: a representation approach, *Comput. J.* **27**, 171–175.

179. M. C. Er [1984], The generalized colour towers of Hanoi: an iterative algorithm, *Comput. J.* **27**, 278–282.

180. M. C. Er [1984], The generalized towers of Hanoi problem, *J. Inform. Optim. Sci.* **5**, 89–94.

181. M. C. Er [1985], The complexity of the generalized cyclic towers of Hanoi problem, *J. Algorithms* **6**, 351–358.

182. M. C. Er [1987], A general algorithm for finding a shortest path between two n-configurations, *Information Sciences* **42**, 137–141.

183. C. Erbas, S. Sarkeshik and M. M. Tanik [1992], Different perspectives of the N-queens problem, in: Proc. ACM Computer Science Conf., Kansas City, MO.

184. C. Erbas and M. M. Tanik [1994], Parallel memory allocation and data alignment in SIMD machines, *Parallel Algorithms and Applications* **4**, 139–151. Preliminary version appeared under the title: Storage schemes for parallel memory systems and the N-queens problem, in: Proc. 15th Ann. Energy Tech. Conf., Houston, TX, Amer. Soc. Mech. Eng., Vol. 43, 1992, pp. 115–120.

185. C. Erbas, M. M. Tanik and Z. Aliyazicioglu [1992], Linear conguence equations for the solutions of the N-queens problem, *Inform. Process. Lett.* **41**, 301–306.

186. P. Erdős and J. L. Selfridge [1973], On a combinatorial game, *J. Combin. Theory* (Ser. A) **14**, 298–301.

187. J. Erickson [1996], Sowing games, in: *Games of No Chance*, Proc. MSRI Workshop on Combinatorial Games, July, 1994, Berkeley, CA (R. J. Nowakowski, ed.), MSRI Publ. Vol. 29, Cambridge University Press, Cambridge, pp. 287–297.

188. J. Erickson [1996], New toads and frogs results, in: *Games of No Chance*, Proc. MSRI Workshop on Combinatorial Games, July, 1994, Berkeley, CA

(R. J. Nowakowski, ed.), MSRI Publ. Vol. 29, Cambridge University Press, Cambridge, pp. 299–310.

189. M. Erickson and F. Harary [1983], Picasso animal achievement games, *Bull. Malaysian Math. Soc.* **6**, 37–44.

190. H. Eriksson [1995], Pebblings, *Electronic J. Combin.* **2**, R7, 18pp., found at http://ejc.math.gatech.edu:8080/Journal/journalhome.html

191. H. Eriksson and B. Lindström [1995], Twin jumping checkers in $\mathbb{Z}^d$, *European J. Combin.* **16**, 153–157.

192. K. Eriksson [1991], No polynomial bound for the chip firing game on directed graphs, *Proc. Amer. Math. Soc.* **112**, 1203–1205.

193. K. Eriksson [1992], Convergence of Mozes' game of numbers, *Linear Algebra Appl.* **166**, 151–165.

194. K. Eriksson [1994], Node firing games on graphs, *Contemp. Math.* **178**, 117–127.

195. K. Eriksson [1994], Reachability is decidable in the numbers game, *Theoret. Comput. Sci. (Math Games)* **131**, 431–439.

196. K. Eriksson [1995], The numbers game and Coxeter groups, *Discrete Math.* **139**, 155–166.

197. R. J. Evans [1974], A winning opening in reverse Hex, *J. Recr. Math.* **7**, 189–192.

198. R. J. Evans [1975–76], Some variants of Hex, *J. Recr. Math.* **8**, 120–122.

199. R. J. Evans [1979], Silverman's game on intervals, *Amer. Math. Monthly* **86**, 277–281.

200. S. Even and R. E. Tarjan [1976], A combinatorial problem which is complete in polynomial space, *J. Assoc. Comput. Mach.* **23**, 710–719. Also appeared in Proc. 7th Ann. ACM Symp. Theory of Computing (Albuquerque, NM, 1975), Assoc. Comput. Mach., New York, NY, 1975, pp. 66–71.

201. E. Falkener [1961], *Games Ancient and Modern*, Dover, New York, NY. (Published previously by Longmans Green, 1897.)

202. B.-J. Falkowski and L. Schmitz [1986], A note on the queens' problem, *Inform. Process. Lett.* **23**, 39–46.

203. U. Feigle and W. Kern [1993], On the game chromatic number of some classes of graphs, *Ars Combin.* **35**, 143–150.

204. T. S. Ferguson [1974], On sums of graph games with last player losing, *Internat. J. Game Theory* **3**, 159–167.

205. T. S. Ferguson [1984], Misère annihilation games, *J. Combin. Theory* (Ser. A) **37**, 205–230.

206. T. S. Ferguson [1989], Who solved the secretary problem? *Statistical Science* **4**, 282–296.

207. T. S. Ferguson [1992], Mate with bishop and knight in kriegspiel, *Theoret. Comput. Sci. (Math Games)* **96**, 389–403.

208. A. S. Finbow and B. L. Hartnell [1983], A game related to covering by stars, *Ars Combinatoria* **16-A**, 189–198.

209. P. C. Fishburn and N. J. A. Sloane [1989], The solution to Berlekamp's switching game, *Discrete Math.* **74**, 263–290.

210. D. C. Fisher and J. Ryan [1992], Optimal strategies for a generalized "scissors, paper, and stone" game, *Amer. Math. Monthly* **99**, 935–942.

211. J. A. Flanigan [1978], Generalized two-pile Fibonacci nim, *Fibonacci Quart.* **16**, 459–469.

212. J. A. Flanigan [1981], On the distribution of winning moves in random game trees, *Bull. Austr. Math. Soc.* **24**, 227–237.

213. J. A. Flanigan [1981], Selective sums of loopy partizan graph games, *Internat. J. Game Theory* **10**, 1–10.

214. J. A. Flanigan [1982], A complete analysis of black-white Hackendot, *Internat. J. Game Theory* **11**, 21–25.

215. J. A. Flanigan [1982], One-pile time and size dependent take-away games, *Fibonacci Quart.* **20**, 51–59.

216. J. A. Flanigan [1983], Slow joins of loopy games, *J. Combin. Theory* (Ser. A) **34**, 46–59.

217. J. O. Flynn [1973], Lion and man: the boundary constraint, *SIAM J. Control* **11**, 397–411.

218. J. O. Flynn [1974], Some results on max-min pursuit, *SIAM J. Control* **12**, 53–69.

219. J. O. Flynn [1974], Lion and man: the general case, *SIAM J. Control* **12**, 581–597.

220. L. R. Foulds and D. G. Johnson [1984], An application of graph theory and integer programming: chessboard non-attacking puzzles, *Math. Mag.* **57**, 95–104.

221. A. S. Fraenkel [1974], Combinatorial games with an annihiliation rule, in: The Influence of Computing on Mathematical Research and Education, *Proc. Symp. Appl. Math.* (J. P. LaSalle, ed.), Vol. 20, Amer. Math. Soc., Providence, RI, pp. 87–91.

222. A. S. Fraenkel [1977], The particles and antiparticles game, *Comput. Math. Appl.* **3**, 327–328.

223. A. S. Fraenkel [1980], From Nim to Go, *Ann. Discrete Math.* **6**: Proc. Symp. on Combinatorial Mathematics, Combinatorial Designs and Their Applications (J. Srivastava, ed.), Colorado State Univ., Fort Collins, CO, June 1978, pp. 137–156.

224. A. S. Fraenkel [1981], Planar kernel and Grundy with $d \leq 3$, $d_{out} \leq 2$, $d_{in} \leq 2$ are NP-complete, *Discrete Appl. Math.* **3**, 257–262.

225. A. S. Fraenkel [1982], How to beat your Wythoff games' opponents on three fronts, *Amer. Math. Monthly* **89**, 353–361.

226. A. S. Fraenkel [1983], 15 Research problems on games, in: "Research Problems", *Discrete Math.* **43–46**.

227. A. S. Fraenkel [1984], Wythoff games, continued fractions, cedar trees and Fibonacci searches, *Theoret. Comput. Sci.* **29**, 49–73. An earlier version appeared in Proc. 10th Internat. Colloq. on Automata, Languages and Programming (J. Diaz, ed.), Barcelona, July 1983, *Lecture Notes in Computer Science* **154**, 203–225, Springer Verlag, Berlin, 1983.

228. A. S. Fraenkel [1991], Complexity of games, in: Combinatorial Games, *Proc. Symp. Appl. Math.* **43** (R. K. Guy, ed.), Amer. Math. Soc., Providence, RI, pp. 111–153.

229. A. S. Fraenkel [1994], Even kernels, *Electronic J. Combinatorics* **1**, R5, 13pp. http://ejc.math.gatech.edu:8080/Journal/Volume_1/volume1.html

230. A. S. Fraenkel [1994], Recreation and depth in combinatorial games, in: The Lighter Side of Mathematics, Proc. E. Strens Memorial Conf. on Recr. Math. and its History (R. K. Guy and R. E. Woodrow, eds.), Calgary, 1986, Spectrum Series, Math. Assoc. of America, Washington, D. C., pp. 159–173.

231. A. S. Fraenkel [1996], Error-correcting codes derived from combinatorial games, in: *Games of No Chance*, Proc. MSRI Workshop on Combinatorial Games, July, 1994, Berkeley, CA (R. J. Nowakowski, ed.), MSRI Publ. Vol. 29, Cambridge University Press, Cambridge, pp. 417–431.

232. A. S. Fraenkel [1996], Scenic trails ascending from sea-level Nim to alpine chess, in: *Games of No Chance*, Proc. MSRI Workshop on Combinatorial Games, July, 1994, Berkeley, CA (R. J. Nowakowski, ed.), MSRI Publ. Vol. 29, Cambridge University Press, Cambridge, pp. 13–42.

233. A. S. Fraenkel and I. Borosh [1973], A generalization of Wythoff's game, *J. Combin. Theory* (Ser. A) **15**, 175–191.

234. A. S. Fraenkel, M. R. Garey, D. S. Johnson, T. Schaefer and Y. Yesha [1978], The complexity of checkers on an $n \times n$ board — preliminary report, Proc. 19th Ann. Symp. Foundations of Computer Science (Ann Arbor, MI, Oct. 1978), IEEE Computer Soc., Long Beach, CA, pp. 55–64.

235. A. S. Fraenkel and E. Goldschmidt [1987], Pspace-hardness of some combinatorial games, *J. Combin. Theory* (Ser. A) **46**, 21–38.

236. A. S. Fraenkel and F. Harary [1989], Geodetic contraction games on graphs, *Internat. J. Game Theory* **18**, 327–338.

237. A. S. Fraenkel and H. Herda [1980], Never rush to be first in playing Nimbi, *Math. Mag.* **53**, 21–26.

238. A. S. Fraenkel, A. Jaffray, A. Kotzig and G. Sabidussi [1995], Modular Nim, *Theoret. Comput. Sci.* (*Math Games*) **143**, 319–333.

239. A. S. Fraenkel and A. Kotzig [1987], Partizan octal games: partizan subtraction games, *Internat. J. Game Theory* **16**, 145–154.

240. A. S. Fraenkel and D. Lichtenstein [1981], Computing a perfect strategy for $n \times n$ chess requires time exponential in n, *J. Combin. Theory* (Ser. A) **31**, 199–214. Preliminary version in Proc. 8th Internat. Colloq. Automata, Languages and Programming (S. Even and O. Kariv, eds.), Acre, Israel, 1981, *Lecture Notes in Computer Science* **115**, 278–293, Springer Verlag, Berlin.

241. A. S. Fraenkel, M. Loebl and J. Nešetřil [1988], Epidemiography, II. Games with a dozing yet winning player, *J. Combin. Theory* (Ser. A) **49**, 129–144.

242. A. S. Fraenkel and M. Lorberbom [1989], Epidemiography with various growth functions, *Discrete Appl. Math.* **25**, 53–71.

243. A. S. Fraenkel and M. Lorberbom [1991], Nimhoff games, *J. Combin. Theory* (Ser. A) **58**, 1–25.

244. A. S. Fraenkel and J. Nešetřil [1985], Epidemiography, *Pacific J. Math.* **118**, 369–381.

245. A. S. Fraenkel and Y. Perl [1975], Constructions in combinatorial games with cycles, *Coll. Math. Soc. János Bolyai*, **10**: Proc. Internat. Colloq. on Infinite and Finite Sets, Vol. 2 (A. Hajnal, R. Rado and V. T. Sós, eds.) Keszthely, Hungary, 1973, North-Holland, pp. 667–699.

246. A. S. Fraenkel and E. R. Scheinerman [1991], A deletion game on hypergraphs, *Discrete Appl. Math.* **30**, 155–162.

247. A. S. Fraenkel, E. R. Scheinerman and D. Ullman [1993], Undirected edge geography, *Theoret. Comput. Sci.* (*Math Games*) **112**, 371–381.

248. A. S. Fraenkel and S. Simonson [1993], Geography, *Theoret. Comput. Sci.* (*Math Games*) **110**, 197–214.

249. A. S. Fraenkel and U. Tassa [1975], Strategy for a class of games with dynamic ties, *Comput. Math. Appl.* **1**, 237–254.

250. A. S. Fraenkel and U. Tassa [1982], Strategies for compounds of partizan games, *Math. Proc. Camb. Phil. Soc.* **92**, 193–204.

251. A. S. Fraenkel, U. Tassa and Y. Yesha [1978], Three annihilation games, *Math. Mag.* **51**, 13–17.

252. A. S. Fraenkel and Y. Yesha [1976], Theory of annihilation games, *Bull. Amer. Math. Soc.* **82**, 775–777.

253. A. S. Fraenkel and Y. Yesha [1979], Complexity of problems in games, graphs and algebraic equations, *Discrete Appl. Math.* **1**, 15–30.

254. A. S. Fraenkel and Y. Yesha [1982], Theory of annihilation games — I, *J. Combin. Theory* (Ser. B) **33**, 60–86.

255. A. S. Fraenkel and Y. Yesha [1986], The generalized Sprague–Grundy function and its invariance under certain mappings, *J. Combin. Theory* (Ser. A) **43**, 165–177.

256. C. N. Frangakis [1981], A backtracking algorithm to generate all kernels of a directed graph, *Intern. J. Comput. Math.* **10**, 35–41.

257. P. Frankl [1987], Cops and robbers in graphs with large girth and Cayley graphs, *Discrete Appl. Math.* **17**, 301–305.

258. P. Frankl [1987], On a pursuit game on Cayley graphs, *Combinatorica* **7**, 67–70.

259. D. Fremlin [1973], Well-founded games, *Eureka* **36**, 33–37.

260. G. H. Fricke, S. M. Hedetniemi, S. T. Hedetniemi, A. A. McRae, C. K. Wallis, M. S. Jacobson, H. W. Martin and W. D. Weakley [1995], Combinatorial problems on chessboards: a brief survey, in: Graph Theory, Combinatorics, and Applications: Proc. 7th Quadrennial Internat. Conf. on the Theory and Applications of Graphs, Vol. 1 (Y. Alavi and A. Schwenk, Eds., Wiley), pp. 507–528.

261. Z. Füredi and Á. Seress [1994], Maximal triangle-free graphs with restrictions on the degrees, *J. Graph Theory* **18**, 11–24.

262. D. Gale [1974], A curious Nim-type game, *Amer. Math. Monthly* **81**, 876–879.

263. D. Gale [1979], The game of Hex and the Brouwer fixed-point theorem, *Amer. Math. Monthly* **86**, 818–827.

264. D. Gale [1986], Problem 1237 (line-drawing game), *Math. Mag.* **59**, 111. Solution by J. Hutchinson and S. Wagon, *ibid.* **60** (1987) 116.

265. D. Gale and A. Neyman [1982], Nim-type games, *Internat. J. Game Theory* **11**, 17–20.

266. D. Gale and F. M. Stewart [1953], Infinite games with perfect information, Contributions to the Theory of Games, *Ann. of Math. Stud.* **2**(28), 245–266, Princeton.

267. H. Galeana-Sánchez [1982], A counterexample to a conjecture of Meyniel on kernel-perfect graphs, *Discrete Math.* **41**, 105–107.

268. H. Galeana-Sánchez [1986], A theorem about a conjecture of Meyniel on kernel-perfect graphs, *Discrete Math.* **59**, 35–41.

269. H. Galeana-Sánchez [1995], B_1 and B_2-orientable graphs in kernel theory, *Discrete Math.* **143**, 269–274.

270. H. Galeana-Sánchez and V. Neuman-Lara [1984], On kernels and semikernels of digraphs, *Discrete Math.* **48**, 67–76.

271. H. Galeana-Sánchez and V. Neuman-Lara [1994], New extensions of kernel perfect digraphs to kernel imperfect critical digraphs, *Graphs Combin.* **10**, 329–336.

272. F. Galvin [1978], Indeterminacy of point-open games. *Bull. de l'Academie Polonaise des Sciences* (math., astr. et phys.) **26**, 445–449.

273. A. Gangolli and T. Plambeck [1989], A note on periodicity in some octal games, *Internat. J. Game Theory* **18**, 311–320.

274. T. E. Gantner [1988], The game of Quatrainment, *Math. Mag.* **61**, 29–34.

275. M. Gardner [1956], *Mathematics, Magic and Mystery*, Dover, New York, NY.

276. M. Gardner, Mathematical Games, a column in Scientific American (Jan. 1957–Dec. 1981).

277. M. Gardner [1959], *Mathematical Puzzles of Sam Loyd*, Dover, New York, NY.

278. M. Gardner [1960], *More Mathematical Puzzles of Sam Loyd*, Dover, New York, NY.

279. M. Gardner [1961], *The Second Scientific American Book of Mathematical Puzzles and Diversions*, Simon and Schuster, NY.

280. M. Gardner [1966], *New Mathematical Diversions from Scientific American*, Simon and Schuster, New York, NY.

281. M. Gardner [1967], *536 Puzzles and Curious Problems* (edited; reissue of H. E. Dudeney's *Modern Puzzles* and *Puzzles and Curious Problems*), Scribner's, NY.

282. M. Gardner [1969], *The Unexpected Hanging and Other Mathematical Diversions*, Simon and Schuster, NY.

283. M. Gardner [1975], *Mathematical Carnival*, Knopf, NY.

284. M. Gardner [1977], *Mathematical Magic Show*, Knopf, NY.

285. M. Gardner [1978], *Aha! Insight*, Freeman, New York, NY.

286. M. Gardner [1979], *Mathematical Circus*, Knopf, NY.

287. M. Gardner [1981], *Science Fiction Puzzle Tales*, Potter.

288. M. Gardner [1982], *Aha! Gotcha!*, Freeman, New York, NY.

289. M. Gardner [1984], *Puzzles from Other Worlds*, Random House.

290. M. Gardner [1984], *The Magic Numbers of Dr. Matrix*, Prometheus.

291. M. Gardner [1984], *The Sixth Book of Mathematical Games*, Univ. of Chicago Press (first appeared in 1971 by Freeman).

292. M. Gardner [1984], *Wheels, Life and Other Mathematical Amusements*, Freeman, New York, NY.

293. M. Gardner [1986], *Knotted Doughnuts and Other Mathematical Entertainments*, Freeman, New York, NY.

294. M. Gardner [1987], *Riddles of the Sphinx*, Math. Assoc. of America.

295. M. Gardner [1987], *Time Travel and Other Mathematical Bewilderments*, Freeman, New York, NY.

296. M. Gardner [1988], *Hexaflexagons and Other Mathematical Diversions*, University of Chicago Press, Chicago, 1988. A first version appeared under the title *The Scientific American Book of Mathematical Puzzles and Diversions*, Simon & Schuster, 1959, NY.

297. M. Gardner [1989], *Penrose Tiles to Trapdoor Ciphers*, Freeman, New York, NY.

298. M. Gardner [1991], *Fractal Music, Hypercards and More*, Freeman, New York, NY.

299. M. R. Garey and D. S. Johnson [1979], *Computers and Intractability: A Guide to the theory of NP-Completeness*, Freeman, San Francisco (Appendix, A8).

300. R. Gasser [1996), Solving nine men's Morris, in: *Games of No Chance*, Proc. MSRI Workshop on Combinatorial Games, July, 1994, Berkeley, CA (R. J. Nowakowski, ed.), MSRI Publ. Vol. 29, Cambridge University Press, Cambridge, pp. 101–114.

301. J. R. Gilbert, T. Lengauer and R. E. Tarjan [1980], The pebbling problem is complete in polynomial space, *SIAM J. Comput.* **9**, 513–524. Preliminary version in Proc. 11th Ann. ACM Symp. Theory of Computing (Atlanta, GA, 1979), Assoc. Comput. Mach., New York, NY, pp. 237–248.

302. J. Ginsburg [1939], Gauss's arithmetization of the problem of 8 queens, *Scripta Math.* **5**, 63–66.

303. A. S. Goldstein and E. M. Reingold [1995], The complexity of pursuit on a graph, *Theoret. Comput. Sci. (Math Games)* **143**, 93–112.

304. E. Goles and M. A. Kiwi [1993], Games on line graphs and sand piles, *Theoret. Comput. Sci. (Math Games)* **115**, 321–349.

305. S. W. Golomb [1966], A mathematical investigation of games of "take-away", *J. Combin. Theory* **1**, 443–458.

306. S. W. Golomb [1994], *Polyominoes: Puzzles, Patterns, Problems, and Packings*, Princeton University Press. Original edition: *Polyominoes*, Scribner's, NY, 1965; Allen and Unwin, London, 1965.

307. D. M. Gordon, R. W. Robinson and F. Harary [1994], Minimum degree games for graphs, *Discrete Math.* **128**, 151–163.

308. E. Grädel [1990], Domino games and complexity, *SIAM J. Comput.* **19**, 787–804.

309. S. B. Grantham [1985], Galvin's "racing pawns" game and a well-ordering of trees, *Memoirs Amer. Math. Soc.* **53**(316), 63 pp.

310. P. M. Grundy [1964], Mathematics and Games, *Eureka* **27**, 9–11; originally published: *ibid.* **2** (1939), 6–8.

311. P. M. Grundy, R. S. Scorer and C. A. B. Smith [1944], Some binary games, *Math. Gaz.* **28**, 96–103.

312. P. M. Grundy and C. A. B. Smith [1956], Disjunctive games with the last player losing, *Proc. Camb. Phil. Soc.* **52**, 527–533.

313. F. Grunfeld and R. C. Bell [1975], *Games of the World*, Holt, Rinehart and Winston.

314. C. D. Grupp [1976], *Brettspiele-Denkspiele*, Humboldt-Taschenbuchverlag, München.

315. S. Gunther [1874], Zur mathematischen Theorie des Schachbretts, *Arch. Math. Physik* **56**, 281–292.

316. R. K. Guy [1976], Packing $[1, n]$ with solutions of $ax + by = cz$; the unity of combinatorics, *Atti Conv. Lincei #17, Accad. Naz. Lincei* Rome, Tomo II, 173–179.

317. R. K. Guy [1976], Twenty questions concerning Conway's sylver coinage, *Amer. Math. Monthly* **83**, 634–637.

318. R. K. Guy [1977], Games are graphs, indeed they are trees, *Proc. 2nd Carib. Conf. Combin. and Comput.*, Letchworth Press, Barbados, 6–18.

319. R. K. Guy [1977], She loves me, she loves me not; relatives of two games of Lenstra, *Een Pak met een Korte Broek* (papers presented to H. W. Lenstra), Mathematisch Centrum, Amsterdam.

320. R. K. Guy [1978], Partizan and impartial combinatorial games, *Colloq. Math. Soc. János Bolyai*, **18**: Proc. 5th Hungar. Conf. Combin., Vol. I (A. Hajnal and V. T. Sós, eds.), Keszthely, Hungary, 1976, North-Holland, pp. 437–461.

321. R. K. Guy [1979], Partizan Games, Colloques Internationaux C. N. R. No. 260 — Problèmes Combinatoires et Théorie des Graphs, 199–205.

322. R. K. Guy [1981], Anyone for twopins?, in: *The Mathematical Gardner* (D. A. Klarner, ed.), Wadsworth Internat., Belmont, CA, pp. 2–15.

323. R. K. Guy [1983], Graphs and games, in: *Selected Topics in Graph Theory*, 2 (L. W. Beineke and R. J. Wilson, eds.), Academic Press, London, pp. 269–295.

324. R. K. Guy [1986], John Isbell's game of beanstalk and John Conway's game of beans-don't-talk, *Math. Mag.* **59**, 259–269.

325. R. K. Guy [1989], *Fair Game*, COMAP Math. Exploration Series, Arlington, MA.

326. R. K. Guy [1990], A guessing game of Bill Sands, and Bernardo Recamán's Barranca, *Amer. Math. Monthly* **97**, 314–315.

327. R. K. Guy [1991], Mathematics from fun & fun from mathematics; an informal autobiographical history of combinatorial games, in: *Paul Halmos: Celebrating 50 Years of Mathematics* (J. H. Ewing and F. W. Gehring, eds.), Springer Verlag, New York, pp. 287–295.

328. R. K. Guy [1996], Combinatorial games, in: *Handbook of Combinatorics* (R. L. Graham, M. Grötschel and L. Lovász, eds.), Vol. II, pp. 2117-2162, North-Holland, Amsterdam.

329. R. K. Guy [1996], Impartial Games, in: *Games of No Chance*, Proc. MSRI Workshop on Combinatorial Games, July, 1994, Berkeley, CA (R. J. Nowakowski, ed.), MSRI Publ. Vol. 29, Cambridge University Press, Cambridge, pp. 61–78. Earlier version in: Combinatorial Games, *Proc. Symp. Appl.*

Math. **43** (R. K. Guy, ed.), Amer. Math. Soc., Providence, RI, pp. 35–55, 1991.

330. R. K. Guy [1996], What is a game? in: *Games of No Chance*, Proc. MSRI Workshop on Combinatorial Games, July, 1994, Berkeley, CA (R. J. Nowakowski, ed.), MSRI Publ. Vol. 29, Cambridge University Press, Cambridge, pp. 43–60. Earlier version in: Combinatorial Games, *Proc. Symp. Appl. Math.* **43** (R. K. Guy, ed.), Amer. Math. Soc., Providence, RI, pp. 1–21, 1991.

331. R. K. Guy [1996], Unsolved problems in combinatorial games, in: *Games of No Chance*, Proc. MSRI Workshop on Combinatorial Games, July, 1994, Berkeley, CA (R. J. Nowakowski, ed.), MSRI Publ. Vol. 29, Cambridge University Press, Cambridge, pp. 475–491. Update with 52 problems of earlier version with 37 problems, in: *Proc. Symp. Appl. Math.* **43** (R. K. Guy, ed.), Amer. Math. Soc., Providence, RI, pp. 183–189, 1991.

332. R. K. Guy and C. A. B. Smith [1956], The G-values of various games, *Proc. Camb. Phil. Soc.* **52**, 514–526.

333. R. K. Guy and R. E. Woodrow, eds. [1994], *The Lighter Side of Mathematics*, Proc. E. Strens Memorial Conf. on Recreational Mathematics and its History, Spectrum Series, Math. Assoc. Amer., Washington, DC.

334. W. Guzicki [1990], Ulam's searching game with two lies, *J. Combin. Theory* (Ser. A) **54**, 1–19.

335. D. R. Hale [1983], A variant of Nim and a function defined by Fibonacci representation, *Fibonacci Quart.* **21**, 139–142.

336. A. W. Hales and R. I. Jewett [1963], Regularity and positional games, *Trans. Amer. Math. Soc.* **106**, 222–229.

337. L. Halpenny and C. Smyth [1992], A classification of minimal standard-path 2×2 switching networks, *Theoret. Comput. Sci. (Math Games)* **102**, 329–354.

338. Y. O. Hamidoune [1987], On a pursuit game of Cayley digraphs, *Europ. J. Combin.* **8**, 289–295.

339. Y. O. Hamidoune and M. Las Vergnas [1985], The directed Shannon switching game and the one-way game, in: *Graph Theory and Its Applications to Algorithms and Computer Science* (Y. Alavi et al., eds.), Wiley, pp. 391–400.

340. Y. O. Hamidoune and M. Las Vergnas [1986], Directed switching games on graphs and matroids, *J. Combin. Theory* (Ser. B) **40**, 237–269.

341. Y. O. Hamidoune and M. Las Vergnas [1987], A solution to the box game, *Discrete Math.* **65**, 157–171.

342. Y. O. Hamidoune and M. Las Vergnas [1988], A solution to the misère Shannon switching game, *Discrete Math.* **72**, 163–166.

343. O. Hanner [1959], Mean play of sums of positional games, *Pacific J. Math.* **9**, 81–99.

344. F. Harary [1982], Achievement and avoidance games for graphs, *Ann. Discrete Math.* **13**, 111–120.

345. F. Harary and K. Plochinski [1987], On degree achievement and avoidance games for graphs, *Math. Mag.* **60**, 316–321.

346. P. J. Hayes [1977], A note on the towers of Hanoi problem, *Computer J.* **20**, 282–285.

347. O. Heden [1992], On the modular n-queen problem, *Discrete Math.* **102**, 155–161.

348. O. Heden [1993], Maximal partial spreads and the modular n-queen problem, *Discrete Math.* **120**, 75–91.

349. O. Heden [1995], Maximal partial spreads and the modular n-queen problem II, *Discrete Math.* **142**, 97–106.

350. P. Hein [1942], Polygon, *Politiken* (description of Hex in this Danish newspaper of Dec. 26).

351. D. Hensley [1988], A winning strategy at Taxman, *Fibonacci Quart.* **26**, 262–270.

352. C. W. Henson [1970], Winning strategies for the ideal game, *Amer. Math. Monthly* **77**, 836–840.

353. G. A. Heuer [1982], Odds versus evens in Silverman-type games, *Internat. J. Game Theory* **11**, 183–194.

354. G. A. Heuer [1989], Reduction of Silverman-like games to games on bounded sets, *Internat. J. Game Theory* **18**, 31–36.

355. G. A. Heuer and W. D. Rieder [1988], Silverman games on disjoint discrete sets, *SIAM J. Disc. Math.* **1**, 485–525.

356. R. Hill and J. P. Karim [1992], Searching with lies: the Ulam problem, *Discrete Math.* **106/107**, 273–283.

357. T. P. Hill and U. Krengel [1991], Minimax-optimal stop rules and distributions in secretary problems, *Ann. Probab.* **19**, 342–353.

358. T. P. Hill and U. Krengel [1992], On the game of Googol, *Internat. J. Game Theory* **21**, 151–160.

359. P. G. Hinman [1972], Finite termination games with tie, *Israel J. Math.* **12**, 17–22.

360. A. M. Hinz [1989], An iterative algorithm for the tower of Hanoi with four pegs, *Computing* **42**, 133–140.

361. A. M. Hinz [1989], The tower of Hanoi, *Enseign. Math.* **35**, 289–321.

362. A. M. Hinz [1992], Pascal's triangle and the tower of Hanoi, *Amer. Math. Monthly* **99**, 538–544.

363. A. M. Hinz [1992], Shortest paths between regular states of the tower of Hanoi, *Inform. Sci.* **63**, 173–181.

364. S. Hitotumatu [1968], Some remarks on nim-like games, *Comment. Math. Univ. St. Paul* **17**, 85–98.

365. E. J. Hoffman, J. C. Loessi and R. C. Moore [1969], Construction for the solution of the *n*-queens problem, *Math. Mag.* **42**, 66–72.

366. J. C. Holladay [1957], Cartesian products of termination games, Contributions to the Theory of Games, *Ann. of Math. Stud.* **3**(39), 189–200, Princeton.

367. J. C. Holladay [1958], Matrix nim, *Amer. Math. Monthly* **65**, 107–109.

368. J. C. Holladay [1966], A note on the game of dots, *Amer. Math. Monthly* **73**, 717–720.

369. K. Igusa [1985], Solution of the Bulgarian solitaire conjecture, *Math. Mag.* **58**, 259–271.

370. J. Isbell [1992], The Gordon game of a finite group, *Amer. Math. Monthly* **99**, 567–569.

371. O. Itzinger [1977], The South American game, *Math. Mag.* **50**, 17–21.

372. S. Iwata and T. Kasai [1994], The Othello game on an $n \times n$ board is PSPACE-complete, *Theoret. Comput. Sci.* (*Math Games*) **123**, 329–340.

373. T. A. Jenkyns and J. P. Mayberry [1980], The skeleton of an impartial game and the Nim-function of Moore's Nim_k, *Internat. J. Game Theory* **9**, 51–63.

374. D. S. Johnson [1983], The NP-Completeness Column: An Ongoing Guide, *J. Algorithms* **4**, 397–411 (9th quarterly column (games); column started in 1981).

375. J. P. Jones [1982], Some undecidable determined games, *Internat. J. Game Theory* **11**, 63–70.

376. J. P. Jones and A. S. Fraenkel [1996], Complexities of winning strategies in diophantine games, *J. Complexity* **11**, 435-455.

377. M. Kac [1974], Hugo Steinhaus, a reminiscence and a tribute, *Amer. Math. Monthly* **81**, 572–581 (p. 577).

378. J. Kahane and A. S. Fraenkel [1987], *k*-Welter — a generalization of Welter's game, *J. Combin. Theory* (Ser. A) **46**, 1–20.

379. J. Kahn, J. C. Lagarias and H. S. Witsenhausen [1987], Single-suit two-person card play, *Internat. J. Game Theory* **16**, 291–320.

380. J. Kahn, J. C. Lagarias and H. S. Witsenhausen [1988], Single-suit two-person card play, II. Dominance, *Order* **5**, 45–60.

381. J. Kahn, J. C. Lagarias and H. S. Witsenhausen [1989], Single-suit two-person card play, III. The misère game, *SIAM J. Disc. Math.* **2**, 329–343.

382. J. Kahn, J. C. Lagarias and H. S. Witsenhausen [1989], On Lasker's card game, in: Differential games and applications (T. S. Başar, P. Bernhard, eds.), *Lecture Notes in Control and Information Sciences* **119**, 1–8, Springer Verlag, Berlin.

383. L. Kalmár [1928], Zur Theorie der abstrakten Spiele, *Acta Sci. Math. Univ. Szeged* **4**, 65–85.

384. B. Kalyanasundram [1991], On the power of white pebbles, *Combinatorica* **11**, 157–171.

385. B. Kalyanasundram and G. Schnitger [1988], On the power of white pebbles, Proc. 20th Ann. ACM Symp. Theory of Computing (Chicago, IL, 1988), Assoc. Comput. Mach., New York, NY, pp. 258–266.

386. M. Kano [1983], Cycle games and cycle cut games, *Combinatorica* **3**, 201–206.

387. R. M. Karp and Y. Zhang [1989], On parallel evaluation of game trees, Proc. ACM Symp. Parallel Algorithms and Architectures, 409–420.

388. T. Kasai, A. Adachi and S. Iwata [1979], Classes of pebble games and complete problems, *SIAM J. Comput.* **8**, 574–586.

389. Y. Kawano [1996], Using similar positions to search game trees, in: *Games of No Chance*, Proc. MSRI Workshop on Combinatorial Games, July, 1994, Berkeley, CA (R. J. Nowakowski, ed.), MSRI Publ. Vol. 29, Cambridge University Press, Cambridge, pp. 193–202.

390. J. C. Kenyon [1967], A Nim-like game with period 349, Univ. of Calgary, Math. Dept. Res. Paper No. 13.

391. J. C. Kenyon [1967], Nim-like games and the Sprague–Grundy theory, M. Sc. Thesis, Univ. of Calgary.

392. B. Kerst [1933], *Mathematische Spiele*, Reprinted by Dr. Martin Sändig oHG, Wiesbaden 1968.

393. Y. Kim [1996], New values in domineering, *Theoret. Comput. Sci. (Math Games)* **156**, 263–280.

394. H. Kinder [1973], Gewinnstrategien des Anziehenden in einigen Spielen auf Graphen, *Arch. Math.* **24**, 332–336.

395. M. M. Klawe [1985], A tight bound for black and white pebbles on the pyramids, *J. Assoc. Comput. Mach.* **32**, 218–228.

396. C. S. Klein and S. Minker [1993], The super towers of Hanoi problem: large rings on small rings, *Discrete Math.* **114**, 283–295.

397. D. J. Kleitman and B. L. Rothschild [1972], A generalization of Kaplansky's game, *Discrete Math.* **2**, 173–178.

398. T. Kløve [1977], The modular n-queen problem, *Discrete Math.* **19**, 289–291.

399. T. Kløve [1981], The modular n-queen problem II, *Discrete Math.* **36**, 33–48.

400. D. E. Knuth [1974], *Surreal Numbers*, Addison-Wesley, Reading, MA.

401. D. E. Knuth [1976], The computer as Master Mind, *J. Recr. Math.* **9**, 1–6.

402. A. Kotzig [1946], O k-posunutiach (On k-translations; in Slovakian), *Časop. pro Pěst. Mat. a Fys.* **71**, 55–61. Extended abstract in French, pp. 62–66.

403. G. Kowalewski [1930], *Alte und neue mathematische Spiele*, Reprinted by Dr. Martin Sändig oHG, Wiesbaden 1968.

404. K. Koyama and T. W. Lai [1993], An optimal Mastermind strategy, *J. Recr. Math.* **25**, 251–256.

405. M. Kraitchik [1953], *Mathematical Recreations*, 2nd ed., Dover, New York, NY.

406. B. Kummer [1980], *Spiele auf Graphen*, Internat. Series of Numerical Mathematics, Birkhäuser Verlag, Basel.

407. R. E. Ladner and J. K. Norman [1985], Solitaire automata, *J. Comput. System Sci.* **30**, 116–129.

408. J. C. Lagarias [1977], Discrete balancing games, *Bull. Inst. Math. Acad. Sinica* **5**, 363-373.

409. S. P. Lalley [1988], A one-dimensional infiltration game, *Naval Research Logistics* **35**, 441–446.

410. H. A. Landman [1996], Eyespace values in Go, in: *Games of No Chance*, Proc. MSRI Workshop on Combinatorial Games, July, 1994, Berkeley, CA (R. J. Nowakowski, ed.), MSRI Publ. Vol. 29, Cambridge University Press, Cambridge, pp. 227–257.

411. L. Larson [1977], A theorem about primes proved on a chessboard, *Math. Mag.* **50**, 69–74.

412. E. Lasker [1931], *Brettspiele der Völker, Rätsel und mathematische Spiele*, Berlin.

413. I. Lavalée [1985], Note sur le problème des tours d'Hanoï, *Rev. Roumaine Math. Pures Appl.* **30**, 433–438.

414. A. J. Lazarus, D. E. Loeb, J. G. Propp and D. Ullman [1996], Richman Games, in: *Games of No Chance*, Proc. MSRI Workshop on Combinatorial Games, July, 1994, Berkeley, CA (R. J. Nowakowski, ed.), MSRI Publ. Vol. 29, Cambridge University Press, Cambridge, pp. 439–449.

415. A. Lehman [1964], A solution to the Shannon switching game, *SIAM J. Appl. Math.* **12**, 687–725.

416. T. Lengauer and R. Tarjan [1980], The space complexity of pebble games on trees, *Inform. Process. Lett.* **10**, 184–188.

417. T. Lengauer and R. Tarjan [1982], Asymptotically tight bounds on time-space trade-offs in a pebble game, *J. Assoc. Comput. Mach.* **29**, 1087–1130.

418. H. W. Lenstra, Jr. [1977], On the algebraic closure of two, *Proc. Kon. Nederl. Akad. Wetensch.* (Ser. A) **80**, 389–396.

419. H. W. Lenstra, Jr. [1977/1978], Nim multiplication, Séminaire de Théorie des Nombres, No. 11, Université de Bordeaux, France.

420. D. N. L. Levy, Editor [1988], *Computer Games I, II*, Springer-Verlag, New York.

421. J. Lewin [1986], The lion and man problem revisited, *J. Optimization Theory and Applications* **49**, 411–430.

422. S.-Y. R. Li [1974], Generalized impartial games, *Internat. J. Game Theory* **3**, 169–184.

423. S.-Y. R. Li [1976], Sums of Zuchswang games, *J. Combin. Theory* (Ser. A) **21**, 52–67.

424. S.-Y. R. Li [1977], N-person nim and N-person Moore's games, *Internat. J. Game Theory* **7**, 31–36.

425. D. Lichtenstein, [1982], Planar formulae and their uses, *SIAM J. Comput.* **11**, 329–343.

426. D. Lichtenstein and M. Sipser [1980], Go is Polynomial-Space hard, *J. Assoc. Comput. Mach.* **27**, 393–401. Earlier draft appeared in Proc. 19th Ann. Symp. Foundations of Computer Science (Ann Arbor, MI, Oct. 1978), IEEE Computer Soc., Long Beach, CA, 1978, pp. 48–54.

427. D. E. Loeb [1996], Stable winning coalitions, in: *Games of No Chance*, Proc. MSRI Workshop on Combinatorial Games, July, 1994, Berkeley, CA (R. J. Nowakowski, ed.), MSRI Publ. Vol. 29, Cambridge University Press, Cambridge, pp. 451–471.

428. A. M. Lopez, Jr. [1991], A prolog Mastermind program, *J. Recr. Math.* **23**, 81–93.

429. S. Loyd [1914], *Cyclopedia of Puzzles and Tricks*, Franklin Bigelow Corporation, Morningside Press, NY. Reissued and edited by M. Gardner under the name *The Mathematical Puzzles of Sam Loyd* (two volumes), Dover, New York, NY, 1959.

430. X. Lu [1991], A matching game, *Discrete Math.* **94**, 199–207.

431. X. Lu [1992], Hamiltonian games, *J. Combin. Theory* (Ser. B) **55**, 18–32.

432. X. Lu [1995], A Hamiltonian game on $K_{n,n}$, *Discrete Math.* **142**, 185–191.

433. X.-M. Lu [1986], Towers of Hanoi graphs, *Intern. J. Comput. Math.* **19**, 23–38.

434. X.-M. Lu [1988], Towers of Hanoi problem with arbitrary $k \geq 3$ pegs, *Intern. J. Comput. Math.* **24**, 39–54.

435. X.-M. Lu [1989], An iterative solution for the 4-peg towers of Hanoi, *Comput. J.* **32**, 187–189.

436. É. Lucas [1960], *Récréations Mathématiques* (four volumes), A. Blanchard, Paris. Previous editions: Gauthier-Villars, Paris, 1891–1894.

437. É. Lucas [1974], *Introduction aux Récréations Mathématiques: L'Arithmétique Amusante*, reprinted by A. Blanchard, Paris. Originally published by A. Blanchard, Paris, 1895.

438. A. L. Ludington [1988], Length of the 7-number game, *Fibonacci Quart.* **26**, 195–204.

439. A. Ludington-Young [1990], Length of the n-number game, *Fibonacci Quart.* **28**, 259–265.

440. M. Maamoun and H. Meyniel [1987], On a game of policemen and robber, *Discrete Appl. Math.* **17**, 307–309.

441. P. A. MacMahon [1921], *New Mathematical Pastimes*, Cambridge University Press, Cambridge.

442. F. Maffray [1986], On kernels in i-triangulated graphs, *Discrete Math.* **61**, 247–251.

443. F. Maffray [1992], Kernels in perfect line-graphs, *J. Combin. Theory* (Ser. B) **55**, 1–8.

444. G. Martin [1991], *Polyominoes: Puzzles and Problems in Tiling*, Math. Assoc. America, Wasington, DC.

445. J. G. Mauldon [1978], Num, a variant of nim with no first player win, *Amer. Math. Monthly* **85**, 575–578.

446. D. P. McIntyre [1942], A new system for playing the game of nim, *Amer. Math. Monthly* **49**, 44–46.

447. E. Mead, A. Rosa and C. Huang [1974], The game of SIM: A winning strategy for the second player, *Math. Mag.* **47**, 243–247.

448. N. Megiddo, S. L. Hakimi, M. R. Garey, D. S. Johnson and C. H. Papadimitriou [1988], The complexity of searching a graph, *J. Assoc. Comput. Mach.* **35**, 18–44.

449. K. Mehlhorn, S. Näher and M. Rauch [1990], On the complexity of a game related to the dictionary problem, *SIAM J. Comput.* **19**, 902–906. Earlier draft appeared in Proc. 30th Ann. Symp. Foundations of Computer Science, pp. 546–548.

450. N. S. Mendelsohn [1946], A psychological game, *Amer. Math. Monthly* **53**, 86–88.

451. C. G. Méndez [1981], On the law of large numbers, infinite games and category, *Amer. Math. Monthly* **88**, 40–42.

452. D. Mey [1994], Finite games for a predicate logic without contractions, *Theoret. Comput. Sci. (Math Games)* **123**, 341–349.

453. F. Meyer auf der Heide [1981], A comparison of two variations of a pebble game on graphs, *Theoret. Comput. Sci.* **13**, 315–322.

454. H. Meyniel and J.-P. Roudneff [1988], The vertex picking game and a variation of the game of dots and boxes, *Discrete Math.* **70**, 311–313.

455. D. Michie and I. Bratko [1987], Ideas on knowledge synthesis stemming from the KBBKN endgame, *Internat. Comp. Chess Assoc. J.* **10**, 3–10.

456. J. Milnor [1953], Sums of positional games, Contributions to the Theory of Games, (H. W. Kuhn and A. W. Tucker, eds.), *Ann. of Math. Stud.* **2**(28), 291–301, Princeton.

457. P.-Min Lin [1982], Principal partition of graphs and connectivity games, *J. Franklin Inst.* **314**, 203–210.

458. S. Minsker [1989], The towers of Hanoi rainbow problem: coloring the rings, *J. Algorithms* **10**, 1–19.

459. S. Minsker [1991], The towers of Antwerpen problem, *Inform. Process. Lett.* **38**, 107–111.

460. D. Moews [1991], Sum of games born on days 2 and 3, *Theoret. Comput. Sci. (Math Games)* **91**, 119–128.

461. D. Moews [1992], Pebbling graphs, *J. Combin. Theory* (Ser. B) **55**, 244–252.

462. D. Moews [1996], Infinitesimals and coin-sliding, in: *Games of No Chance*, Proc. MSRI Workshop on Combinatorial Games, July, 1994, Berkeley, CA (R. J. Nowakowski, ed.), MSRI Publ. Vol. 29, Cambridge University Press, Cambridge, pp. 315–327.

463. D. Moews [1996], Loopy games and Go, in: *Games of No Chance*, Proc. MSRI Workshop on Combinatorial Games, July, 1994, Berkeley, CA (R. J. Nowakowski, ed.), MSRI Publ. Vol. 29, Cambridge University Press, Cambridge, pp. 259–272.

464. D. Moews [≥1996], Coin-sliding and Go, to appear in *Theoret. Comput. Sci. (Math Games)*.

465. E. H. Moore [1909–1910], A generalization of the game called nim, *Ann. of Math.* **11** (Ser. 2), 93–94.

466. F. L. Morris [1981], Playing disjunctive sums is polynomial space complete, *Internat. J. Game Theory* **10**, 195–205.

467. M. Müller and R. Gasser [1996], Experiments in computer Go endgames, in: *Games of No Chance*, Proc. MSRI Workshop on Combinatorial Games, July, 1994, Berkeley, CA (R. J. Nowakowski, ed.), MSRI Publ. Vol. 29, Cambridge University Press, Cambridge, pp. 273–284.

468. H. J. R. Murray [1952], *A History of Board Games Other Than Chess*, Oxford University Press.

469. B. Nadel [1990], Representation selection for constraint satisfaction: a case study using n-queens, *IEEE Expert*, 16–23.

470. A. Napier [1970], A new game in town, *Empire Mag., Denver Post*, May 2.

471. A. Negro and M. Sereno [1992], Solution of Ulam's problem on binary search with three lies, *J. Combin. Theory* (Ser. A) **59**, 149–154.

472. J. Nešetřil and R. Thomas [1987], Well quasi ordering, long games and combinatorial study of undecidability, *Contemp. Math.* **65**, 281–293.

473. S. Neufeld and R. J. Nowakowski [1993], A vertex-to-vertex pursuit game played with disjoint sets of edges, in: *Finite and Infinite Combinatorics in Sets and Logic* (N. W. Sauer et al., eds.), Kluwer, Dordrecht, pp. 299–312.

474. R. J. Nowakowski [1991], . . .,Welter's game, Sylver coinage, dots-and-boxes, . . ., in: Combinatorial Games, *Proc. Symp. Appl. Math.* **43** (R. K. Guy, ed.), Amer. Math. Soc., Providence, RI, pp. 155–182.

475. R. J. Nowakowski and D. G. Poole [1996], Geography played on products of directed cycles, in: *Games of No Chance*, Proc. MSRI Workshop on Combinatorial Games, July, 1994, Berkeley, CA (R. J. Nowakowski, ed.), MSRI Publ. Vol. 29, Cambridge University Press, Cambridge, pp. 329–337.

476. R. Nowakowski and P. Winkler [1983], Vertex-to-vertex pursuit in a graph, *Discrete Math.* **43**, 235–239.

477. S. P. Nudelman [1995], The modular n-queens problem in higher dimensions, *Discrete Math.* **146**, 159–167.

478. T. H. O'Beirne [1984], *Puzzles and Paradoxes*, Dover, New York, NY. (Appeared previously by Oxford University Press, London, 1965.)

479. H. K. Orman [1996], Pentominoes: a first player win, in: *Games of No Chance*, Proc. MSRI Workshop on Combinatorial Games, July, 1994, Berkeley, CA (R. J. Nowakowski, ed.), MSRI Publ. Vol. 29, Cambridge University Press, Cambridge, pp. 339–344.

480. E. W. Packel [1987], The algorithm designer versus nature: a game-theoretic approach to information-based complexity, *J. Complexity* **3**, 244–257.

481. C. H. Papadimitriou [1985], Games against nature, *J. Comput. System Sci.* **31**, 288–301.

482. A. Papaioannou [1982], A Hamiltonian game, *Ann. Discrete Math.* **13**, 171–178.

483. T. D. Parsons [1978], Pursuit-evasion in a graph, in: *Theory and Applications of Graphs* (Y. Alavi and D. R. Lick, eds.), Springer-Verlag, 426–441.

484. T. D. Parsons [1978], The search number of a connected graph, *Proc. 9th South-Eastern Conf. on Combinatorics, Graph Theory, and Computing*, 549–554.

485. O. Patashnik [1980], Qubic: $4 \times 4 \times 4$ Tic-Tac-Toe, *Math. Mag.* **53**, 202–216.

486. J. L. Paul [1978], Tic-Tac-Toe in n dimensions, *Math. Mag.* **51**, 45–49.

487. W. J. Paul, E. J. Prauss and R. Reischuk [1980], On alternation, *Acta Informatica* **14**, 243–255.

488. W. J. Paul and R. Reischuk [1980], On alternation, II, *Acta Informatica* **14**, 391–403.

489. J. Pearl [1980], Asymptotic properties of minimax trees and game-searching procedures, *Artificial Intelligence* **14**, 113–138.

490. J. Pearl [1984], *Heuristics: Intelligent Search Strategies for Computer Problem Solving*, Addison-Wesley, Reading, MA.

491. A. Pelc [1987], Solution of Ulam's problem on searching with a lie, *J. Combin. Theory* (Ser. A) **44**, 129–140.

492. A. Pelc [1988], Prefix search with a lie, *J. Combin. Theory* (Ser. A) **48**, 165–173.

493. A. Pelc [1989], Detecting errors in searching games, *J. Combin. Theory* (Ser. A) **51**, 43–54.

494. D. H. Pelletier [1987], Merlin's magic square, *Amer. Math. Monthly* **94**, 143–150.

495. G. L. Peterson, Press-Ups is Pspace-complete, Unpublished manuscript.

496. G. L. Peterson and J. H. Reif [1979], Multiple-person alternation, Proc. 20th Ann. Symp. Foundations Computer Science (San Juan, Puerto Rico, Oct. 1979), IEEE Computer Soc., Long Beach, CA, pp. 348–363.

497. N. Pippenger [1980], Pebbling, Proc. 5th IBM Symp. Math. Foundations of Computer Science, IBM, Japan (19 pp.)

498. N. Pippenger [1982], Advances in pebbling, in: Proc. 9th Internat. Colloq. Automata, Languages and Programming (M. Nielson and E. M. Schmidt, eds.), *Lecture Notes in Computer Science* **140**, 407–417, Springer Verlag, New York, NY.

499. T. E. Plambeck [1992], Daisies, Kayles, and the Sibert–Conway decomposition in misère octal games, *Theoret. Comput. Sci.* (*Math Games*) **96**, 361–388.

500. V. Pless [1991], Games and codes, in: Combinatorial Games, *Proc. Symp. Appl. Math.* **43** (R. K. Guy, ed.), Amer. Math. Soc., Providence, RI, pp. 101–110.

501. D. Poole [1992], The bottleneck towers of Hanoi problem, *J. Recr. Math.* **24**, 203–207.

502. D. G. Poole [1994], The towers and triangles of Professor Claus (or, Pascal knows Hanoi), *Math. Mag.* **67**, 323–344.

503. J. Propp [1994], A new take-away game, in: The Lighter Side of Mathematics, Proc. E. Strens Memorial Conf. on Recr. Math and its History (R. K. Guy and R. E. Woodrow, eds.), Calgary, 1986, Spectrum Series, Math. Assoc. of America, Washington, DC, pp. 212–221.

504. J. Propp [1996], About David Richman, Prologue to the paper by J. D. Lazarus et al. [1996], in: *Games of No Chance*, Proc. MSRI Workshop on Combinatorial Games, July, 1994, Berkeley, CA (R. J. Nowakowski, ed.), MSRI Publ. Vol. 29, Cambridge University Press, Cambridge, p. 437.

505. J. Propp [≥1996], Three-person impartial games, preprint.

506. J. Propp and D. Ullman [1992], On the cookie game, *Internat. J. Game Theory* **20**, 313–324.

507. A. Pultr and F. L. Morris [1984], Prohibiting repetitions makes playing games substantially harder, *Internat. J. Game Theory* **13**, 27–40.

508. A. Quilliot [1982], Discrete pursuit games, Proc. 13th Conference on Graphs and Combinatorics, Boca Raton, FL.

509. A. Quilliot [1985], A short note about pursuit games played on a graph with a given genus, *J. Combin. Theory* (Ser. B) **38**, 89–92.

510. M. O. Rabin [1957], Effective computability of winning strategies, Contributions to the Theory of Games, *Ann. of Math. Stud.* **3**(39), 147–157, Princeton.

511. M. O. Rabin [1976], Probabilistic algorithms, Proc. Symp. on New Directions and Recent Results in Algorithms and Complexity (J. F. Traub, ed.), Carnegie-Mellon, Academic Press, New York, NY, pp. 21–39.

512. B. Ravikumar and K. B. Lakshmanan [1984], Coping with known patterns of lies in a search game, *Theoret. Comput. Sci.* **33**, 85–94.

513. M. Reichling [1987], A simplified solution of the N queens' problem, *Inform. Process. Lett.* **25**, 253–255.

514. J. H. Reif [1984], The complexity of two-player games of incomplete information, *J. Comput. System Sci.* **29**, 274–301. Earlier draft entitled Universal games of incomplete information, appeared in Proc. 11th Ann. ACM Symp. Theory of Computing (Atlanta, GA, 1979), Assoc. Comput. Mach., New York, NY, pp. 288–308.

515. S. Reisch [1980], Gobang ist PSPACE-vollständig, *Acta Informatica* **13**, 59–66.

516. S. Reisch [1981], Hex ist PSPACE-vollständig, *Acta Informatica* **15**, 167–191.

517. M. Richardson [1953], Extension theorems for solutions of irreflexive relations, *Proc. Nat. Acad. Sci. U. S. A.* **39**, 649.

518. M. Richardson [1953], Solutions of irreflexive relations, *Ann. of Math.* **58**, 573–590.

519. R. D. Ringeisen [1974], Isolation, a game on a graph, *Math. Mag.* **47**, 132–138.

520. R. L. Rivest, A. R. Meyer, D. J. Kleitman, K. Winklman and J. Spencer [1980], Coping with errors in binary search procedures, *J. Comput. System Sci.* **20**, 396–404.

521. I. Rivin, I. Vardi and P. Zimmermann[1994], The n-queens problem, *Amer. Math. Monthly* **101**, 629–638.

522. I. Rivin and R. Zabih [1992], A dynamic programming solution to the N-queens problem, *Inform. Process. Lett.* **41**, 253–256.

523. E. Robertson and I. Munro [1978], NP-completeness, puzzles and games, *Utilitas Math.* **13**, 99–116.

524. A. G. Robinson and A. J. Goldman [1989], The set coincidence game: complexity, attainability, and symmetric strategies, *J. Comput. System Sci.* **39**, 376–387.

525. A. G. Robinson and A. J. Goldman [1990], On Ringeisen's isolation game, *Discrete Math.* **80**, 297–312.

526. A. G. Robinson and A. J. Goldman [1990], On the set coincidence game, *Discrete Math.* **84**, 261–283.

527. A. G. Robinson and A. J. Goldman [1991], On Ringeisen's isolation game, II, *Discrete Math.* **90**, 153–167.

528. A. G. Robinson and A. J. Goldman [1993], The isolation game for regular graphs, *Discrete Math.* **112**, 173–184.

529. J. M. Robson [1983], The complexity of Go, Proc. Information Processing 83 (R. E. A. Mason, ed.), Elsevier, Amsterdam, pp. 413–417.

530. J. M. Robson [1984], Combinatorial games with exponential space complete decision problems, Proc. 11th Symp. Math. Foundations of Computer Science (M. P. Chytie and V. Koubek, eds.), Praha, Czechoslovakia, 1984, *Lecture Notes in Computer Science* **176**, 498–506, Springer, Berlin.

531. J. M. Robson [1984], N by N checkers is Exptime complete, *SIAM J. Comput.* **13**, 252–267.

532. J. M. Robson [1985], Alternation with restrictions on looping, *Inform. and Control* **67**, 2–11.

533. E. Y. Rodin [1989], A pursuit-evasion bibliography – version 2, *Comput. Math. Appl.* **18**, 245–250.

534. J. S. Rohl [1983], A faster lexicographical n-queens algorithm, *Inform. Process. Lett.* **17**, 231–233.

535. I. Roizen and J. Pearl [1983], A minimax algorithm better than alpha-beta? Yes and no, *Artificial Intelligence* **21**, 199–220.

536. I. Rosenholtz [1993], Solving some variations on a variant of Tic-Tac-Toe using invariant subsets, *J. Recr. Math.* **25**, 128-135.

537. A. S. C. Ross [1953], The name of the game of Nim, Note 2334, *Math. Gaz.* **37**, 119–120.

538. A. E. Roth [1978], A note concerning asymmetric games on graphs, *Naval Res. Logist. Quart.* **25**, 365–367.

539. A. E. Roth [1978], Two-person games on graphs, *J. Combin. Theory* (Ser. B) **24**, 238–241.

540. T. Roth [1974], The tower of Brahma revisited, *J. Recr. Math.* **7**, 116–119.

541. E. M. Rounds and S. S. Yau [1974], A winning strategy for SIM, *J. Recr. Math.* **7**, 193–202.

542. W. L. Ruzzo [1980], Tree-size bounded alternation, *J. Comput. Systems Sci.* **21** (1980), 218–235.

543. S. Sackson [1946], *A Gamut of Games*, Random House.

544. M. Saks and A. Wigderson [1986], Probabilistic Boolean decision trees and the complexity of evaluating game trees, 27th Ann. Symp. Foundations of Computer Science (Toronto, Ont., Canada), IEEE Computer Soc., Washington, DC, pp. 29–38.

545. M. Sato [1972], Grundy functions and linear games, Publ. Res. Inst. Math. Sciences, Kyoto Univ., Vol. 7, 645–658.

546. W. L. Schaaf [1955, 1970, 1973, 1978], *A Bibliography of Recreational Mathematics* (four volumes), Nat'l. Council of Teachers of Mathematics, Reston, VA.

547. T. J. Schaefer [1976], Complexity of decision problems based on finite two-person perfect information games, 8th Ann. ACM Symp. Theory of Computing (Hershey, PA, 1976), Assoc. Comput. Mach., New York, NY, pp. 41–49.

548. T. J. Schaefer [1978], On the complexity of some two-person perfect-information games, *J. Comput. System Sci.* **16**, 185–225.

549. J. Schaeffer [1996], Solving the game of checkers, in: *Games of No Chance*, Proc. MSRI Workshop on Combinatorial Games, July, 1994, Berkeley, CA (R. J. Nowakowski, ed.), MSRI Publ. Vol. 29, Cambridge University Press, Cambridge, pp. 119–134.

550. M. Scheepers [1994], Variations on a game of Gale (II): Markov strategies, *Theoret. Comput. Sci. (Math Games)* **129**, 385–396.

551. G. Schmidt and T. Ströhlein [1985], On kernels of graphs and solutions of games: a synopsis based on relations and fixpoints, *SIAM J. Alg. Disc. Math.* **6**, 54–65.

552. G. Schrage [1985], A two-dimensional generalization of Grundy's game, *Fibonacci Quart.* **23**, 325–329.

553. H. Schubert [1953], *Mathematische Mussestunden*, Neubearbeitet von F. Fitting, Elfte Auflage, De Gruyter, Berlin.

554. F. Schuh [1952], Spel van delers (The game of divisors), *Nieuw Tijdschrift voor Wiskunde* **39**, 299–304.

555. F. Schuh [1968], *The Master Book of Mathematical Recreations*, translated by F. Göbel, edited by T. H. O'Beirne, Dover, New York, NY.

556. B. L. Schwartz [1971], Some extensions of Nim, *Math. Mag.* **44**, 252–257.

557. A. J. Schwenk [1970], Take-away games, *Fibonacci Quart.* **8**, 225–234.

558. R. S. Scorer, P. M. Grundy and C. A. B. Smith [1944], Some binary games, *Math. Gaz.* **28**, 96–103.

559. Á. Seress [1992], On Hajnal's triangle-free game, *Graphs Combin.* **8**, 75–79.

560. L. E. Shader [1978], Another strategy for SIM, *Math. Mag.* **51**, 60–64.

561. A. S. Shaki [1979], Algebraic solutions of partizan games with cycles, *Math. Proc. Camb. Phil. Soc.* **85**, 227–246.

562. A. Shamir, R. L. Rivest and L. M. Adleman [1981], Mental Poker, in: *The Mathematical Gardner* (D. A. Klarner, ed.), Wadsworth Internat., Belmont, CA, pp. 37–43.

563. G. J. Sherman [1978], A child's game with permutations, *Math. Mag.* **51**, 67–68.

564. W. L. Sibert and J. H. Conway [1992], Mathematical Kayles, *Internat. J. Game Theory* **20**, 237–246.

565. R. Silber [1976], A Fibonacci property of Wythoff pairs, *Fibonacci Quart.* **14**, 380–384.

566. R. Silber [1977], Wythoff's Nim and Fibonacci representations, *Fibonacci Quart.* **15**, 85–88.

567. J.-N. O. Silva [1993], Some game bounds depending on birthdays, *Portugaliae Math.* **3**, 353–358.

568. R. Silver [1967], The group of automorphisms of the game of 3-dimensional ticktacktoe, *Amer. Math. Monthly* **74**, 247–254.

569. D. L. Silverman [1971], *Your Move*, McGraw-Hill.

570. G. J. Simmons [1969], The game of SIM, *J. Recr. Math.* **2**, 193–202.

571. D. Singmaster [1981], Almost all games are first person games, *Eureka* **41**, 33–37.

572. D. Singmaster [1982], Almost all partizan games are first person and almost all impartial games are maximal, *J. Combin. Inform. System Sci.* **7**, 270–274.

573. C. A. B. Smith [1966], Graphs and composite games, *J. Combin. Theory* **1**, 51–81. Reprinted in slightly modified form in: *A Seminar on Graph Theory* (F. Harary, ed.), Holt, Rinehart and Winston, New York, NY, 1967.

574. C. A. B. Smith [1968], Compound games with counters, *J. Recr. Math.* **1**, 67–77.

575. C. A. B. Smith [1971], Simple game theory and its applications, *Bull. Inst. Math. Appl.* **7**, 352–357.

576. D. E. Smith and C. C. Eaton [1911], Rithmomachia, the great medieval number game, *Amer. Math. Montly* **18**, 73–80.

577. R. Sosic and J. Gu [1990], A polynomial time algorithm for the n-queens problem, *SIGART* **1**, 7–11.

578. J. Spencer [1977], Balancing games, *J. Combin. Theory* (Ser. B) **23**, 68–74.

579. J. Spencer [1984], Guess a number with lying, *Math. Mag.* **57**, 105–108.

580. J. Spencer [1986], Balancing vectors in the max norm, *Combinatorica* **6**, 55–65.

581. J. Spencer [1991], Threshold spectra via the Ehrenfeucht game, *Discrete App. Math.* **30**, 235–252.

582. J. Spencer [1992], Ulam's searching game with a fixed number of lies, *Theoret. Comput. Sci.* (*Math Games*) **95**, 307–321.

583. J. Spencer [1994], Randomization, derandomization and antirandomization: three games, *Theoret. Comput. Sci.* (*Math Games*) **131**, 415–429.

584. E. L. Spitznagel, Jr. [1973], Properties of a game based on Euclid's algorithm, *Math. Mag.* **46**, 87–92.

585. R. Sprague [1935–36], Über mathematische Kampfspiele, *Tôhoku Math. J.* **41**, 438–444.

586. R. Sprague [1937], Über zwei Abarten von Nim, *Tôhoku Math. J.* **43**, 351–359.

587. R. Sprague [1947], Bemerkungen über eine spezielle Abelsche Gruppe, *Math. Z.* **51**, 82–84.

588. R. Sprague [1961], *Unterhaltsame Mathematik*, Vieweg and Sohn, Braunschweig, Paperback reprint, translation by T. H. O'Beirne: *Recreations in Mathematics*, Blackie, 1963.

589. H. Steinhaus [1960], Definitions for a theory of games and pursuit, *Naval Res. Logist. Quart.* **7**, 105–108.

590. V. N. Stepanenko [1975], Grundy games under conditions of semidefiniteness, *Cybernetics* **11**, 167–172 (trans. of *Kibernetika* **11** (1975) 145–149).

591. B. M. Stewart [1939], Problem 3918 (*k*-peg tower of Hanoi), *Amer. Math. Monthly* **46**, 363. Solution by J. S. Frame, *ibid.* **48** (1941) 216–217; by the proposer, *ibid.* 217–219.

592. L. Stiller [1988], Massively parallel retrograde analysis. Tech. Report BU-CS TR88–014, Comp. Sci. Dept., Boston University.

593. L. Stiller [1989], Parallel analysis of certain endgames, *Internat. Comp. Chess Assoc. J.* **12**, 55–64.

594. L. Stiller [1991], Group graphs and computational symmetry on massively parallel architecture, *J. Supercomputing* **5**, 99–117.

595. L. Stiller [1996], Multilinear algebra and chess endgames in: *Games of No Chance*, Proc. MSRI Workshop on Combinatorial Games, July, 1994, Berkeley, CA (R. J. Nowakowski, ed.), MSRI Publ. Vol. 29, Cambridge University Press, Cambridge, pp. 151–192.

596. D. L. Stock [1989], Merlin's magic square revisited, *Amer. Math. Monthly* **96**, 608–610.

597. L. J. Stockmeyer and A. K. Chandra [1979], Provably difficult combinatorial games, *SIAM J. Comput.* **8**, 151–174.

598. J. A. Storer [1983], On the complexity of chess, *J. Comput. System Sci.* **27**, 77–100.

599. P. D. Straffin, Jr. [1985], Three-person winner-take-all games with McCarthy's revenge rule, *College J. Math.* **16**, 386–394.

600. Th. Ströhlein and L. Zagler [1977], Analyzing games by Boolean matrix iteration, *Discrete Math.* **19**, 183–193.

601. W. Stromquist and D. Ullman [1993], Sequential compounds of combinatorial games, *Theoret. Comput. Sci. (Math Games)* **119**, 311–321.

602. K. Sugihara and I. Suzuki [1989], Optimal algorithms for a pursuit-evasion problem in grids, *SIAM J. Disc. Math.* **1**, 126–143.

603. K. Sutner [1988], On σ-automata, *Complex Systems* **2**, 1-28.

604. K. Sutner [1989], Linear cellular automata and the Garden-of-Eden, *Math. Intelligencer* **11**, 49–53.

605. K. Sutner [1990], The σ-game and cellular automata, *Amer. Math. Monthly* **97**, 24–34.

606. J. L. Szwarcfiter and G. Chaty [1994], Enumerating the kernels of a directed graph with no odd circuits, *Inform. Process. Lett.* **51**, 149–153.

607. G. Tardos [1988], Polynomial bound for a chip firing game on graphs, *SIAM J. Disc. Math.* **1**, 397–398.

608. M. Tarsi [1983], Optimal search on some game trees, *J. Assoc. Comput. Mach.* **30**, 389–396.

609. R. Telgársky [1987], Topological games: on the 50th anniversary of the Banach-Mazur game, *Rocky Mountain J. Math.* **17**, 227–276.

610. K. Thompson [1986], Retrograde analysis of certain engames, *Internat. Comp. Chess Assoc. J.* **9**, 131–139.

611. I. Tomescu [1990], Almost all digraphs have a kernel, *Discrete Math.* **84**, 181–192.

612. R. Tošić and S. Šćekić [1983], An analysis of some partizan graph games, *Proc. 4th Yugoslav Seminar on Graph Theory, Novi Sad*, pp. 311–319.

613. A. M. Turing, M. A. Bates, B. V. Bowden and C. Strachey [1953], Digital computers applied to games, in: *Faster Than Thought* (B. V. Bowden, ed.), Pitman, London, pp. 286–310.

614. R. Uehara and S. Iwata [1990], Generalized Hi-Q is NP-complete, *Trans. IEICE* **E73**, 270–273.

615. J. Úlehla [1980], A complete analysis of von Neumann's Hackendot, *Internat. J. Game Theory* **9**, 107–113.

616. D. Ullman [1992], More on the four-numbers game, *Math. Mag.* **65**, 170–174.

617. S. Vajda [1992], *Mathematical Games and How to Play Them*, Ellis Horwood Series in Mathematics and its Applications, Chichester, England.

618. H. J. van den Herik and I. S. Herschberg [1985], The construction of an omniscient endgame database, *Internat. Comp. Chess Assoc. J.* **8**, 66–87.

619. J. van Leeuwen [1976], Having a Grundy-numbering is NP-complete, Report No. 207, Computer Science Dept., Pennsylvania State University, University Park, PA.

620. A. J. van Zanten [1990], The complexity of an optimal algorithm for the generalized tower of Hanoi problem, *Intern. J. Comput. Math.* **36**, 1–8.

621. A. J. van Zanten [1991], An iterative optimal algorithm for the generalized tower of Hanoi problem, *Intern. J. Comput. Math.* **39**, 163–168.

622. I. Vardi [1990], *Computational Recreations in Mathematica*, Addison Wesley.

623. I. P. Varvak [1968], Games on the sum of graphs, *Cybernetics* **4**, 49–51 (trans. of *Kibernetika* **4** (1968) 63–66).

624. J. Veerasamy and I. Page [1994], On the towers of Hanoi problem with multiple spare pegs, *Intern. J. Comput. Math.* **52**, 17–22.

625. H. Venkateswaran and M. Tompa [1989], A new pebble game that characterizes parallel complexity classes, *SIAM J. Comput.* **18**, 533–549.

626. D. Viaud [1987], Une stratégie générale pour jouer au Master-Mind, *RAIRO Recherche opérationelle/Operations Research* **21**, 87–100.

627. J. von Neumann [1928], Zur Theorie der Gesellschaftsspiele, *Math. Ann.* **100**, 295–320.

628. J. von Neumann and O. Morgenstern [1953], *Theory of Games and Economic Behaviour*, 3rd ed., Princeton University Press, Princeton, NJ.

629. J. L. Walsh [1953], The name of the game of Nim, Letter to the Editor, *Math. Gaz.* **37**, 290.

630. T. R. Walsh [1982], The towers of Hanoi revisited: moving the rings by counting the moves, *Inform. Process. Lett.* **15**, 64–67.

631. T. R. Walsh [1983], Iteration strikes back at the cyclic towers of Hanoi, *Inform. Process. Lett.* **16**, 91–93.

632. A. Washburn [1990], Deterministic graphical games, *J. Math. Anal. Appl.* **153**, 84–96.

633. W. A. Webb [1982], The length of the four-number game, *Fibonacci Quart.* **20**, 33–35.

634. C. P. Welter [1952], The advancing operation in a special abelian group, *Nederl. Akad. Wetensch. Proc.* (Ser. A) **55** = *Indag. Math.* **14**, 304–314.

635. C. P. Welter [1954], The theory of a class of games on a sequence of squares, in terms of the advancing operation in a special group, *Nederl. Akad. Wetensch. Proc.* (Ser. A) **57** = *Indag. Math.* **16**, 194–200.

636. J. West [1996], Champion-level play of domineering, in: *Combinatorial Games*, Proc. MSRI Workshop on Combinatorial Games, July, 1994, Berkeley, CA (R. J. Nowakowski, ed.), MSRI Publ. Vol. 29, Cambridge University Press, Cambridge, pp. 85–91.

637. J. West [1996], Champion-level play of dots-and-boxes, in: *Games of No Chance*, Proc. MSRI Workshop on Combinatorial Games, July, 1994, Berkeley, CA (R. J. Nowakowski, ed.), MSRI Publ. Vol. 29, Cambridge University Press, Cambridge, pp. 79–84.

638. J. West [1996], New values for Top Entails, in: *Games of No Chance*, Proc. MSRI Workshop on Combinatorial Games, July, 1994, Berkeley, CA (R. J. Nowakowski, ed.), MSRI Publ. Vol. 29, Cambridge University Press, Cambridge, pp. 345–350.

639. M. J. Whinihan [1963], Fibonacci Nim, *Fibonacci Quart.* **1**(4), 9–12.

640. R. Wilber [1988], White pebbles help, *J. Comput. System Sci.* **36**, 108–124.

641. R. M. Wilson [1974], Graph puzzles, homotopy and the alternating group, *J. Combin. Theory* (Ser. B) **16**, 86–96.

642. D. Wolfe [1993], Snakes in domineering games, *Theoret. Comput. Sci.* (*Math Games*) **119**, 323–329.

643. D. Wolfe [1996], The gamesman's toolkit, in: *Games of No Chance*, Proc. MSRI Workshop on Combinatorial Games, July, 1994, Berkeley, CA (R. J. Nowakowski, ed.), MSRI Publ. Vol. 29, Cambridge University Press, Cambridge, pp. 93–98.

644. D. Wood [1981], The towers of Brahma and Hanoi revisited, *J. Recr. Math.* **14**, 17–24.

645. D. Wood [1983], Adjudicating a towers of Hanoi contest, *Intern. J. Comput. Math.* **14**, 199–207.

646. J.-S. Wu and R.-J. Chen [1992], The towers of Hanoi problem with parallel moves, *Inform. Process. Lett.* **44**, 241–243.

647. J.-S. Wu and R.-J. Chen [1993], The towers of Hanoi problem with cyclic parallel moves, *Inform. Process. Lett.* **46**, 1–6.

648. W. A. Wythoff [1907], A modification of the game of Nim, *Nieuw Arch. Wisk.* **7**, 199–202.

649. A. M. Yaglom and I. M. Yaglom [1967], *Challenging Mathematical Problems with Elementary Solutions*, translated by J. McCawley, Jr., revised and edited by B. Gordon, Vol. II, Holden-Day, San Francisco.

650. Y. Yamasaki [1978], Theory of division games, Publ. Res. Inst. Math. Sciences, Kyoto Univ., Vol. 14, pp. 337–358.

651. Y. Yamasaki [1980], On misère Nim-type games, *J. Math. Soc. Japan* **32**, 461–475.

652. Y. Yamasaki [1981], The projectivity of *Y*-games, Publ. Res. Inst. Math. Sciences, Kyoto Univ., Vol. 17, pp. 245–248.

653. Y. Yamasaki [1981], Theory of Connexes I, Publ. Res. Inst. Math. Sciences, Kyoto Univ., Vol. 17, pp. 777–812.

654. Y. Yamasaki [1985], Theory of connexes II, Publ. Res. Inst. Math. Sciences, Kyoto Univ., Vol. 21, 403–410.

655. Y. Yamasaki [1989], *Combinatorial Games: Back and Front* (in Japanese), Springer Verlag, Tokyo.

656. Y. Yamasaki [1991], A difficulty in particular Shannon-like games, *Discrete Appl. Math.* **30**, 87–90.

657. Y. Yamasaki [1993], Shannon-like games are difficult, *Discrete Math.* **111**, 481–483.

658. L. J. Yedwab [1985], On playing well in a sum of games, M. Sc. thesis, MIT, MIT/LCS/TR-348.

659. Y. Yesha [1978], Theory of annihilation games, Ph. D. thesis, Weizmann Institute of Science, Rehovot, Israel.

660. Y. K. Yu and R. B. Banerji [1982], Periodicity of Sprague–Grundy function in graphs with decomposable nodes, *Cybernetics and Systems: An Internat. J.* **13**, 299–310.

661. S. Zachos [1988], Probabilistic quantifiers and games, *J. Comput. System Sci.* **36**, 433–451.

662. E. Zermelo [1912], Über eine Anwendung der Mengenlehre auf die Theorie des Schachspiels, Proc. 5th Int. Cong. Math., Cambridge, Cambridge University Press, 1913, Vol. II, pp. 501–504.

663. M. Zieve [1996], Take-away games, in: *Games of No Chance*, Proc. MSRI Workshop on Combinatorial Games, July, 1994, Berkeley, CA (R. J. Nowakowski, ed.), MSRI Publ. Vol. 29, Cambridge University Press, Cambridge, pp. 351–361.

664. U. Zwick and M. S. Paterson [1993], The memory game, *Theoret. Comput. Sci. (Math Games)* **110**, 169–196.

665. U. Zwick and M. S. Paterson [1996], The complexity of mean payoff games on graphs, *Theoret. Comput. Sci. (Math Games)* **158**, 343–359.

666. W. S. Zwicker [1987], Playing games with games: the hypergame paradox, *Amer. Math. Monthly* **94**, 507–514.

AVIEZRI S. FRAENKEL
DEPARTMENT OF APPLIED MATHEMATICS AND COMPUTER SCIENCE
WEIZMANN INSTITUTE OF SCIENCE
REHOVOT 76100, ISRAEL
    fraenkel@wisdom.weizmann.ac.il
    http://www.wisdom.weizmann.ac.il/~fraenkel/fraenkel.html